Implications of the COVID-19 Pandemic and the Russia-Ukraine Crisis on the Agricultural Sector

Vítor João Pereira Domingues Martinho
ESAV, CERNAS-IPV Research Centre, Polytechnic Institute of Viseu, Portugal

IGI Global
PUBLISHER of TIMELY KNOWLEDGE

A volume in the Practice, Progress, and Proficiency in Sustainability (PPPS) Book Series

Published in the United States of America by
IGI Global
Engineering Science Reference (an imprint of IGI Global)
701 E. Chocolate Avenue
Hershey PA, USA 17033
Tel: 717-533-8845
Fax: 717-533-8661
E-mail: cust@igi-global.com
Web site: http://www.igi-global.com

Library of Congress Cataloging-in-Publication Data

Names: Martinho, Vítor João Pereira Domingues, editor.
Title: Implications of the Covid-19 pandemic and the Russia-Ukraine crisis on the agricultural sector / authored by: Vítor João Pereira Domingues Martinho.
Description: Hershey, PA : Engineering Science Reference, [2023] | Includes bibliographical references and index. | Summary: "The main objectives of this publication are to assess the real impacts of the Covid-19 pandemic and Russia- Ukraine crisis on the world agricultural sector. These objectives and the methodologies considered bring novelty to the scientific community, with added value for several stakeholders, including international organisations and policymakers"-- Provided by publisher.
Identifiers: LCCN 2023008187 (print) | LCCN 2023008188 (ebook) | ISBN 9781668489239 (hardcover) | ISBN 9781668489246 (paperback) | ISBN 9781668489253 (ebook)
Subjects: LCSH: Agriculture--Economic aspects. | COVID-19 Pandemic, 2020---Economic aspects. | Ukraine--History--Russian Invasion, 2022---Economic aspects.
Classification: LCC HD1415 .I473 2023 (print) | LCC HD1415 (ebook) | DDC 338.10947--dc23/eng/20230412
LC record available at https://lccn.loc.gov/2023008187
LC ebook record available at https://lccn.loc.gov/2023008188

This book is published in the IGI Global book series Practice, Progress, and Proficiency in Sustainability (PPPS) (ISSN: 2330-3271; eISSN: 2330-328X)

British Cataloguing in Publication Data
A Cataloguing in Publication record for this book is available from the British Library.

For electronic access to this publication, please contact: eresources@igi-global.com.

Practice, Progress, and Proficiency in Sustainability (PPPS) Book Series

Ayman Batisha
International Sustainability Institute, Egypt

ISSN:2330-3271
EISSN:2330-328X

Mission

In a world where traditional business practices are reconsidered and economic activity is performed in a global context, new areas of economic developments are recognized as the key enablers of wealth and income production. This knowledge of information technologies provides infrastructures, systems, and services towards sustainable development.

The **Practices, Progress, and Proficiency in Sustainability (PPPS) Book Series** focuses on the local and global challenges, business opportunities, and societal needs surrounding international collaboration and sustainable development of technology. This series brings together academics, researchers, entrepreneurs, policy makers and government officers aiming to contribute to the progress and proficiency in sustainability.

Coverage

- Global Business
- Intellectual Capital
- Knowledge clusters
- Green Technology
- Innovation Networks
- Environmental informatics
- Outsourcing
- Socio-Economic
- Eco-Innovation
- Strategic Management of IT

IGI Global is currently accepting manuscripts for publication within this series. To submit a proposal for a volume in this series, please contact our Acquisition Editors at acquisitions@igi-global.com or visit: https://www.igi-global.com/publish/.

The Practice, Progress, and Proficiency in Sustainability (PPPS) Book Series (ISSN 2330-3271) is published by IGI Global, 701 E. Chocolate Avenue, Hershey, PA 17033-1240, USA, www.igi-global.com. This series is composed of titles available for purchase individually; each title is edited to be contextually exclusive from any other title within the series. For pricing and ordering information please visit https://www.igi-global.com/book-series/practice-progress-proficiency-sustainability/73810. Postmaster: Send all address changes to above address.

Titles in this Series

For a list of additional titles in this series, please visit: https://www.igi-global.com/book-series/practice-progress-proficiency-sustainability/73810

Sustainable Approaches and Strategies for E-Waste Management and Utilization
A. M. Rawani (National Institute of Technology, Raipur, India) Mithilesh Kumar Sahu (O.P. Jindal University, India) Siddharth S. Chakarabarti (O.P. Jindal University, India) and Ajit Kumar Singh (GLA Universit, India)
Engineering Science Reference • © 2023 • 300pp • H/C (ISBN: 9781668475737) • US $240.00

Handbook of Research on Solving Societal Challenges Through Sustainability-Oriented Innovation
Luísa Cagica Carvalho (Instituto Politécnico de Setúbal, Portugal & CEFAGE, University of Évora, Portugal) Paulo Bogas (Instituto Politécnico de Setúbal, Portugal) Jordana Kneipp (Federal University of Santa Maria, Brazil) Lucas Avila (Federal University of Santa Maria, Brazil) and Elis Ossmane (Universidade Aberta, Portugal)
Engineering Science Reference • © 2023 • 419pp • H/C (ISBN: 9781668461235) • US $295.00

Handbook of Research on Socio-Economic Sustainability in the Post-Pandemic Era
Jozef Oleński (High School Technology and Economics, Jaroslaw, Poland) Jeffrey Sachs (Columbia University, USA) Masayuki Susai (Nagasaki University, Japan) Yannis Tsekouras (University of Macedonia, Greece) and Arjan Gjonça (London School of Economics and Political Science, UK)
Information Science Reference • © 2023 • 400pp • H/C (ISBN: 9781799897606) • US $270.00

Handbook of Research on Promoting Sustainable Public Transportation Strategies in Urban Environments
Zafer Yilmaz (TED University, Turkey) Silvia Golem (University of Split, Croatia) and Dorinela Costescu (Polytechnic University of Bucharest, Romania)
Engineering Science Reference • © 2023 • 401pp • H/C (ISBN: 9781668459966) • US $295.00

Energy Transition in the African Economy Post 2050
Olayinka Ohunakin (Covenant University, Nigeria)
Engineering Science Reference • © 2023 • 300pp • H/C (ISBN: 9781799886389) • US $215.00

Implications of Industry 5.0 on Environmental Sustainability
Muhammad Jawad Sajid (Xuzhou University of Technology, China) Syed Abdul Rehman Khan (Xuzhou University of Technology, China) and Zhang Yu (ILMA University, Pakistan)
Business Science Reference • © 2023 • 328pp • H/C (ISBN: 9781668461136) • US $250.00

IGI Global
PUBLISHER of TIMELY KNOWLEDGE

701 East Chocolate Avenue, Hershey, PA 17033, USA
Tel: 717-533-8845 x100 • Fax: 717-533-8661
E-Mail: cust@igi-global.com • www.igi-global.com

Table of Contents

Preface

The COVID-19 pandemic and the Russia-Ukraine conflict brought about several changes in the organisation of economic sectors, including agriculture. These changes occurred in the markets and, specifically, in the distribution chains, as a result of the lockdowns and the increase in resource prices, namely energy. However, the real impacts of these frameworks deserve special assessment.

The main objectives of this publication are to assess the real impacts of the COVID-19 pandemic and Russia-Ukraine crisis on the world agricultural sector. These objectives, the methodologies and approaches considered bring novelty to the scientific community, with added value for several stakeholders, including international organisations and policymakers.

The topics to be covered in this publication are the following: Agricultural contexts before and after the COVID-19 pandemic; Main factors that contributed to the changes identified; Suggestions for the future. The target audiences for this book are students, professors, researchers, farmers, policymakers, and national and international organisations.

This publication is divided into 15 chapters. The first two chapters give a perspective on the contributions from the scientific literature, in terms of insights into the interrelationships between the COVID-19 pandemic and the Russia-Ukraine conflict and the agricultural sector, namely bibliometric analysis and systematic review. Chapter 3 highlights the implications of the current context on the world food markets. Chapters 4-6 address the most recent frameworks worldwide related to agricultural output, agricultural policy, and sustainable agriculture. The remaining chapters present the agricultural conjunctures in recent years for the European Union and Portugal. The choice of these two contexts is associated with the importance of the European Union in the international agricultural realities and the economic specificities of countries, such as Portugal.

The research carried out in this book shows that the topics related to agriculture and the recent international contexts have motivated the researchers to bring more scientific contributions to these fields. On the other hand, the impacts on the world food markets seem to be greater from the Russia-Ukraine crisis than from the pandemic. The implications of these shocks on the agricultural output worldwide were reduced over the period considered (until September 2022). The expectations about the contributions from the impacts, specifically, of the pandemic to rethink agricultural practices and improve sustainability worldwide seem to be exaggerated. The agricultural policies have relevant impacts on the evolution of the sector, and they could have been more effective, in some circumstances. Considering data until September/October 2022, the assessments carried out reveal that the pandemic and the conflict had a reduced impact on the European Union dairy sector, poultry farming and meat sector, specifically in the dynamics related to production. The implications from the pandemic, for example, on the European Union producer prices of fruits and vegetables were relevant, however with differences among productions. In the Portuguese context, the impacts on the agricultural producer prices have been reduced in some cases and greater from the Russia-Ukraine crisis than from the pandemic in other cases.

A brief description of the importance of each chapter will be presented following. In any case, namely for the public institutions and policymakers, it is suggested to design more effective instruments and measures to avoid speculation in some cases over the agricultural chains and adjusted policies to support the farmers dealing with the increasing in the costs of production in some circumstances.

CHAPTER 1. INSIGHTS FROM THE LITERATURE ON COVID-19 AND THE AGRICULTURAL SECTOR

This part of the book intends to bring insights from the literature about the relationships among the pandemic and the agricultural sector. Information from the Web of Science (Core Collection) was taken into account, for 1614 documents, on a search carried out on 04 January 2023 for the topics "Covid-19" and "agricultur*". These data were assessed through bibliometric analysis, systematic review and factor analysis. The results show the attention dedicated, by the scientific community, to the relationships between the pandemic and agriculture, in countries such as the US, China, India and England.

CHAPTER 2. SYSTEMATIC REVIEW ON THE IMPACTS ON AGRICULTURE OF THE RUSSIA-UKRAINE SCENARIO

This chapter aims to analyse the literature insights about the several linkages among the agricultural sector and the Russia-Ukraine conflict. For that, 106 documents from the Web of Science (Core Collection) were taken into account for the topics "Russia-Ukraine conflict", or "Russia-Ukraine crisis", or "Russia-Ukraine war" and "agricultur*", on a search carried out on 04 January 2023. This information was analysed through bibliometric analysis, systematic review and factor-cluster analyses. The topics related to the Russia-Ukraine conflict and the farming sectors have attracted the attention of researchers, but there is still a field to be discovered.

CHAPTER 3. CONSEQUENCES OF THE CURRENT CONJUNCTURE ON WORLD FOOD MARKETS

In this research, it was aimed to assess the evolution worldwide of food prices over the last few years and to analyse the eventual implications of the pandemic and Russia-Ukraine conflict on these frameworks. Statistical information from the FAOSTAT was taken into account for the period January 2016-September 2022 and the following indicators: Food Price Inflation (%); Consumer Prices, Food Indices (2015 = 100). The implications on food prices were greater from the Russia-Ukraine crisis, than from the Covid-19 pandemic. This is principally worrying in more vulnerable nations.

CHAPTER 4. EVOLUTION OF GLOBAL AGRICULTURAL OUTPUT SINCE THE PANDEMIC

The objective of this study is to analyse the consequences of the most recent international occurrences on agricultural output worldwide. Data from the OECD (Organisation for Economic Co-operation and Development) databases were considered for the period 2012-2022 and the following indicators: crop production (thousand hectares; tonnes/ha; thousand tonnes); meat consumption (Kg/capita; thousand tonnes). The statistical information seems to show no relevant consequences from the international occurrences since 2019/2020, for instance, in crop production and meat consumption worldwide.

CHAPTER 5. GLOBAL AGRICULTURAL POLICY AFTER COVID-19

The objective of this research is to carry out an analysis of the agricultural policies' implications from, namely, the Covid-19 pandemic. Statistical information from the OECD (Organisation for Economic Co-operation and Development) database for the period 2018-2021 was taken into account for agricultural support and producer protection. It seems that, in specific frameworks, agricultural policies could have a more relevant involvement to support the farmers dealing with the challenges generated by the pandemic.

CHAPTER 6. THE REALITY OF SUSTAINABLE AGRICULTURE IN RECENT YEARS

This study intends to assess the consequences of recent events, specifically the Covid-19 pandemic, on the sustainability of agriculture worldwide. Data from OECD (Organisation for Economic Co-operation and Development) were taken into account for the agricultural land and nutrient balance over the period 2019-2021. There is a long way to run to increase the sustainability of agriculture worldwide and the contributions of the pandemic to open opportunities to mitigate the environmental implications from the agricultural sector are not yet visible.

CHAPTER 7. IMPLICATIONS OF THE PANDEMIC AND RECENT CONFLICTS ON THE EUROPEAN UNION DAIRY SECTOR

This chapter has as its principal objective to highlight impacts from the most recent international occurrences in the European Union dairy sector. Statistical information from the Eurostat database was taken into account over the period January 2019-October 2022 for indicators related to milk production and dairy products: raw cows' milk delivered to dairies; raw cream delivered to dairies (in milk equivalent); drinking milk; cream for direct consumption; milk and cream powders, excluding skimmed milk powders; skimmed milk powder; concentrated milk; acidified milk (yoghurts and other); butter (including dehydrated butter and ghee, and other fats and oils derived from milk; dairy spreads); and cheese from cows' milk (pure). These indicators were not significantly altered by the pandemic and recent crises.

CHAPTER 8. THE EVOLUTION OF POULTRY FARMING IN THE EUROPEAN UNION

This chapter intends to analyse the implications of the pandemic and recent crisis in the European Union poultry farming. Statistical information from the Eurostat was taken into account for the period January 2019 – September 2022 and for the following indicators related to the chicks of laying hen breeds (laying): chicks hatched; eggs placed in incubation; exports intra-EU; exports – total; exports extra-EU; imports intra-EU; imports – total; imports extra-EU. The consequences from the most recent occurrences seem modest in this sector, including in the international conjuncture.

CHAPTER 9. CHARACTERISATION OF THE EUROPEAN UNION MEAT SECTOR BEFORE AND AFTER THE PANDEMIC

This study aims to show the evolution of this sector after the Covid-19 pandemic. Data from the Eurostat were taken into account over the period January 2017 – September 2022 for the following indicators related to slaughtering in slaughterhouses in the European Union countries: bovine meat; pigmeat; meat of sheep and goats; and poultry meat. The greatest average values for the European Union context over the period considered were found for the pigmeat, followed by the poultry meat, bovine meat and meat of sheep and goats, respectively. The pandemic and the Russia-Ukraine crisis seem to have here reduced consequences.

CHAPTER 10. EVOLUTION OF THE AGRICULTURAL PRODUCER MARKETS IN THE EUROPEAN UNION

This chapter aims to analyse the European Union agricultural producer markets before and after the pandemic. Data from the Eurostat were taken into account over the period 2016-2021 for the following producer prices (euros): soft wheat - prices per 100 kg; durum wheat - prices per 100 kg; maize - prices per 100 kg; rice - prices per 100 kg; soya - prices per 100 kg; tomatoes in the open: all qualities - prices per 100 kg; lettuce in the open: all qualities - prices per 100 kg; carrots: all qualities - prices per 100 kg; onions: all qualities - prices per 100 kg; and dessert apples: all varieties - prices per 100 kg. There are visible implications from the pandemic on the producer prices, but they are not identical for the different production.

CHAPTER 11. CURRENT REALITIES OF PORTUGUESE ORGANIC MARKETS

This research assesses producer prices in the Portuguese organic markets. Statistical information was taken into account from the SIMA (Sistema de Informação de Mercados Agrícolas) database for the weeks from 2017 to 2022, namely for the producer prices of blueberries and strawberries across the diverse Portuguese regions and markets. The implications of the Covid-19 pandemic and Russia-Ukraine on these Portuguese markets seem not important in some cases and transitory in other cases.

CHAPTER 12. PORTUGUESE DAIRY MARKETS AFTER THE PANDEMIC

This chapter intends to assess the prices of milk production (at the farm level) and dairy products (at the factory level) over the period 2017-2022 in the Portuguese Mainland and Azores region. Data from the SIMA (Sistema de Informação de Mercados Agrícolas) database were taken into account. The implications on the producer prices of these sectors were bigger on the dairy products prices than on the milk prices and particularly from the Russia-Ukraine crisis.

CHAPTER 13. COVID-19 AND CONFLICT'S IMPACT ON PORTUGUESE LIVESTOCK PRODUCTION MARKETS

This study aims to analyse the consequences of the Covid-19 pandemic and the Russia-Ukraine crisis on Portuguese livestock producer prices. Data were taken into account from SIMA (Sistema de Informação de Mercados Agrícolas) for the weeks of 2017-2022, specifically for cow and country chicken carcasses. There is some evidence of implications on the producer prices from the most recent international occurrences, principally the Russia-Ukraine conflict.

CHAPTER 14. IMPLICATIONS OF RECENT EVENTS ON PORTUGUESE FRUIT PRODUCTION MARKETS

The progress of the fresh fruit producer prices in the Portuguese regions and markets was analysed. A special emphasis was given to the apple markets, considering their relevance for the diet of the Portuguese citizens. For that, data from the SIMA (Sistema de Informação de Mercados Agrícolas) were taken into account for the weeks from 2017 to 2022. The consequences from the most recent events, if, in fact, had importance, they seem to have been slight.

CHAPTER 15. CURRENT REALITIES FOR PORTUGUESE VEGETABLE AND CEREAL PRODUCTION MARKETS

It was assessed the potential implications of the pandemic and recent crises on the potatoes and wheat producer prices, taking into account statistical information for the weeks of 2017-2022 from the SIMA (Sistema de Informação de Mercados Agrícolas). The consequences seem to be greater on the wheat prices than on the potatoes ones and these consequences came specifically in 2022.

Vítor João Pereira Domingues Martinho
ESAV, CERNAS-IPV Research Centre, Polytechnic Institute of Viseu, Portugal

Chapter 1
Insights From the Literature on COVID-19 and the Agricultural Sector

ABSTRACT

The agricultural sector has an important contribution to the socio-economic dynamics worldwide, but it is also vulnerable to internal particularities and external conjunctures. The frameworks created by the Covid-19 pandemic had implications for several economic sectors and agriculture was no exception. This chapter intends to highlight insights from the literature about the interrelationships between the pandemic and the agricultural sector. Information was considered from the Web of Science (Core Collection) for 1614 documents on a search carried out on 04 January 2023 for the topics "Covid-19" and "agricultur". These data were analysed through bibliometric analysis, systematic review, and factor analysis. As insights, it is worth noting the attention dedicated by the scientific community to the interrelationships among the pandemic and agriculture; namely, for example, in countries such as the US, China, India, and England.*

INTRODUCTION

The Covid-19 pandemic topic has been addressed by the scientific community and researchers from diverse fields of science, some with perspectives about new technologies and sustainability (Ayan et al., 2022), others about the environment (Casado-Aranda et al., 2021), the agri-food chains (Das & Roy, 2022), food security (Martinho et al., 2022), agricultural production (Okolie & Ogundeji, 2022), socioeconomic dimensions (Patil et al., 2021), geospatial platforms (Pérez-Cutillas et al., 2023), digital transition (Rejeb, Rejeb, Abdollahi, et al., 2022) and food supply (Rejeb, Rejeb, Appolloni, et al., 2022). The bibliometric analysis was an approach considered in these researches.

The impacts of the Covid-19 pandemic have been enormous in several sustainability dimensions (Joshi et al., 2022), including in social dynamics (Quigley et al., 2022). The agricultural sector was no an exception, with diverse disturbances in the dairy sector (Abhijit et al., 2021) and meat markets

DOI: 10.4018/978-1-6684-8923-9.ch001

(Zielińska-Chmielewska et al., 2021), for example. The new contexts created by the pandemic brought additional concerns about the world population demand for food (Ahmad Rizal & Md Nordin, 2022) in quality (Éliás & Jámbor, 2021) and quantity (Rad et al., 2021). This is particularly worrying in countries with historical problems of poverty (Apaliya et al., 2022) and other difficulties (Rukasha et al., 2021). The new technologies and developments bring relevant contributions to the assessments of these recent challenges (Bahrami et al., 2022). Including to better respond to the difficulties created (Botene et al., 2021), detection (Patel et al., 2022) and carrying out predictions (Keshavamurthy et al., 2022), but promote other risks (Consavage Stanley et al., 2022). In fact, the events created by the pandemic are additional frameworks to those already existing and promoted by climate change, for instance (Belsey-Priebe et al., 2021). Nonetheless, the pandemic opened new opportunities and promoted new knowledge (Chandorkar et al., 2021).

The emergence of novel diseases is a health risk (White & Razgour, 2020), they are a consequence of increased human-animal interactions (Debnath et al., 2021) and require special attention from international organisations, governments and the scientific community (Dhama et al., 2021). The scientific community has an important role to produce knowledge, namely in the health domains (Heaney et al., 2021) and resource preservation (Hernández Castaño, 2021).

This literature review shows that has been brought relevant contributions by the scientific community about the several interrelationships between the Covid-19 pandemic and agriculture topics, considering bibliometric analysis and systematic reviews, but highlights also that there is a field to be explored. From this perspective, this research aims to bring more insights into these relationships through a systematic review based on bibliometric analysis (Martinho, 2022). For that, 1614 documents were considered from the Web of Science Core Collection (2023) in a search carried out on 4 January 2023 for the topics "Covid-19" and "agricultur*". For the bibliometric analysis, the procedures proposed by the software VOSviewer (van Eck & Waltman, 2010, 2022; VOSviewer, 2023) were followed. The Scopus (2023) database was also consulted. To better explore the different interrelations, a factor analysis (Kim & Mueller, 1978a, 1978b; Vincent, 1971; Torres-Reyna, n.d.) was carried out following Stata (StataCorp, 2017a, 2017b; Stata, 2023) procedures.

BIBLIOGRAPHIC DATA ANALYSIS

The top networked authors with the highest total link strength, for the topics "Covid-19" and "agricultur*" are the following (figure 1 and table 1): El Bilali, Hamid; Idris, Zainab; Loh, Soh Kheang; Parveez, Ghulam Kadir Ahmad; Iranmanesh, Mohammad. Some of the top 50 authors with the greatest total link strength have also high scores for the other metrics (documents, citations, normalised citation, average citations and average normalised citations). In general, the average publication year of the documents published by these researchers ranges between 2020 and 2022.

Figure 1. Network map for authors as items, bibliographic coupling links, fractional counting and 1 as the minimum number of documents of an author

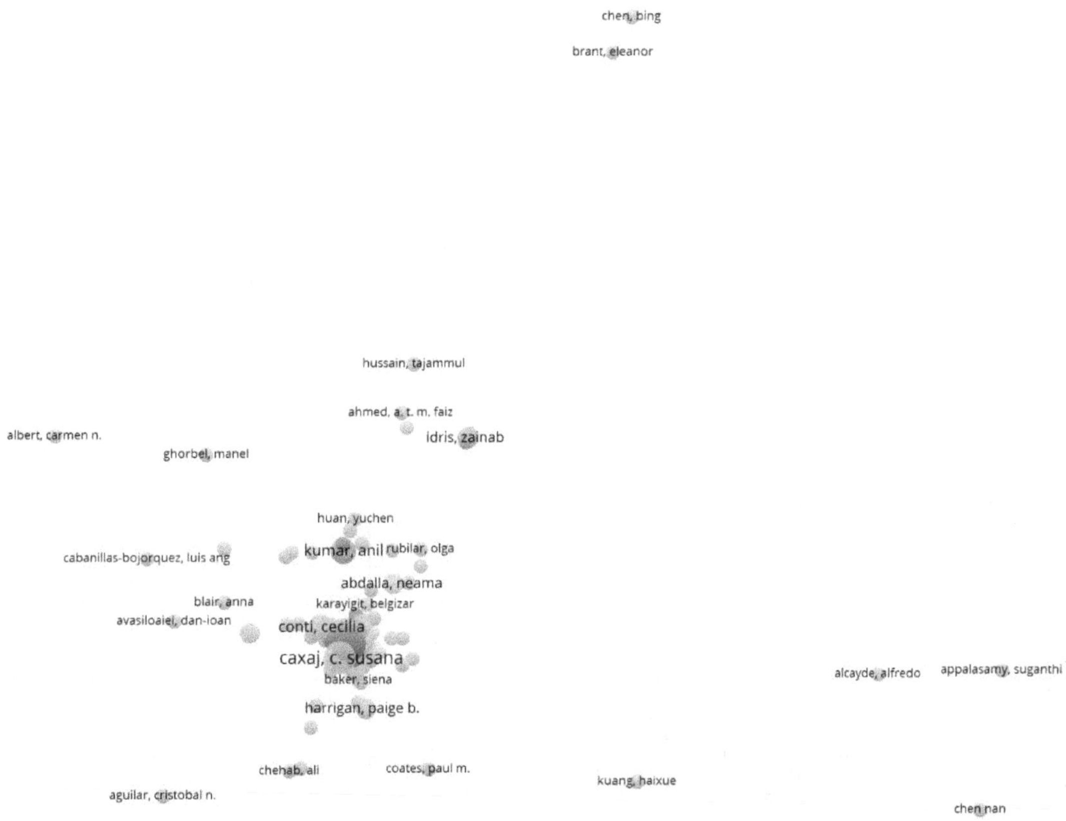

Table 1. Top 50 authors with the highest total link strength for bibliographic coupling links, fractional counting and 1 as the minimum number of documents of an author

Authors	Total link strength	Documents	Citations	Normalised citations	Average publication year	Average citations	Average normalised citations
El Bilali, Hamid	431	5	40	12	2022	8	2
Idris, Zainab	396	2	17	3	2022	9	2
Loh, Soh Kheang	396	2	17	3	2022	9	2
Parveez, Ghulam Kadir Ahmad	396	2	17	3	2022	9	2
Iranmanesh, Mohammad	393	2	11	6	2022	6	3
Rejeb, Abderahman	393	2	11	6	2022	6	3
Rejeb, Karim	393	2	11	6	2022	6	3
Haridy, Mohie A. M.	392	1	7	1	2021	7	1
Metwally, Asmaa A.	392	1	7	1	2021	7	1
Mohamed, Hams M. A.	392	1	7	1	2021	7	1
Saied, Abdulrahman A.	392	1	7	1	2021	7	1
Garg, Dixit	366	3	68	26	2022	23	9
Luthra, Sunil	366	3	68	26	2022	23	9
Yadav, Sanjeev	366	3	68	26	2022	23	9
Chehab, Ali	364	1	26	15	2022	26	15
Noura, Hassan N.	364	1	26	15	2022	26	15
Salman, Ola	364	1	26	15	2022	26	15
Yaacoub, Jean-Paul A.	364	1	26	15	2022	26	15
Springmann, Marco	341	2	6	2	2022	3	1
Coates, Paul M.	339	1	24	1	2020	24	1
Flaherty, Grace E.	339	1	24	1	2020	24	1
Fleischhacker, Sheila E.	339	1	24	1	2020	24	1
Glickman, Daniel R.	339	1	24	1	2020	24	1
Harkin, Thomas R.	339	1	24	1	2020	24	1
Hubbard, Van S.	339	1	24	1	2020	24	1
Kessler, David	339	1	24	1	2020	24	1
Li, William W.	339	1	24	1	2020	24	1
Loscalzo, Joseph	339	1	24	1	2020	24	1
Mozaffarian, Dariush	339	1	24	1	2020	24	1
Parekh, Anand	339	1	24	1	2020	24	1
Rowe, Sylvia	339	1	24	1	2020	24	1
Stover, Patrick J.	339	1	24	1	2020	24	1
Tagtow, Angie	339	1	24	1	2020	24	1
Woteki, Catherine E.	339	1	24	1	2020	24	1
Yun, Anthony Joon	339	1	24	1	2020	24	1
Morens, David M.	337	1	2	1	2022	2	1
Muller, Adrian	337	1	2	1	2022	2	1
Murray, Kris A.	337	1	2	1	2022	2	1

Continued on following page

Table 1. Continued

Authors	Total link strength	Documents	Citations	Normalised citations	Average publication year	Average citations	Average normalised citations
Saylors, Karen	337	1	2	1	2022	2	1
Sokolow, Susanne H.	337	1	2	1	2022	2	1
Wegner, Giulia, I	337	1	2	1	2022	2	1
Nordin, Shahrina Md	334	2	11	1	2022	6	1
Abdalla, Neama	304	2	16	3	2022	8	2
El-Ramady, Hassan	304	2	16	3	2022	8	2
Brant, Eleanor	300	1	49	6	2021	49	6
Budak, Hikmet	300	1	49	6	2021	49	6
Li, Chao	300	1	49	6	2021	49	6
Zhang, Baohong	300	1	49	6	2021	49	6
Jagtap, Sandeep	297	2	11	1	2022	6	1
Kumar, Vinod	293	2	35	12	2022	18	6

United States, China, India, England, Australia, Italy, Germany, France, Malaysia, Pakistan, Saudi Arabia, Canada, Spain, South Africa, Turkey, Netherlands, Nigeria, Egypt, Bangladesh and Iran are the networked countries of affiliation with the greatest total link strength and in some cases with great values for other metrics (figure 2 and table 2). Generally, the respective documents have 2021 as the average publication year.

Figure 2. Network map for countries as items, bibliographic coupling links, fractional counting and 1 as the minimum number of documents of a country

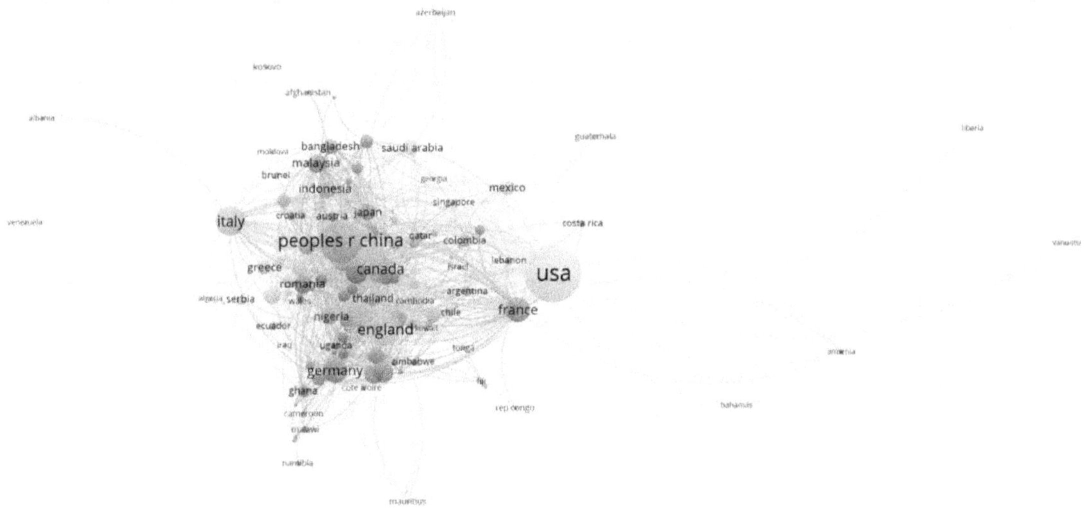

Table 2. Top 50 countries with the highest total link strength for bibliographic coupling links, fractional counting and 1 as the minimum number of documents of a country

Countries	Total link strength	Documents	Citations	Normalised citations	Average publication year	Average citations	Average normalised citations
USA	10647	360	3776	369	2021	10	1
Peoples R China	7063	186	1956	226	2021	11	1
India	6706	181	1454	258	2021	8	1
England	5653	109	1520	178	2021	14	2
Australia	4823	87	1011	135	2021	12	2
Italy	4254	95	1104	105	2021	12	1
Germany	3727	66	667	78	2021	10	1
France	3420	62	777	62	2021	13	1
Malaysia	2618	37	617	79	2021	17	2
Pakistan	2464	38	414	56	2021	11	1
Saudi Arabia	2315	29	305	49	2021	11	2
Canada	1969	78	1019	83	2021	13	1
Spain	1889	51	488	64	2021	10	1
South Africa	1885	45	207	35	2021	5	1
Turkey	1856	41	481	62	2022	12	2
Netherlands	1794	34	478	48	2021	14	1
Nigeria	1776	29	438	56	2021	15	2
Egypt	1680	26	281	34	2021	11	1
Bangladesh	1647	25	332	45	2021	13	2
Iran	1613	22	178	31	2021	8	1
U Arab Emirates	1520	19	459	89	2022	24	5
Brazil	1362	39	171	15	2021	4	0
Thailand	1354	29	163	20	2021	6	1
Switzerland	1320	19	218	26	2021	11	1
Indonesia	1316	38	110	17	2021	3	0
Poland	1285	34	158	18	2021	5	1
Kenya	1245	28	365	36	2021	13	1
Scotland	1207	22	207	20	2021	9	1
Mexico	1179	25	180	14	2021	7	1
Vietnam	1178	27	146	31	2021	5	1
Japan	1141	28	451	50	2021	16	2
Russia	1128	37	194	35	2021	5	1
South Korea	1118	21	256	33	2022	12	2
Portugal	1053	14	88	13	2021	6	1
Hungary	1020	16	108	20	2021	7	1
Tunisia	895	12	44	10	2021	4	1

Continued on following page

Table 2. Continued

Countries	Total link strength	Documents	Citations	Normalised citations	Average publication year	Average citations	Average normalised citations
Austria	858	12	118	14	2021	10	1
Greece	856	22	158	14	2021	7	1
New Zealand	846	13	142	13	2021	11	1
Sweden	775	13	267	29	2021	21	2
Czech Republic	745	19	36	11	2021	2	1
Ghana	712	16	97	16	2021	6	1
Qatar	701	8	15	8	2022	2	1
Romania	699	36	156	32	2021	4	1
Philippines	668	17	125	16	2021	7	1
Lebanon	664	6	28	16	2022	5	3
Morocco	660	12	119	6	2021	10	1
Belgium	646	15	242	26	2021	16	2
Singapore	636	8	71	5	2021	9	1
Denmark	634	8	152	16	2021	19	2

The top 10 organisations with the biggest total strength of the links between each other are the following (figure 3 and table 3): Chinese Acad Sci; Univ Calif Davis; Cornell Univ; Univ Edinburgh; South Valley Univ; Chinese Acad Agr Sci; Univ Sao Paulo; Wageningen Univ; Univ Exeter; Univ Oxford. These are some of the most important world organisations for the topics here addressed.

Figure 3. Network map for organisations as items, bibliographic coupling links, fractional counting and 1 as the minimum number of documents of an organisation

Table 3. Top 50 organisations with the highest total link strength for bibliographic coupling links, fractional counting and 1 as the minimum number of documents of an organisation

Organisations	Total link strength	Documents	Citations	Normalised citations	Average publication year	Average citations	Average normalised citations
Chinese Acad Sci	1118	17	75	16	2022	4	1
Univ Calif Davis	879	14	198	30	2021	14	2
Cornell Univ	847	16	223	18	2021	14	1
Univ Edinburgh	705	13	90	11	2021	7	1
South Valley Univ	686	3	34	4	2021	11	1
Chinese Acad Agr Sci	681	9	154	16	2021	17	2
Univ Sao Paulo	649	6	60	6	2021	10	1
Wageningen Univ	648	8	162	13	2021	20	2
Univ Exeter	625	9	266	19	2021	30	2
Univ Oxford	607	8	198	17	2021	25	2
Univ Calif Berkeley	578	15	343	31	2021	23	2
King Saud Univ	575	8	73	8	2021	9	1
Qatar Univ	569	7	15	8	2022	2	1
Univ Montpellier	560	11	92	6	2021	8	1
China Agr Univ	557	10	124	15	2021	12	1
Univ Chinese Acad Sci	548	7	32	8	2022	5	1
Banaras Hindu Univ	541	7	83	15	2021	12	2
Univ Guelph	527	14	232	17	2021	17	1
Univ Agr Faisalabad	522	7	46	4	2021	7	1
Australian Natl Univ	519	8	30	4	2021	4	0
South Ural State Univ	504	7	97	28	2022	14	4
Univ Sheffield	504	5	266	36	2021	53	7
Natl Inst Technol	502	4	85	26	2021	21	6
Univ Kebangsaan Malaysia	498	4	247	31	2021	62	8
Purdue Univ	497	13	153	16	2021	12	1
Michigan State Univ	485	10	111	13	2021	11	1
Edith Cowan Univ	479	4	16	7	2022	4	2
York Univ	476	7	16	5	2022	2	1
Univ Melbourne	474	8	76	13	2021	10	2
Univ Maryland	468	10	122	8	2021	12	1
North Carolina State Univ	451	5	15	2	2022	3	0
Univ Bourgogne Franche Comte	449	2	26	15	2022	13	7
Univ Malaysia Sabah	445	5	44	6	2022	9	1
Univ Aveiro	444	2	30	4	2021	15	2
Univ Michigan	442	7	43	5	2021	6	1
Nih	441	2	30	2	2021	15	1

Continued on following page

Table 3. Continued

Organisations	Total link strength	Documents	Citations	Normalised citations	Average publication year	Average citations	Average normalised citations
Int Livestock Res Inst	440	8	149	8	2021	19	1
Georgetown Univ	438	4	33	2	2021	8	1
Swedish Univ Agr Sci	435	4	44	5	2021	11	1
Amer Univ Beirut	432	3	28	16	2022	9	5
Stellenbosch Univ	431	8	37	6	2021	5	1
Natl Univ Singapore	426	3	59	4	2021	20	1
Kafrelsheikh Univ	423	3	17	4	2022	6	1
Leibniz Ctr Agr Landscape Res Zalf	419	5	19	2	2021	4	0
King Abdulaziz Univ	415	5	109	9	2021	22	2
Kansas State Univ	415	9	166	21	2021	18	2
Kharazmi Univ	414	3	44	18	2022	15	6
Univ Adelaide	409	4	14	1	2021	4	0
Penn State Univ	403	9	69	4	2021	8	0
Zhejiang Univ	402	7	108	8	2021	15	1

Some of the top documents with the highest metrics are the following (table 4): Wegner et al. (2022); Ekstrand et al. (2021); Rejeb, Rejeb, Appolloni, et al. (2022); Cardoso et al. (2021); Meekins et al. (2021).

Table 4. Top 20 documents with the highest total link strength for bibliographic coupling links, fractional counting and 0 as the minimum number of citations of a document

Documents	url	Total link strength	Citations	Normalised citations	Publication year
(Wegner et al., 2022)	https://doi.org/10.1016/j.eclinm.2022.101386	111	2	1	2022
(Ekstrand et al., 2021)	https://doi.org/10.3390/ani11072044	106	3	0	2021
(Rejeb, Rejeb, Appolloni, et al., 2022)	https://doi.org/10.3390/su141912437	101	0	0	2022
(Cardoso et al., 2021)	https://doi.org/10.3390/su13179799	101	10	1	2021
(Meekins et al., 2021)	https://doi.org/10.3390/v13101993	96	13	2	2021
(Baležentis et al., 2021)	https://doi.org/10.3390/su132111899	86	4	0	2021
(Rivera-Ferre et al., 2021)	https://doi.org/10.1016/j.agsy.2021.103134	83	21	3	2021
(Amahmid et al., 2022)	https://doi.org/10.1080/09603123.2021.1901865	83	5	3	2022
(Lang, Behl, Dong, Thu, et al., 2022)	https://doi.org/10.1108/ijlm-01-2021-0039	81	14	8	2022
(Talukder et al., 2022)	https://doi.org/10.1016/j.onehlt.2022.100416	80	1	1	2022
(Lang, Behl, Dong, Temouri, et al., 2022)	https://doi.org/10.1108/jic-01-2021-0039	79	4	2	2022
(Chen et al., 2022)	https://doi.org/10.1111/rode.12885	79	0	0	2022
(Aday & Aday, 2020)	https://doi.org/10.1093/fqsafe/fyaa024	78	226	9	2020
(Balcilar et al., 2021)	https://doi.org/10.1016/j.resourpol.2021.102219	77	44	5	2021
(Adelodun et al., 2021)	https://doi.org/10.1016/j.jclepro.2021.128451	75	18	2	2021
(Štreimikienė et al., 2022)	https://doi.org/10.1080/1331677x.2021.1919542	71	12	7	2022
(Haji Ali et al., 2021)	https://doi.org/10.1016/j.jece.2021.106201	71	5	1	2021
(Rejeb, Rejeb, Abdollahi, et al., 2022)	https://doi.org/10.3390/su14010083	70	11	6	2022
(Cao & Cheng, 2021)	https://doi.org/10.1016/j.resourpol.2021.102364	70	12	1	2021
(Priyadarshini & Abhilash, 2021)	https://doi.org/10.1016/j.gfs.2021.100537	67	15	2	2021

FACTOR ANALYSIS

Considering the all information obtained for the authors' network and taking into account factor analysis, three factors were obtained (table 5). Citations, normalised citations, average citations and average normalised citations define and are correlated with factor1 (table 6). On the other hand, factor2 is defined by the average publication year and factor3 by the documents. The total link strength has a great percentage of variance that is not shared by other variables (uniqueness). The Kaiser-Meyer-Olkin (kmo) measure shows the sampling adequacy (table 7). The overall results for the kmo are acceptable.

Table 5. Factor analysis with principal-component factors and orthogonal varimax, for authors and all networked data

Factor	Variance	Difference	Proportion	Cumulative
Factor1	3.490	2.256	0.499	0.499
Factor2	1.234	0.172	0.176	0.675
Factor3	1.062	.	0.152	0.827

Table 6. Rotated factor loadings, for authors and all networked data

Variable	Factor1	Factor2	Factor3	Uniqueness
Total link strength	0.351	0.402	-0.267	0.645
Documents	0.068	0.030	0.963	0.066
Citations	0.893	-0.364	0.083	0.063
Normalised Citations	0.923	0.155	0.199	0.084
Average Publication Year	-0.154	0.879	0.054	0.200
Average Citations	0.888	-0.364	-0.104	0.069
Average Normalised Citations	0.949	0.098	-0.049	0.087

Table 7. Sampling adequacy assessment through the Kaiser-Meyer-Olkin measure, for authors and all networked data

Variable	kmo
Total link strength	0.669
Documents	0.133
Citations	0.584
Normalised Citations	0.566
Average Publication Year	0.861
Average Citations	0.562
Average Normalised Citations	0.590
Overall	0.569

For the countries of affiliation were obtained two factors (table 8). Factor1 is defined by total link strength, documents, citations and normalised citations (table 9). The remaining variables (average publication year, average citations and average normalised citation) are correlated with factor2. Nonetheless, the uniqueness of the average publication year is greater than 0.5. The results for the kmo are in this case much better (table 10).

Table 8. Factor analysis with principal-component factors and orthogonal varimax, for countries and all networked data

Factor	Variance	Difference	Proportion	Cumulative
Factor1	3.909	1.951	0.558	0.558
Factor2	1.958	.	0.280	0.838

Table 9. Rotated factor loadings, for countries and all networked data

Variable	Factor1	Factor2	Uniqueness
Total link strength	0.982	0.063	0.032
Documents	0.987	0.008	0.025
Citations	0.983	0.095	0.024
Normalised Citations	0.987	0.096	0.017
Average Publication Year	-0.009	-0.691	0.522
Average Citations	0.066	0.940	0.113
Average Normalised Citations	0.158	0.758	0.400

Table 10. Sampling adequacy assessment through the Kaiser-Meyer-Olkin measure, for countries and all networked data

Variable	kmo
Total link strength	0.906
Documents	0.825
Citations	0.849
Normalised Citations	0.810
Average Publication Year	0.453
Average Citations	0.486
Average Normalised Citations	0.437
Overall	0.759

Three factors were found for the organisations of affiliation (table 11). Factor1 is defined by the citations, normalised citations, average citations and average normalised citations (table 12). Total link strength and documents are correlated with factor2 and the average publication year with factor3. The results for the kmo are acceptable (table 13).

Table 11. Factor analysis with principal-component factors and orthogonal varimax, for organisations and all networked data

Factor	Variance	Difference	Proportion	Cumulative
Factor1	2.723	0.499	0.389	0.389
Factor2	2.224	0.999	0.318	0.707
Factor3	1.225	.	0.175	0.882

Table 12. Rotated factor loadings, for organisations and all networked data

Variable	Factor1	Factor2	Factor3	Uniqueness
Total link strength	0.100	0.861	-0.019	0.249
Documents	0.006	0.936	-0.057	0.121
Citations	0.667	0.562	-0.305	0.147
Normalised Citations	0.783	0.534	0.077	0.096
Average Publication Year	-0.057	-0.056	0.975	0.044
Average Citations	0.868	-0.025	-0.400	0.086
Average Normalised Citations	0.948	-0.051	0.111	0.086

Table 13. Sampling adequacy assessment through the Kaiser-Meyer-Olkin measure, for organisations and all networked data

Variable	kmo
Total link strength	0.757
Documents	0.694
Citations	0.546
Normalised Citations	0.563
Average Publication Year	0.459
Average Citations	0.465
Average Normalised Citations	0.430
Overall	0.537

Table 14 presents the two factors found for the networked documents and table 15 shows that citations and normalised citations define factor1 and total link strength and publication year are correlated with factor2. The values for the kmo are very low (table 16).

Table 14. Factor analysis with principal-component factors and orthogonal varimax, for documents and all networked data

Factor	Variance	Difference	Proportion	Cumulative
Factor1	1.786	0.650	0.447	0.447
Factor2	1.136	.	0.284	0.731

Table 15. Rotated factor loadings, for documents and all networked data

Variable	Factor1	Factor2	Uniqueness
Total link strength	0.331	0.701	0.399
Citations	0.899	-0.236	0.135
Normalised Citations	0.840	0.224	0.245
Publication Year	-0.404	0.734	0.298

Table 16. Sampling adequacy assessment through the Kaiser-Meyer-Olkin measure, for documents and all networked data

Variable	kmo
Total link strength	0.651
Citations	0.444
Normalised Citations	0.423
Publication Year	0.323
Overall	0.418

LITERATURE REVIEW BASED ON BIBLIOMETRIC ANALYSIS

Table 17 presents insights from the literature review based on the top 20 documents identified through bibliometric analysis, considering bibliographic coupling links and bibliographic data from the Web of Science.

Table 17. Insights from the literature

Documents	Insights
(Wegner et al., 2022)	Discussion about the mitigation of zoonosis threats.
(Ekstrand et al., 2021)	Dogs and pigs seem to have low susceptibility to Covid-19.
(Rejeb, Rejeb, Appolloni, et al., 2022)	Contributions to better understand the food supply chains.
(Cardoso et al., 2021)	Policies are crucial to deal with the difficulties created by the pandemic.
(Meekins et al., 2021)	Understanding animal susceptibility to Covid-19 is crucial for policymakers.
(Baležentis et al., 2021)	The policies related to alleviation of the financial obligations seem to be the more effective.
(Rivera-Ferre et al., 2021)	Food system and food security have been impacted by the pandemic.
(Amahmid et al., 2022)	Sewage monitoring, water analysis and adjusted practices to control the dissemination of Covid-19.
(Lang, Behl, Dong, Thu, et al., 2022)	Social capital and agribusiness diversification have causal interrelationships.
(Talukder et al., 2022)	Impacts on the sustainable development goals.
(Lang, Behl, Dong, Temouri, et al., 2022)	Social capital is interrelated with the agribusiness diversification.
(Chen et al., 2022)	Contagion risks between crude oil and agricultural markets.
(Aday & Aday, 2020)	Covid-19 impacted the food supply chains since the producer until the consumer.
(Balcilar et al., 2021)	Crude oil affects other commodity markets.
(Adelodun et al., 2021)	Circular economy is a way for the sustainability in the agri-food systems.
(Štreimikienė et al., 2022)	Strategies to improve the agricultural resilience.
(Haji Ali et al., 2021)	More studies needed on wastewater aerosols in dissemination of Covid-19.
(Rejeb, Rejeb, Abdollahi, et al., 2022)	Contributions to understand the digital transition in the agri-food sector.
(Cao & Cheng, 2021)	For food and crude oil markets, the spillovers in the pandemic are weaker than under the financial crisis.
(Priyadarshini & Abhilash, 2021)	Suggestions for food security.

CONCLUSION

This chapter aimed to highlight the main insights from the scientific literature about the relations among the Covid-19 pandemic and agriculture. To achieve these objectives, data from the Web of Science Core Collection were considered on a search carried out on 04 January 2023. From this scientific database, 1614 documents were obtained for the topics "Covid-19" and "agricultur*". The topic "agricultur*" was considered in this way for a broader search and to identify documents were terms such as "agriculture" and "agricultural", for example. These documents were assessed through bibliometric analysis (bibliographic data and bibliographic coupling links) and systematic review. For a deeper assessment, factor analysis was also taken into account.

The literature review shows that the topics related with Covid-19 pandemic were already addressed abundantly, however, there are still subjects to be explored. The pandemic topic was considered in studies related to sustainability, food chains, food security and digital transition. Some of these dimensions have been impacted by the Covid-19 context. This is particularly serious in countries with old problems of development and already affected by other challenges. Some of these difficult frameworks are associated with climate change and global warming. In any case, these scenarios also create new opportunities and open doors for innovation. Digital technologies may bring here important contributions, as well as scientific research.

The bibliometric analysis highlights that the authors more strongly linked (with the highest total link strength) for the topics here considered are: El Bilali, Hamid; Idris, Zainab; Loh, Soh Kheang; Parveez, Ghulam Kadir Ahmad; Iranmanesh, Mohammad. For this metric (total link strength) the top countries are the United States, China, India, England, Australia, Italy, Germany, France, Malaysia, Pakistan, Saudi Arabia, Canada, Spain, South Africa, Turkey, Netherlands, Nigeria, Egypt, Bangladesh and Iran. The top organisations are: Chinese Acad Sci; Univ Calif Davis; Cornell Univ; Univ Edinburgh; South Valley Univ; Chinese Acad Agr Sci; Univ Sao Paulo; Wageningen Univ; Univ Exeter; Univ Oxford. These findings highlight the importance of these authors, countries and organisations of affiliation for the network analysed.

The factor analysis presents that the interrelationships between the metrics used (total link strength, documents, citations, normalised citations, average publication year, average citations and average normalised citations) are different among the three items assessed (authors, countries and organisations), showing that the network is distinct.

The literature survey based on the bibliometric analysis shows the following focus of the researchers on the topics addressed:

- Zoonosis threats;
- Animal susceptibility to Covid-19;
- Food chains;
- Importance of the policies;
- Food security;
- Agribusiness diversification;
- Contagion among commodity markets;
- Agricultural sustainability and resilience;
- Digital transition.

In terms of practical implications, this study reveals that there are subjects that may be further analysed, namely those related to the producer prices and those relationships with the prices in the retailers' chains. On the other hand, there is a field to be explored about the impacts of the Covid-19 pandemic on other countries, some of them with specific particularities. For policy recommendations, it is suggested to better understand the implication of the pandemic on the agricultural sector and, in consequence, to design more adjusted policies to support the farmers, particularly the most vulnerable. For future research, it could be interesting to analyse the consequences of the Russia-Ukraine crisis in the sector and bring more insights into the producer markets.

ACKNOWLEDGMENT

This work is funded by National Funds through the FCT - Foundation for Science and Technology, I.P., within the scope of the project Refª UIDB/00681/2020. Furthermore we would like to thank the CERNAS Research Centre and the Polytechnic Institute of Viseu for their support.

REFERENCES

Abhijit, D. A. S., Sivaram, M., & Thejesh, S. (2021). Economic impact of COVID-19 pandemic on dairy sector: A meta-analysis. [Scopus.]. *The Indian Journal of Animal Sciences*, *91*(7), 582–594.

Aday, S., & Aday, M. S. (2020). Impact of COVID-19 on the food supply chain. *Food Quality and Safety*, *4*(4), 167–180. doi:10.1093/fqsafe/fyaa024

Adelodun, B., Kareem, K. Y., Kumar, P., Kumar, V., Choi, K. S., Yadav, K. K., Yadav, A., El-Denglawey, A., Cabral-Pinto, M., Son, C. T., Krishnan, S., & Khan, N. A. (2021). Understanding the impacts of the COVID-19 pandemic on sustainable agri-food system and agroecosystem decarbonization nexus: A review. *Journal of Cleaner Production*, *318*, 128451. doi:10.1016/j.jclepro.2021.128451 PMID:36570877

Ahmad Rizal, A. R., & Md Nordin, S. (2022). Getting ahead of the pandemic curve: A systematic review of critical determining factors for innovation adoption in ensuring food security. *Frontiers in Nutrition, 9*. Scopus. doi:10.3389/fnut.2022.986324

Amahmid, O., El Guamri, Y., Rakibi, Y., Ouizat, S., Yazidi, M., Razoki, B., Kaid Rassou, K., Asmama, S., Bouhoum, K., & Belghyti, D. (2022). Occurrence of SARS-CoV-2 in excreta, sewage, and environment: Epidemiological significance and potential risks. *International Journal of Environmental Health Research*, *32*(8), 1686–1706. doi:10.1080/09603123.2021.1901865 PMID:33752527

Apaliya, M. T., Kwaw, E., Osae, R., Alolga, R. N., Aikins, A. S. S., Otoo, G. S., Kaburi, S. A., Lamptey, F. P., & Amo-Broni, C. (2022). THE IMPACT OF COVID-19 ON FOOD SECURITY: GHANA IN REVIEW. *Journal of Food Technology Research, 9*(3), 160–175. Scopus. doi:10.18488/jftr.v9i3.3228

Ayan, B., Güner, E., & Son-Turan, S. (2022). Blockchain Technology and Sustainability in Supply Chains and a Closer Look at Different Industries: A Mixed Method Approach. *Logistics, 6*(4). Scopus. doi:10.3390/logistics6040085

Bahrami, H., McNairn, H., Mahdianpari, M., & Homayouni, S. (2022). A Meta-Analysis of Remote Sensing Technologies and Methodologies for Crop Characterization. *Remote Sensing, 14*(22). Scopus. doi:10.3390/rs14225633

Balcilar, M., Gabauer, D., & Umar, Z. (2021). Crude Oil futures contracts and commodity markets: New evidence from a TVP-VAR extended joint connectedness approach. *Resources Policy*, *73*, 102219. doi:10.1016/j.resourpol.2021.102219

Baležentis, T., Morkūnas, M., Žičkienė, A., Volkov, A., Ribašauskienė, E., & Štreimikienė, D. (2021). Policies for Rapid Mitigation of the Crisis' Effects on Agricultural Supply Chains: A Multi-Criteria Decision Support System with Monte Carlo Simulation. *Sustainability*, *13*(21), 21. doi:10.3390u132111899

Belsey-Priebe, M., Lyons, D., & Buonocore, J. J. (2021). COVID-19′ s impact on american women's food insecurity foreshadows vulnerabilities to climate change. *International Journal of Environmental Research and Public Health, 18*(13). Scopus. doi:10.3390/ijerph18136867

Botene, P. H. R., de Azevedo, A. T., & de Arruda Ignácio, P. S. (2021). Blockchain as an enabling technology in the COVID-19 pandemic: A systematic review. *Health and Technology, 11*(6), 1369–1382. Scopus. doi:10.1007/s12553-021-00593-z

Cao, Y., & Cheng, S. (2021). Impact of COVID-19 outbreak on multi-scale asymmetric spillovers between food and oil prices. *Resources Policy, 74*, 102364. doi:10.1016/j.resourpol.2021.102364 PMID:34584328

Cardoso, B., Cunha, L., Leiras, A., Gonçalves, P., Yoshizaki, H., de Brito, I. Junior, & Pedroso, F. (2021). Causal Impacts of Epidemics and Pandemics on Food Supply Chains: A Systematic Review. *Sustainability, 13*(17), 17. doi:10.3390u13179799

Casado-Aranda, L.-A., Sánchez-Fernández, J., & Viedma-del-Jesús, M. I. (2021). Analysis of the scientific production of the effect of COVID-19 on the environment: A bibliometric study. *Environmental Research, 193*. Scopus. doi:10.1016/j.envres.2020.110416

Chandorkar, N., Tambe, S., Amin, P., & Madankar, C. (2021). *A systematic and comprehensive review on current understanding of the pharmacological actions, molecular mechanisms, and clinical implications of the genus Eucalyptus. Phytomedicine Plus, 1(4)*. Scopus., doi:10.1016/j.phyplu.2021.100089

Chen, Z., Yan, B., & Kang, H. (2022). Dynamic correlation between crude oil and agricultural futures markets. *Review of Development Economics, 26*(3), 1798–1849. doi:10.1111/rode.12885

Consavage Stanley, K., Harrigan, P. B., Serrano, E. L., & Kraak, V. I. (2022). A systematic scoping review of the literacy literature to develop a digital food and nutrition literacy model for low-income adults to make healthy choices in the online food retail ecosystem to reduce obesity risk. *Obesity Reviews, 23*(4). Scopus. doi:10.1111/obr.13414

Das, N. K., & Roy, A. (2022). COVID-19 and agri-food value chain: A systematic review and bibliometric mapping. *Journal of Agribusiness in Developing and Emerging Economies, 12*(3), 442–462. Scopus. doi:10.1108/JADEE-07-2021-0188

Debnath, F., Chakraborty, D., Deb, A., Saha, M., & Dutta, S. (2021). Increased human-animal interface & emerging zoonotic diseases: An enigma requiring multi-sectoral efforts to address. *Indian Journal of Medical Research, 153*(5), 577–584. Scopus. doi:10.4103/ijmr.IJMR_2971_20

Dhama, K., Patel, S. K., Yatoo, M. I., Tiwari, R., Sharun, K., Dhama, J., Natesan, S., Malik, Y. S., Singh, K. P., & Harapan, H. (2021). SARS-CoV-2 existence in sewage and wastewater: A global public health concern? *Journal of Environmental Management, 280*. Scopus. doi:10.1016/j.jenvman.2020.111825

Ekstrand, K., Flanagan, A. J., Lin, I. E., Vejseli, B., Cole, A., Lally, A. P., Morris, R. L., & Morgan, K. N. (2021). Animal Transmission of SARS-CoV-2 and the Welfare of Animals during the COVID-19 Pandemic. *Animals (Basel), 11*(7), 7. Advance online publication. doi:10.3390/ani11072044 PMID:34359172

Éliás, B. A., & Jámbor, A. (2021). Food security and covid-19: A systematic review of the first-year experience. *Sustainability (Switzerland), 13*(9). Scopus. doi:10.3390/su13095294

Haji Ali, B., Shahin, M. S., Masoumi Sangani, M. M., Faghihinezhad, M., & Baghdadi, M. (2021). Wastewater aerosols produced during flushing toilets, WWTPs, and irrigation with reclaimed municipal wastewater as indirect exposure to SARS-CoV-2. *Journal of Environmental Chemical Engineering, 9*(5), 106201. doi:10.1016/j.jece.2021.106201 PMID:34405082

Heaney, A. K., Head, J. R., Broen, K., Click, K., Taylor, J., Balmes, J. R., Zelner, J., & Remais, J. V. (2021). Coccidioidomycosis and COVID-19 co-infection, United States, 2020. *Emerging Infectious Diseases, 27*(5), 1266–1273. Scopus. doi:10.3201/eid2705.204661

Hernández Castaño, C. (2021). Advances in the production of knowledge on biofertilizers and new economic scenarios for Peru. *Revista Venezolana de Gerencia, 26*(Special Issue 5), 708–721. Scopus. doi:10.52080/rvgluz.26.e5.45

Joshi, T., Poudel, R. P., Kafle, K., Bhattarai, B., Prasai, B. P., & Adhikari, S. (2022). Assessing the Impacts of the COVID-19 Pandemics on Sustainable Development Goals in Nepal. *Frontiers in Sustainable Food Systems, 6*. Scopus. doi:10.3389/fsufs.2022.852759

Keshavamurthy, R., Dixon, S., Pazdernik, K. T., & Charles, L. E. (2022). Predicting infectious disease for biopreparedness and response: A systematic review of machine learning and deep learning approaches. *One Health, 15*. Scopus. doi:10.1016/j.onehlt.2022.100439

Kim, J.-O., & Mueller, C. W. (1978a). Factor Analysis: Statistical Methods and Practical Issues. *Sage (Atlanta, Ga.).*

Kim, J.-O., & Mueller, C. W. (1978b). *Introduction to Factor Analysis: What It Is and How To Do It.* SAGE Publications. doi:10.4135/9781412984652

Lang, L. D., Behl, A., Dong, N. T., Temouri, Y., & Thu, N. H. (2022). Effect of social capital on agribusiness diversification intention in the emerging market. *Journal of Intellectual Capital, 23*(1), 56–84. doi:10.1108/JIC-01-2021-0039

Lang, L. D., Behl, A., Dong, N. T., Thu, N. H., & Dewani, P. P. (2022). Social capital in agribusiness: An exploratory investigation from a supply chain perspective during the COVID-19 crisis. *International Journal of Logistics Management, 33*(4), 1437–1473. doi:10.1108/IJLM-01-2021-0039

Martinho, V. J. P. D. (2022). Impacts of the COVID-19 Pandemic and the Russia–Ukraine Conflict on Land Use across the World. *Land (Basel), 11*(10), 10. Advance online publication. doi:10.3390/land11101614

Martinho, V. J. P. D., Cunha, C. A. D. S., Pato, M. L., Costa, P. J. L., Sánchez-Carreira, M. C., Georgantzís, N., Rodrigues, R. N., & Coronado, F. (2022). Machine Learning and Food Security: Insights for Agricultural Spatial Planning in the Context of Agriculture 4.0. *Applied Sciences (Switzerland), 12*(22). Scopus. doi:10.3390/app122211828

Meekins, D. A., Gaudreault, N. N., & Richt, J. A. (2021). Natural and Experimental SARS-CoV-2 Infection in Domestic and Wild Animals. *Viruses, 13*(10), 10. Advance online publication. doi:10.3390/v13101993 PMID:34696423

Okolie, C. C., & Ogundeji, A. A. (2022). *Effect of COVID-19 on agricultural production and food security: A scientometric analysis. Humanities and Social Sciences Communications, 9(1)*. Scopus., doi:10.105741599-022-01080-0

Patel, S. K., Surve, J., Parmar, J., Ahmed, K., Bui, F. M., & Al-Zahrani, F. A. (2022). Recent Advances in Biosensors for Detection of COVID-19 and Other Viruses. *IEEE Reviews in Biomedical Engineering*, 1–16. Scopus. doi:10.1109/RBME.2022.3212038

Patil, S., Verma, A., Patil, I., Singh, R., & Gaur, R. (2021). Bibliometric Survey on Effect of Socio-Economic Factors on Spread of Corona virus (COVID-19). [Scopus.]. *Library Philosophy and Practice*, *2021*, 1–16.

Pérez-Cutillas, P., Pérez-Navarro, A., Conesa-García, C., Zema, D. A., & Amado-Álvarez, J. P. (2023). *What is going on within google earth engine? A systematic review and meta-analysis. Remote Sensing Applications: Society and Environment, 29*. Scopus., doi:10.1016/j.rsase.2022.100907

Priyadarshini, P., & Abhilash, P. C. (2021). Agri-food systems in India: Concerns and policy recommendations for building resilience in post COVID-19 pandemic times. *Global Food Security*, *29*, 100537. doi:10.1016/j.gfs.2021.100537 PMID:35155096

Quigley, D. D., Qureshi, N., Gahlon, G., & Gidengil, C. (2022). Worker and employer experiences with COVID-19 and the California Workers' Compensation System: A review of the literature. *American Journal of Industrial Medicine, 65*(3), 203–213. Scopus. doi:10.1002/ajim.23326

Rad, A. K., Shamshiri, R. R., Azarm, H., Balasundram, S. K., & Sultan, M. (2021). Effects of the covid-19 pandemic on food security and agriculture in iran: A survey. *Sustainability (Switzerland), 13*(18). Scopus. doi:10.3390/su131810103

Rejeb, A., Rejeb, K., Abdollahi, A., Zailani, S., Iranmanesh, M., & Ghobakhloo, M. (2022). Digitalization in food supply chains: A bibliometric review and key-route main path analysis. *Sustainability (Switzerland), 14*(1). Scopus. doi:10.3390/su14010083

Rejeb, A., Rejeb, K., Appolloni, A., Iranmanesh, M., Treiblmaier, H., & Jagtap, S. (2022). Exploring Food Supply Chain Trends in the COVID-19 Era: A Bibliometric Review. *Sustainability (Switzerland), 14*(19). Scopus. doi:10.3390/su141912437

Rivera-Ferre, M. G., López-i-Gelats, F., Ravera, F., Oteros-Rozas, E., di Masso, M., Binimelis, R., & El Bilali, H. (2021). The two-way relationship between food systems and the COVID19 pandemic: Causes and consequences. *Agricultural Systems*, *191*, 103134. doi:10.1016/j.agsy.2021.103134

Rukasha, T., Nyagadza, B., Pashapa, R., & Muposhi, A. (2021). Covid-19 impact on Zimbabwean agricultural supply chains and markets: A sustainable livelihoods perspective. *Cogent Social Sciences, 7*(1). Scopus. doi:10.1080/23311886.2021.1928980

Scopus. (2023). *Scopus Database*. Scopus. https://www.scopus.com/search/form.uri?display=basic#basic

Stata. (2023). *Statistical software for data science | Stata*. https://www.stata.com/

StataCorp. (2017a). *Stata 15 Base Reference Manual*. Stata Press.

StataCorp. (2017b). *Stata Statistical Software: Release 15*. StataCorp LLC.

Štreimikienė, D., Baležentis, T., Volkov, A., Ribašauskienė, E., Morkūnas, M., & Žičkienė, A. (2022). Negative effects of covid-19 pandemic on agriculture: Systematic literature review in the frameworks of vulnerability, resilience and risks involved. *Economic Research-Ekonomska Istraživanja, 35*(1), 529–545. doi:10.1080/1331677X.2021.1919542

Talukder, B., vanLoon, G. W., & Hipel, K. W. (2022). Planetary health & COVID-19: A multi-perspective investigation. *One Health, 15*, 100416. doi:10.1016/j.onehlt.2022.100416 PMID:35892119

Torres-Reyna, O. (n.d.). *Getting Started in Factor Analysis (using Stata) (ver. 1.0 beta/draft).* https://www.princeton.edu/~otorres/Stata/Factor

van Eck, N. J., & Waltman, L. (2010). Software survey: VOSviewer, a computer program for bibliometric mapping. *Scientometrics, 84*(2), 523–538. doi:10.100711192-009-0146-3 PMID:20585380

van Eck, N. J., & Waltman, L. (2022). *Manual for VOSviewer version 1. 6.18.* CWTS, Universiteit Leiden. https://www.vosviewer.com/documentation/Manual_VOSviewer_1.6.18.pdf

Vincent, J. E. (1971). *Factor Analysis in International Relations: Interpretation, Problem Areas, and an Application.* University of Florida Press.

VOSviewer. (2023). *VOSviewer—Visualizing scientific landscapes—Version 1.6.18.* VOSviewer. https://www.vosviewer.com//

Web of Science. (2023). *Web of Science Core Collection.* https://www.webofscience.com/wos/woscc/basic-search

Wegner, G. I., Murray, K. A., Springmann, M., Muller, A., Sokolow, S. H., Saylors, K., & Morens, D. M. (2022). Averting wildlife-borne infectious disease epidemics requires a focus on socio-ecological drivers and a redesign of the global food system. *EClinicalMedicine, 47*, 101386. doi:10.1016/j.eclinm.2022.101386 PMID:35465645

White, R. J., & Razgour, O. (2020). Emerging zoonotic diseases originating in mammals: A systematic review of effects of anthropogenic land-use change. *Mammal Review, 50*(4), 336–352. Scopus. doi:10.1111/mam.12201

Zielińska-Chmielewska, A., Mruk-Tomczak, D., & Wielicka-Regulska, A. (2021). Qualitative research on solving difficulties in maintaining continuity of food supply chain on the meat market during the COVID-19 pandemic. *Energies, 14*(18). Scopus. doi:10.3390/en14185634

Chapter 2
A Systematic Review of the Impacts on Agriculture of the Russia-Ukraine Scenario

ABSTRACT

When the world began to be prepared to return to normalcy after months of disturbances with the pandemic, a new worrying context appeared with the Russia-Ukraine crisis. This conflict brought implications in the different socioeconomic dimensions, specifically those related to energy prices and food chains. This research aims to analyse the literature highlights about the several interlinkages between the farming sector and the Russia-Ukraine conflict. For that, 106 documents from the Web of Science (Core Collection) were considered for the topics "Russia-Ukraine conflict," or "Russia-Ukraine crisis," or "Russia-Ukraine war," and "agricultur," on a search carried out on 04 January 2023. This information was assessed through bibliometric analysis, systematic review, and factor-cluster analyses. The topics associated with the Russia-Ukraine conflict and the agricultural sectors have attracted the attention of researchers, but there is still a field to be explored.*

INTRODUCTION

The Russia-Ukraine conflict appears as an additional exogenous shock for the socioeconomic context worldwide (Abu Hatab, 2022) with risks of systemic contamination (Qureshi et al., 2022). Inflation seems to be one of the major problems, namely for the most vulnerable economies (Aharon & Qadan, 2022), as well as the volatility in the financial (Beraich et al., 2022) and commodity markets (Fang & Shao, 2022). The stock market efficiency (Gaio et al., 2022), international monetary system stability (Miao & Fei, 2022), international stock markets volatility (Mu et al., 2022), global economic uncertainty (Shen & Hong, 2023) and the spillovers among agricultural commodity markets (Just & Echaust, 2022) are also challenges.

These scenarios jointly with the supply chain constraints are threats to the international conjuncture, namely to food security (Ben Hassen & El Bilali, 2022), global food crisis (Elbasuney et al., 2022), food

DOI: 10.4018/978-1-6684-8923-9.ch002

chains (Jagtap et al., 2022), including in China (Li & Song, 2022), energy markets dynamics (Mo & Wang, 2022), including for US (Shepard & Pratson, 2022), investors priorities (Singh et al., 2022) and trade sustainability (Cabelkova et al., 2022). There are specific impacts, such as those related to broiler producers (Myeki et al., 2022).

The consequences of this crisis are economic, but have other dimensions, such as those related to health (Chaaya, Thambi, et al., 2022), environment (Rawtani et al., 2022), refugees (Lewtak et al., 2022), migrations (Subramaniam & Rajoo, 2022), dietary patterns (Yazbeck et al., 2022), opinions (Ibar-Alonso et al., 2022) and comments (Lewandowska-Tomaszczyk, 2022) about current topics. This conflict promote also discussions of the laws, specifically the naval ones (Fink, 2022), corporate decisions (Tosun & Eshraghi, 2022), sustainability of the agricultural systems (Yang & Liu, 2022) and food resilience (Sun et al., 2022). In any case, these events open always alternatives (Huang et al., 2022), opportunities and challenges for the investors (Huang et al., 2023) in an asymmetric way (Nerlinger & Utz, 2022).

Another question is about the perception of the citizens about the news related to this conflict, which is an additional ingredient in these frameworks (Alyukov, 2022), about the role of the states in these processes (Dylan & Maguire, 2022) and the roots of the crisis (Henrikson, 2022), including the religious (Jha et al., 2022).

International organisations have a crucial role in world stability (Anil & Kumar, 2022), however not always without failures (Murithi, 2022). Other organisations have also their influence (Deniz & Ozgur, 2022). The policy instruments may have here a fundamental contribution, namely in the European Union in terms of energy and climate measures (Costantini et al., 2022). The digital transition may contribute significantly to deal with the difficulties created by recent events (Nakandala et al., 2022), including in the food sector (Martinho et al., 2022).

Considering the framework described before, this study intends to highlight the several dimensions associated with the Russia-Ukraine conflict and the agricultural sector. Bibliometric analysis and systematic review were carried out. For that, 106 documents were considered from the Web of Science Core Collection (2023) in a search performed on 4 January 2023 for the following topics: "Russia-Ukraine conflict", or "Russia-Ukraine crisis", or "Russia-Ukraine war" and "agricultur*". For the bibliometric assessment, the VOSviewer (van Eck & Waltman, 2010, 2022; VOSviewer, 2023) software procedures were followed, as well as the Martinho (2022) developments. To complement these insights a cluster analysis (Vincent, 1971; Kim & Mueller, 1978a, 1978b; Torres-Reyna, n.d.) was taken into account following Stata (StataCorp, 2017a, 2017b; Stata, 2023) software procedures.

BIBLIOGRAPHIC DATA ASSESSMENT

The top 10 networked authors with the biggest total link strength, for the topics here addressed ("Russia-Ukraine conflict", or "Russia-Ukraine crisis", or "Russia-Ukraine war" and "agricultur*"), are the following (figure 1 and table 1): Dylan, Huw; Maguire, Thomas J.; Cross, Mai'a K. Davis; Karolewski, Ireneusz Pawel; Kormych, Borys; Malyarenko, Tetyana; Liu, Weidong; Yang, Xinyu; Bak, Iwona; Singh, Guru Ashish. Some of these items have also great results for other metrics.

Figure 1. Networked authors for bibliographic coupling links, fractional counting and 1 as the minimum number of documents of an author

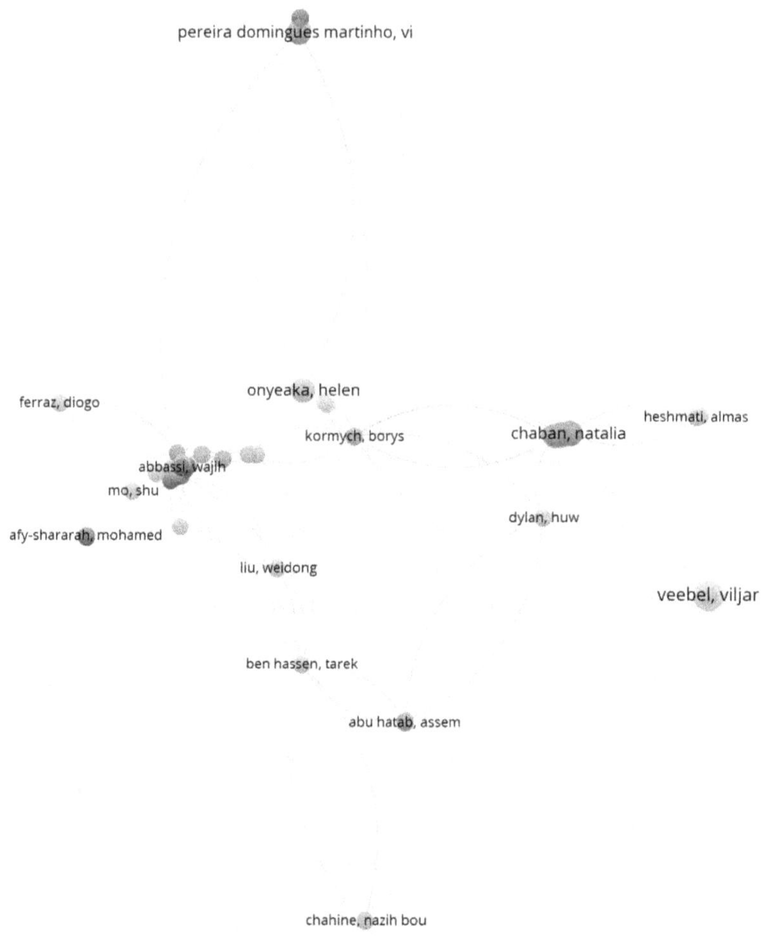

Table 1. Top 50 items (authors) with the greatest values for the total link strength, for bibliographic coupling links, fractional counting and 1 as the minimum number of documents of an author

Authors	Total link strength	Documents	Citations	Normalised citations	Average publication year	Average citations	Average normalised citations
Dylan, Huw	150	1	0	0	2022	0	0
Maguire, Thomas J.	150	1	0	0	2022	0	0
Cross, Mai'a K. Davis	140	2	31	2	2017	16	1
Karolewski, Ireneusz Pawel	140	2	31	2	2017	16	1
Kormych, Borys	134	1	1	0		1	
Malyarenko, Tetyana	134	1	1	0		1	
Liu, Weidong	117	1	0	0	2022	0	0
Yang, Xinyu	117	1	0	0	2022	0	0
Bak, Iwona	109	1	0	0	2022	0	0
Singh, Guru Ashish	109	1	0	0	2022	0	0
Singh, Uma Shankar	109	1	0	0	2022	0	0
Tarczynska-Luniewska, Malgorzata	109	1	0	0	2022	0	0
Li, Jingdong	102	1	0	0	2022	0	0
Song, Zhouying	102	1	0	0	2022	0	0
Afy-Shararah, Mohamed	101	1	7	4	2022	7	4
Duong, Linh	101	1	7	4	2022	7	4
Garcia-Garcia, Guillermo	101	1	7	4	2022	7	4
Hassoun, Abdo	101	1	7	4	2022	7	4
Hdaifeh, Ammar	101	1	7	4	2022	7	4
Jagtap, Sandeep	101	1	7	4	2022	7	4
Lorenzo, Jose M.	101	1	7	4	2022	7	4
Martindale, Wayne	101	1	7	4	2022	7	4
Munekata, Paulo E. S.	101	1	7	4	2022	7	4
Parra-Lopez, Carlos	101	1	7	4	2022	7	4
Salonitis, Konstantinos	101	1	7	4	2022	7	4
Trollman, Frank	101	1	7	4	2022	7	4
Trollman, Hana	101	1	7	4	2022	7	4
Pereira Domingues Martinho, Vitor Joao	89	2	1	1	2022	1	0
Ben Hassen, Tarek	89	1	12	6	2022	12	6
El Bilali, Hamid	89	1	12	6	2022	12	6
Ferraz, Diogo	87	1	0	0		0	
Mohammed, Kamel Si	87	1	0	0		0	
Shahzadi, Irum	87	1	0	0		0	
Tiwari, Sunil	87	1	0	0		0	
Chaban, Natalia	85	2	17	4	2019	9	2

Continued on following page

Table 1. Continued

Authors	Total link strength	Documents	Citations	Normalised citations	Average publication year	Average citations	Average normalised citations
Elgstrom, Ole	85	2	17	4	2019	9	2
Knodt, Michele	85	2	17	4	2019	9	2
Coronado, Freddy	85	1	0	0	2022	0	0
Da Silva Cunha, Carlos Augusto	85	1	0	0	2022	0	0
Georgantzis, Nikolaos	85	1	0	0	2022	0	0
Lourenco Costa, Paulo Jorge	85	1	0	0	2022	0	0
Pato, Maria Lucia	85	1	0	0	2022	0	0
Rodrigues, Raimundo Nonato	85	1	0	0	2022	0	0
Sanchez-Carreira, Maria Carmen	85	1	0	0	2022	0	0
Islam, Md Monirul	76	1	0	0		0	
Mansour, Hoda	76	1	0	0		0	
Sohag, Kazi	76	1	0	0		0	
Zikovic, Ivana Tomas	76	1	0	0		0	
Gupta, Gunjan	65	1	3	2	2022	3	2
Hussain, Chaudhery Mustansar	65	1	3	2	2022	3	2

England, Germany, India, France, USA, Poland, Norway, Spain Netherlands, Russia, Brazil, Turkey, China, Italy and Portugal are some of the countries of researchers' affiliation with the greatest total strength for the links among items (figure 2 and table 2). The average publication year ranges between 2015 and 2022.

Figure 2. Networked countries for bibliographic coupling links, fractional counting and 1 as the minimum number of documents of a country

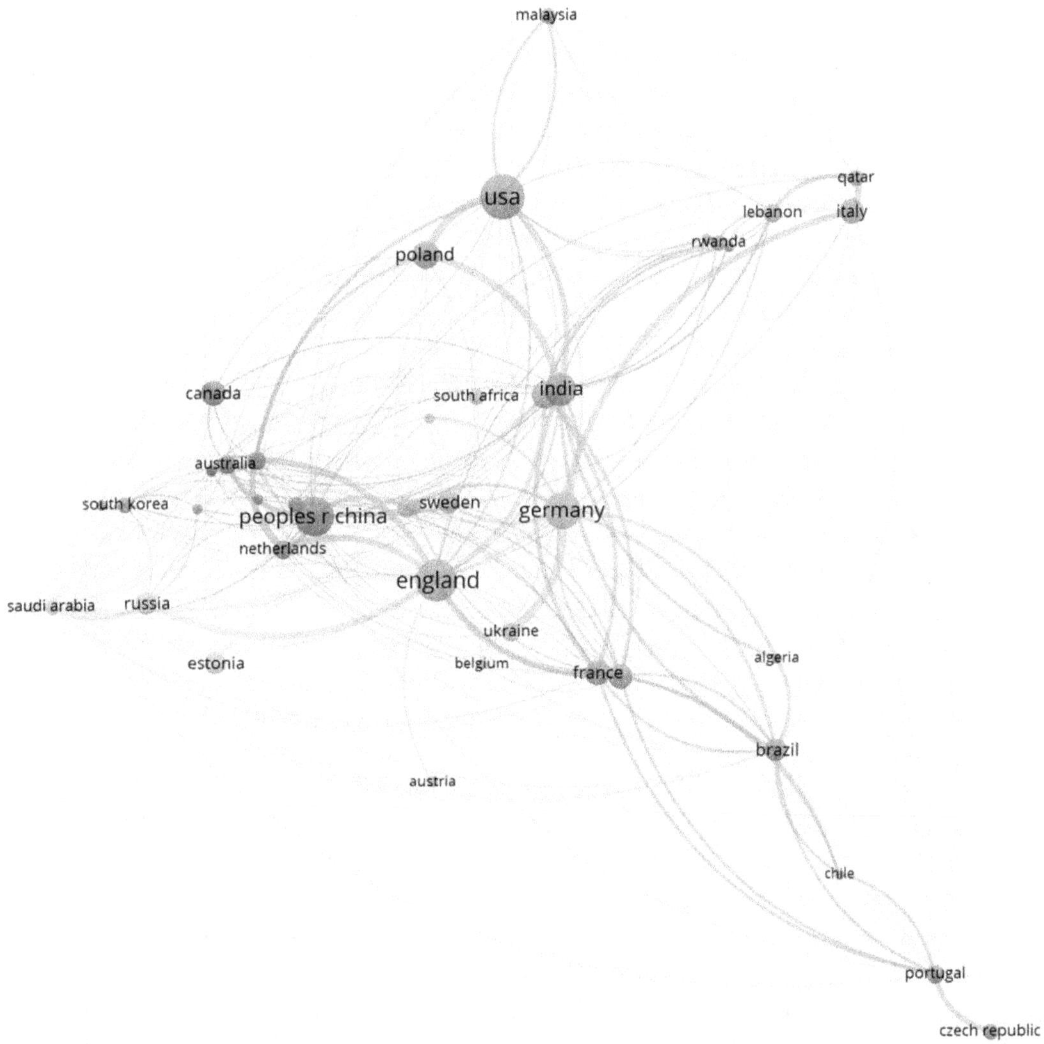

Table 2. Networked countries for bibliographic coupling links, fractional counting and 1 as the minimum number of documents of a country

Countries	Total link strength	Documents	Citations	Normalised citations	Average publication year	Average citations	Average normalised citations
England	475	14	32	18	2021	2	1
Germany	401	11	55	7	2017	5	1
India	384	9	18	6	2022	2	1
France	303	5	16	5	2022	3	1
Usa	296	16	96	14	2019	6	1
Poland	251	6	34	4	2020	6	1
Norway	248	3	35	2	2020	12	1
Spain	241	5	16	5	2021	3	1
Netherlands	201	3	0	0	2022	0	0
Russia	187	4	23	12	2022	6	3
Brazil	182	4	6	3	2022	2	1
Turkey	180	6	39	20	2021	7	3
Peoples R China	175	13	7	2	2022	1	0
Italy	163	5	13	7	2022	3	1
Portugal	143	3	3	1	2021	1	0
Ukraine	139	3	2	1	2019	1	0
Qatar	130	2	12	6	2022	6	3
New Zealand	129	3	22	4	2019	7	1
Australia	102	3	5	0	2022	2	0
Lebanon	97	3	8	4	2022	3	1
Saudi Arabia	88	2	8	4	2022	4	2
Sweden	87	4	21	5	2019	5	1
Algeria	87	1	0	0		0	
Chile	85	1	0	0	2022	0	0
Croatia	76	1	0	0		0	
South Korea	60	2	20	10	2020	10	5
Rwanda	55	2	8	4	2022	4	2
Czech Republic	54	2	2	0	2020	1	0
Israel	53	2	8	1	2020	4	1
Malaysia	44	2	0	0	2022	0	0
Yemen	44	1	0	0		0	
Bangladesh	38	1	5	0		5	
Finland	38	1	5	0		5	
Iran	36	1	2	0	2017	2	0
Nigeria	29	1	3	2	2022	3	2
Tanzania	29	1	3	2	2022	3	2

Continued on following page

Table 2. Continued

Countries	Total link strength	Documents	Citations	Normalised citations	Average publication year	Average citations	Average normalised citations
Sudan	27	1	5	3	2022	5	3
U Arab Emirates	24	1	18	10	2022	18	10
Canada	22	5	16	1	2019	3	0
Switzerland	22	1	2	1	2022	2	1
Egypt	17	2	4	2	2022	2	1
Pakistan	12	1	8	4	2022	8	4
Wales	7	1	9	5	2022	9	5
Estonia	6	4	14	6	2019	4	2
Belgium	5	1	1	1	2022	1	1
Austria	3	1	22	5	2016	22	5
South Africa	2	2	1	1	2022	1	0
Slovakia	1	1	0	0	2015	0	0

Figure 3. Networked organisations for bibliographic coupling links, fractional counting and 1 as the minimum number of documents of an organisation

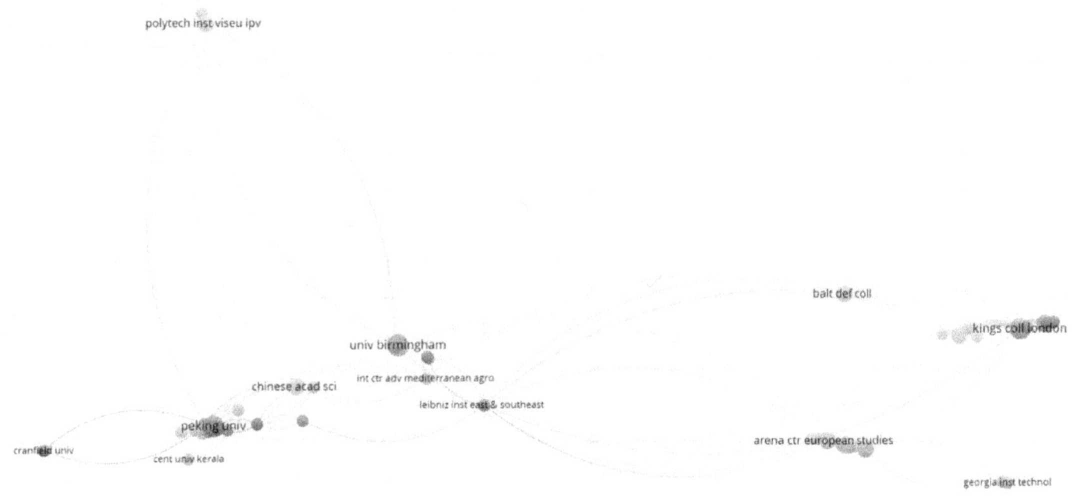

The top organizations with the highest results for the total link strength are the following (figure 3 and table 3): Kings Coll London; Leiden Univ; Norwegian Intelligence Sch; Univ Tyumen; Arena Ctr European Studies; Northeastern Univ; Univ Wroclaw; Leibniz Inst East & Southeast European Studies Re; Natl Univ Odesa Law Acad; Chinese Acad Sci; Flame Univ; Univ Szczecin; West Pomeranian Univ Technol; Cranfield Univ.

Table 3. Top 50 items (organisations) with the greatest values for the total link strength, for bibliographic coupling links, fractional counting and 1 as the minimum number of documents of an organisation

Organisations	Total link strength	Documents	Citations	Normalised citations	Average publication year	Average citations	Average normalised citations
Kings Coll London	243	3	6	4	2022	2	1
Leiden Univ	198	2	0	0	2022	0	0
Norwegian Intelligence Sch	151	1	0	0	2022	0	0
Univ Tyumen	145	2	5	2	2022	3	1
Arena Ctr European Studies	140	2	31	2	2017	16	1
Northeastern Univ	140	2	31	2	2017	16	1
Univ Wroclaw	140	2	31	2	2017	16	1
Leibniz Inst East & Southeast European Studies Re	134	1	1	0		1	
Natl Univ Odesa Law Acad	134	1	1	0		1	
Chinese Acad Sci	120	2	0	0	2022	0	0
Univ Chinese Acad Sci	117	1	0	0	2022	0	0
Flame Univ	109	1	0	0	2022	0	0
Univ Szczecin	109	1	0	0	2022	0	0
West Pomeranian Univ Technol	109	1	0	0	2022	0	0
Cranfield Univ	101	1	7	4	2022	7	4
Ctr Tecnol Carne Galicia	101	1	7	4	2022	7	4
Inst Agr & Fisheries Res & Training Ifapa	101	1	7	4	2022	7	4
Nhs Trust	101	1	7	4	2022	7	4
Sustainable Agrifoodtech Innovat & Res Safir	101	1	7	4	2022	7	4
Syrian Acad Expertise Sae	101	1	7	4	2022	7	4
Univ Leicester	101	1	7	4	2022	7	4
Univ Lincoln	101	1	7	4	2022	7	4
Univ Lorraine	101	1	7	4	2022	7	4
Univ Vigo	101	1	7	4	2022	7	4
Univ West England	101	1	7	4	2022	7	4
Int Ctr Adv Mediterranean Agron Studies Ciheam Ba	89	1	12	6	2022	12	6
Qatar Univ	89	1	12	6	2022	12	6
Polytech Inst Viseu Ipv	89	2	1	1	2022	1	0
Cent Univ Kerala	87	1	0	0		0	
Fed Univ Ouro Preto Deeco Ufop	87	1	0	0		0	
Sao Paulo State Univ Unesp	87	1	0	0		0	
Univ Ain Temouchent	87	1	0	0		0	
Univ Hohenheim	87	1	0	0		0	
Lund Univ	86	2	17	4	2019	9	2
Tech Univ Darmstadt	86	2	17	4	2019	9	2
Univ Canterbury	86	2	17	4	2019	9	2
Univ Bourgogne Franche Comte	85	1	0	0	2022	0	0
Univ Chile	85	1	0	0	2022	0	0
Univ Fed Pernambuco	85	1	0	0	2022	0	0
Univ Santiago De Compostela	85	1	0	0	2022	0	0

Continued on following page

Table 3. Continued

Organisations	Total link strength	Documents	Citations	Normalised citations	Average publication year	Average citations	Average normalised citations
Univ Business & Technol	76	1	0	0		0	
Univ Rijeka	76	1	0	0		0	
Ural Fed Univ	76	1	0	0		0	
Ctr Independent Social Res	71	1	1	0		1	
Lebanese Univ	69	2	5	3	2022	3	1
Univ Free State	69	2	1	1	2022	1	0
Gujarat Environm Management Inst Gemi	65	1	3	2	2022	3	2
Natl Forens Sci Univ	65	1	3	2	2022	3	2
New Jersey Inst Technol	65	1	3	2	2022	3	2
Univ Oslo	59	1	30	2	2017	30	2

The following documents are those with high metrics, namely total link strength (table 4): Sjursen & Rosén (2017); Cross & Karolewski (2017); Sohag et al. (2022); Mahran (2022); Chaban et al. (2019).

Table 4. Top 20 items (documents) with the greatest values for the total link strength, for bibliographic coupling links, fractional counting and 0 as the minimum number of citations of a document

Documents	url	Total link strength	Citations	Normalised citations	Publication year
(Sjursen & Rosén, 2017)	https://doi.org/10.1111/jcms.12443	59	30	2	2017
(Cross & Karolewski, 2017)	https://doi.org/10.1111/jcms.12442	58	20	2	2017
(Sohag et al., 2022)	https://doi.org/10.1108/bfj-09-2022-0793	17	0	0	
(Mahran, 2022)	https://doi.org/10.1108/jrf-06-2022-0163	17	0	0	
(Chaban et al., 2019)	https://doi.org/10.1177/0010836718823813	16	9	3	2019
(Elgström et al., 2018)	https://doi.org/10.1163/15718069-23021154	16	8	1	2018
(Abbassi et al., 2022)	https://doi.org/10.1108/jrf-05-2022-0108	15	7	0	
(Umar et al., 2022)	https://doi.org/10.1016/j.frl.2022.102976	12	18	10	2022
(Schilde, 2017)	https://doi.org/10.1111/jcms.12444	11	11	1	2017
(Ahmed et al., 2022)	https://doi.org/10.1111/eufm.12386	10	5	0	
(Chaaya, Devi Thambi, et al., 2022)	https://doi.org/10.1016/j.amsu.2022.104033	10	5	3	2022
(Huang et al., 2021)	https://doi.org/10.1080/10242694.2021.2007333	10	5	0	
(Clem, 2017)	https://doi.org/10.1080/15387216.2018.1424006	10	5	1	2018
(Fang & Shao, 2022)	https://doi.org/10.1016/j.frl.2022.103264	9	0	0	2022
(Ben Hassen & El Bilali, 2022)	https://doi.org/10.3390/foods11152301	9	12	6	2022
(Kormych & Malyarenko, 2022)	https://doi.org/10.1080/09592318.2022.2122278	9	1	0	
(Karolewski & Cross, 2017)	https://doi.org/10.1111/jcms.12446	8	11	1	2017
(Gaio et al., 2022)	https://doi.org/10.1016/j.frl.2022.103302	8	1	1	2022
(Le et al., 2022)	https://doi.org/10.1108/jrf-06-2022-0168	8	0	0	
(Nerlinger & Utz, 2022)	https://doi.org/10.1016/j.frl.2022.103243	7	2	1	2022

FACTOR-CLUSTER ANALYSES

Through factor analysis and with information for the top 50 networked authors were obtained two factors (table 5). Some metrics were removed because of the lack of data. Factor1 is defined by documents, citations, normalised citations and average citations. Total link strength is correlated with factor2. The results for kmo (Kaiser-Meyer-Olkin) are low (table 6). With factor1 and factor2 and through cluster analysis, the clusters presented in table 7 were found.

Table 5. Rotated factor analysis, for the top 50 networked authors

Variable	Factor1	Factor2	Uniqueness
Total link strength	0.061	0.907	0.173
Documents	0.645	0.349	0.462
Citations	0.937	0.263	0.052
Normalised Citations	0.836	-0.326	0.195
Average Citations	0.975	0.059	0.045

Table 6. Kaiser-Meyer-Olkin to assess sampling adequacy, for the top 50 networked authors

Variable	kmo
Total link strength	0.302
Documents	0.359
Citations	0.433
Normalised Citations	0.351
Average Citations	0.435
Overall	0.400

Table 7. Cluster analysis for the top 50 networked authors

Authors	Clusters
Dylan, Huw	1
Maguire, Thomas J.	1
Cross, Mai'a K. Davis	2
Karolewski, Ireneusz Pawel	2
Kormych, Borys	1
Malyarenko, Tetyana	1
Liu, Weidong	1
Yang, Xinyu	1
Bak, Iwona	1
Singh, Guru Ashish	1
Singh, Uma Shankar	1
Tarczynska-Luniewska, Malgorzata	1
Li, Jingdong	1
Song, Zhouying	1
Afy-Shararah, Mohamed	4
Duong, Linh	4
Garcia-Garcia, Guillermo	4
Hassoun, Abdo	4
Hdaifeh, Ammar	4
Jagtap, Sandeep	4
Lorenzo, Jose M.	4
Martindale, Wayne	4
Munekata, Paulo E. S.	4
Parra-Lopez, Carlos	4
Salonitis, Konstantinos	4
Trollman, Frank	4
Trollman, Hana	4
Pereira Domingues Martinho, Vitor Joao	1
Ben Hassen, Tarek	4
El Bilali, Hamid	4
Ferraz, Diogo	3
Mohammed, Kamel Si	3
Shahzadi, Irum	3
Tiwari, Sunil	3
Chaban, Natalia	4
Elgstrom, Ole	4
Knodt, Michele	4
Coronado, Freddy	3

Continued on following page

Table 7. Continued

Authors	Clusters
Da Silva Cunha, Carlos Augusto	3
Georgantzis, Nikolaos	3
Lourenco Costa, Paulo Jorge	3
Pato, Maria Lucia	3
Rodrigues, Raimundo Nonato	3
Sanchez-Carreira, Maria Carmen	3
Islam, Md Monirul	3
Mansour, Hoda	3
Sohag, Kazi	3
Zikovic, Ivana Tomas	3
Gupta, Gunjan	3
Hussain, Chaudhery Mustansar	3

For the networked countries were identified also two factors (table 8) with better results for the kmo (table 9). All variables define factor1, with the exception of average citations that are correlated with factor2. With these two factors and cluster analysis, the clusters presented in table 10 were obtained.

Table 8. Rotated factor analysis, for networked countries

Variable	Factor1	Factor2	Uniqueness
Total link strength	0.907	-0.097	0.168
Documents	0.949	-0.036	0.098
Citations	0.783	0.482	0.155
Normalised Citations	0.675	0.552	0.240
Average Citations	-0.061	0.960	0.075

Table 9. Kaiser-Meyer-Olkin to assess sampling adequacy, for networked countries

Variable	kmo
Total link strength	0.806
Documents	0.645
Citations	0.703
Normalised Citations	0.853
Average Citations	0.345
Overall	0.692

Table 10. Cluster analysis for networked countries

Countries	Clusters
England	1
Germany	1
India	1
France	1
Usa	1
Poland	1
Norway	2
Spain	1
Netherlands	4
Russia	2
Brazil	4
Turkey	2
Peoples R China	1
Italy	1
Portugal	4
Ukraine	4
Qatar	2
New Zealand	2
Australia	4
Lebanon	4
Saudi Arabia	4
Sweden	2
Algeria	4
Chile	4
Croatia	4
South Korea	2
Rwanda	4
Czech Republic	4
Israel	4
Malaysia	4
Yemen	4
Bangladesh	4
Finland	4
Iran	4
Nigeria	4
Tanzania	4
Sudan	4
U Arab Emirates	3

Continued on following page

Table 10. Continued

Countries	Clusters
Canada	4
Switzerland	4
Egypt	4
Pakistan	2
Wales	2
Estonia	2
Belgium	4
Austria	3
South Africa	4
Slovakia	4

Considering the top 50 networked organisations of affiliation as items, through factor analysis were identified two factors (table 11). Factor 1 is defined by citations, normalised citations and average citations and factor2 is defined by total link strength and documents. The value for the kmo is low (table 12). With these factors, the clusters shown in table 13 were obtained.

Table 11. Rotated factor analysis, for the top 50 networked organisations

Variable	Factor1	Factor2	Uniqueness
Total link strength	-0.057	0.884	0.215
Documents	0.226	0.852	0.224
Citations	0.913	0.219	0.118
Normalised Citations	0.759	-0.009	0.424
Average Citations	0.961	-0.010	0.077

Table 12. Kaiser-Meyer-Olkin to assess sampling adequacy, for the top 50 networked organisations

Variable	kmo
Total link strength	0.620
Documents	0.370
Citations	0.492
Normalised Citations	0.637
Average Citations	0.476
Overall	0.492

Table 13. Cluster analysis for the top 50 networked organisations

Organisations	Clusters
Kings Coll London	1
Leiden Univ	1
Norwegian Intelligence Sch	4
Univ Tyumen	1
Arena Ctr European Studies	2
Northeastern Univ	2
Univ Wroclaw	2
Leibniz Inst East & Southeast European Studies Re	4
Natl Univ Odesa Law Acad	4
Chinese Acad Sci	1
Univ Chinese Acad Sci	4
Flame Univ	4
Univ Szczecin	4
West Pomeranian Univ Technol	4
Cranfield Univ	3
Ctr Tecnol Carne Galicia	3
Inst Agr & Fisheries Res & Training Ifapa	3
Nhs Trust	3
Sustainable Agrifoodtech Innovat & Res Safir	3
Syrian Acad Expertise Sae	3
Univ Leicester	3
Univ Lincoln	3
Univ Lorraine	3
Univ Vigo	3
Univ West England	3
Int Ctr Adv Mediterranean Agron Studies Ciheam Ba	3
Qatar Univ	3
Polytech Inst Viseu Ipv	4
Cent Univ Kerala	4
Fed Univ Ouro Preto Deeco Ufop	4
Sao Paulo State Univ Unesp	4
Univ Ain Temouchent	4
Univ Hohenheim	4
Lund Univ	2
Tech Univ Darmstadt	2
Univ Canterbury	2
Univ Bourgogne Franche Comte	4
Univ Chile	4

Continued on following page

Table 13. Continued

Organisations	Clusters
Univ Fed Pernambuco	4
Univ Santiago De Compostela	4
Univ Business & Technol	4
Univ Rijeka	4
Ural Fed Univ	4
Ctr Independent Social Res	4
Lebanese Univ	3
Univ Free State	4
Gujarat Environm Management Inst Gemi	4
Natl Forens Sci Univ	4
New Jersey Inst Technol	4
Univ Oslo	3

For the top 20 documents, one factor was obtained and is that presented in table 14 (normalised citations has a uniqueness greater than 0.5) and the kmo results are also low (table 15). The clusters identified with this factor are presented in table 16.

Table 14. Rotated factor analysis, for the top 20 networked documents

Variable	Factor1	Uniqueness
Total link strength	0.811	0.343
Citations	0.971	0.058
Normalised Citations	0.580	0.664

Table 15. Kaiser-Meyer-Olkin to assess sampling adequacy, for the top 20 networked documents

Variable	kmo
Total link strength	0.346
Citations	0.401
Normalised Citations	0.234
Overall	0.343

Table 16. Cluster analysis for the top 20 networked documents

Documents	Clusters
Sjursen (2017)	1
Cross (2017)	1
Sohag	2
Mahran	2
Chaban (2019)	3
Elgstrom (2018)	3
Abbassi	3
Umar (2022)	1
Schilde (2017)	3
Ahmed	2
Chaaya (2022)	3
Huang	2
Clem (2018)	2
Fang (2022)	2
Ben Hassen (2022)	3
Kormych	2
Karolewski (2017)	3
Gaio (2022)	2
Le	2
Nerlinger (2022)	2

INSIGHTS FROM THE LITERATURE BASED ON BIBLIOGRAPHIC DATA

Insights from the literature review based on bibliometric analysis are presented in table 17. This table summarises contributions from the literature for the topics here addressed and related to the Russia-Ukraine conflict and the agricultural sector dimensions and dynamics worldwide.

Table 17. Insights from the literature review

Documents	Contributions from the literature
(Abbassi et al., 2022)	Stock prices are vulnerable to trade dependence.
(Ahmed et al., 2022)	European stock markets are impacted by the Russia-Ukraine conflicts.
(Ben Hassen & El Bilali, 2022)	Reforms and policies are required to deal with the impacts on food security.
(Chaaya, Devi Thambi, et al., 2022)	The conflict has implication on mental health of Ukrainian young people.
(Chaban et al., 2019)	European Union has been a relevant role of the crisis mediation.
(Clem, 2017)	The information and its assessment are important for the context understanding.
(Cross & Karolewski, 2017)	Contributions for the understanding of the European Union role on the crisis.
(Elgström et al., 2018)	Perceptions about the European Union role on the conflict.
(Fang & Shao, 2022)	Commodity markets present volatility in the context of Russia-Ukraine crisis.
(Gaio et al., 2022)	The conflict has impacts on the stock market efficiency.
(Huang et al., 2021)	Geopolitical events affect the volatility of gold markets.
(Karolewski & Cross, 2017)	Contributions for the understanding of the European Union as a policy actor.
(Kormych & Malyarenko, 2022)	Perspectives about the future of the conflict.
(Le et al., 2022)	Impacts on airline and defense markets.
(Mahran, 2022)	The average connectedness between sectors in Egypt was 85.63% during the crisis.
(Nerlinger & Utz, 2022)	The consequences on energy firms were asymmetric.
(Schilde, 2017)	The recent events changed the European military viewpoint.
(Sjursen & Rosén, 2017)	There was an integration of the European external policy.
(Sohag et al., 2022)	Inflation is a consequence.
(Umar et al., 2022)	European equities and Russian bonds are spreaders of shocks in financial markets.

CONCLUSION

In this chapter was intended to explore the scientific literature about the interrelationships among the Russia-Ukraine conflict and the agricultural sector. In this way, 106 documents were considered for the topics "Russia-Ukraine conflict", or "Russia-Ukraine crisis", or "Russia-Ukraine war" and "agricultur*". These documents were obtained on a search carried out on 04 January 2023 on the Web of Science Core Collection database. These studies were assessed through bibliometric analysis (bibliographic coupling links and authors, countries and organisations as items) and analysed deeper considering systematic review based on bibliographic data. For further assessment, cluster analysis was also taken into account.

From the literature review, it is important to refer to the relevant impacts of the Russia-Ukraine crisis on inflation, market volatility and spillover effects between the commodity markets. The main concerns from these frameworks are associated with food security, agri-food chains and international trade dynamics. In the discussions about these topics, the resilience of the agricultural sector and the respective markets appears often as a fundamental factor to take into account by several stakeholders in future strategies. The perception of the citizens about the implications of the conflict on the several dimensions of their lives is another question to deeper addressed. Governments and international organisations play here a determinant role in the world stability, resilience of agriculture and food sector, and promote food security.

The authors with the highest total strength of the links (including other metrics) are: Dylan, Huw; Maguire, Thomas J.; Cross, Mai'a K. Davis; Karolewski, Ireneusz Pawel; Kormych, Borys; Malyarenko, Tetyana; Liu, Weidong; Yang, Xinyu; Bak, Iwona; Singh, Guru Ashish. England, Germany, India, France, USA, Poland, Norway, Spain Netherlands, Russia, Brazil, Turkey, China, Italy and Portugal are some of the countries with the biggest scores for this metric. The top organisations are: Kings Coll London; Leiden Univ; Norwegian Intelligence Sch; Univ Tyumen; Arena Ctr European Studies; Northeastern Univ; Univ Wroclaw; Leibniz Inst East & Southeast European Studies Re; Natl Univ Odesa Law Acad; Chinese Acad Sci; Flame Univ; Univ Szczecin; West Pomeranian Univ Technol; Cranfield Univ. These results show that there is a field to be explored by other countries and organisations, for example.

The factor-cluster analyses highlight that the network has different characteristics for the items considered (authors, countries and organisations). The literature survey based on bibliometric analysis reveals that the following topics are among the addressed by the researchers with the greatest total link strength:

- Stock markets;
- Policies and impacts on food security;
- The mental health of the population;
- Mediation role of the European Union;
- Importance of the information;
- Perceptions;
- Commodity markets volatility;
- Changes in the military perspectives in the European Union;
- Inflation;
- Connectedness between sectors and markets.

This research reveals the impacts of the conflict on the markets' volatility and prices. This needs deeper assessment and adjusted policies. Other important implications of the crisis are those associated with the mental health of the citizens and how the perceptions may be impacted by social communication media. Another question is how these perceptions are interrelated with the instability of the markets. In terms of policy recommendation, it is suggested to identify the groups, sectors and markets more vulnerable and better support these dimensions with adjusted measures and policy instruments. It would be important to give special focus to the market dynamics and the main factors that affect them. For future research, the dynamics of the producer markets could be better assessed.

ACKNOWLEDGMENT

This work is funded by National Funds through the FCT - Foundation for Science and Technology, I.P., within the scope of the project Refª UIDB/00681/2020. Furthermore we would like to thank the CERNAS Research Centre and the Polytechnic Institute of Viseu for their support.

REFERENCES

Abbassi, W., Kumari, V., & Pandey, D. K. (2022). What makes firms vulnerable to the Russia–Ukraine crisis? *The Journal of Risk Finance*. doi:10.1108/JRF-05-2022-0108

Abu Hatab, A. (2022). Africa's Food Security under the Shadow of the Russia-Ukraine Conflict. *Strategic Review for Southern Africa*, *44*(1), 37–46. doi:10.35293rsa.v44i1.4083

Aharon, D. Y., & Qadan, M. (2022). Infection, invasion, and inflation: Recent lessons. *Finance Research Letters*, *50*, 103307. doi:10.1016/j.frl.2022.103307

Ahmed, S., Hasan, M. M., & Kamal, M. R. (2022). Russia–Ukraine crisis: The effects on the European stock market. *European Financial Management,* 1–41. doi:10.1111/eufm.12386

Alyukov, M. (2022). Making Sense of the News in an Authoritarian Regime: Russian Television Viewers' Reception of the Russia-Ukraine Conflict. *Europe-Asia Studies*, *74*(3), 337–359. doi:10.1080/09668136.2021.2016633

Anil, S. M., & Kumar, K. A. (2022). India and the UN. *International Journal of Early Childhood Special Education*, *14*(5), 1083–1092. doi:10.9756/INTJECSE/V14I5.107

Ben Hassen, T., & El Bilali, H. (2022). Impacts of the Russia-Ukraine War on Global Food Security: Towards More Sustainable and Resilient Food Systems? *Foods*, *11*(15), 2301. doi:10.3390/foods11152301 PMID:35954068

Beraich, M., Amzile, K., Laamire, J., Zirari, O., & Fadali, M. A. (2022). Volatility Spillover Effects of the US, European and Chinese Financial Markets in the Context of the Russia-Ukraine Conflict. *International Journal of Financial Studies*, *10*(4), 95. doi:10.3390/ijfs10040095

Cabelkova, I., Smutka, L., Rotterova, S., Zhytna, O., Kluger, V., & Mares, D. (2022). The Sustainability of International Trade: The Impact of Ongoing Military Conflicts, Infrastructure, Common Language, and Economic Wellbeing in Post-Soviet Region. *Sustainability*, *14*(17), 10840. doi:10.3390u141710840

Chaaya, C., Thambi, V. D., Sabuncu, O., Abedi, R., Osman, A. O. A., Uwishema, O., & Onyeaka, H. (2022). Ukraine—Russia crisis and its impacts on the mental health of Ukrainian young people during the COVID-19 pandemic. *Annals of Medicine and Surgery (London)*, *79*, 104033. doi:10.1016/j.amsu.2022.104033 PMID:35765517

Chaban, N., Elgström, O., & Knodt, M. (2019). Perceptions of EU mediation and mediation effectiveness: Comparing perspectives from Ukraine and the EU. *Cooperation and Conflict*, *54*(4), 506–523. doi:10.1177/0010836718823813

Clem, R. S. (2017). Clearing the Fog of War: Public versus official sources and geopolitical storylines in the Russia-Ukraine conflict. *Eurasian Geography and Economics*, *58*(6), 592–612. doi:10.1080/15387216.2018.1424006

Costantini, V., Morando, V., Olk, C., & Tausch, L. (2022). Fuelling the Fire: Rethinking European Policy in Times of Energy and Climate Crises. *Energies*, *15*(20), 7781. doi:10.3390/en15207781

Cross, M. K. D., & Karolewski, I. P. (2017). What Type of Power has the EU Exercised in the Ukraine–Russia Crisis? A Framework of Analysis. *Journal of Common Market Studies, 55*(1), 3–19. doi:10.1111/jcms.12442

Deniz, A., & Ozgur, E. M. (2022). The establishment of the Ukrainian diaspora in Turkey through migrant associations: 'We feel as part of the diaspora from now on.'. *Southeast European and Black Sea Studies, 22*(2), 243–261. doi:10.1080/14683857.2022.2037820

Dylan, H., & Maguire, T. J. (2022). Secret Intelligence and Public Diplomacy in the Ukraine War. *Survival, 64*(4), 33–74. doi:10.1080/00396338.2022.2103257

Elbasuney, S., El-Sayyad, G. S., Attia, M. S., & Abdelaziz, A. M. (2022). Ferric Oxide Colloid: Towards Green Nano-Fertilizer for Tomato Plant with Enhanced Vegetative Growth and Immune Response Against Fusarium Wilt Disease. *Journal of Inorganic and Organometallic Polymers and Materials, 32*(11), 4270–4283. doi:10.100710904-022-02442-6 PMID:35910584

Elgström, O., Chaban, N., Knodt, M., Müller, P., & Pardo, S. (2018). Perceptions of the EU's Role in the Ukraine-Russia and the Israel-Palestine Conflicts: A Biased Mediator? *International Negotiation, 23*(2), 299–318. doi:10.1163/15718069-23021154

Fang, Y., & Shao, Z. (2022). The Russia-Ukraine conflict and volatility risk of commodity markets. *Finance Research Letters, 50*, 103264. doi:10.1016/j.frl.2022.103264

Fink, M. (2022). Naval Blockade and the Russia-Ukraine Conflict. *Netherlands International Law Review, 69*(3), 411–437. doi:10.100740802-022-00228-8

Gaio, L. E., Stefanelli, N. O., Pimenta, T. Junior, Bonacim, C. A. G., & Gatsios, R. C. (2022). The impact of the Russia-Ukraine conflict on market efficiency: Evidence for the developed stock market. *Finance Research Letters, 50*, 103302. doi:10.1016/j.frl.2022.103302

Henrikson, A. K. (2022). The Trauma of Territorial Break-up: The Russia-Ukraine Conflict and Its International Management-Geopolitical Strategy and Diplomatic Therapy. *Geopoliticas-Revista De Estudios Sobre Espacio Y Poder, 13*(1), 15–40. doi:10.5209/geop.81880

Huang, J., Cao, Y., & Zhong, P. (2022). Searching for a safe haven to crude oil: Green bond or precious metals? *Finance Research Letters, 50*, 103303. doi:10.1016/j.frl.2022.103303

Huang, J., Li, Y., Suleman, M. T., & Zhang, H. (2021). Effects of Geopolitical Risks on Gold Market Return Dynamics: Evidence from a Nonparametric Causality-in-quantiles Approach. *Defence and Peace Economics*, 1–15. doi:10.1080/10242694.2021.2007333

Huang, M., Shao, W., & Wang, J. (2023). Correlations between the crude oil market and capital markets under the Russia-Ukraine conflict: A perspective of crude oil importing and exporting countries. *Resources Policy, 80*, 103233. doi:10.1016/j.resourpol.2022.103233

Ibar-Alonso, R., Quiroga-Garcia, R., & Arenas-Parra, M. (2022). Opinion Mining of Green Energy Sentiment: A Russia-Ukraine Conflict Analysis. *Mathematics, 10*(14), 2532. doi:10.3390/math10142532

Jagtap, S., Trollman, H., Trollman, F., Garcia-Garcia, G., Parra-Lopez, C., Duong, L., Martindale, W., Munekata, P. E. S., Lorenzo, J. M., Hdaifeh, A., Hassoun, A., Salonitis, K., & Afy-Shararah, M. (2022). The Russia-Ukraine Conflict: Its Implications for the Global Food Supply Chains. *Foods*, *11*(14), 2098. doi:10.3390/foods11142098 PMID:35885340

Jha, A., Bhatt, N., Dixit, A. K., Singh, U. K., & Choudhary, L. R. (2022). Y Role of Religion and Politics in War Between Russia and Ukraine: A Study. *International Journal of Early Childhood Special Education*, *14*(4), 1181–1185. doi:10.9756/INT-JECSE/V14I4.156

Just, M., & Echaust, K. (2022). Dynamic spillover transmission in agricultural commodity markets: What has changed after the COVID-19 threat? *Economics Letters*, *217*, 110671. doi:10.1016/j.econlet.2022.110671

Karolewski, I. P., & Cross, M. K. D. (2017). The EU's Power in the Russia–Ukraine Crisis: Enabled or Constrained? *Journal of Common Market Studies*, *55*(1), 137–152. doi:10.1111/jcms.12446

Kim, J.-O., & Mueller, C. W. (1978a). Factor Analysis: Statistical Methods and Practical Issues. *Sage (Atlanta, Ga.)*.

Kim, J.-O., & Mueller, C. W. (1978b). *Introduction to Factor Analysis: What It Is and How To Do It*. SAGE Publications. doi:10.4135/9781412984652

Kormych, B., & Malyarenko, T. (2022). From gray zone to conventional warfare: The Russia-Ukraine conflict in the Black Sea. *Small Wars & Insurgencies*, *0*(0), 1–36. doi:10.1080/09592318.2022.2122278

Le, V. H., von Mettenheim, H.-J., Goutte, S., & Liu, F. (2022). News-based sentiment: Can it explain market performance before and after the Russia–Ukraine conflict? *The Journal of Risk Finance,* doi:10.1108/JRF-06-2022-0168

Lewandowska-Tomaszczyk, B. (2022). Online comments as inwards and outwards-directed acts in Polish and English The Russia-Ukraine conflict. *International Review of Pragmatics*, *14*(2), 252–275. doi:10.1163/18773109-01402005

Lewtak, K., Kanecki, K., Tyszko, P., Gorynski, P., Kosinska, I., Poznanska, A., Rzad, M., & Nitsch-Osuch, A. (2022). Hospitalizations of Ukrainian Migrants and Refugees in Poland in the Time of the Russia-Ukraine Conflict. *International Journal of Environmental Research and Public Health*, *19*(20), 13350. doi:10.3390/ijerph192013350 PMID:36293932

Li, J., & Song, Z. (2022). Dynamic Impacts of External Uncertainties on the Stability of the Food Supply Chain: Evidence from China. *Foods*, *11*(17), 2552. doi:10.3390/foods11172552 PMID:36076737

Mahran, H. A. (2022). The impact of the Russia–Ukraine conflict (2022) on volatility connectedness between the Egyptian stock market sectors: Evidence from the DCC-GARCH-CONNECTEDNESS approach. *The Journal of Risk Finance*. doi:10.1108/JRF-06-2022-0163

Martinho, V. J. P. D. (2022). Impacts of the COVID-19 Pandemic and the Russia–Ukraine Conflict on Land Use across the World. *Land (Basel)*, *11*(10), 10. doi:10.3390/land11101614

Martinho, V. J. P. D., da Silva Cunha, C. A., Pato, M. L., Lourenco Costa, P. J., Sanchez-Carreira, M. C., Georgantzis, N., Rodrigues, R. N., & Coronado, F. (2022). Machine Learning and Food Security: Insights for Agricultural Spatial Planning in the Context of Agriculture 4.0. *Applied Sciences-Basel*, *12*(22), 11828. doi:10.3390/app122211828

Miao, Y., & Fei, X. (2022). The international monetary system: Evolution and revolution. *China Economic Journal*, *15*(3), 235–252. doi:10.1080/17538963.2022.2117193

Mo, S., & Wang, T. (2022). Synergistic Effects of International Oil Price Fluctuations and Carbon Tax Policies on the Energy-Economy-Environment System in China. *International Journal of Environmental Research and Public Health*, *19*(21), 14177. doi:10.3390/ijerph192114177 PMID:36361057

Mu, S., Huang, G., Li, P., & Hou, Y. (2022). A Study on Volatility Spillovers among International Stock Markets during the Russia-Ukraine Conflict. *Discrete Dynamics in Nature and Society*, *4948444*, 1–8. doi:10.1155/2022/4948444

Murithi, T. (2022). The Failure of the United Nations Security Council in Creating the Framework Conditions for Mediation in the Russia-Ukraine Crisis. *Strategic Review for Southern Africa*, *44*(1), 65–92. doi:10.35293rsa.v44i1.4082

Myeki, L. W., Nengovhela, N. B., Mudau, L., Nakana, E., & Ngqangweni, S. (2022). Estimation of Technical, Allocative, and Economic Efficiencies for Smallholder Broiler Producers in South Africa. *Agriculture-Basel*, *12*(10), 1601. doi:10.3390/agriculture12101601

Nakandala, D., Tsang, Y. P., Lau, H., & Lee, C. K. M. (2022). An Industrial Blockchain-Based Multi-Criteria Decision Framework for Global Freight Management in Agricultural Supply Chains. *Mathematics*, *10*(19), 3550. doi:10.3390/math10193550

Nerlinger, M., & Utz, S. (2022). The impact of the Russia-Ukraine conflict on energy firms: A capital market perspective. *Finance Research Letters*, *50*, 103243. doi:10.1016/j.frl.2022.103243

Qureshi, A., Rizwan, M. S., Ahmad, G., & Ashraf, D. (2022). Russia-Ukraine war and systemic risk: Who is taking the heat? *Finance Research Letters*, *48*, 103036. doi:10.1016/j.frl.2022.103036

Rawtani, D., Gupta, G., Khatri, N., Rao, P. K., & Hussain, C. M. (2022). Environmental damages due to war in Ukraine: A perspective. *The Science of the Total Environment*, *850*, 157932. doi:10.1016/j.scitotenv.2022.157932 PMID:35952889

Schilde, K. (2017). European Military Capabilities: Enablers and Constraints on EU Power? *Journal of Common Market Studies*, *55*(1), 37–53. doi:10.1111/jcms.12444

Shen, L., & Hong, Y. (2023). Can geopolitical risks excite Germany economic policy uncertainty: Rethinking in the context of the Russia-Ukraine conflict. *Finance Research Letters*, *51*, 103420. doi:10.1016/j.frl.2022.103420

Shepard, J. U., & Pratson, L. F. (2022). The myth of US energy independence. *Nature Energy*, *7*(6), 462–464. doi:10.103841560-022-01053-2

Singh, A., Patel, R., & Singh, H. (2022). Recalibration of priorities: Investor preference and Russia-Ukraine conflict. *Finance Research Letters*, *50*, 103294. doi:10.1016/j.frl.2022.103294

Sjursen, H., & Rosén, G. (2017). Arguing Sanctions. On the EU's Response to the Crisis in Ukraine. *Journal of Common Market Studies*, *55*(1), 20–36. doi:10.1111/jcms.12443

Sohag, K., Islam, M. M., Tomas Žiković, I., & Mansour, H. (2022). Food inflation and geopolitical risks: Analyzing European regions amid the Russia-Ukraine war. *British Food Journal*. doi:10.1108/BFJ-09-2022-0793

Stata. (2023). *Statistical software for data science | Stata*. https://www.stata.com/

StataCorp. (2017a). *Stata 15 Base Reference Manual*. Stata Press.

StataCorp. (2017b). *Stata Statistical Software: Release 15*. StataCorp LLC.

Subramaniam, S., & Rajoo, M. (2022). Review on Vietnamese Refugees, Resettlement and Mental Health: From Pulau Bidong, a Malaysian Experience. *Clinical Social Work and Health Intervention*, *13*(2), 38–44. doi:10.22359/cswhi_13_2_10

Sun, Z., Scherer, L., Zhang, Q., & Behrens, P. (2022). Adoption of plant-based diets across Europe can improve food resilience against the Russia-Ukraine conflict. *Nature Food*, *3*(11), 905–910. doi:10.103843016-022-00634-4

Torres-Reyna, O. (n.d.). *Getting Started in Factor Analysis (using Stata) (ver. 1.0 beta/draft)*. Princeton. https://www.princeton.edu/~otorres/Stata/Factor

Tosun, O. K., & Eshraghi, A. (2022). Corporate decisions in times of war: Evidence from the Russia-Ukraine conflict. *Finance Research Letters*, *48*, 102920. doi:10.1016/j.frl.2022.102920

Umar, Z., Polat, O., Choi, S.-Y., & Teplova, T. (2022). The impact of the Russia-Ukraine conflict on the connectedness of financial markets. *Finance Research Letters*, *48*, 102976. doi:10.1016/j.frl.2022.102976

van Eck, N. J., & Waltman, L. (2010). Software survey: VOSviewer, a computer program for bibliometric mapping. *Scientometrics*, *84*(2), 523–538. doi:10.100711192-009-0146-3 PMID:20585380

van Eck, N. J., & Waltman, L. (2022). *Manual for VOSviewer version 1. 6.18*. CWTS, Universiteit Leiden. https://www.vosviewer.com/documentation/Manual_VOSviewer_1.6.18.pdf

Vincent, J. E. (1971). *Factor Analysis in International Relations: Interpretation, Problem Areas, and an Application*. University of Florida Press.

VOSviewer. (2023). *VOSviewer—Visualizing scientific landscapes—Version 1.6.18*. VOSviewer. https://www.vosviewer.com//

Web of Science. (2023). *Web of Science Core Collection*. WOS. https://www.webofscience.com/wos/woscc/basic-search

Yang, X., & Liu, W. (2022). Agricultural Production Networks and Upgrading from a Global-Local Perspective: A Review. *Land (Basel)*, *11*(10), 1864. doi:10.3390/land11101864

Yazbeck, N., Mansour, R., Salame, H., Chahine, N. B., & Hoteit, M. (2022). The Ukraine-Russia War Is Deepening Food Insecurity, Unhealthy Dietary Patterns and the Lack of Dietary Diversity in Lebanon: Prevalence, Correlates and Findings from a National Cross-Sectional Study. *Nutrients*, *14*(17), 3504. doi:10.3390/nu14173504 PMID:36079761

Chapter 3
Consequences of the Current Conjuncture on World Food Markets

ABSTRACT

The current conjunctures, namely the Covid-19 pandemic and the Russia-Ukraine crisis, created new challenges in the food markets worldwide, because of the pressures in the prices and constraints in the food chains. These frameworks brought additional pressures for already existing problems related to global warming and climate change. In this study, it was intended to analyse the evolution worldwide of food prices over the last few years and to assess the eventual impacts of the pandemic and Russia-Ukraine conflict on these contexts. Statistical information from the FAOSTAT was considered for the period January 2016-September 2022 and the following variables: food price inflation (%); consumer prices, food indices (2015 = 100). The impacts on food prices were greater from the Russia-Ukraine crisis than from the Covid-19 pandemic. This is particularly worrying in more vulnerable countries.

INTRODUCTION

The recent international events have had multidimensional consequences (Ahoya et al., 2022). The implications on the food markets, including the online ones (Macias et al., 2022), are no an exception (Akber & Paltasingh, 2022). These impacts are true for the following countries, for instance: Bangladesh (Akter et al., 2022); Indonesia (Amin et al., 2022); Nigeria (Ayo-Lawal et al., 2022); India (Bairagi et al., 2022); Mozambique (Bliznashka et al., 2022); Uganda (Buzigi & Onakuse, 2023); BRICST (Brazil, Russia, India, China, South Africa and Turkey) countries (Chang et al., 2022); Zimbabwe (Chari et al., 2022); US (Codjia & Saghaian, 2022); Iran (Ghalibaf et al., 2022); Myanmar (Goeb, Maredia, et al., 2022; Goeb, Zone, et al., 2022); South Africa (Hart et al., 2022); Lao PDR (Head et al., 2022); Caribbean countries (Henry et al., 2022); Latin America and the Caribbean (Hernández-Vásquez et al., 2022); Malaysia (Kh'ng et al., 2022); Turkey (Korkmaz Aslan et al., 2022); Russian context (Lialina & Morachevskaya, 2022); China (Liang & Zhong, 2023); Brazil (Lopes et al., 2022); Democratic Republic

DOI: 10.4018/978-1-6684-8923-9.ch003

of the Congo (Manyong et al., 2022); Yemen (Rahmat et al., 2022); France (Recchia et al., 2022); Togo (Sodji, 2022). This is a reality also for the broiler meat markets (Firmansyah et al., 2022), potato (Rajpoot et al., 2022), maize (Nasir et al., 2022) and wheat prices (Martin & Minot, 2022).

These consequences on food prices have occurred during the Russia-Ukraine crisis (Ben Hassen & El Bilali, 2022), but also with the Covid-19 pandemic (Bai et al., 2022). The roots of these food markets implications are in food chains (Carrière-Swallow et al., 2023) and transport routes (Dasgupta & Robinson, 2022) disruptions, suggesting the need for more resilient agri-food supply chains (Das & Roy, 2022). Agricultural production and the accessibilities have here a crucial role (Iheme et al., 2022), as well as the following dimensions: new technologies (Jagtap et al., 2022); public policies (Kaicker et al., 2022); food and agricultural strategies (Munonye et al., 2022); scientific contributions (Nekmahmud, 2022), changes in the consumer behaviour (O'Meara et al., 2022), tourism (Phukamchanoada, 2022); resources markets (Shahini et al., 2022); and food tracking (Mendi, 2022).

The digital transition is also in the order of the day (Nikonenko et al., 2022) and public interventions are specifically important in more vulnerable contexts (Omotayo et al., 2022). Nonetheless, the pressures on the food markets are not new (Kyfyak et al., 2022) and came from other events (Higashi et al., 2022), such as, for example, climate changes (Cheng et al., 2023), agroterrorism (Djurle et al., 2022) and energy markets (Guo & Tanaka, 2022). These pressures are not equal across several world markets (Dietrich et al., 2022) and, in some circumstances, have implications on food security (Ibiyemi & Oldewage-Theron, 2022) worldwide (Jacobs et al., 2022), social conflicts (Gutiérrez-Romero, 2022) and consumer options (Yang et al., 2022). Some of these consequences may persist if not well addressed (Wegerif, 2022). This is particularly worrying in low and middle-income nations (Mueller et al., 2022).

Following these trends, the present study aims to analyse the impacts of the recent international events on Food Price Inflation (%) and Consumer Prices, Food Indices (2015 = 100), considering data from FAOSTAT (2023) for the period 2016-2022. Some results were obtained with the Stata (StataCorp, 2017a, 2017b; Stata, 2023) software. The Scopus (2023) database was surveyed for the literature review. For a better understanding of the context over the period considered, normalised values were calculated $((x_i-x_{minimum})/(x_{maximum}-x_{minimum}))$.

FOOD PRICE INFLATION ANALYSIS

The consumer food price inflation (in percentage for the world monthly homologous rate) results reveal, that over the period 2016-2019, this variable changed relatively to the previous month of the same year, on average, 0.066% points (table 1). The standard deviation was 0.250% points and the sample had 44 observations. In levels, this variable had an average of 2.791%, a standard deviation of 1.058% and 48 observations.

Over the period January 2020-September 2022, the food price inflation changed by 0.551% points, on average, namely because of the values of the year 2022 (table 2), with a standard deviation of 0.749% points and 30 observations. The values in levels had an average of 9.711%, a standard deviation of 5.609% and 33 observations were considered. On average, the prices decreased in 2020, increased slightly in 2021 and increased more strongly in 2022.

On the other hand, these frameworks were not similar worldwide, such as tables 3, 4 and 5 highlight for the period January 2016-September 2022 and tables 6, 7 and 8 for the year 2022 (until September).

Table 1. Consumer food price inflation (%) for the world monthly homologous rate (2016-2019)

Year	Months	Value	Normalised value for all period	Changes (% points)
2016	January	2.247	0.037	
2016	February	2.514	0.049	0.267
2016	March	2.388	0.043	-0.126
2016	April	2.350	0.042	-0.038
2016	May	2.029	0.027	-0.321
2016	June	1.906	0.021	-0.123
2016	July	2.006	0.026	0.100
2016	August	1.617	0.008	-0.389
2016	September	1.490	0.002	-0.127
2016	October	1.484	0.002	-0.006
2016	November	1.702	0.012	0.218
2016	December	1.642	0.009	-0.060
2017	January	1.790	0.016	
2017	February	1.451	0.000	-0.339
2017	March	1.708	0.012	0.257
2017	April	1.841	0.018	0.133
2017	May	2.206	0.035	0.365
2017	June	2.169	0.033	-0.037
2017	July	2.201	0.035	0.032
2017	August	2.345	0.041	0.144
2017	September	2.348	0.042	0.003
2017	October	2.407	0.044	0.059
2017	November	2.433	0.045	0.026
2017	December	2.695	0.058	0.262
2018	January	2.680	0.057	
2018	February	2.578	0.052	-0.102
2018	March	2.328	0.041	-0.250
2018	April	2.169	0.033	-0.159
2018	May	2.016	0.026	-0.153
2018	June	2.409	0.044	0.393
2018	July	2.506	0.049	0.097
2018	August	2.857	0.065	0.351
2018	September	3.324	0.087	0.467
2018	October	3.387	0.090	0.063
2018	November	3.211	0.081	-0.176
2018	December	3.117	0.077	-0.094
2019	January	3.193	0.081	
2019	February	3.600	0.099	0.407

Continued on followiing page

Table 1. Continued

Year	Months	Value	Normalised value for all period	Changes (% points)
2019	March	4.202	0.127	0.602
2019	April	4.379	0.136	0.177
2019	May	4.691	0.150	0.312
2019	June	4.438	0.138	-0.253
2019	July	4.398	0.136	-0.040
2019	August	4.307	0.132	-0.091
2019	September	4.139	0.124	-0.168
2019	October	4.552	0.144	0.413
2019	November	5.159	0.172	0.607
2019	December	5.351	0.181	0.192

Table 2. Consumer food price inflation (%) for the world monthly homologous rate (2020-2022)

Year	Months	Value	Normalised value for all period	Changes (% points)
2020	January	5.799	0.201	
2020	February	5.946	0.208	0.147
2020	March	5.738	0.198	-0.208
2020	April	6.917	0.253	1.179
2020	May	6.657	0.241	-0.260
2020	June	6.894	0.252	0.237
2020	July	6.709	0.243	-0.185
2020	August	6.538	0.235	-0.171
2020	September	6.301	0.225	-0.237
2020	October	6.214	0.220	-0.087
2020	November	5.801	0.201	-0.413
2020	December	5.753	0.199	-0.048
2021	January	5.648	0.194	
2021	February	5.618	0.193	-0.030
2021	March	5.645	0.194	0.027
2021	April	4.955	0.162	-0.690
2021	May	5.368	0.181	0.413
2021	June	5.662	0.195	0.294
2021	July	6.516	0.234	0.854
2021	August	7.253	0.269	0.737
2021	September	7.990	0.303	0.737
2021	October	8.709	0.336	0.719
2021	November	9.894	0.391	1.185
2021	December	11.022	0.443	1.128
2022	January	11.864	0.482	
2022	February	12.742	0.523	0.878
2022	March	14.738	0.615	1.996
2022	April	17.021	0.721	2.283
2022	May	18.372	0.783	1.351
2022	June	19.959	0.857	1.587
2022	July	21.269	0.917	1.310
2022	August	21.906	0.947	0.637
2022	September	23.053	1.000	1.147

Table 3. Decreasing order for the consumer food price inflation (%), on average over the period January 2016-September 2022, for the countries with the highest values

Country	Value
Venezuela (Bolivarian Republic of)	29850.407
Zimbabwe	203.710
Lebanon	120.202
South Sudan	118.153
Sudan	95.340
Syrian Arab Republic	49.824
Argentina	40.189
Suriname	35.279
Iran (Islamic Republic of)	34.717
Angola	25.380
Türkiye	24.715
Ethiopia	19.793
Haiti	19.111
Nigeria	16.829
Malawi	16.777
Sierra Leone	16.446
Zambia	15.108
Uzbekistan	15.071
Egypt	14.574
Libya	13.369
Sri Lanka	13.293
Guinea	13.059
Liberia	12.639
Ukraine	11.487
Mozambique	11.474
Ghana	11.121
Kazakhstan	10.532
Republic of Moldova	10.429
Azerbaijan	9.538
Pakistan	9.157
Sao Tome and Principe	9.025
Uruguay	8.767
Burundi	8.581
Gambia	8.485
Kenya	8.461
Belarus	8.454
Lesotho	8.385

Continued on followiing page

Table 3. Continued

Country	Value
Colombia	8.377
Tajikistan	8.342
Mongolia	8.197
Georgia	8.171
Rwanda	8.104
Guatemala	7.749
Brazil	7.584
Yemen	7.399
Myanmar	7.352
Afghanistan	7.316
Madagascar	7.258
Russian Federation	6.730
Mexico	6.649
Jamaica	6.608
Barbados	6.595
Bhutan	6.442
South Africa	6.298
Hungary	6.207
Tunisia	6.091
Namibia	5.981
Bangladesh	5.974
Paraguay	5.819
Dominican Republic	5.704
Vanuatu	5.668
Chile	5.614
Kyrgyzstan	5.614
Bulgaria	5.481
United Republic of Tanzania	5.401
Nepal	5.388
Somalia	5.351
Curaçao	5.241
Lithuania	5.162
Nicaragua	5.053

Table 4. Decreasing order for the consumer food price inflation (%), on average over the period January 2016-September 2022, for the countries with the intermediate values

Country	Value
Tonga	5.046
Burkina Faso	5.037
Eswatini	4.925
Algeria	4.868
Mauritania	4.830
China, Hong Kong SAR	4.755
Poland	4.681
Latvia	4.676
India	4.631
Estonia	4.543
Mauritius	4.481
Djibouti	4.429
Albania	4.417
Samoa	4.356
San Marino	4.321
Serbia	4.194
Indonesia	4.172
Uganda	4.159
Montenegro	4.152
Romania	4.055
Slovakia	4.034
Armenia	4.002
Trinidad and Tobago	3.979
Senegal	3.937
Honduras	3.925
Czechia	3.860
Togo	3.821
Cambodia	3.777
Peru	3.753
Côte d'Ivoire	3.662
Malta	3.654
Botswana	3.646
Lao People's Democratic Republic	3.620
Philippines	3.542
Bosnia and Herzegovina	3.541
Papua New Guinea	3.487
Saudi Arabia	3.483

Continued on followiing page

Table 4. Continued

Country	Value
Kuwait	3.464
Republic of Korea	3.460
North Macedonia	3.358
Cameroon	3.337
Guam	3.291
Aruba	3.272
Fiji	3.271
Germany	3.235
Democratic Republic of the Congo	3.218
Cayman Islands	3.121
Viet Nam	3.049
Palau	3.014
Bahrain	2.997
Guinea-Bissau	2.997
Slovenia	2.935
Croatia	2.898
Sweden	2.865
Iceland	2.848
Bermuda	2.848
Spain	2.789
Malaysia	2.781
Antigua and Barbuda	2.773
Costa Rica	2.742
Niger	2.741
China, mainland	2.725
Netherlands	2.714
Seychelles	2.683
Andorra	2.626
Canada	2.574
Cook Islands	2.521
Cabo Verde	2.506
Austria	2.503
Luxembourg	2.474

Table 5. Decreasing order for the consumer food price inflation (%), on average over the period January 2016-September 2022, for the countries with the lowest values

Country	Value
United States of America	2.431
Bahamas	2.422
China, Macao SAR	2.380
Saint Vincent and the Grenadines	2.368
Portugal	2.346
Bolivia (Plurinational State of)	2.340
Belgium	2.293
El Salvador	2.285
Mali	2.229
Denmark	2.223
Singapore	2.144
Timor-Leste	2.112
Anguilla	2.101
Australia	2.087
United Arab Emirates	2.065
France	2.022
New Zealand	2.003
French Polynesia	1.984
Morocco	1.968
Italy	1.937
Greece	1.899
Maldives	1.885
Greenland	1.876
Norway	1.834
Guadeloupe	1.832
United Kingdom of Great Britain and Northern Ireland	1.790
Puerto Rico	1.772
Gabon	1.754
Åland Islands	1.727
New Caledonia	1.710
Finland	1.678
Réunion	1.651
Thailand	1.647
Congo	1.625
Comoros	1.604
French Guyana	1.584
Belize	1.579

Continued on followiing page

Table 5. Continued

Country	Value
Brunei Darussalam	1.515
Martinique	1.495
Kiribati	1.448
Cyprus	1.386
Solomon Islands	1.343
Iraq	1.308
Japan	1.282
Dominica	1.275
Benin	1.182
Equatorial Guinea	1.145
Israel	1.056
Montserrat	1.031
Oman	0.977
Qatar	0.970
Saint Kitts and Nevis	0.881
Ecuador	0.872
Palestine	0.820
Jordan	0.812
Panama	0.755
Grenada	0.729
British Virgin Islands	0.646
Chad	0.479
Micronesia (Federated States of)	0.429
Switzerland	0.235
Saint Lucia	0.190
Ireland	-0.361

Table 6. Decreasing order for the consumer food price inflation (%), on average over the period January-September 2022, for the countries with the highest values

Country	Value
Lebanon	331.827
Venezuela (Bolivarian Republic of)	192.433
Zimbabwe	188.094
Sudan	126.335
Türkiye	82.557
Argentina	66.205
Iran (Islamic Republic of)	60.355
Sri Lanka	56.722
Suriname	51.988
Ethiopia	38.876
Syrian Arab Republic	37.640
Republic of Moldova	31.142
Haiti	28.974
Ghana	26.995
Angola	26.379
Sierra Leone	25.752
Burkina Faso	25.556
Malawi	24.718
Colombia	24.092
Ukraine	23.560
Lithuania	23.346
Afghanistan	22.432
Pakistan	21.453
Hungary	21.437
Egypt	20.734
Bulgaria	20.419
Burundi	20.389
Bosnia and Herzegovina	20.354
Montenegro	20.202
Azerbaijan	19.871
Nigeria	19.761
Mongolia	19.431
Rwanda	19.082
Latvia	18.942
Georgia	18.479
Myanmar	18.360
Russian Federation	18.329

Continued on followiing page

Table 6. Continued

Country	Value
North Macedonia	17.840
Sao Tome and Principe	17.535
Belarus	17.348
Serbia	17.273
Kazakhstan	17.049
Estonia	16.875
Paraguay	16.680
Kyrgyzstan	16.099
Samoa	16.014
Chile	15.670
Slovakia	15.608
Nicaragua	15.314
Cabo Verde	15.270
Uzbekistan	14.833
Croatia	14.639
Somalia	14.455
Mauritania	14.298
Mali	14.269
Palau	14.235
Costa Rica	14.137
Barbados	14.122
Mauritius	14.120
Algeria	13.990
Czechia	13.750
Romania	13.681
Armenia	13.555
Zambia	13.547
Guinea	13.339
Senegal	13.307
Mexico	13.263
Poland	13.243
Mozambique	13.099
Gambia	13.005

Table 7. Decreasing order for the consumer food price inflation (%), on average over the period January-September 2022, for the countries with the intermediate values

Country	Value
Jamaica	12.817
Togo	12.738
Guam	12.655
Peru	12.552
Bahamas	12.524
Honduras	12.440
Brazil	12.262
Kenya	12.170
El Salvador	12.111
Dominican Republic	11.402
United States of America	11.189
Uruguay	10.926
Albania	10.891
Djibouti	10.875
Cameroon	10.840
Portugal	10.784
San Marino	10.703
Greece	10.630
Germany	10.429
Curaçao	10.421
Uganda	10.392
Spain	10.314
Tonga	10.237
Denmark	10.210
Slovenia	10.205
Morocco	10.165
Antigua and Barbuda	10.087
Bahrain	10.044
Chad	9.946
Tunisia	9.717
Malta	9.613
Botswana	9.408
Saint Kitts and Nevis	9.346
Canada	9.276
Guinea-Bissau	9.255
Austria	9.217
Netherlands	9.161

Continued on followiing page

Table 7. Continued

Country	Value
Aruba	9.154
Côte d'Ivoire	9.121
Sweden	9.108
United Kingdom of Great Britain and Northern Ireland	9.005
Trinidad and Tobago	8.939
Solomon Islands	8.838
Cook Islands	8.735
Saint Vincent and the Grenadines	8.727
Finland	8.682
Madagascar	8.526
Anguilla	8.522
Niger	8.446
Lesotho	8.436
South Africa	8.233
Guatemala	8.200
Puerto Rico	8.184
Tajikistan	8.150
Italy	7.690
Andorra	7.669
Timor-Leste	7.656
Åland Islands	7.596
India	7.596
Kuwait	7.589
Bangladesh	7.586
Cyprus	7.329
New Zealand	7.240
United Arab Emirates	6.982
French Polynesia	6.947
Belgium	6.923
Nepal	6.921
Iraq	6.918
Namibia	6.835
Palestine	6.787

Table 8. Decreasing order for the consumer food price inflation (%), on average over the period January-September 2022, for the countries with the lowest values

Country	Value
Bermuda	6.699
United Republic of Tanzania	6.604
Belize	6.567
Congo	6.404
Cayman Islands	6.344
Iceland	6.316
Thailand	6.231
Equatorial Guinea	6.060
Indonesia	5.949
Republic of Korea	5.932
Vanuatu	5.893
Luxembourg	5.799
Australia	5.615
Papua New Guinea	5.579
Ireland	5.459
Cambodia	5.444
New Caledonia	5.394
France	5.392
Oman	5.362
Fiji	5.351
Malaysia	5.302
Qatar	5.048
Réunion	4.982
Brunei Darussalam	4.973
Micronesia (Federated States of)	4.964
Montserrat	4.963
Bhutan	4.844
Singapore	4.760
China, Hong Kong SAR	4.736
Libya	4.662
Norway	4.589
Ecuador	4.503
Israel	4.333
Maldives	4.187
French Guyana	4.087
Kiribati	4.054
Martinique	3.993

Continued on followiing page

Table 8. Continued

Country	Value
Greenland	3.993
Dominica	3.690
Jordan	3.685
Guadeloupe	3.656
Panama	3.649
Saudi Arabia	3.621
Japan	3.377
Philippines	3.324
Lao People's Democratic Republic	3.274
Grenada	3.267
Gabon	3.151
Yemen	2.734
Eswatini	2.547
Saint Lucia	2.257
Liberia	2.134
China, mainland	1.987
Comoros	1.832
China, Macao SAR	1.769
Viet Nam	1.697
Democratic Republic of the Congo	1.480
British Virgin Islands	1.061
Bolivia (Plurinational State of)	0.906
Switzerland	0.814
Seychelles	0.763
Benin	-0.700
South Sudan	-3.420

FOOD INDICES (2015 = 100) ASSESSMENT

On average the food indices (2015=100) presented monthly growth rates (relatively to the previous month of the same year), over the period 2016-2019, of 0.216% (table 9), with a standard deviation of 0.249% and 44 observations. The food indices had an average of 106.167 and a standard deviation of 3.770. For these results, 48 observations were taken into account.

These growth rates had an average of 0.958% in the period January 2020-September 2022 (table 10) and 1.885% in the year 2022 (until September). The standard deviation for the period January 2020-September 2022 was 0.709% (30 observations). The values in levels had an average of 129.588, a standard deviation of 12.651 and 33 observations. Again, the pandemic and, particularly, the Russia-Ukraine crisis had impacts on food prices.

These realities diverge across the different countries, such as revealed in tables 11, 12 and 13 for the period January 2016-September 2022 and tables 14, 15 and 16 for the period January-September 2022.

Table 9. Consumer prices, food indices (2015 = 100) for world monthly values (2016-2019)

Year	Months	Value	Growth rates (%)
2016	January	101.426	
2016	February	102.159	0.723
2016	March	101.849	-0.303
2016	April	101.940	0.089
2016	May	101.725	-0.211
2016	June	101.626	-0.097
2016	July	101.827	0.198
2016	August	101.878	0.050
2016	September	102.090	0.208
2016	October	102.209	0.117
2016	November	102.273	0.063
2016	December	102.425	0.149
2017	January	103.242	
2017	February	103.642	0.387
2017	March	103.589	-0.051
2017	April	103.817	0.220
2017	May	103.970	0.147
2017	June	103.830	-0.135
2017	July	104.069	0.230
2017	August	104.267	0.190
2017	September	104.487	0.211
2017	October	104.668	0.173
2017	November	104.761	0.089
2017	December	105.185	0.405
2018	January	106.009	
2018	February	106.313	0.287
2018	March	106.000	-0.294
2018	April	106.069	0.065
2018	May	106.066	-0.003
2018	June	106.332	0.251
2018	July	106.677	0.324
2018	August	107.245	0.532
2018	September	107.960	0.667
2018	October	108.213	0.234
2018	November	108.126	-0.080
2018	December	108.463	0.312
2019	January	109.395	
2019	February	110.141	0.682

Continued on followiing page

Table 9. Continued

Year	Months	Value	Growth rates (%)
2019	March	110.454	0.284
2019	April	110.714	0.235
2019	May	111.041	0.295
2019	June	111.051	0.009
2019	July	111.368	0.285
2019	August	111.864	0.445
2019	September	112.428	0.504
2019	October	113.139	0.632
2019	November	113.704	0.499
2019	December	114.266	0.494

Table 10. Consumer prices food indices (2015 = 100) for world monthly values (2020-2022)

Year	Months	Value	Growth rates (%)
2020	January	115.739	
2020	February	116.690	0.822
2020	March	116.792	0.087
2020	April	118.372	1.353
2020	May	118.434	0.052
2020	June	118.707	0.231
2020	July	118.839	0.111
2020	August	119.178	0.285
2020	September	119.512	0.280
2020	October	120.170	0.551
2020	November	120.300	0.108
2020	December	120.840	0.449
2021	January	122.276	
2021	February	123.245	0.792
2021	March	123.385	0.114
2021	April	124.238	0.691
2021	May	124.791	0.445
2021	June	125.429	0.511
2021	July	126.584	0.921
2021	August	127.821	0.977
2021	September	129.062	0.971
2021	October	130.635	1.219
2021	November	132.202	1.200
2021	December	134.159	1.480
2022	January	136.783	
2022	February	138.949	1.584
2022	March	141.570	1.886
2022	April	145.385	2.695
2022	May	147.718	1.605
2022	June	150.463	1.858
2022	July	153.506	2.022
2022	August	155.822	1.509
2022	September	158.814	1.920

Table 11. Decreasing order for the consumer prices food indices (2015 = 100), on average over the period January 2016-September 2022, for the countries with the highest values

Country	Value
Venezuela (Bolivarian Republic of)	67800000000.000
Zimbabwe	4470.643
South Sudan	4037.763
Sudan	1684.186
Lebanon	978.184
Syrian Arab Republic	457.642
Argentina	376.178
Suriname	329.825
Iran (Islamic Republic of)	287.235
Angola	263.339
Türkiye	194.357
Haiti	192.767
Libya	190.144
Nigeria	188.027
Malawi	187.840
Egypt	184.499
Ethiopia	184.219
Sierra Leone	182.318
Uzbekistan	176.346
Zambia	169.296
Liberia	168.552
Guinea	164.402
Mozambique	160.492
Ukraine	148.488
Kazakhstan	145.435
Azerbaijan	144.231
Ghana	143.578
Yemen	138.916
Kenya	138.135
Lesotho	137.409
Uruguay	137.115
Guatemala	136.837
Republic of Moldova	136.261
Gambia	135.859
Tajikistan	134.742
Belarus	134.107
Madagascar	133.265

Continued on followiing page

Table 11. Continued

Country	Value
Sri Lanka	133.026
Sao Tome and Principe	131.917
Rwanda	130.216
Burundi	129.854
South Africa	129.058
Brazil	128.087
Colombia	128.074
Pakistan	127.987
Eswatini	127.437
Namibia	127.390
Mongolia	126.937
Georgia	126.523
Myanmar	126.382
Bhutan	126.204
United Republic of Tanzania	125.349
Mexico	125.128
Barbados	125.027
Bangladesh	124.850
Tunisia	123.729
Afghanistan	123.459
Jamaica	122.844
Vanuatu	122.452
Russian Federation	121.319
Paraguay	120.581
Nepal	120.088
Tonga	119.638
China, Hong Kong SAR	118.950
Somalia	118.529
Indonesia	117.718
Democratic Republic of the Congo	116.931
Dominican Republic	116.853
Hungary	116.771
Curaçao	116.727

Table 12. Decreasing order for the consumer prices food indices (2015 = 100), on average over the period January 2016-September 2022, for the countries with the intermediate values

Country	Value
Chile	116.596
Uganda	116.393
India	116.207
Djibouti	116.001
Albania	115.796
Cambodia	115.557
Trinidad and Tobago	114.973
Algeria	114.890
Poland	114.816
Samoa	114.682
Bulgaria	114.377
Mauritania	114.364
San Marino	114.255
Mauritius	114.030
Nicaragua	113.903
Philippines	113.841
Papua New Guinea	113.508
Estonia	113.429
Lao People's Democratic Republic	113.073
Latvia	112.800
Lithuania	112.488
Bolivia (Plurinational State of)	112.355
Malta	112.309
Republic of Korea	112.022
China, mainland	111.901
Fiji	111.892
Malaysia	111.434
Viet Nam	111.126
Botswana	110.999
Honduras	110.960
Czechia	110.652
Côte d'Ivoire	110.559
Serbia	110.384
China, Macao SAR	110.383
Peru	110.275
Romania	110.204
Slovakia	109.859

Continued on followiing page

Table 12. Continued

Country	Value
Saudi Arabia	109.816
Bermuda	109.746
Aruba	109.686
Senegal	109.542
Cayman Islands	109.225
Germany	109.192
Slovenia	108.909
Armenia	108.843
Togo	108.814
Luxembourg	108.814
Sweden	108.809
Burkina Faso	108.329
Netherlands	108.315
Bahrain	108.159
Andorra	107.991
Kuwait	107.846
Cameroon	107.834
Comoros	107.796
Belgium	107.741
Montenegro	107.472
Singapore	107.451
Greenland	107.448
Antigua and Barbuda	107.285
Spain	107.186
Iceland	107.182
Kyrgyzstan	107.173
Guinea-Bissau	107.030
Austria	106.998
Norway	106.924
Guadeloupe	106.380
Guam	106.242
Maldives	106.240
France	106.227
Croatia	106.076

Table 13. Decreasing order for the consumer prices food indices (2015 = 100), on average over the period January 2016-September 2022, for the countries with the lowest values

Country	Value
Saint Vincent and the Grenadines	106.047
United Arab Emirates	106.042
Canada	105.968
Cook Islands	105.881
Australia	105.875
Palau	105.835
Seychelles	105.787
Denmark	105.308
French Polynesia	105.281
Bosnia and Herzegovina	105.230
Portugal	105.220
Martinique	105.218
Réunion	105.217
Gabon	105.197
New Zealand	105.037
Italy	105.030
Costa Rica	104.850
Japan	104.822
North Macedonia	104.803
Morocco	104.536
Anguilla	104.535
Thailand	104.502
New Caledonia	104.396
French Guyana	104.387
Niger	104.016
Kiribati	103.936
Bahamas	103.533
El Salvador	103.488
Puerto Rico	103.483
Benin	103.481
Dominica	103.409
Timor-Leste	103.384
United Kingdom of Great Britain and Northern Ireland	103.348
United States of America	103.254
Åland Islands	103.224
Brunei Darussalam	103.021
Cabo Verde	102.942

Continued on followiing page

Table 13. Continued

Country	Value
Ecuador	102.711
Greece	102.639
Solomon Islands	102.611
Finland	102.152
Congo	102.100
Cyprus	101.884
Belize	101.634
Switzerland	101.563
Panama	101.408
Israel	101.381
British Virgin Islands	101.300
Montserrat	101.175
Oman	100.732
Mali	100.663
Qatar	100.571
Grenada	100.378
Equatorial Guinea	100.350
Jordan	100.041
Iraq	99.704
Palestine	99.067
Saint Lucia	98.992
Micronesia (Federated States of)	98.158
Saint Kitts and Nevis	97.605
Ireland	95.341
Chad	93.675

Table 14. Decreasing order for the consumer prices food indices (2015 = 100), on average over the period January-September 2022, for the countries with the highest values

Country	Value
Venezuela (Bolivarian Republic of)	334000000000.000
Zimbabwe	20653.280
Sudan	6894.484
South Sudan	6748.994
Lebanon	4864.737
Syrian Arab Republic	1177.564
Argentina	982.027
Suriname	736.555
Iran (Islamic Republic of)	710.714
Angola	471.398
Türkiye	417.031
Ethiopia	341.215
Haiti	330.259
Nigeria	293.232
Sierra Leone	285.488
Malawi	281.744
Uzbekistan	263.707
Zambia	253.359
Egypt	244.039
Guinea	233.414
Liberia	221.212
Sri Lanka	220.116
Libya	215.140
Ghana	207.686
Ukraine	206.405
Mozambique	205.264
Kazakhstan	197.399
Republic of Moldova	192.913
Azerbaijan	186.602
Pakistan	181.343
Sao Tome and Principe	181.052
Uruguay	177.436
Mongolia	174.854
Gambia	174.836
Tajikistan	173.703
Belarus	173.693
Kenya	173.510

Continued on followiing page

Table 14. Continued

Country	Value
Lesotho	173.168
Georgia	170.595
Colombia	170.063
Burundi	169.022
Guatemala	166.193
Brazil	163.371
Rwanda	162.351
Madagascar	162.082
Myanmar	161.250
Afghanistan	160.363
Yemen	160.228
Mexico	155.919
Russian Federation	155.613
Jamaica	153.501
Barbados	153.337
Bhutan	153.205
South Africa	151.874
Hungary	150.693
Tunisia	150.633
Namibia	148.821
Bangladesh	148.671
Paraguay	148.425
Vanuatu	147.586
Dominican Republic	144.525
Chile	144.420
Kyrgyzstan	143.757
Bulgaria	143.678
Somalia	143.323
United Republic of Tanzania	143.091
Curaçao	142.422
Tonga	141.889
Nicaragua	140.162
Nepal	140.158

Table 15. Decreasing order for the consumer prices food indices (2015 = 100), on average over the period January-September 2022, for the countries with the intermediate values

Country	Value
Lithuania	140.151
Eswatini	138.192
Algeria	137.874
Mauritania	137.802
Poland	137.246
China, Hong Kong SAR	137.229
Latvia	136.526
Burkina Faso	136.409
Mauritius	136.122
Estonia	135.919
India	135.349
Albania	135.278
Djibouti	134.776
San Marino	133.485
Samoa	132.878
Serbia	132.808
Romania	132.446
Indonesia	132.154
Armenia	131.948
Montenegro	131.278
Slovakia	131.156
Honduras	130.182
Czechia	130.078
Uganda	130.005
Trinidad and Tobago	129.867
Cambodia	128.799
Togo	128.698
Côte d'Ivoire	128.254
Botswana	128.022
Peru	127.986
Malta	127.889
Senegal	127.682
Philippines	127.173
Lao People's Democratic Republic	127.054
Republic of Korea	126.851
Bosnia and Herzegovina	126.442
Saudi Arabia	126.245

Continued on followiing page

Table 15. Continued

Country	Value
Papua New Guinea	125.855
Kuwait	125.498
North Macedonia	124.997
Cameroon	124.925
Germany	124.600
Guam	124.124
Aruba	124.041
Democratic Republic of the Congo	124.032
Fiji	123.836
Cayman Islands	123.685
Palau	122.846
Viet Nam	122.762
Slovenia	122.221
Croatia	121.633
Bermuda	121.519
Sweden	121.388
Guinea-Bissau	121.320
Iceland	121.078
Bahrain	120.906
Spain	120.496
Malaysia	120.478
Antigua and Barbuda	120.357
Netherlands	120.171
Costa Rica	119.989
China, mainland	119.414
Andorra	119.379
Seychelles	119.080
Canada	118.743
Austria	118.655
Cook Islands	118.624
Luxembourg	118.388
Niger	118.267
Saint Vincent and the Grenadines	117.979
United States of America	117.706

Table 16. Decreasing order for the consumer prices food indices (2015 = 100), on average over the period January-September 2022, for the countries with the lowest values

Country	Value
Cabo Verde	117.511
China, Macao SAR	117.353
Bahamas	117.304
Portugal	117.197
Belgium	116.828
Bolivia (Plurinational State of)	116.734
El Salvador	116.513
Denmark	116.267
Singapore	115.867
Timor-Leste	115.719
Anguilla	115.495
New Zealand	115.380
Australia	115.344
Mali	115.163
France	114.823
French Polynesia	114.621
United Arab Emirates	114.499
Italy	114.022
Morocco	113.812
Greenland	113.326
Greece	113.289
Maldives	113.211
Norway	113.189
United Kingdom of Great Britain and Northern Ireland	113.144
Guadeloupe	113.118
Gabon	112.533
Åland Islands	112.483
New Caledonia	112.405
Réunion	112.120
Puerto Rico	112.065
Finland	112.038
Thailand	111.739
French Guyana	111.283
Belize	111.148
Brunei Darussalam	110.808
Congo	110.756
Martinique	110.702

Continued on followiing page

Table 16. Continued

Country	Value
Comoros	110.576
Kiribati	110.107
Japan	109.135
Cyprus	109.094
Dominica	108.953
Solomon Islands	108.364
Iraq	108.128
Equatorial Guinea	107.627
Benin	107.215
Montserrat	107.186
Israel	106.807
Oman	106.758
Saint Kitts and Nevis	106.493
Qatar	106.352
Ecuador	105.848
Jordan	105.457
Grenada	105.390
Panama	105.059
Palestine	104.605
British Virgin Islands	103.942
Micronesia (Federated States of)	103.021
Chad	101.961
Switzerland	101.771
Saint Lucia	101.652
Ireland	97.333

TESTING FOR STRUCTURAL BREAKS

Figures 1 and 2 present the results of the QLR (Quandt Likelihood Ratio) test for unknown structural breaks in the data (Quandt, 1960), since the beginning of 2020. The two figures highlight that the strongest structural impacts occurred at the beginning of 2022 after the start of the Russia-Ukraine conflict. These results confirm that the prices were more affected by the conflict than by the pandemic.

Figure 1. QLR test for unknown structural breaks in food price inflation (%) over the period January 2020-September 2022

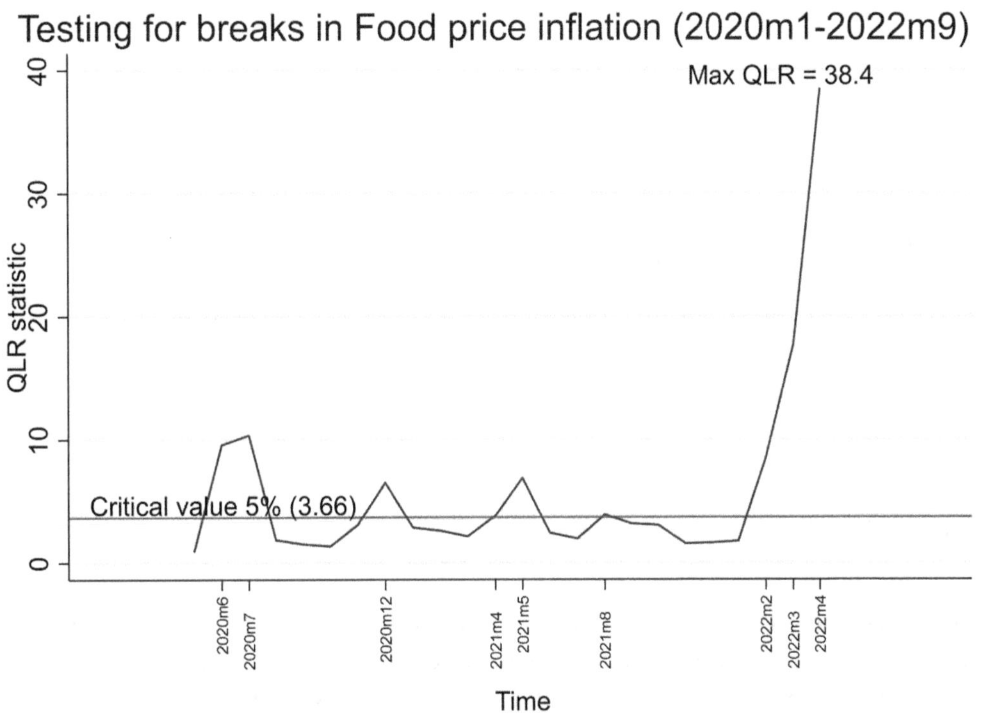

Figure 2. QLR test for unknown structural breaks in consumer prices, food indices (2015 = 100) over the period January 2020-September 2022

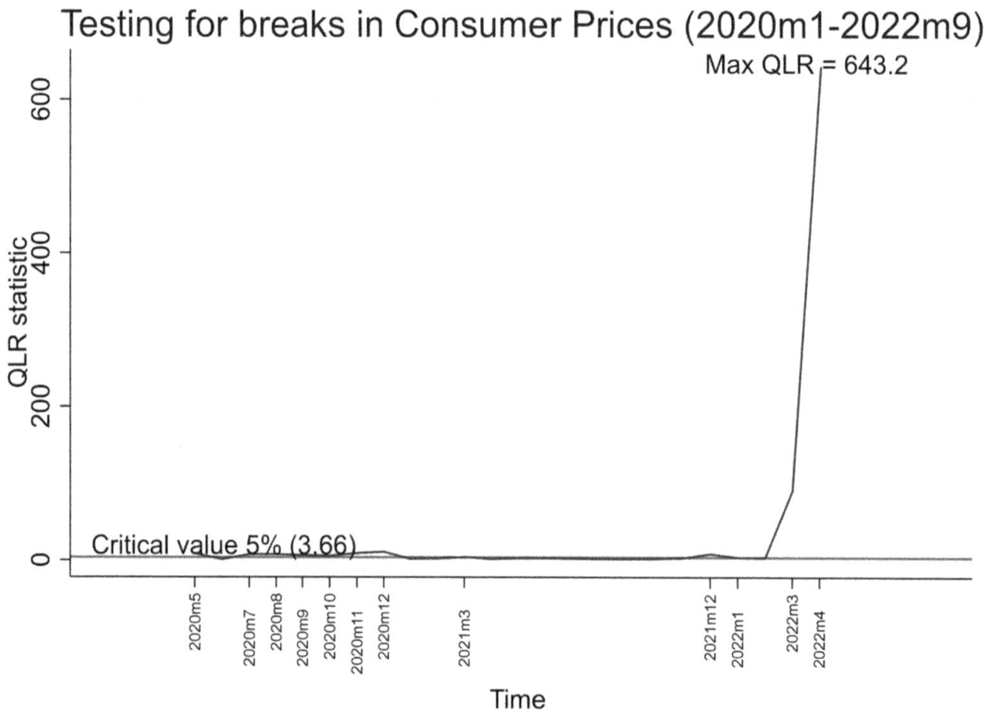

CONCLUSION

This contribution proposed to analyse the impacts from, namely, the Covid-19 pandemic and the Russia-Ukraine conflict on the food prices evolution worldwide. For that, data from the FAOSTAT database were considered for the months from the beginning of 2016 (January) until the end of 2022 (September). The statistical information was taken into account for the food price inflation (%) and consumer prices (food indices, 2015=100). For a deeper assessment, normalised values were found, as well as changes in the percentages, growth rates and QLR tests for unknown structural breaks.

The literature shows that these two exogenous shocks (the pandemic and the Russia-Ukraine crisis) had impacts on the different food markets worldwide. The roots of these consequences are related to food chain disruptions, transport route disturbances and resource prices, namely energy ones. The new technologies and public policies may have here an important contribution. These two shocks only brought additional difficulties to those created by climate change, animal diseases and agroterrorism, for example. The different world food markets were not affected similarly and, in some cases, there are risks to food security and social stability. This is a concern, specifically for low and middle-income countries and may persist if not properly considered by governments and international organisations.

The data analysis for the food price inflation (%) and food indices (2015=100) reveals that, indeed, the pandemic and Russia-Ukraine conflict had implications on the food market prices. Nonetheless, the strongest impacts came from the conflict. On the other hand, the pandemic seems to have promoted decreases in food price inflation in the months of 2020. In addition, this data analysis confirms the literature insights about the distinct reality of the food markets worldwide and about the different ways these markets were affected by international shocks. The unknown structural breaks assessment also confirms these findings about the strongest consequences on the food prices from the Russia-Ukraine crisis.

From a perspective of practical implications, this chapter highlights the distinct reality of the food markets worldwide and the different patterns of reaction of these markets to shocks, calling for analysis and policy instruments that take into account the specific particularities of each case, namely to deal with the consequences on the prices from the Russia-Ukraine crisis. For policy recommendations, it is suggested to publically present information from the markets in the producer and over the remaining chain. For future research, it would be important to analyse also the supply and demand changes, in the food markets, since 2020.

ACKNOWLEDGMENT

This work is funded by National Funds through the FCT - Foundation for Science and Technology, I.P., within the scope of the project Refª UIDB/00681/2020. Furthermore we would like to thank the CERNAS Research Centre and the Polytechnic Institute of Viseu for their support.

REFERENCES

Ahoya, B., Kavle, J. A., Kiige, L., Gathi, C., Samburu, B., Maina, L., Ramirez, L., Wambu, R., & Codjia, P. (2022). *How COVID-19 affected food systems, health service delivery and maternal and infant nutrition practices: Implications for moving forward in Kenya. Maternal and Child Nutrition*. Scopus., doi:10.1111/mcn.13466

Akber, N., & Paltasingh, K. R. (2022). COVID-19 lockdown and prices of essential food items in India: Examining law of one price. *Journal of Agribusiness in Developing and Emerging Economies, 12*(3), 510–530. Scopus. doi:10.1108/JADEE-12-2021-0316

Akter, M. S., Talsma, E. F., Feskens, E. J. M., Thilsted, S. H., & Rasheed, S. (2022). How did the COVID-19 pandemic affect food environment, food purchase, and fish consumption among low-income urban households in Bangladesh—A path analysis. *Frontiers in Public Health, 10*. Scopus. doi:10.3389/fpubh.2022.994236

Amin, F., Poespito Hadi, W., Zauhar, S., & Santoso Haryono, B. (2022). Determinants of post COVID-19 food security policy success. *International Journal of Disaster Resilience in the Built Environment, 13*(4), 440–450. Scopus. doi:10.1108/IJDRBE-11-2020-0118

Ayo-Lawal, R. A., Ilevbare, O. E., Omotoso, K. O., Omimakinde, E. A., & Ukwuoma, O. (2022). Social assistance and food security during covid-19 pandemic lock-down: Insights from Nigeria. *Future of Food: Journal on Food, Agriculture and Society, 10*(1), 1–14. Scopus. doi:10.17170/kobra-202110144902

Bai, Y., Costlow, L., Ebel, A., Laves, S., Ueda, Y., Volin, N., Zamek, M., & Masters, W. A. (2022). Retail prices of nutritious food rose more in countries with higher COVID-19 case counts. *Nature Food, 3*(5), 325–330. Scopus. doi:10.1038/s43016-022-00502-1

Bairagi, S., Mishra, A. K., & Mottaleb, K. A. (2022). Impacts of the COVID-19 pandemic on food prices: Evidence from storable and perishable commodities in India. *PLoS ONE, 17*(3 March). Scopus. doi:10.1371/journal.pone.0264355

Ben Hassen, T., & El Bilali, H. (2022). Impacts of the Russia-Ukraine War on Global Food Security: Towards More Sustainable and Resilient Food Systems? *Foods, 11*(15). Scopus. doi:10.3390/foods11152301

Bliznashka, L., Ahun, M. N., Velthausz, D., Donco, R., Karuskina-Drivdale, S., Pinto, J., Yousafzai, A. K., & Jeong, J. (2022). Effects of COVID-19 on child health services utilization and delivery in rural Mozambique: A qualitative study. *Health Policy and Planning, 37*(6), 737–746. Scopus. doi:10.1093/heapol/czac009

Buzigi, E., & Onakuse, S. (2023). Food price volatility and socio-economic inequalities in poor food consumption status during coronavirus disease-2019 lockdown among slum and non-slum households in urban Nansana municipality, Uganda. *Nutrition Journal, 22*(1). Scopus. doi:10.1186/s12937-023-00836-x

Carrière-Swallow, Y., Deb, P., Furceri, D., Jiménez, D., & Ostry, J. D. (2023). Shipping costs and inflation. *Journal of International Money and Finance, 130*. Scopus. doi:10.1016/j.jimonfin.2022.102771

Chang, B. H., Gohar, R., Derindag, O. F., & Uche, E. (2022). COVID-19, lockdown measures and their impact on food and healthcare prices: Empirical evidence using a dynamic ARDL model. *Journal of Economic Studies*. Scopus. doi:10.1108/JES-12-2021-0627

Chari, F., Muzinda, O., Novukela, C., & Ngcamu, B. S. (2022). *Pandemic outbreaks and food supply chains in developing countries: A case of COVID-19 in Zimbabwe. Cogent Business and Management, 9(1)*. Scopus., doi:10.1080/23311975.2022.2026188

Cheng, S., Li, X., & Cao, Y. (2023). Global evidence of the exposure-lag-response associations between temperature anomalies and food markets. *Journal of Environmental Management, 325*. Scopus. doi:10.1016/j.jenvman.2022.116592

Codjia, C. O., & Saghaian, S. H. (2022). Determinants of Food Expenditure Patterns: Evidence from U.S. Consumers in the Context of the COVID-19 Price Shocks. *Sustainability (Switzerland), 14*(13). Scopus. doi:10.3390/su14138156

Das, N. K., & Roy, A. (2022). COVID-19 and agri-food value chain: A systematic review and bibliometric mapping. *Journal of Agribusiness in Developing and Emerging Economies, 12*(3), 442–462. Scopus. doi:10.1108/JADEE-07-2021-0188

Dasgupta, S., & Robinson, E. J. Z. (2022). Impact of COVID-19 on food insecurity using multiple waves of high frequency household surveys. *Scientific Reports, 12*(1). Scopus. doi:10.1038/s41598-022-05664-3

Dietrich, S., Giuffrida, V., Martorano, B., & Schmerzeck, G. (2022). COVID-19 policy responses, mobility, and food prices. *American Journal of Agricultural Economics, 104*(2), 569–588. Scopus. doi:10.1111/ajae.12278

Djurle, A., Young, B., Berlin, A., Vågsholm, I., Blomström, A.-L., Nygren, J., & Kvarnheden, A. (2022). Addressing biohazards to food security in primary production. *Food Security, 14*(6), 1475–1497. Scopus. doi:10.1007/s12571-022-01296-7

FAOSTAT. (2023). *Several Statistics and Information*. FAO. https://www.fao.org/faostat/en/#home

Firmansyah, Afriani, H., & Hoesni, F. (2022). The Volatility Of Broiler Meat Prices In Traditional And Modern Markets Before, At The Beginning Of, And During The Covid-19 Pandemic In Jambi City. *Xinan Jiaotong Daxue Xuebao/Journal of Southwest Jiaotong University, 57*(3), 138–151. Scopus. doi:10.35741/issn.0258-2724.57.3.12

Ghalibaf, M. B., Gholami, M., & Mohammadian, N. (2022). Stability of Food Security in Iran; Challenges and Ways Forward: A Narrative Review. [Scopus.]. *Iranian Journal of Public Health, 51*(12), 2654–2663. doi:10.18502/ijph.v51i12.11456 PMID:36742249

Goeb, J., Maredia, M., Win, K. Z., Masias, I., Lambrecht, I., Boughton, D., & Minten, B. (2022). Urban food prices under lockdown: Evidence from Myanmar's traditional food retail sector during COVID-19. *Applied Economics, 54*(47), 5412–5441. Scopus. doi:10.1080/00036846.2022.2044999

Goeb, J., Zone, P. P., Kham Synt, N. L., Zu, A. M., Tang, Y., & Minten, B. (2022). Food prices, processing, and shocks: Evidence from rice and COVID-19. *Journal of Agricultural Economics, 73*(2), 338–355. Scopus. doi:10.1111/1477-9552.12461

Guo, J., & Tanaka, T. (2022). Do biofuel production and financial speculation in agricultural commodities influence African food prices? New evidence from a TVP-VAR extended joint connectedness approach. *Energy Economics, 116*. Scopus. doi:10.1016/j.eneco.2022.106422

Gutiérrez-Romero, R. (2022). Conflicts increased in Africa shortly after COVID-19 lockdowns, but welfare assistance reduced fatalities. *Economic Modelling, 116*. Scopus. doi:10.1016/j.econmod.2022.105991

Hart, T. G., Davids, Y. D., Rule, S., Tirivanhu, P., & Mtyingizane, S. (2022). The COVID-19 pandemic reveals an unprecedented rise in hunger: The South African Government was ill-prepared to meet the challenge. *Scientific African, 16*. Scopus. 10.1016/j.sciaf.2022.e01169

Head, J. R., Chanthavilay, P., Catton, H., Vongsitthi, A., Khamphouxay, K., & Simphaly, N. (2022). Changes in household food security, access to health services and income in northern Lao PDR during the COVID-19 pandemic: A cross-sectional survey. *BMJ Open, 12*(6). Scopus. doi:10.1136/bmjopen-2021-055935

Henry, F. J., Lawrence, B., & Nelson, M. (2022). *Comparative cost of diets for low-income families in the Caribbean. Revista Panamericana de Salud Publica/Pan American Journal of Public Health, 46*. Scopus., doi:10.26633/RPSP.2022.120

Hernández-Vásquez, A., Visconti-Lopez, F. J., & Vargas-Fernández, R. (2022). Factors Associated with Food Insecurity in Latin America and the Caribbean Countries: A Cross-Sectional Analysis of 13 Countries. *Nutrients, 14*(15). Scopus. doi:10.3390/nu14153190

Higashi, R. T., Sood, A., Conrado, A. B., Shahan, K. L., Leonard, T., & Pruitt, S. L. (2022). Experiences of increased food insecurity, economic and psychological distress during the COVID-19 pandemic among Supplemental Nutrition Assistance Program-enrolled food pantry clients. *Public Health Nutrition, 25*(4), 1027–1037. Scopus. doi:10.1017/S1368980021004717

Ibiyemi, T. E., & Oldewage-Theron, W. (2022). A review of food and nutrition security in nigeria. *African Journal of Food, Agriculture, Nutrition and Development*, *22*(1).

Iheme, G. O., Adile, A. D., Egechizuorom, I. M., Kupoluyi, O. E., Ogbonna, O. C., Olah, L. E., Enuka, H. C., Idris, H., Asouzu, N. C., & Oyebamiji, E. A. (2022). Impact of COVID-19 pandemic on food price index in Nigeria. *Future of Food: Journal on Food, Agriculture and Society, 10*(3), 1–9. Scopus. doi:10.17170/kobra-202204136008

Jacobs, M., McDade, T. R., Chaparro, M. V., & Corea, M. (2022). Sick, Hungry, and Vulnerable: Federal Stimulus and Food Security on Marginalized Populations During the COVID-19 Pandemic. *Journal of Racial and Ethnic Health Disparities*. Scopus. doi:10.1007/s40615-022-01447-8

Jagtap, S., Trollman, H., Trollman, F., Garcia-Garcia, G., Parra-López, C., Duong, L., Martindale, W., Munekata, P. E. S., Lorenzo, J. M., Hdaifeh, A., Hassoun, A., Salonitis, K., & Afy-Shararah, M. (2022). The Russia-Ukraine Conflict: Its Implications for the Global Food Supply Chains. *Foods, 11*(14). Scopus. doi:10.3390/foods11142098

Kaicker, N., Gupta, A., & Gaiha, R. (2022). Covid-19 pandemic and food security in India: Can authorities alleviate the disproportionate burden on the disadvantaged? *Journal of Policy Modeling, 44*(5), 963–980. Scopus. doi:10.1016/j.jpolmod.2022.08.001

Kh'ng, K., Chang, C.-C., & Hsu, S.-H. (2022). Implications of food and nutrition security on household food expenditure: The case of Malaysia. *Agriculture and Food Security, 11*(1). Scopus. doi:10.1186/s40066-022-00367-4

Korkmaz Aslan, G., Kılınç, E., & Kartal, A. (2022). The effect of COVID-19 pandemic on lifestyle-related behaviours in Turkey: A web-based cross-sectional study. *International Journal of Nursing Practice, 28*(5). Scopus. doi:10.1111/ijn.13053

Kyfyak, V., Verbivska, L., Alioshkina, L., Galunets, N., Kucher, L., & Skrypnyk, S. (2022). The Influence of the Social and Economic Situation on Agribusiness. *WSEAS Transactions on Environment and Development, 18*, 1021–1035. Scopus. doi:10.37394/232015.2022.18.98

Lialina, A. V., & Morachevskaya, K. A. (2022). Economic Access to Food and COVID-19: New Challenges for the Russian Exclave. *Regional Research of Russia, 12*(3), 335–349. Scopus. doi:10.1134/S2079970522700289

Liang, Y., & Zhong, T. (2023). Impacts of community-level grassroots organizations on household food security during the COVID-19 epidemic period in China. *International Journal of Disaster Risk Reduction, 85*. Scopus. doi:10.1016/j.ijdrr.2022.103490

Lopes, M. S., de Freitas, P. P., de Carvalho, M. C. R., Silva, U. M., & Lopes, A. C. S. (2022). The COVID-19 pandemic in a Brazilian metropolis: Repercussion on food prices. *Cadernos de Saude Publica, 38*(4). Scopus. doi:10.1590/0102-311XEN166721

Macias, P., Stelmasiak, D., & Szafranek, K. (2022). Nowcasting food inflation with a massive amount of online prices. *International Journal of Forecasting*. Scopus. doi:10.1016/j.ijforecast.2022.02.007

Manyong, V., Bokanga, M., Akonkwa Nyamuhirwa, D.-M., Bamba, Z., Adeoti, R., Mwepu, G., Cole, S. M., & Dontsop Nguezet, P. M. (2022). *COVID-19 outbreak and rural household food security in the Western Democratic Republic of the Congo. World Development Perspectives, 28*. Scopus., doi:10.1016/j.wdp.2022.100469

Martin, W., & Minot, N. (2022). The impacts of price insulation on world wheat markets during the 2022 food price crisis. *Australian Journal of Agricultural and Resource Economics, 66*(4), 753–774. Scopus. doi:10.1111/1467-8489.12498

Mendi, A. F. (2022). Blockchain for Food Tracking. *Electronics (Switzerland), 11*(16). Scopus. doi:10.3390/electronics11162491

Mueller, V., Grépin, K. A., Rabbani, A., Navia, B., Ngunjiri, A. S. W., & Wu, N. (2022). Food insecurity and COVID-19 risk in low- and middle-income countries. *Applied Economic Perspectives and Policy, 44*(1), 92–109. Scopus. doi:10.1002/aepp.13200

Munonye, J., Osuji, E., Olaolu, M., Okoisu, A., Obi, J., Eze, G., Ibrahim-Olesin, S., Njoku, L., Amadi, M., Izuogu, C., & Azuamairo, G. (2022). Perceived Effects of COVID-19 Pandemic on Food Security in Southeast Nigeria. *Frontiers in Sustainable Food Systems, 6*. Scopus. doi:10.3389/fsufs.2022.936157

Nasir, M. A., Nugroho, A. D., & Lakner, Z. (2022). Impact of the Russian–Ukrainian Conflict on Global Food Crops. *Foods, 11*(19). Scopus. doi:10.3390/foods11192979

Nekmahmud, M. (2022). Food consumption behavior, food supply chain disruption, and food security crisis during the COVID-19: The mediating effect of food price and food stress. *Journal of Foodservice Business Research*. Scopus. doi:10.1080/15378020.2022.2090802

Nikonenko, U., Shtets, T., Kalinin, A., Dorosh, I., & Sokolik, L. (2022). Assessing the Policy of Attracting Investments in the Main Sectors of the Economy in the Context of Introducing Aspects of Industry 4.0. *International Journal of Sustainable Development and Planning, 17*(2), 497–505. Scopus. doi:10.18280/ijsdp.170214

O'Meara, L., Turner, C., Coitinho, D. C., & Oenema, S. (2022). Consumer experiences of food environments during the Covid-19 pandemic: Global insights from a rapid online survey of individuals from 119 countries. *Global Food Security, 32*. Scopus. doi:10.1016/j.gfs.2021.100594

Omotayo, A. O., Omotoso, A. B., Daud, S. A., Omotayo, O. P., & Adeniyi, B. A. (2022). Rising Food Prices and Farming Households Food Insecurity during the COVID-19 Pandemic: Policy Implications from SouthWest Nigeria. *Agriculture (Switzerland), 12*(3). Scopus. doi:10.3390/agriculture12030363

Phukamchanoada, P. (2022). Community-based tourism product development based on the community identity of Klong-Lad-Ma-Yom floating market Bangkok, Thailand. *Turyzm/Tourism, 32*(1), 59–76. Scopus. doi:10.18778/0867-5856.32.1.03

Quandt, R. E. (1960). Tests of the Hypothesis that a Linear Regression System Obeys Two Separate Regimes. *Journal of the American Statistical Association, 55*(290), 324–330. doi:10.1080/01621459.1960.10482067

Rahmat, Z. S., Islam, Z., Mohanan, P., Kokash, D. M., Essar, M. Y., Hasan, M. M., Hashim, H. T., & Basalilah, A. F. M. (2022). Food Insecurity during COVID-19 in Yemen. *American Journal of Tropical Medicine and Hygiene, 106*(6), 1589–1592. Scopus. doi:10.4269/ajtmh.22-0059

Rajpoot, K., Singla, S., Singh, A., & Shekhar, S. (2022). Impact of COVID-19 lockdown on prices of potato and onion in metropolitan cities of India. *Journal of Agribusiness in Developing and Emerging Economies, 12*(3), 386–399. Scopus. doi:10.1108/JADEE-04-2021-0099

Recchia, D., Rollet, P., Perignon, M., Bricas, N., Vonthron, S., Perrin, C., Méjean, C., & Surfood-Foodscapes Working Group. (2022). Changes in Food Purchasing Practices of French Households During the First COVID-19 Lockdown and Associated Individual and Environmental Factors. *Frontiers in Nutrition, 9*. Scopus. doi:10.3389/fnut.2022.828550

Scopus. (2023). *Scopus Database*. https://www.scopus.com/search/form.uri?display=basic#basic

Shahini, E., Skuraj, E., Sallaku, F., & Shahini, S. (2022). The Supply Shock in Organic Fertilizers for Agriculture Caused by the Effect of Russia-Ukraine War. *Scientific Horizons, 25*(2), 97–103. Scopus. doi:10.48077/scihor.25(2).2022.97-103

Sodji, K. (2022). Estimating the link between trade uncertainty, pandemic uncertainty and food price stability in Togo: New evidence for an asymmetric analysis. *Review of Development Economics*. Scopus. doi:10.1111/rode.12963

Stata. (2023). *Statistical software for data science*. Stata. https://www.stata.com/

StataCorp. (2017a). *Stata 15 Base Reference Manual*. Stata Press.

StataCorp. (2017b). *Stata Statistical Software: Release 15*. StataCorp LLC.

Wegerif, M. (2022). The impact of Covid-19 on black farmers in South Africa. *Agrekon, 61*(1), 52–66. Scopus. doi:10.1080/03031853.2021.1971097

Yang, B., Asche, F., & Li, T. (2022). Consumer behavior and food prices during the COVID-19 pandemic: Evidence from Chinese cities. *Economic Inquiry, 60*(3), 1437–1460. Scopus. doi:10.1111/ecin.13067

Chapter 4
Evolution of Global Agricultural Output Since the Pandemic

ABSTRACT

One of the main concerns worldwide is food security, considering the challenges created by climate change, the Covid-19 pandemic, and the Russia-Ukraine crisis, among others. In these scenarios, some of the main discussions by international organisations and governments are related to the design of the most adjusted policies to deal with food security and food sovereignty. The objective of this chapter is to assess the consequences of the most recent events on agricultural output worldwide. Data from the OECD (Organisation for Economic Co-operation and Development) databases were considered for the period 2012-2022 and the following variables: crop production (thousand hectares; tonnes/ha; thousand tonnes); and meat consumption (Kg/capita; thousand tonnes). The statistical information seems to highlight no relevant impacts from the events since 2019/2020, for example, in crop production and meat consumption worldwide.

INTRODUCTION

In globalised supply chains (Hamilton et al., 2020), the current international shocks create challenges for several stakeholders (Biadgilign & Yigzaw, 2020), including farmers (Burrone et al., 2022). They promote serious threats to the different economic sectors (Haque et al., 2021), workers (Lusk & Chandra, 2021), including migrants (Molinero-Gerbeau, 2021), rice cultivators (Niaz et al., 2022), human health (Madzorera et al., 2021), wildlife (van Vliet et al., 2022), commercialisation channels (Vargas et al., 2021) and environment (Liu et al., 2022). There is still a field to be explored regarding the environmental repercussions (Mahmud et al., 2021). The smaller farmers are generally the most affected by these events (Haqiqi & Bahalou Horeh, 2021). But, these events bring also opportunities (Usman et al., 2021) to deal, for example, with greenhouse gas emissions (Andrieu et al., 2021), carry out alternative assessments (Dentener et al., 2020) and promote eco-friendly productions (Arshad et al., 2022).

The impacts of these recent events have been unequal across countries, sectors, groups (Gatto & Islam, 2021) and farming systems (Tripathi et al., 2021), namely in terms of severity (Asegie et al., 2021), and,

DOI: 10.4018/978-1-6684-8923-9.ch004

in some cases, the recovery will be long (Boughton et al., 2021). The consequences on cereals production and trade are particularly worrying in the most vulnerable countries (Selim & Eltarabily, 2022) and populations (Jha et al., 2021), specifically in rural areas (Sun et al., 2022). These events impacted the agricultural production and the decisions about the use of some resources (Johnson et al., 2022), as well as the farmers' perceptions (Yuan et al., 2022).

Some countries are crucial for the cereal supply (Araujo-Enciso & Fellmann, 2020) and food security (Lioubimtseva & Henebry, 2012) worldwide (Meyfroidt et al., 2016), as well as specific regions and lands (Arsene & Nkulu Mwine Fyama, 2021). Food security is a current problem (Muthamilarasan & Prasad, 2021), namely in specific frameworks (Okoli, 2021), and it will continue to be among the most relevant research topics (Bartol, 2023). Wheat is one of the most important cereals for the diet of the world population (Gutiérrez-Moya et al., 2021). Rice has also its importance for world food security (Zhao et al., 2022).

Agricultural policies have a fundamental role in these frameworks (Pan et al., 2020), particularly the fiscal and monetary ones (Temesgen et al., 2022), however not always with the most adjusted instruments (Belasco & Smith, 2022). The policymakers and other stakeholders will have an important role to increase the resilience of the agricultural and food sectors (Vicente-Vicente et al., 2021), to better deal with unexpected shocks (Darnhofer, 2021), that may increase in the future (Karoliina et al., 2023). The promotion of shorter supply chains may be between better solutions (Yovchevska, 2021). Policy design and planning are critical for effective implementation (Sangha et al., 2022).

The new technologies may provide relevant support for assessments (Fouotsa Manfouo & Von Fintel, 2022) and simulations related to agriculture (Meunier et al., 2022) and the environment (Mohite et al., 2022), such as to predict the wheat yield (Franch et al., 2021), in a context of smart farming (Koo et al., 2021). The biotechnology progresses may bring also important contributions, specifically to improve farming productivity and efficiency (Gaffney et al., 2022).

Considering the context presented before, this research intends to analyse the implications of the Covid-19 pandemic and the Russia-Ukraine crisis in agricultural output. For that, statistical information was considered, for the period from 2012 until 2022, from the OECD (2023) database. Some findings were obtained following Stata (StataCorp, 2017b, 2017a; Stata, 2023) software procedures. For the literature survey, Scopus (2023) was consulted. Normalised values $((x_i-x_{minimum})/(x_{maximum}-x_{minimum}))$, growth rates and logarithms results were taken into account.

CROP PRODUCTION

The crop production (OECD countries and world), in thousand hectares, presented the highest growth rates (%) in 2016 and 2020, for rice, and 2020 for wheat (table 1). The greatest growth rates appear in 2016 for maize, and in 2013 (world), 2014 (world and OECD countries) and 2017 (OECD countries) for soybean (table 2). On the other hand, the lowest growth rates appear, generally, before 2020. Does not seem visible an impact from the Covid-19 and Russia-Ukraine crisis on these data.

Table 1. Crop production (thousand hectares) of rice and wheat for OECD countries and the world

Location	Crop	Year	Value	Growth rates (%)
OECD	Rice	2012	4740	
OECD	Rice	2013	4636	-2.194
OECD	Rice	2014	4711	1.618
OECD	Rice	2015	4508	-4.309
OECD	Rice	2016	4838	7.320
OECD	Rice	2017	4515	-6.676
OECD	Rice	2018	4595	1.772
OECD	Rice	2019	4453	-3.090
OECD	Rice	2020	4747	6.602
OECD	Rice	2021	4541	-4.340
OECD	Rice	2022	4545	0.088
World	Rice	2012	161010	
World	Rice	2013	163386	1.476
World	Rice	2014	162488	-0.550
World	Rice	2015	161302	-0.730
World	Rice	2016	163687	1.479
World	Rice	2017	162521	-0.712
World	Rice	2018	163447	0.570
World	Rice	2019	161190	-1.381
World	Rice	2020	164558	2.089
World	Rice	2021	164635	0.047
World	Rice	2022	164301	-0.203
World	Wheat	2012	219819	
World	Wheat	2013	221141	0.601
World	Wheat	2014	222818	0.758
World	Wheat	2015	221967	-0.382
World	Wheat	2016	221690	-0.125
World	Wheat	2017	217831	-1.741
World	Wheat	2018	215516	-1.063
World	Wheat	2019	216956	0.668
World	Wheat	2020	221967	2.310
World	Wheat	2021	223156	0.536
World	Wheat	2022	222513	-0.288
OECD	Wheat	2012	76939	
OECD	Wheat	2013	76330	-0.792
OECD	Wheat	2014	76816	0.637
OECD	Wheat	2015	76210	-0.789
OECD	Wheat	2016	75032	-1.546

Continued on following page

Table 1. Continued

Location	Crop	Year	Value	Growth rates (%)
OECD	Wheat	2017	70173	-6.476
OECD	Wheat	2018	70808	0.905
OECD	Wheat	2019	69995	-1.148
OECD	Wheat	2020	71277	1.832
OECD	Wheat	2021	71938	0.927
OECD	Wheat	2022	71670	-0.373

Table 2. Crop production (thousand hectares) of maize and soybean for OECD countries and the world

Location	Crop	Year	Value	Growth rates (%)
World	Maize	2012	185166	
World	Maize	2013	189944	2.580
World	Maize	2014	186806	-1.652
World	Maize	2015	186016	-0.423
World	Maize	2016	194393	4.503
World	Maize	2017	189334	-2.602
World	Maize	2018	187684	-0.871
World	Maize	2019	191121	1.831
World	Maize	2020	192749	0.852
World	Maize	2021	194045	0.672
World	Maize	2022	194629	0.301
World	Soybean	2012	108704	
World	Soybean	2013	114566	5.393
World	Soybean	2014	119251	4.089
World	Soybean	2015	121212	1.644
World	Soybean	2016	120660	-0.455
World	Soybean	2017	124593	3.260
World	Soybean	2018	126354	1.413
World	Soybean	2019	123218	-2.482
World	Soybean	2020	127835	3.747
World	Soybean	2021	129206	1.072
World	Soybean	2022	129658	0.350
OECD	Maize	2012	54969	
OECD	Maize	2013	55184	0.391
OECD	Maize	2014	52930	-4.085
OECD	Maize	2015	51709	-2.307
OECD	Maize	2016	53905	4.247

Continued on following page

Table 2. Continued

Location	Crop	Year	Value	Growth rates (%)
OECD	Maize	2017	51774	-3.953
OECD	Maize	2018	50973	-1.547
OECD	Maize	2019	51149	0.345
OECD	Maize	2020	52162	1.980
OECD	Maize	2021	52444	0.541
OECD	Maize	2022	52616	0.328
OECD	Soybean	2012	33387	
OECD	Soybean	2013	33664	0.830
OECD	Soybean	2014	36745	9.152
OECD	Soybean	2015	36706	-0.106
OECD	Soybean	2016	37090	1.046
OECD	Soybean	2017	40672	9.658
OECD	Soybean	2018	39424	-3.068
OECD	Soybean	2019	33866	-14.098
OECD	Soybean	2020	36721	8.430
OECD	Soybean	2021	37527	2.195
OECD	Soybean	2022	37551	0.064

Table 3 highlights the increases in rice productivity (tonnes/ha) in 2021 (table 3). On the other hand, 2020 presents some of the worse growth rates for wheat productivity. In turn, 2019 was one of the years with the worse growth rates for maize and soybean productivity (table 4). Maybe, the pandemic, in the beginning, had an impact on the wheat area productivity, nonetheless, these findings deserve deeper analysis in future research.

Table 3. Crop production (tonnes/ha) of rice and wheat for OECD countries and the world

Location	Crop	Year	Value	Growth rates (%)
World	Rice	2012	3.018	
World	Rice	2013	3.006	-0.398
World	Rice	2014	3.021	0.499
World	Rice	2015	3.033	0.397
World	Rice	2016	3.043	0.330
World	Rice	2017	3.079	1.183
World	Rice	2018	3.104	0.812
World	Rice	2019	3.123	0.612
World	Rice	2020	3.143	0.640
World	Rice	2021	3.189	1.464
World	Rice	2022	3.221	1.003
World	Wheat	2012	3.001	
World	Wheat	2013	3.231	7.664
World	Wheat	2014	3.291	1.857
World	Wheat	2015	3.316	0.760
World	Wheat	2016	3.440	3.739
World	Wheat	2017	3.492	1.512
World	Wheat	2018	3.394	-2.806
World	Wheat	2019	3.503	3.212
World	Wheat	2020	3.455	-1.370
World	Wheat	2021	3.478	0.666
World	Wheat	2022	3.507	0.834
OECD	Rice	2012	4.915	
OECD	Rice	2013	4.938	0.468
OECD	Rice	2014	4.999	1.235
OECD	Rice	2015	4.979	-0.400
OECD	Rice	2016	5.036	1.145
OECD	Rice	2017	4.932	-2.065
OECD	Rice	2018	4.973	0.831
OECD	Rice	2019	4.863	-2.212
OECD	Rice	2020	4.902	0.802
OECD	Rice	2021	5.008	2.162
OECD	Rice	2022	5.028	0.399
OECD	Wheat	2012	3.535	
OECD	Wheat	2013	3.841	8.656
OECD	Wheat	2014	3.790	-1.328
OECD	Wheat	2015	3.890	2.639
OECD	Wheat	2016	3.979	2.288

Continued on following page

Table 3. Continued

Location	Crop	Year	Value	Growth rates (%)
OECD	Wheat	2017	3.968	-0.276
OECD	Wheat	2018	3.741	-5.721
OECD	Wheat	2019	4.014	7.298
OECD	Wheat	2020	3.875	-3.463
OECD	Wheat	2021	3.882	0.181
OECD	Wheat	2022	3.917	0.902

Table 4. Crop production (tonnes/ha) of maize and soybean for OECD countries and the world

Location	Crop	Year	Value	Growth rates (%)
World	Maize	2012	4.968	
World	Maize	2013	5.553	11.775
World	Maize	2014	5.757	3.674
World	Maize	2015	5.568	-3.283
World	Maize	2016	5.933	6.555
World	Maize	2017	5.845	-1.483
World	Maize	2018	6.117	4.654
World	Maize	2019	6.010	-1.749
World	Maize	2020	6.005	-0.083
World	Maize	2021	6.099	1.565
World	Maize	2022	6.173	1.213
World	Soybean	2012	2.497	
World	Soybean	2013	2.505	0.320
World	Soybean	2014	2.706	8.024
World	Soybean	2015	2.633	-2.698
World	Soybean	2016	2.920	10.900
World	Soybean	2017	2.780	-4.795
World	Soybean	2018	2.893	4.065
World	Soybean	2019	2.756	-4.736
World	Soybean	2020	2.842	3.120
World	Soybean	2021	2.886	1.548
World	Soybean	2022	2.910	0.832
OECD	Maize	2012	6.852	
OECD	Maize	2013	8.419	22.869
OECD	Maize	2014	9.145	8.623
OECD	Maize	2015	8.758	-4.232
OECD	Maize	2016	9.265	5.789

Continued on following page

Table 4. Continued

Location	Crop	Year	Value	Growth rates (%)
OECD	Maize	2017	9.408	1.543
OECD	Maize	2018	9.482	0.787
OECD	Maize	2019	9.099	-4.039
OECD	Maize	2020	9.073	-0.286
OECD	Maize	2021	9.238	1.819
OECD	Maize	2022	9.291	0.574
OECD	Soybean	2012	2.687	
OECD	Soybean	2013	2.936	9.267
OECD	Soybean	2014	3.151	7.323
OECD	Soybean	2015	3.177	0.825
OECD	Soybean	2016	3.427	7.869
OECD	Soybean	2017	3.233	-5.661
OECD	Soybean	2018	3.339	3.279
OECD	Soybean	2019	3.140	-5.960
OECD	Soybean	2020	3.338	6.306
OECD	Soybean	2021	3.372	1.019
OECD	Soybean	2022	3.399	0.801

Table 5 reveals that 2020 appears also among the years with the greatest growth rates for rice production in thousand tonnes. Again, it is not visible, from these data, that the Covid-19 pandemic and the Russia-Ukraine conflicts have had an impact on rice and wheat production in thousand of tonnes around the world and the OECD countries. The same for the maize and soybean (table 6).

Table 5. Crop production (thousand tonnes) of rice and wheat for OECD countries and the world

Location	Crop	Year	Value	Growth rates (%)
World	Rice	2012	485937	
World	Rice	2013	491159	1.075
World	Rice	2014	490905	-0.052
World	Rice	2015	489204	-0.347
World	Rice	2016	498144	1.827
World	Rice	2017	500429	0.459
World	Rice	2018	507343	1.382
World	Rice	2019	503317	-0.794
World	Rice	2020	517210	2.760
World	Rice	2021	525068	1.519
World	Rice	2022	529211	0.789
OECD	Rice	2012	23301	
OECD	Rice	2013	22892	-1.755
OECD	Rice	2014	23552	2.883
OECD	Rice	2015	22445	-4.700
OECD	Rice	2016	24363	8.545
OECD	Rice	2017	22271	-8.587
OECD	Rice	2018	22852	2.609
OECD	Rice	2019	21657	-5.229
OECD	Rice	2020	23270	7.448
OECD	Rice	2021	22742	-2.269
OECD	Rice	2022	22852	0.484
World	Wheat	2012	659768	
World	Wheat	2013	714545	8.302
World	Wheat	2014	733359	2.633
World	Wheat	2015	736096	0.373
World	Wheat	2016	762579	3.598
World	Wheat	2017	760578	-0.262
World	Wheat	2018	731389	-3.838
World	Wheat	2019	759896	3.898
World	Wheat	2020	766823	0.912
World	Wheat	2021	776177	1.220
World	Wheat	2022	780372	0.540
OECD	Wheat	2012	271971	
OECD	Wheat	2013	293151	7.788
OECD	Wheat	2014	291125	-0.691
OECD	Wheat	2015	296454	1.830
OECD	Wheat	2016	298555	0.709

Continued on following page

Table 5. Continued

Location	Crop	Year	Value	Growth rates (%)
OECD	Wheat	2017	278418	-6.745
OECD	Wheat	2018	264877	-4.864
OECD	Wheat	2019	280991	6.084
OECD	Wheat	2020	276201	-1.705
OECD	Wheat	2021	279226	1.095
OECD	Wheat	2022	280727	0.538

Table 6. Crop production (thousand tonnes) of maize and soybean for OECD countries and the world

Location	Crop	Year	Value	Growth rates (%)
World	Soybean	2012	271440	
World	Soybean	2013	286966	5.720
World	Soybean	2014	322687	12.448
World	Soybean	2015	319140	-1.099
World	Soybean	2016	352316	10.395
World	Soybean	2017	346373	-1.687
World	Soybean	2018	365512	5.526
World	Soybean	2019	339541	-7.105
World	Soybean	2020	363345	7.011
World	Soybean	2021	372859	2.618
World	Soybean	2022	377343	1.203
World	Maize	2012	919951	
World	Maize	2013	1054736	14.651
World	Maize	2014	1075411	1.960
World	Maize	2015	1035711	-3.692
World	Maize	2016	1153396	11.363
World	Maize	2017	1106702	-4.048
World	Maize	2018	1148124	3.743
World	Maize	2019	1148622	0.043
World	Maize	2020	1157534	0.776
World	Maize	2021	1183381	2.233
World	Maize	2022	1201435	1.526
OECD	Maize	2012	376650	
OECD	Maize	2013	464621	23.356
OECD	Maize	2014	484040	4.180
OECD	Maize	2015	452855	-6.443
OECD	Maize	2016	499431	10.285

Continued on following page

Table 6. Continued

Location	Crop	Year	Value	Growth rates (%)
OECD	Maize	2017	487081	-2.473
OECD	Maize	2018	483332	-0.770
OECD	Maize	2019	465425	-3.705
OECD	Maize	2020	473290	1.690
OECD	Maize	2021	484459	2.360
OECD	Maize	2022	488874	0.911
OECD	Soybean	2012	89707	
OECD	Soybean	2013	98841	10.182
OECD	Soybean	2014	115779	17.137
OECD	Soybean	2015	116606	0.714
OECD	Soybean	2016	127088	8.989
OECD	Soybean	2017	131476	3.453
OECD	Soybean	2018	131642	0.126
OECD	Soybean	2019	106328	-19.229
OECD	Soybean	2020	122558	15.264
OECD	Soybean	2021	126529	3.240
OECD	Soybean	2022	127635	0.874

The top countries for production (thousand tonnes on average from 2012 to 2022) are presented in table 7 for wheat, table 8 for maize, table 9 for rice and table 10 for soybean. These tables show that each country has particular conditions for the production of each one of these crops. Of highlighting, the importance of the BRICS countries, for example, for these crop productions.

Table 7. Crop production (thousand tonnes on average decreasing order over 2012-2022) of wheat

Location	Value	Normalised value
World	743780	1.000
BRICS	306835	0.413
OECD	282881	0.380
China	130541	0.176
India	99255	0.133
Russian Federation	69561	0.094
USA	54324	0.073
Canada	32156	0.043
Pakistan	25258	0.034
Ukraine	25232	0.034
Australia	23465	0.032
Turkey	20633	0.028
Argentina	15967	0.021
United Kingdom	14335	0.019
Kazakhstan	13523	0.018
Iran, Islamic Rep.	12731	0.017
Egypt, Arab Rep.	9016	0.012
Brazil	5654	0.008
Ethiopia (excludes Eritrea)	4637	0.006
Mexico	3331	0.004
South Africa	1824	0.002
Chile	1401	0.002
Paraguay	1144	0.002
Japan	891	0.001
Switzerland	486	0.001
New Zealand	439	0.001
Norway	426	0.001
Saudi Arabia	382	0.001
Peru	202	0.000
Israel	120	0.000
Nigeria	70	0.000
Korea, Rep.	28	0.000
Colombia	11	0.000
Thailand	1	0.000
Indonesia	0	0.000
Malaysia	0	0.000
Philippines	0	0.000
Vietnam	0	0.000

Table 8. Crop production (thousand tonnes on average decreasing order over 2012-2022) of maize

Location	Value	Normalised value
World	1107728	1.000
OECD	469096	0.423
BRICS	402623	0.363
USA	354377	0.320
China	257634	0.233
Brazil	91266	0.082
Argentina	47507	0.043
Ukraine	30535	0.028
India	26666	0.024
Mexico	26110	0.024
Indonesia	21413	0.019
South Africa	14079	0.013
Canada	13362	0.012
Russian Federation	12980	0.012
Nigeria	11220	0.010
Ethiopia (excludes Eritrea)	8346	0.008
Philippines	7806	0.007
Egypt, Arab Rep.	7754	0.007
Turkey	6022	0.005
Pakistan	5981	0.005
Paraguay	5267	0.005
Vietnam	4993	0.005
Thailand	4726	0.004
Colombia	1583	0.001
Peru	1568	0.001
Iran, Islamic Rep.	1364	0.001
Chile	997	0.001
Kazakhstan	795	0.001
Australia	368	0.000
New Zealand	201	0.000
Switzerland	148	0.000
Israel	97	0.000
Saudi Arabia	85	0.000
Malaysia	78	0.000
Korea, Rep.	76	0.000
United Kingdom	0	0.000
Japan	0	0.000
Norway	0	0.000

Table 9. Crop production (thousand tonnes on average decreasing order over 2012-2022) of rice

Location	Value	Normalised value
World	503448	1.000
BRICS	268035	0.532
China	145158	0.288
India	113970	0.226
Indonesia	36855	0.073
Vietnam	28456	0.057
OECD	22927	0.046
Thailand	21040	0.042
Philippines	12370	0.025
Brazil	7841	0.016
Japan	7596	0.015
Pakistan	7352	0.015
USA	6386	0.013
Nigeria	4580	0.009
Egypt, Arab Rep.	4576	0.009
Korea, Rep.	4060	0.008
Peru	2173	0.004
Iran, Islamic Rep.	2044	0.004
Colombia	1694	0.003
Malaysia	1659	0.003
Russian Federation	1064	0.002
Argentina	947	0.002
Paraguay	644	0.001
Turkey	569	0.001
Australia	515	0.001
Kazakhstan	313	0.001
Mexico	261	0.001
Chile	110	0.000
Ethiopia (excludes Eritrea)	98	0.000
Ukraine	52	0.000
South Africa	2	0.000
Switzerland	0	0.000
Canada	0	0.000
United Kingdom	0	0.000
Israel	0	0.000
Norway	0	0.000
New Zealand	0	0.000
Saudi Arabia	0	0.000

Table 10. Crop production (thousand tonnes on average decreasing order over 2012-2022) of soybean

Location	Value	Normalised value
World	337956	1.000
BRICS	145891	0.432
OECD	117654	0.348
Brazil	113513	0.336
USA	107831	0.319
Argentina	51831	0.153
China	15799	0.047
India	12037	0.036
Paraguay	9555	0.028
Canada	6527	0.019
Ukraine	3487	0.010
Russian Federation	3435	0.010
South Africa	1108	0.003
Indonesia	684	0.002
Nigeria	672	0.002
Mexico	339	0.001
Kazakhstan	246	0.001
Japan	233	0.001
Iran, Islamic Rep.	200	0.001
Turkey	147	0.000
Vietnam	118	0.000
Korea, Rep.	106	0.000
Ethiopia (excludes Eritrea)	99	0.000
Colombia	72	0.000
Thailand	49	0.000
Australia	43	0.000
Egypt, Arab Rep.	31	0.000
Peru	5	0.000
Switzerland	5	0.000
Pakistan	3	0.000
Philippines	1	0.000
Malaysia	0	0.000
Chile	0	0.000
United Kingdom	0	0.000
Israel	0	0.000
Norway	0	0.000
New Zealand	0	0.000
Saudi Arabia	0	0.000

MEAT CONSUMPTION

Meat consumption (Kg/capita) does not seem to be affected by the recent events worldwide, with the exception, maybe, of poultry meat consumption (world) which presents the worse growth rates for the years 2021 and 2022. In turn, pig meat consumption decreased significantly in 2020 and increased in 2021 and 2022 (table 11). For the OECD countries (table 12) a significant decrease in beef consumption for 2022 was observed and a relevant reduction in pig meat consumption in 2020 was verified. In addition, it was observed in 2021 one of the lowest growth rates for poultry meat consumption. It seems that the pandemic and the conflict between Russia and Ukraine had impacts on meat consumption with changes in the preferences of consumers.

For the meat consumption in thousand tonnes (world), poultry meat appears again with the lowest growth rates in 2021 and 2022 (table 13). In the OECD countries, the years 2020, 2021 and 2022 present some of the lowest growth rates for beef, pig meat and poultry meat consumption (table 14).

Table 11. Meat consumption (Kg/capita) for the world

Meat	Year	Value	Logarithm	Growth rates (%)
Beef	2012	6.433	1.861	
Beef	2013	6.472	1.867	0.606
Beef	2014	6.362	1.850	-1.700
Beef	2015	6.308	1.842	-0.849
Beef	2016	6.357	1.850	0.777
Beef	2017	6.375	1.852	0.283
Beef	2018	6.401	1.856	0.408
Beef	2019	6.421	1.860	0.312
Beef	2020	6.391	1.855	-0.467
Beef	2021	6.377	1.853	-0.219
Beef	2022	6.330	1.845	-0.737
Pig	2012	12.439	2.521	
Pig	2013	12.577	2.532	1.109
Pig	2014	12.723	2.543	1.161
Pig	2015	12.632	2.536	-0.715
Pig	2016	12.435	2.521	-1.560
Pig	2017	12.394	2.517	-0.330
Pig	2018	12.327	2.512	-0.541
Pig	2019	11.091	2.406	-10.027
Pig	2020	10.678	2.368	-3.724
Pig	2021	10.817	2.381	1.302
Pig	2022	11.085	2.406	2.478
Poultry	2012	13.299	2.588	
Poultry	2013	13.344	2.591	0.338

Continued on following page

Table 11. Continued

Meat	Year	Value	Logarithm	Growth rates (%)
Poultry	2014	13.527	2.605	1.371
Poultry	2015	13.876	2.630	2.580
Poultry	2016	14.101	2.646	1.622
Poultry	2017	14.130	2.648	0.206
Poultry	2018	14.334	2.663	1.444
Poultry	2019	14.748	2.691	2.888
Poultry	2020	14.875	2.700	0.861
Poultry	2021	14.858	2.699	-0.114
Poultry	2022	14.852	2.698	-0.040
Sheep	2012	1.649	0.500	
Sheep	2013	1.675	0.516	1.577
Sheep	2014	1.698	0.529	1.373
Sheep	2015	1.737	0.552	2.297
Sheep	2016	1.734	0.550	-0.173
Sheep	2017	1.743	0.556	0.519
Sheep	2018	1.745	0.557	0.115
Sheep	2019	1.750	0.560	0.287
Sheep	2020	1.759	0.565	0.514
Sheep	2021	1.766	0.569	0.398
Sheep	2022	1.774	0.573	0.453

Table 12. Meat consumption (Kg/capita) for OECD countries

Meat	Year	Value	Logarithm	Growth rates (%)
Beef	2012	14.204	2.654	
Beef	2013	14.048	2.642	-1.098
Beef	2014	13.822	2.626	-1.609
Beef	2015	13.841	2.628	0.137
Beef	2016	14.140	2.649	2.160
Beef	2017	14.259	2.657	0.842
Beef	2018	14.521	2.676	1.837
Beef	2019	14.472	2.672	-0.337
Beef	2020	14.425	2.669	-0.325
Beef	2021	14.406	2.668	-0.132
Beef	2022	14.244	2.656	-1.125
Pig	2012	21.583	3.072	
Pig	2013	21.518	3.069	-0.301

Continued on following page

Table 12. Continued

Meat	Year	Value	Logarithm	Growth rates (%)
Pig	2014	21.698	3.077	0.837
Pig	2015	22.515	3.114	3.765
Pig	2016	22.545	3.116	0.133
Pig	2017	22.651	3.120	0.470
Pig	2018	23.014	3.136	1.603
Pig	2019	23.022	3.136	0.035
Pig	2020	22.856	3.129	-0.721
Pig	2021	22.848	3.129	-0.035
Pig	2022	22.820	3.128	-0.123
Poultry	2012	26.975	3.295	
Poultry	2013	27.045	3.298	0.259
Poultry	2014	27.746	3.323	2.592
Poultry	2015	29.033	3.368	4.639
Poultry	2016	29.776	3.394	2.559
Poultry	2017	30.056	3.403	0.940
Poultry	2018	30.686	3.424	2.096
Poultry	2019	31.323	3.444	2.076
Poultry	2020	31.708	3.457	1.229
Poultry	2021	31.816	3.460	0.341
Poultry	2022	31.984	3.465	0.528
Sheep	2012	1.277	0.245	
Sheep	2013	1.292	0.256	1.175
Sheep	2014	1.298	0.261	0.464
Sheep	2015	1.360	0.307	4.777
Sheep	2016	1.333	0.287	-1.985
Sheep	2017	1.305	0.266	-2.101
Sheep	2018	1.317	0.275	0.920
Sheep	2019	1.279	0.246	-2.885
Sheep	2020	1.269	0.238	-0.782
Sheep	2021	1.268	0.237	-0.079
Sheep	2022	1.267	0.237	-0.079

Table 13. Meat consumption (thousand tonnes) for the world

Meat	Year	Value	Logarithm	Growth rates (%)
Beef	2012	65185.000	11.085	
Beef	2013	66372.000	11.103	1.821
Beef	2014	66024.000	11.098	-0.524
Beef	2015	66223.000	11.101	0.301
Beef	2016	67506.000	11.120	1.937
Beef	2017	68462.000	11.134	1.416
Beef	2018	69500.000	11.149	1.516
Beef	2019	70474.000	11.163	1.401
Beef	2020	70882.000	11.169	0.579
Beef	2021	71451.000	11.177	0.803
Beef	2022	71637.000	11.179	0.260
Pig	2012	113120.000	11.636	
Pig	2013	115748.000	11.659	2.323
Pig	2014	118492.000	11.683	2.371
Pig	2015	119015.000	11.687	0.441
Pig	2016	118506.000	11.683	-0.428
Pig	2017	119446.000	11.691	0.793
Pig	2018	120115.000	11.696	0.560
Pig	2019	109245.000	11.601	-9.050
Pig	2020	106279.000	11.574	-2.715
Pig	2021	108775.000	11.597	2.349
Pig	2022	112585.000	11.631	3.503
Poultry	2012	107201.000	11.582	
Poultry	2013	108860.000	11.598	1.548
Poultry	2014	111664.000	11.623	2.576
Poultry	2015	115884.000	11.660	3.779
Poultry	2016	119118.000	11.688	2.791
Poultry	2017	120705.000	11.701	1.332
Poultry	2018	123795.000	11.726	2.560
Poultry	2019	128758.000	11.766	4.009
Poultry	2020	131230.000	11.785	1.920
Poultry	2021	132426.000	11.794	0.911
Poultry	2022	133696.000	11.803	0.959
Sheep	2012	13291.000	9.495	
Sheep	2013	13666.000	9.523	2.821
Sheep	2014	14020.000	9.548	2.590
Sheep	2015	14502.000	9.582	3.438
Sheep	2016	14644.000	9.592	0.979

Continued on following page

Table 13. Continued

Meat	Year	Value	Logarithm	Growth rates (%)
Sheep	2017	14891.000	9.609	1.687
Sheep	2018	15069.000	9.620	1.195
Sheep	2019	15279.000	9.634	1.394
Sheep	2020	15520.000	9.650	1.577
Sheep	2021	15742.000	9.664	1.430
Sheep	2022	15965.000	9.678	1.417

Table 14. Meat consumption (thousand tonnes) for OECD countries

Meat	Year	Value	Logarithm	Growth rates (%)
Beef	2012	27129.000	10.208	
Beef	2013	26992.000	10.203	-0.505
Beef	2014	26714.000	10.193	-1.030
Beef	2015	26901.000	10.200	0.700
Beef	2016	27631.000	10.227	2.714
Beef	2017	28011.000	10.240	1.375
Beef	2018	28668.000	10.264	2.346
Beef	2019	28702.000	10.265	0.119
Beef	2020	28731.000	10.266	0.101
Beef	2021	28801.000	10.268	0.244
Beef	2022	28575.000	10.260	-0.785
Pig	2012	36994.000	10.519	
Pig	2013	37106.000	10.522	0.303
Pig	2014	37634.000	10.536	1.423
Pig	2015	39271.000	10.578	4.350
Pig	2016	39538.000	10.585	0.680
Pig	2017	39932.000	10.595	0.997
Pig	2018	40774.000	10.616	2.109
Pig	2019	40978.000	10.621	0.500
Pig	2020	40853.000	10.618	-0.305
Pig	2021	40994.000	10.621	0.345
Pig	2022	41083.000	10.623	0.217
Poultry	2012	40983.000	10.621	
Poultry	2013	41337.000	10.630	0.864
Poultry	2014	42655.000	10.661	3.188
Poultry	2015	44885.000	10.712	5.228
Poultry	2016	46285.000	10.743	3.119

Continued on following page

Table 14. Continued

Meat	Year	Value	Logarithm	Growth rates (%)
Poultry	2017	46966.000	10.757	1.471
Poultry	2018	48186.000	10.783	2.598
Poultry	2019	49416.000	10.808	2.553
Poultry	2020	50234.000	10.824	1.655
Poultry	2021	50597.000	10.832	0.723
Poultry	2022	51038.000	10.840	0.872
Sheep	2012	1940.000	7.570	
Sheep	2013	1974.000	7.588	1.753
Sheep	2014	1996.000	7.599	1.114
Sheep	2015	2102.000	7.651	5.311
Sheep	2016	2073.000	7.637	-1.380
Sheep	2017	2039.000	7.620	-1.640
Sheep	2018	2068.000	7.634	1.422
Sheep	2019	2018.000	7.610	-2.418
Sheep	2020	2010.000	7.606	-0.396
Sheep	2021	2017.000	7.609	0.348
Sheep	2022	2022.000	7.612	0.248

CORRELATIONS ASSESSMENT

Table 15, with the results for the Spearman's rank correlation (Spearman, 1904) coefficients (among the crop production worldwide), over the period 2012-2022, highlights strong correlations. These correlations present the highest coefficient between rice production and soybean production, among rice production and maize production, and between soybean production and maize production. This reveals that the trends followed over the period considered were similar for these crop productions. The lowest coefficients appear among wheat production and other crop productions.

Table 15. Spearman's rank correlation coefficients for crop production worldwide (1000 tonnes, 2012-2022)

	Crop production-World (Rice, 1000 tonnes)	Crop production-World (Wheat, 1000 tonnes)	Crop production-World (Soybean, 1000 tonnes)	Crop production-World (Maize, 1000 tonnes)
Crop production-World (Rice, 1000 tonnes)	1.000			
Crop production-World (Wheat, 1000 tonnes)	0.7727*	1.000		
	(0.005)			
Crop production-World (Soybean, 1000 tonnes)	0.9273*	0.8000*	1.000	
	(0.000)	(0.003)		
Crop production-World (Maize, 1000 tonnes)	0.9273*	0.8909*	0.9182*	1.000
	(0.000)	(0.000)	(0.000)	

Note: *, statistically significant at 5%.

CONCLUSION

This chapter intended to investigate the impacts of the two recent international events (the Covid-19 pandemic and the Russia-Ukraine crisis) on crop production and meat consumption worldwide. To achieve these objectives, statistical information from the OECD database was considered for the years 2012-2022. Variables related to the crop output, crop area productivity, total meat consumption and meat consumption per capita were considered. For deeper assessments, normalised values, growth rates and logarithms results were found.

The literature analysis reveals that the recent international shocks brought disturbances for the economic sectors, supply chains and workers, including migrants, and sustainability. The smaller economic agents are, in general, the most affected by these disruptions. There are also opportunities to mitigate greenhouse gas emissions and promote eco-friendly productions. The rural areas in some countries will find serious difficulties to deal with the new frameworks created by these exogenous shocks. Wheat and rice are among the most important cereals for the human diet worldwide. The policymakers (creating adjusted policy instruments to better deal with the new challenges), short supply chains (to increase the resilience of the food and agricultural sectors) and the new technologies (particularly to predict cereals yield) may play here a fundamental role.

The findings obtained from the statistical information show that it is not visible the impact of the pandemic and the Russia-Ukraine crisis on crop production worldwide and in the OECD countries (number of hectares and output). On the other hand, there are some signs that the wheat area productivity was affected at the beginning of the pandemic. In any case, the results highlight the importance of the BRICS countries for these crop productions (rice, wheat, maize and soybean). In meat consumption, the impacts of these two international events are more visible, as well as changes in the preferences of the consumers. An assessment of the crop production evolution, over the period taken into account, reveals a strong correlation between these productions.

From a practical implications viewpoint about the dimensions here analysed, the focus of the different stakeholders should be on the impacts of the pandemic and international crisis on consumption, assessing how the respective changes impact the food chains and the human diet. For policy recommendations, it is suggested measures that take into account the changes verified in the demand and that be able to mitigate the consequent implications on the farmers. For future research, it would be important to analyse the reaction of the policymakers to these new contexts.

ACKNOWLEDGMENT

This work is funded by National Funds through the FCT - Foundation for Science and Technology, I.P., within the scope of the project Refª UIDB/00681/2020. Furthermore we would like to thank the CERNAS Research Centre and the Polytechnic Institute of Viseu for their support.

REFERENCES

Andrieu, N., Hossard, L., Graveline, N., Dugue, P., Guerra, P., & Chirinda, N. (2021). Covid-19 management by farmers and policymakers in Burkina Faso, Colombia and France: Lessons for climate action. *Agricultural Systems, 190*. Scopus. doi:10.1016/j.agsy.2021.103092

Araujo-Enciso, S. R., & Fellmann, T. (2020). Yield Variability and Harvest Failures in Russia, Ukraine and Kazakhstan and Their Possible Impact on Food Security in the Middle East and North Africa. *Journal of Agricultural Economics, 71*(2), 493–516. Scopus. doi:10.1111/1477-9552.12367

Arsene, M. B., & Nkulu Mwine Fyama, J. (2021). Potential threats to agricultural food production and farmers' coping strategies in the marshlands of Kabare in the Democratic Republic of Congo. *Cogent Food and Agriculture, 7*(1). Scopus. doi:10.1080/23311932.2021.1933747

Arshad, M. U., Zhao, Y., Hanif, O., & Fatima, F. (2022). Evolution of Overall Cotton Production and Its Determinants: Implications for Developing Countries Using Pakistan Case. *Sustainability (Switzerland), 14*(2). Scopus. doi:10.3390/su14020840

Asegie, A. M., Adisalem, S. T., & Eshetu, A. A. (2021). The effects of COVID-19 on livelihoods of rural households: South Wollo and Oromia Zones, Ethiopia. *Heliyon, 7*(12). Scopus. doi:10.1016/j.heliyon.2021.e08550

Bartol, T. (2023). Smallholders and small-scale agriculture: Mapping and visualization of knowledge domains and research trends. *Cogent Social Sciences, 9*(1). Scopus. doi:10.1080/23311886.2022.2161778

Belasco, E. J., & Smith, V. (2022). The impact of policy design on payment concentration in Ad-hoc disaster Relief: Lessons from the Market Facilitation and Coronavirus food Assistance programs. *Food Policy, 106*. Scopus. doi:10.1016/j.foodpol.2021.102189

Biadgilign, S., & Yigzaw, M. (2020). Covid-19 in ethiopia: Current situation, missed opportunities, and the risk of health system disruptions. *Pan African Medical Journal, 35*, 1–2. Scopus. doi:10.11604/pamj.supp.2020.35.2.23906

Boughton, D., Goeb, J., Lambrecht, I., Headey, D., Takeshima, H., Mahrt, K., Masias, I., Goudet, S., Ragasa, C., Maredia, M. K., Minten, B., & Diao, X. (2021). Impacts of COVID-19 on agricultural production and food systems in late transforming Southeast Asia: The case of Myanmar. *Agricultural Systems, 188*. Scopus. doi:10.1016/j.agsy.2020.103026

Burrone, S., Dingacci, G., Dia, M., Bamba, B., Tarchiani, V., Grieco, E., Zini, C., Di Vecchia, A., & Vignaroli, P. (2022). *The role of staple crop production during the Covid-19 outbreak. Evidence for women small producers in Senegal. Applied Economics*. Scopus., doi:10.1080/00036846.2022.2108749

Darnhofer, I. (2021). Resilience or how do we enable agricultural systems to ride the waves of unexpected change? *Agricultural Systems, 187*. Scopus. doi:10.1016/j.agsy.2020.102997

Dentener, F., Emberson, L., Galmarini, S., Cappelli, G., Irimescu, A., Mihailescu, D., Van Dingenen, R., & Van Den Berg, M. (2020). Lower air pollution during COVID-19 lock-down: Improving models and methods estimating ozone impacts on crops: O3-crop impact during COVID-19. *Philosophical Transactions of the Royal Society A: Mathematical, Physical and Engineering Sciences, 378*(2183). Scopus. doi:10.1098/rsta.2020.0188

Fouotsa Manfouo, N. C., & Von Fintel, D. (2022). Investigating the effects of drought and lockdowns on smallholder and commercial agricultural production in KwaZulu-Natal using remotely sensed data. *Heliyon, 8*(11). Scopus. doi:10.1016/j.heliyon.2022.e11637

Franch, B., Vermote, E., Skakun, S., Santamaria-Artigas, A., Kalecinski, N., Roger, J.-C., Becker-Reshef, I., Barker, B., Justice, C., & Sobrino, J. A. (2021). The ARYA crop yield forecasting algorithm: Application to the main wheat exporting countries. *International Journal of Applied Earth Observation and Geoinformation, 104*. Scopus. doi:10.1016/j.jag.2021.102552

Gaffney, J., Girma, D., Kane, N. A., Llaca, V., Mace, E., Taylor, N., & Tibebu, R. (2022). Maximizing value of genetic sequence data requires an enabling environment and urgency. *Global Food Security, 33*. Scopus. doi:10.1016/j.gfs.2022.100619

Gatto, M., & Islam, A. H. M. S. (2021). Impacts of COVID-19 on rural livelihoods in Bangladesh: Evidence using panel data. *PLoS ONE, 16*(11 November). Scopus. doi:10.1371/journal.pone.0259264

Gutiérrez-Moya, E., Adenso-Díaz, B., & Lozano, S. (2021). Analysis and vulnerability of the international wheat trade network. *Food Security, 13*(1), 113–128. Scopus. doi:10.1007/s12571-020-01117-9

Hamilton, H., Henry, R., Rounsevell, M., Moran, D., Cossar, F., Allen, K., Boden, L., & Alexander, P. (2020). Exploring global food system shocks, scenarios and outcomes. *Futures, 123*. Scopus. doi:10.1016/j.futures.2020.102601

Haqiqi, I., & Bahalou Horeh, M. (2021). Assessment of COVID-19 impacts on U.S. counties using the immediate impact model of local agricultural production (IMLAP). *Agricultural Systems, 190*. Scopus. doi:10.1016/j.agsy.2021.103132

Haque, M. H., Islam, M. A., Karim, M. R., Kayesh, M. E. H., Sarker, S., Nazir, K. H. M. N. H., & Anwer, M. S. (2021). Coronavirus disease 2019 and future pandemics: Impacts on livestock health and production and possible mitigation measures. *Veterinary World, 14*(9), 2434–2443. Scopus. doi:10.14202/vetworld.2021.2434-2443

Jha, P. K., Araya, A., Stewart, Z. P., Faye, A., Traore, H., Middendorf, B. J., & Prasad, P. V. V. (2021). Projecting potential impact of COVID-19 on major cereal crops in Senegal and Burkina Faso using crop simulation models. *Agricultural Systems, 190*. Scopus. doi:10.1016/j.agsy.2021.103107

Johnson, R. A. B., Hann, K., Leno, A., Timire, C., Bangura, A. J. A., George, M. I. Z., Davtyan, H., Satyanarayana, S., Nair, D., Mansaray, A. H. D., Bangura, F. I., Kanu, J. S., & Edwards, J. K. (2022). Pesticide Importation in Sierra Leone, 2010–2021: Implications for Food Production and Antimicrobial Resistance. *International Journal of Environmental Research and Public Health, 19*(8). Scopus. doi:10.3390/ijerph19084792

Karoliina, R., Jyrki, A., Kalle, A., & Pasi, R. (2023). *The elements of resilience in the food system and means to enhance the stability of the food supply. Environment Systems and Decisions*. Scopus., doi:10.100710669-022-09889-5

Koo, J. H., Lee, S. W., Lee, G. E., & Pyeon, M. W. (2021). A Study on Smart Farmer Service Using Community Mapping. *Journal of the Korean Society of Surveying, Geodesy, Photogrammetry and Cartography, 39*(6), 419–427. Scopus. doi:10.7848/ksgpc.2021.39.6.419

Lioubimtseva, E., & Henebry, G. M. (2012). Grain production trends in Russia, Ukraine and Kazakhstan: New opportunities in an increasingly unstable world? *Frontiers of Earth Science, 6*(2), 157–166. Scopus. doi:10.1007/s11707-012-0318-y

Liu, M., Yao, H., & Matsui, H. (2022). Marked rebound of agricultural fire emissions in Asia after the outbreak of COVID-19. *Environmental Research Letters, 17*(11). Scopus. doi:10.1088/1748-9326/ac9e69

Lusk, J. L., & Chandra, R. (2021). Farmer and farm worker illnesses and deaths from COVID-19 and impacts on agricultural output. *PLoS ONE, 16*(4 April 2021). Scopus. doi:10.1371/journal.pone.0250621

Madzorera, I., Ismail, A., Hemler, E. C., Korte, M. L., Olufemi, A. A., Wang, D., Assefa, N., Workneh, F., Lankoande, B., Chukwu, A., Ourohire, M., Mattei, J., Soura, A., Berhane, Y., Sie, A., Oduola, A., & Fawzi, W. W. (2021). Impact of COVID-19 on nutrition, food security, and dietary diversity and quality in Burkina Faso, Ethiopia and Nigeria. *American Journal of Tropical Medicine and Hygiene, 105*(2), 295–309. Scopus. doi:10.4269/ajtmh.20-1617

Mahmud, K., Panday, D., Mergoum, A., & Missaoui, A. (2021). Nitrogen losses and potential mitigation strategies for a sustainable agroecosystem. *Sustainability (Switzerland), 13*(4), 1–23. Scopus. doi:10.3390/su13042400

Meunier, C., Casagrande, M., Rosiès, B., Bedoussac, L., Topp, C. F. E., Walker, R. L., Watson, C. A., & Martin, G. (2022). Interplay: A game for the participatory design of locally adapted cereal–legume intercrops. *Agricultural Systems, 201*. Scopus. doi:10.1016/j.agsy.2022.103438

Meyfroidt, P., Schierhorn, F., Prishchepov, A. V., Müller, D., & Kuemmerle, T. (2016). Drivers, constraints and trade-offs associated with recultivating abandoned cropland in Russia, Ukraine and Kazakhstan. *Global Environmental Change, 37*, 1–15. Scopus. doi:10.1016/j.gloenvcha.2016.01.003

Mohite, J., Sawant, S., Pandit, A., & Pappula, S. (2022). Impact of lockdown and crop stubble burning on air quality of India: A case study from wheat-growing region. *Environmental Monitoring and Assessment, 194*(2). Scopus. doi:10.1007/s10661-021-09723-6

Molinero-Gerbeau, Y. (2021). The Problem is not Covid-19, it's the Model! Industrial Agriculture and Migrant Farm Labour in the EU. *EuroChoices, 20*(3), 69–74. Scopus. doi:10.1111/1746-692X.12308

Muthamilarasan, M., & Prasad, M. (2021). Small Millets for Enduring Food Security Amidst Pandemics. *Trends in Plant Science, 26*(1), 33–40. Scopus. doi:10.1016/j.tplants.2020.08.008

Niaz, A., Asad, M. M., Abdulmuhsin, A. A., Shavanov, M., & Churi, P. (2022). Risk factors for the rice crop farming community in China: A documentary analysis of the challenges during post COVID-19. *Asian Education and Development Studies, 11*(2), 298–310. Scopus. doi:10.1108/AEDS-11-2020-0257

OECD. (2023). *Agriculture—OECD Data*. TheOECD. https://data.oecd.org/agriculture.htm

Okoli, T. O. (2021). Coping strategies adopted by female-headed households in combating food insecurity during corona virus-19 (Covid-19) lockdown in rural areas in anambra state. [Scopus.]. *Journal of Home Economics Research, 28*(2), 132–141.

Pan, D., Yang, J., Zhou, G., & Kong, F. (2020). The influence of COVID-19 on agricultural economy and emergency mitigation measures in China: A text mining analysis. *PLoS ONE, 15*(10). Scopus. doi:10.1371/journal.pone.0241167

Sangha, K. K., Ahammad, R., Mazahar, M. S., Hall, M., Owens, G., Kruss, L., Verrall, G., Moro, J., & Dickinson, G. (2022). An Integrated Assessment of the Horticulture Sector in Northern Australia to Inform Future Development. *Sustainability (Switzerland), 14*(18). Scopus. doi:10.3390/su141811647

Scopus. (2023). *Scopus Database*. Scopus. https://www.scopus.com/search/form.uri?display=basic#basic

Selim, T., & Eltarabily, M. G. (2022). Impact of COVID-19 lockdown on small-scale farming in Northeastern Nile Delta of Egypt and learned lessons for water conservation potentials. *Ain Shams Engineering Journal, 13*(4). Scopus. doi:10.1016/j.asej.2021.11.018

Spearman, C. (1904). The Proof and Measurement of Association between Two Things. *The American Journal of Psychology, 15*(1), 72–101. doi:10.2307/1412159 PMID:3322052

Stata. (2023). *Statistical software for data science.* Stata. https://www.stata.com/

StataCorp. (2017a). *Stata 15 Base Reference Manual.* Stata Press.

StataCorp. (2017b). *Stata Statistical Software: Release 15.* StataCorp LLC.

Sun, F., Qian, P., Cao, S., Chen, Y., & Feng, Z. (2022). The impact of crop specialization on nutritional intake: Evidence from farm households in China. *PLoS ONE, 17*(8 August). Scopus. doi:10.1371/journal.pone.0272347

Temesgen, F., Wakjira, M., & Abirham, A. (2022). Assessing the Economic Impact of the Coronavirus Pandemic on the Production and Distribution of Livestock Across Value Chain Approach: The Case of Kellem Wollega Zone, Oromia National Regional State, Ethiopia. *International Journal of Rural Management, 18*(1), 39–55. Scopus. doi:10.1177/0973005221993369

Tripathi, H. G., Smith, H. E., Sait, S. M., Sallu, S. M., Whitfield, S., Jankielsohn, A., Kunin, W. E., Mazibuko, N., & Nyhodo, B. (2021). Impacts of covid-19 on diverse farm systems in tanzania and south africa. *Sustainability (Switzerland), 13*(17). Scopus. doi:10.3390/su13179863

Usman, M., Farooq, M., Jellali, S., Charabi, Y., Al-Sadi, A. M., & Al-Badi, A. (2021). COVID-19: The Challenges and Opportunities for Water, Air, Agriculture and Energy Sectors. *International Journal of Agriculture and Biology, 25*(5), 1085–1095. Scopus. doi:10.17957/IJAB/15.1767

van Vliet, N., Sirén, A., Paemelaere, E. A. D., Rios, E. Y. C., Correa, M. S. A., Williams, T. B., & Quintero, S. (2022). Immediate effects of the COVID-19 pandemic on the use of wildlife as food among indigenous people and local communities in South America. *Ecology and Society, 27*(3). Scopus. doi:10.5751/ES-13570-270342

Vargas, R., Fonseca, C., Hareau, G., Ordinola, M., Pradel, W., Robiglio, V., & Suarez, V. (2021). Health crisis and quarantine measures in Peru: Effects on livelihoods of coffee and potato farmers. *Agricultural Systems, 187*. Scopus. doi:10.1016/j.agsy.2020.103033

Vicente-Vicente, J. L., Doernberg, A., Zasada, I., Ludlow, D., Staszek, D., Bushell, J., Hainoun, A., Loibl, W., & Piorr, A. (2021). Exploring alternative pathways toward more sustainable regional food systems by foodshed assessment – City region examples from Vienna and Bristol. *Environmental Science and Policy, 124*, 401–412. Scopus. doi:10.1016/j.envsci.2021.07.013

Yovchevska, P. (2021). Bulgarian agriculture in the focus of representative statistical surveys. [Scopus.]. *Bulgarian Journal of Agricultural Science*, *27*(5), 859–864.

Yuan, Y.-S., Cao, Z.-Y., Chen, Y.-T., Gong, P.-L., Huang, G.-H., & He, L. (2022). Efficiency Analysis of the Crop Production in China in 2019 and 2020: Role of Uncertainty Perceptions in COVID-19. *Discrete Dynamics in Nature and Society, 2022*. Scopus. doi:10.1155/2022/7044474

Zhao, W., Chou, J., Li, J., Xu, Y., Li, Y., & Hao, Y. (2022). Impacts of Extreme Climate Events on Future Rice Yields in Global Major Rice-Producing Regions. *International Journal of Environmental Research and Public Health, 19*(8). Scopus. doi:10.3390/ijerph19084437

Chapter 5
Global Agricultural Policy After COVID-19

ABSTRACT

Agriculture has relevant particularities that make this sector different from other economic activities. These frameworks imply the need for adjusted strategies and make the agricultural policy among the most important public interventions to correct, namely market failures. This is one explanation of why, for example, the CAP (common agricultural policy) is the most famous policy in the European Union context. The aim of this chapter is to carry out an assessment of the agricultural policies' impacts from, specifically, the Covid-19 pandemic. Statistical information from the OECD (Organisation for Economic Co-operation and Development) database for the period 2018-2021 was considered for agricultural support and producer protection. It seems that in specific contexts agricultural policies could have a more relevant contribution to support the farmers dealing with the challenges created by the pandemic.

INTRODUCTION

The frameworks created by the Covid-19 pandemic and the Russia-Ukraine crisis have had consequences on several domains of economies and societies, such as agriculture (Cao et al., 2021) and respective strategies (Patyka et al., 2021). But they bring also relevant insights for several stakeholders, including agricultural policymakers (Adhikari et al., 2021), for a more balanced development (Kumar et al., 2021), to rethink the farming systems (Popescu & Popescu, 2022) and the supply chains (Yovchevska, 2021).

The recent international events added difficulties to old problems faced by governments, public institutions and international organisations (Anderson, 2022). These shocks promote changes in the policies (Ewert & Loer, 2022), require modifications in the agricultural initiatives paradigm (Cheng et al., 2022) and involve adjustments to the new realities (Ker, 2020), through more robust measures (Munonye et al., 2022). In any case, the implementation of regulations is always subject to several constraints, including institutional and organisational (Gascón, 2020).

Agricultural policies have a crucial role in the current dynamics of the agricultural sector (Baležentis et al., 2021) and rural areas (Espinoza Lastra et al., 2021). Particularly to support the associated activities

DOI: 10.4018/978-1-6684-8923-9.ch005

(Gurbuz & Ozkan, 2021) to deal with the respective impacts (Varshney et al., 2021) and avoid setbacks in the achievements attained (Harris et al., 2020).

However, the design of the respective instruments is not an easy task (Vik, 2020), namely in the contexts such as those from the European Union with a great heterogeneity of realities (Alfaro-Navarro & Andrés-Martínez, 2021). To deal with the impacts on the food and agricultural sectors, OECD nations implemented economic support and emerging countries adopted market and trade strategies (Gruère & Brooks, 2021). Additionally, in the policy design and implementation processes, choices have to be made (Sánchez et al., 2022), but these decisions generally do not satisfy all stakeholders (Belasco & Smith, 2022).

The effectiveness and adequacy of the interventions on agriculture are generally a challenge for the institutions involved (Jiménez & Saldarriaga-Isaza, 2022). This happens because many factors often the results and effects obtained are not equal to those expected and projected ex-ante (Ali et al., 2020) by the related institutions (Srairi, 2021), or at least are not considered socially fairer (Hoanga et al., 2021).

For effective implementation of agricultural policies, the process of design is decisive and here the availability of information (Ali & Imran, 2021) and scientific research insights (Neganova & Chistyakov, 2020) are crucial, namely for the European Union (Scown et al., 2020) and Chinese (Wang et al., 2020) contexts. The consideration of local and specific particularities and bottom-up approaches (Bisht et al., 2020) are fundamental for the policy success (Bezáková & Bezák, 2022) and a more adjusted development (Leakey et al., 2021).

Agroecology, among other approaches, may offer a set of opportunities for policymakers to promote a more sustainable and resilient food supply and farming systems (Altieri & Nicholls, 2020). The resilience of the agri-food chains and food security (Jouili & Elloumi, 2022), the digital transition (Kondratieva, 2021) and the respective approaches (Zysk et al., 2022) are discussions in the order of the day.

The future agricultural policies will be called to deal with new dimensions (Sama-Berrocal & Martínez-Azúa, 2022), where the health and social risks could be integrated (Bitsoumanou Nkounkou & Temple, 2021) together in a multidimensional approach (Karagöl et al., 2021), where solidarity should have its place (Kurczewska, 2021). It will be important that the supply and demand sides be addressed (Détang-Dessendre et al., 2020).

Following these viewpoints, the present study has as its objective to assess the impacts on agricultural policy from the recent international events, namely Covid-19. For that, data from the OECD (OECD, 2023) database was considered for the following variables: TSE, total support estimate (total support to the agricultural sector); PSE, producer support estimate (includes market price funding, budgetary expenses and the cost of income foregone); CSE, consumer support estimate (if negative, the CSE quantifies the burden on consumers through market price support); GSSE, general services support estimate (is related to instruments promoting enabling conditions). The Stata (StataCorp, 2017a, 2017b; Stata, 2023) software was also considered. The literature review was based on Scopus (2023).

AGRICULTURAL SUPPORT FOR OECD CONTEXT

In general, for the OECD countries, it was observed a decrease in agricultural support (million US dollars) in 2021 (table 1). The PSE, on average, increased by 0.803% (the standard deviation of the sample was 3.245%), the TSE increased by 2.106% (standard deviation equal 4.223%), CSE with negative values increased by 18.937% (the standard deviation was 15.017%) and GSSE decreased, on average, -0.999%

(the standard deviation was equal to 4.377%). In percentage of the GDP (Gross Domestic Product) the conclusions are similar (table 2).

These findings may be a coincidence from economic cycles, for example, or can be an impact of the Covid-19 pandemic with some delay. In any case, this needs deeper assessment in future research, namely to better understand what in times of crisis for the several economic sectors and the different social dimensions, promoted reductions in the support to the farming sector.

Table 1. Agricultural support (million US dollars), for the different measures

Location	Support	Year	Value	Growth rates (%)
OECD	PSE	2018	239809	
OECD	PSE	2019	243906	1.708
OECD	PSE	2020	252441	3.499
OECD	PSE	2021	245375	-2.799
OECD	TSE	2018	281282	
OECD	TSE	2019	298517	6.127
OECD	TSE	2020	305936	2.485
OECD	TSE	2021	298918	-2.294
OECD	CSE	2018	-77400	
OECD	CSE	2019	-74274	4.039
OECD	CSE	2020	-60383	18.702
OECD	CSE	2021	-39810	34.071
OECD	GSSE	2018	45202	
OECD	GSSE	2019	44425	-1.719
OECD	GSSE	2020	46066	3.694
OECD	GSSE	2021	43776	-4.971

Note: TSE, total support estimate; PSE, producer support estimate; CSE, consumer support estimate; GSSE, general services support estimate.

Table 2. Agricultural support (% of GDP), for total support estimate

Location	Support	Year	Value	Growth rates (%)
OECD	TSE	2018	0.621	
OECD	TSE	2019	0.619	-0.322
OECD	TSE	2020	0.665	7.431
OECD	TSE	2021	0.612	-7.970

Note: TSE, total support estimate.

AGRICULTURAL SUPPORT BY COUNTRY

The different realities around the world for the total support estimate in millions of US dollars for the years 2018-2021 are presented in tables 3 and 4. On average, the TSE had growth rates of -42% (standard deviation of 262%) for the year 2019, 10% (standard deviation of 78%) in 2020 and 4% (standard deviation of 38%) in 2021. For the producer support estimate (million US dollars), the average growth rates are, respectively, 9% (38% for the standard deviation), 1% (30% for the standard deviation) and 13% (49% for the standard deviation) for the years 2019, 2020 and 2021 (tables 5 and 6). The frameworks for the consumer support estimate (million US dollars) are those described in tables 7 and 8. China, European Union, India, Japan and the USA are between the countries (or groups of countries) with higher general services support estimate in millions of US dollars (tables 9 and 10).

Table 3. Total support estimate (million US dollars), for 2018 and 2019

Location	Year	Value	Growth rates (%)
Argentina	2018	-10002	
Australia	2018	2177	
Brazil	2018	5143	
Canada	2018	4617	
Switzerland	2018	5787	
Chile	2018	721	
China	2018	187883	
Colombia	2018	3820	
Costa Rica	2018	213	
EU	2018	103728	
United Kingdom	2018	6876	
Indonesia	2018	22510	
India	2018	-683	
Iceland	2018	219	
Israel	2018	981	
Japan	2018	39041	
Kazakhstan	2018	1085	
Korea, Rep.	2018	23001	
Mexico	2018	6229	
Norway	2018	3055	
New Zealand	2018	411	
Philippines	2018	8038	
Russian Federation	2018	9333	
Turkey	2018	8917	
Ukraine	2018	753	
USA	2018	83294	

Continued on following page

Table 3. Continued

Location	Year	Value	Growth rates (%)
Vietnam	2018	-3440	
South Africa	2018	1132	
Argentina	2019	-10245	2.430
Australia	2019	2166	-0.505
Brazil	2019	4694	-8.730
Canada	2019	5364	16.179
Switzerland	2019	6104	5.478
Chile	2019	736	2.080
China	2019	204498	8.843
Colombia	2019	3587	-6.099
Costa Rica	2019	292	37.089
EU	2019	105581	1.786
United Kingdom	2019	7056	2.618
Indonesia	2019	25016	11.133
India	2019	8724	-1377.306
Iceland	2019	208	-5.023
Israel	2019	1178	20.082
Japan	2019	42577	9.057
Kazakhstan	2019	1820	67.742
Korea, Rep.	2019	21424	-6.856
Mexico	2019	6415	2.986
Norway	2019	2882	-5.663
New Zealand	2019	551	34.063
Philippines	2019	7857	-2.252
Russian Federation	2019	8691	-6.879
Turkey	2019	9724	9.050
Ukraine	2019	1108	47.145
USA	2019	95135	14.216
Vietnam	2019	-2972	-13.605
South Africa	2019	804	-28.975

Table 4. Total support estimate (million US dollars), for 2020 and 2021

Location	Year	Value	Growth rates (%)
Argentina	2020	-3818	-62.733
Australia	2020	2631	21.468
Brazil	2020	4113	-12.378
Canada	2020	5441	1.435
Switzerland	2020	6960	14.024
Chile	2020	615	-16.440
China	2020	251062	22.770
Colombia	2020	2737	-23.697
Costa Rica	2020	302	3.425
EU	2020	104561	-0.966
United Kingdom	2020	6649	-5.768
Indonesia	2020	19395	-22.470
India	2020	43038	393.329
Iceland	2020	189	-9.135
Israel	2020	1504	27.674
Japan	2020	46447	9.089
Kazakhstan	2020	1629	-10.495
Korea, Rep.	2020	22013	2.749
Mexico	2020	5523	-13.905
Norway	2020	2713	-5.864
New Zealand	2020	546	-0.907
Philippines	2020	8904	13.326
Russian Federation	2020	6611	-23.933
Turkey	2020	14632	50.473
Ukraine	2020	627	-43.412
USA	2020	94299	-0.879
Vietnam	2020	-2001	-32.672
South Africa	2020	750	-6.716
Argentina	2021	-8781	129.990
Australia	2021	2768	5.207
Brazil	2021	7648	85.947
Canada	2021	8267	51.939
Switzerland	2021	6271	-9.899
Chile	2021	708	15.122
China	2021	276364	10.078
Colombia	2021	2198	-19.693
Costa Rica	2021	226	-25.166
EU	2021	95479	-8.686

Continued on following page

Table 4. Continued

Location	Year	Value	Growth rates (%)
United Kingdom	2021	8213	23.522
Indonesia	2021	20028	3.264
India	2021	14601	-66.074
Iceland	2021	201	6.349
Israel	2021	1488	-1.064
Japan	2021	35835	-22.848
Kazakhstan	2021	1688	3.622
Korea, Rep.	2021	22675	3.007
Mexico	2021	5579	1.014
Norway	2021	2718	0.184
New Zealand	2021	541	-0.916
Philippines	2021	9510	6.806
Russian Federation	2021	4771	-27.832
Turkey	2021	7480	-48.879
Ukraine	2021	725	15.630
USA	2021	107459	13.956
Vietnam	2021	-1337	-33.183
South Africa	2021	771	2.800

Table 5. Producer support estimate (million US dollars), for 2018 and 2019

Location	Year	Value	Growth rates (%)
Argentina	2018	-12178	
Australia	2018	1122	
Brazil	2018	2674	
Canada	2018	3766	
Switzerland	2018	6052	
Chile	2018	387	
China	2018	185899	
Colombia	2018	3943	
Costa Rica	2018	186	
EU	2018	109172	
United Kingdom	2018	7257	
Indonesia	2018	24157	
India	2018	-33092	
Iceland	2018	247	

Continued on following page

Table 5. Continued

Location	Year	Value	Growth rates (%)
Israel	2018	932	
Japan	2018	36921	
Kazakhstan	2018	443	
Korea, Rep.	2018	23083	
Mexico	2018	6335	
Norway	2018	3331	
New Zealand	2018	92	
Philippines	2018	7787	
Russian Federation	2018	8892	
Turkey	2018	8338	
Ukraine	2018	668	
USA	2018	41594	
Vietnam	2018	-4817	
South Africa	2018	1009	
Argentina	2019	-11760	-3.432
Australia	2019	1414	26.025
Brazil	2019	2282	-14.660
Canada	2019	4351	15.534
Switzerland	2019	6076	0.397
Chile	2019	369	-4.651
China	2019	199382	7.253
Colombia	2019	3584	-9.105
Costa Rica	2019	265	42.473
EU	2019	105117	-3.714
United Kingdom	2019	6937	-4.410
Indonesia	2019	24489	1.374
India	2019	-23079	-30.258
Iceland	2019	222	-10.121
Israel	2019	1074	15.236
Japan	2019	37245	0.878
Kazakhstan	2019	1201	171.106
Korea, Rep.	2019	19945	-13.594
Mexico	2019	6421	1.358
Norway	2019	2946	-11.558
New Zealand	2019	134	45.652
Philippines	2019	7406	-4.893
Russian Federation	2019	7599	-14.541
Turkey	2019	9795	17.474

Table 5. Continued

Location	Year	Value	Growth rates (%)
Ukraine	2019	1018	52.395
USA	2019	50270	20.859
Vietnam	2019	-4315	-10.421
South Africa	2019	594	-41.130

Table 6. Producer support estimate (million US dollars), for 2020 and 2021

Location	Year	Value	Growth rates (%)
Argentina	2020	-4614	-60.765
Australia	2020	1600	13.154
Brazil	2020	2362	3.506
Canada	2020	4331	-0.460
Switzerland	2020	7108	16.985
Chile	2020	310	-15.989
China	2020	250737	25.757
Colombia	2020	2812	-21.540
Costa Rica	2020	281	6.038
EU	2020	103988	-1.074
United Kingdom	2020	6662	-3.964
Indonesia	2020	18846	-23.043
India	2020	-41481	79.735
Iceland	2020	205	-7.658
Israel	2020	1410	31.285
Japan	2020	40331	8.286
Kazakhstan	2020	931	-22.481
Korea, Rep.	2020	21810	9.351
Mexico	2020	5587	-12.989
Norway	2020	2797	-5.058
New Zealand	2020	199	48.507
Philippines	2020	8424	13.746
Russian Federation	2020	5739	-24.477
Turkey	2020	15655	59.826
Ukraine	2020	490	-51.866
USA	2020	48864	-2.797
Vietnam	2020	-3337	-22.665
South Africa	2020	553	-6.902
Argentina	2021	-10614	130.039

Continued on following page

Table 6. Continued

Location	Year	Value	Growth rates (%)
Australia	2021	1747	9.188
Brazil	2021	6798	187.807
Canada	2021	7942	83.376
Switzerland	2021	6574	-7.513
Chile	2021	390	25.806
China	2021	289007	15.263
Colombia	2021	2130	-24.253
Costa Rica	2021	196	-30.249
EU	2021	97980	-5.778
United Kingdom	2021	8940	34.194
Indonesia	2021	20623	9.429
India	2021	-41231	-0.603
Iceland	2021	228	11.220
Israel	2021	1488	5.532
Japan	2021	33634	-16.605
Kazakhstan	2021	1065	14.393
Korea, Rep.	2021	23205	6.396
Mexico	2021	5852	4.743
Norway	2021	2880	2.967
New Zealand	2021	142	-28.643
Philippines	2021	9652	14.577
Russian Federation	2021	3657	-36.278
Turkey	2021	7466	-52.309
Ukraine	2021	590	20.408
USA	2021	53045	8.556
Vietnam	2021	-2807	-15.883
South Africa	2021	589	6.510

Table 7. Consumer support estimate (million US dollars), for 2018 and 2019

Location	Year	Value	Growth rates (%)
Argentina	2018	9142	
Australia	2018	0	
Brazil	2018	913	
Canada	2018	-2213	
Switzerland	2018	-4000	
Chile	2018	-32	
China	2018	-134201	
Colombia	2018	-3994	
Costa Rica	2018	-203	
EU	2018	-20456	
United Kingdom	2018	-2539	
Indonesia	2018	-17555	
India	2018	79009	
Iceland	2018	-127	
Israel	2018	-1465	
Japan	2018	-43164	
Kazakhstan	2018	818	
Korea, Rep.	2018	-27401	
Mexico	2018	-1795	
Norway	2018	-1647	
New Zealand	2018	-70	
Philippines	2018	-7987	
Russian Federation	2018	-6684	
Turkey	2018	-4885	
Ukraine	2018	-270	
USA	2018	33239	
Vietnam	2018	1608	
South Africa	2018	-528	
Argentina	2019	8377	-8.368
Australia	2019	0	0.000
Brazil	2019	872	-4.491
Canada	2019	-2480	-12.065
Switzerland	2019	-4127	-3.175
Chile	2019	-26	18.750
China	2019	-158714	-18.266
Colombia	2019	-4375	-9.539
Costa Rica	2019	-287	-41.379
EU	2019	-18524	9.445

Continued on following page

Table 7. Continued

Location	Year	Value	Growth rates (%)
United Kingdom	2019	-2004	21.071
Indonesia	2019	-18059	-2.871
India	2019	75351	-4.630
Iceland	2019	-106	16.535
Israel	2019	-1283	12.423
Japan	2019	-43115	0.114
Kazakhstan	2019	419	-48.778
Korea, Rep.	2019	-24275	11.408
Mexico	2019	-2128	-18.552
Norway	2019	-1102	33.090
New Zealand	2019	-107	-52.857
Philippines	2019	-7791	2.454
Russian Federation	2019	-4396	34.231
Turkey	2019	-6107	-25.015
Ukraine	2019	-694	-157.037
USA	2019	32971	-0.806
Vietnam	2019	104	-93.532
South Africa	2019	-465	11.932

Table 8. Consumer support estimate (million US dollars), for 2020 and 2021

Location	Year	Value	Growth rates (%)
Argentina	2020	4138	-50.603
Australia	2020	0	0.000
Brazil	2020	-4	-100.459
Canada	2020	-2751	-10.927
Switzerland	2020	-5050	-22.365
Chile	2020	-23	11.538
China	2020	-241348	-52.065
Colombia	2020	-3609	17.509
Costa Rica	2020	-316	-10.105
EU	2020	-14677	20.768
United Kingdom	2020	-1567	21.806
Indonesia	2020	-11984	33.640
India	2020	150272	99.429
Iceland	2020	-94	11.321
Israel	2020	-1534	-19.564

Continued on following page

Table 8. Continued

Location	Year	Value	Growth rates (%)
Japan	2020	-43692	-1.338
Kazakhstan	2020	832	98.568
Korea, Rep.	2020	-24085	0.783
Mexico	2020	-2306	-8.365
Norway	2020	-1097	0.454
New Zealand	2020	-143	-33.645
Philippines	2020	-8566	-9.947
Russian Federation	2020	-3152	28.298
Turkey	2020	-6182	-1.228
Ukraine	2020	-192	72.334
USA	2020	44674	35.495
Vietnam	2020	-39	-137.500
South Africa	2020	-325	30.108
Argentina	2021	8000	93.330
Australia	2021	0	0.000
Brazil	2021	-2023	-50475.000
Canada	2021	-3753	-36.423
Switzerland	2021	-4547	9.960
Chile	2021	-133	-478.261
China	2021	-263527	-9.190
Colombia	2021	-2201	39.014
Costa Rica	2021	-208	34.177
EU	2021	-12100	17.558
United Kingdom	2021	-4463	-184.812
Indonesia	2021	-11900	0.701
India	2021	121293	-19.284
Iceland	2021	-109	-15.957
Israel	2021	-1795	-17.014
Japan	2021	-37120	15.042
Kazakhstan	2021	1098	31.971
Korea, Rep.	2021	-24471	-1.603
Mexico	2021	-1308	43.278
Norway	2021	-1002	8.660
New Zealand	2021	-72	49.650
Philippines	2021	-9643	-12.573
Russian Federation	2021	-614	80.520
Turkey	2021	-5484	11.291
Ukraine	2021	-214	-11.458

Continued on following page

Table 8. Continued

Location	Year	Value	Growth rates (%)
USA	2021	58971	32.003
Vietnam	2021	-547	-1302.564
South Africa	2021	-325	0.000

Table 9. General services support estimate (million US dollars), for 2018 and 2019

Location	Year	Value	Growth rates (%)
Argentina	2018	374	
Australia	2018	1447	
Brazil	2018	2221	
Canada	2018	1683	
Switzerland	2018	774	
Chile	2018	464	
China	2018	35838	
Colombia	2018	566	
Costa Rica	2018	66	
EU	2018	12847	
United Kingdom	2018	853	
Indonesia	2018	1060	
India	2018	16243	
Iceland	2018	11	
Israel	2018	226	
Japan	2018	9149	
Kazakhstan	2018	313	
Korea, Rep.	2018	4025	
Mexico	2018	535	
Norway	2018	171	
New Zealand	2018	393	
Philippines	2018	1699	
Russian Federation	2018	1891	
Turkey	2018	2185	
Ukraine	2018	221	
USA	2018	10884	
Vietnam	2018	758	
South Africa	2018	327	
Argentina	2019	292	-21.925
Australia	2019	1011	-30.131

Continued on following page

Table 9. Continued

Location	Year	Value	Growth rates (%)
Brazil	2019	1969	-11.346
Canada	2019	1633	-2.971
Switzerland	2019	752	-2.842
Chile	2019	454	-2.155
China	2019	29530	-17.601
Colombia	2019	431	-23.852
Costa Rica	2019	62	-6.061
EU	2019	12783	-0.498
United Kingdom	2019	958	12.309
Indonesia	2019	2082	96.415
India	2019	16014	-1.410
Iceland	2019	10	-9.091
Israel	2019	245	8.407
Japan	2019	10410	13.783
Kazakhstan	2019	286	-8.626
Korea, Rep.	2019	4001	-0.596
Mexico	2019	485	-9.346
Norway	2019	159	-7.018
New Zealand	2019	483	22.901
Philippines	2019	1389	-18.246
Russian Federation	2019	1912	1.111
Turkey	2019	1089	-50.160
Ukraine	2019	222	0.452
USA	2019	11248	3.344
Vietnam	2019	988	30.343
South Africa	2019	306	-6.422

Table 10. General services support estimate (million US dollars), for 2020 and 2021

Location	Year	Value	Growth rates (%)
Argentina	2020	264	-9.589
Australia	2020	1398	38.279
Brazil	2020	1487	-24.479
Canada	2020	1720	5.328
Switzerland	2020	805	7.048
Chile	2020	390	-14.097
China	2020	35360	19.743
Colombia	2020	307	-28.770
Costa Rica	2020	63	1.613
EU	2020	14676	14.809
United Kingdom	2020	891	-6.994
Indonesia	2020	1840	-11.623
India	2020	16524	3.185
Iceland	2020	9	-10.000
Israel	2020	304	24.082
Japan	2020	12592	20.961
Kazakhstan	2020	335	17.133
Korea, Rep.	2020	3245	-18.895
Mexico	2020	548	12.990
Norway	2020	151	-5.031
New Zealand	2020	404	-16.356
Philippines	2020	1723	24.046
Russian Federation	2020	1620	-15.272
Turkey	2020	1018	-6.520
Ukraine	2020	224	0.901
USA	2020	9476	-15.754
Vietnam	2020	1057	6.984
South Africa	2020	301	-1.634
Argentina	2021	230	-12.879
Australia	2021	1527	9.227
Brazil	2021	1447	-2.690
Canada	2021	1712	-0.465
Switzerland	2021	836	3.851
Chile	2021	447	14.615
China	2021	37837	7.005
Colombia	2021	469	52.769
Costa Rica	2021	72	14.286
EU	2021	14511	-1.124

Continued on following page

Table 10. Continued

Location	Year	Value	Growth rates (%)
United Kingdom	2021	771	-13.468
Indonesia	2021	1663	-9.620
India	2021	18499	11.952
Iceland	2021	9	0.000
Israel	2021	272	-10.526
Japan	2021	8740	-30.591
Kazakhstan	2021	267	-20.299
Korea, Rep.	2021	3571	10.046
Mexico	2021	579	5.657
Norway	2021	180	19.205
New Zealand	2021	476	17.822
Philippines	2021	1595	-7.429
Russian Federation	2021	1555	-4.012
Turkey	2021	1380	35.560
Ukraine	2021	268	19.643
USA	2021	10464	10.426
Vietnam	2021	1226	15.989
South Africa	2021	323	7.309

Table 11. Total support estimate (% of GDP), for 2018 and 2019

Location	Year	Value	Growth rates (%)
Argentina	2018	-2.281	
Australia	2018	0.177	
Brazil	2018	0.317	
Canada	2018	0.316	
Switzerland	2018	0.929	
Chile	2018	0.286	
China	2018	1.596	
Colombia	2018	1.350	
Costa Rica	2018	0.403	
EU	2018	0.649	
United Kingdom	2018	0.280	
Indonesia	2018	2.548	
India	2018	-0.030	
Iceland	2018	0.984	
Israel	2018	0.310	

Continued on following page

Table 11. Continued

Location	Year	Value	Growth rates (%)
Japan	2018	0.915	
Kazakhstan	2018	0.714	
Korea, Rep.	2018	1.573	
Mexico	2018	0.599	
Norway	2018	0.825	
New Zealand	2018	0.231	
Philippines	2018	2.735	
Russian Federation	2018	0.666	
Turkey	2018	1.355	
Ukraine	2018	0.679	
USA	2018	0.477	
Vietnam	2018	-1.686	
South Africa	2018	0.363	
Argentina	2019	-2.579	13.064
Australia	2019	0.176	-0.565
Brazil	2019	0.280	-11.672
Canada	2019	0.345	9.177
Switzerland	2019	0.934	0.538
Chile	2019	0.295	3.147
China	2019	1.597	0.063
Colombia	2019	1.242	-8.000
Costa Rica	2019	0.51	26.551
EU	2019	0.637	-1.849
United Kingdom	2019	0.274	-2.143
Indonesia	2019	2.503	-1.766
India	2019	0.340	-1233.333
Iceland	2019	0.937	-4.776
Israel	2019	0.331	6.774
Japan	2019	0.926	1.202
Kazakhstan	2019	1.122	57.143
Korea, Rep.	2019	1.452	-7.692
Mexico	2019	0.564	-5.843
Norway	2019	0.797	-3.394
New Zealand	2019	0.293	26.840
Philippines	2019	2.334	-14.662
Russian Federation	2019	0.576	-13.514
Turkey	2019	1.431	5.609

Continued on following page

Table 11. Continued

Location	Year	Value	Growth rates (%)
Ukraine	2019	0.806	18.704
USA	2019	0.497	4.193
Vietnam	2019	-1.280	-24.081
South Africa	2019	0.256	-29.477

Table 12. Total support estimate (% of GDP), for 2020 and 2021

Location	Year	Value	Growth rates (%)
Argentina	2020	-1.137	-55.913
Australia	2020	0.211	19.886
Brazil	2020	0.324	15.714
Canada	2020	0.377	9.275
Switzerland	2020	1.055	12.955
Chile	2020	0.277	-6.102
China	2020	1.944	21.728
Colombia	2020	1.149	-7.488
Costa Rica	2020	0.556	9.020
EU	2020	0.661	3.768
United Kingdom	2020	0.274	0.000
Indonesia	2020	2.090	-16.500
India	2020	1.855	445.588
Iceland	2020	0.990	5.656
Israel	2020	0.421	27.190
Japan	2020	1.049	13.283
Kazakhstan	2020	1.086	-3.209
Korea, Rep.	2020	1.531	5.441
Mexico	2020	0.577	2.305
Norway	2020	0.853	7.026
New Zealand	2020	0.298	1.706
Philippines	2020	2.806	20.223
Russian Federation	2020	0.509	-11.632
Turkey	2020	2.318	61.985
Ukraine	2020	0.459	-43.052
USA	2020	0.514	3.421
Vietnam	2020	-0.842	-34.219
South Africa	2020	0.283	10.547
Argentina	2021	-2.123	86.719
Australia	2021	0.202	-4.265

Continued on following page

Table 12. Continued

Location	Year	Value	Growth rates (%)
Brazil	2021	0.559	72.531
Canada	2021	0.494	31.034
Switzerland	2021	0.922	-12.607
Chile	2021	0.262	-5.415
China	2021	1.848	-4.938
Colombia	2021	0.826	-28.111
Costa Rica	2021	0.430	-22.662
EU	2021	0.665	0.605
United Kingdom	2021	0.304	10.949
Indonesia	2021	2.017	-3.493
India	2021	0.538	-70.997
Iceland	2021	0.936	-5.455
Israel	2021	0.373	-11.401
Japan	2021	0.853	-18.684
Kazakhstan	2021	1.018	-6.262
Korea, Rep.	2021	1.498	-2.155
Mexico	2021	0.514	-10.919
Norway	2021	0.676	-20.750
New Zealand	2021	0.259	-13.087
Philippines	2021	2.840	1.212
Russian Federation	2021	0.333	-34.578
Turkey	2021	1.155	-50.173
Ukraine	2021	0.492	7.190
USA	2021	0.554	7.782
Vietnam	2021	-0.552	-34.442
South Africa	2021	0.221	-21.908

Table 13. Producer support estimate (% of gross farm receipts), for 2018 and 2019

Location	Year	Value	Growth rates (%)
Argentina	2018	-25.079	
Australia	2018	2.406	
Brazil	2018	1.784	
Canada	2018	7.768	
Switzerland	2018	47.973	
Chile	2018	2.568	
China	2018	13.005	
Colombia	2018	13.541	
Costa Rica	2018	3.715	
EU	2018	19.689	
United Kingdom	2018	19.668	
Indonesia	2018	21.599	
India	2018	-6.848	
Iceland	2018	57.526	
Israel	2018	10.708	
Japan	2018	41.369	
Kazakhstan	2018	3.199	
Korea, Rep.	2018	48.397	
Mexico	2018	10.529	
Norway	2018	58.051	
New Zealand	2018	0.468	
OECD	2018	17.613	
Philippines	2018	25.747	
Russian Federation	2018	11.466	
Turkey	2018	15.182	
Ukraine	2018	2.101	
USA	2018	10.491	
Vietnam	2018	-10.921	
South Africa	2018	4.677	
Argentina	2019	-28.366	13.107
Australia	2019	3.206	33.250
Brazil	2019	1.520	-14.798
Canada	2019	8.721	12.268
Switzerland	2019	47.296	-1.411
Chile	2019	2.707	5.413
China	2019	13.311	2.353
Colombia	2019	12.200	-9.903
Costa Rica	2019	5.401	45.384

Continued on following page

Table 13. Continued

Location	Year	Value	Growth rates (%)
EU	2019	19.540	-0.757
United Kingdom	2019	19.167	-2.547
Indonesia	2019	20.991	-2.815
India	2019	-4.713	-31.177
Iceland	2019	56.501	-1.782
Israel	2019	12.146	13.429
Japan	2019	41.934	1.366
Kazakhstan	2019	8.214	156.768
Korea, Rep.	2019	44.753	-7.529
Mexico	2019	10.137	-3.723
Norway	2019	55.357	-4.641
New Zealand	2019	0.675	44.231
OECD	2019	18.131	2.941
Philippines	2019	25.208	-2.093
Russian Federation	2019	9.032	-21.228
Turkey	2019	17.449	14.932
Ukraine	2019	2.847	35.507
USA	2019	12.667	20.742
Vietnam	2019	-9.752	-10.704
South Africa	2019	2.904	-37.909

Table 14. Producer support estimate (% of gross farm receipts), for 2020 and 2021

Location	Year	Value	Growth rates (%)
Argentina	2020	-10.468	-63.097
Australia	2020	3.290	2.620
Brazil	2020	1.562	2.763
Canada	2020	8.176	-6.249
Switzerland	2020	52.654	11.329
Chile	2020	2.429	-10.270
China	2020	14.794	11.141
Colombia	2020	10.162	-16.705
Costa Rica	2020	5.715	5.814
EU	2020	19.252	-1.474
United Kingdom	2020	18.922	-1.278
Indonesia	2020	16.037	-23.601
India	2020	-8.707	84.744
Iceland	2020	57.219	1.271
Israel	2020	15.356	26.428
Japan	2020	43.496	3.725
Kazakhstan	2020	5.735	-30.180
Korea, Rep.	2020	47.913	7.061
Mexico	2020	9.518	-6.106
Norway	2020	53.401	-3.533
New Zealand	2020	0.978	44.889
OECD	2020	18.227	0.529
Philippines	2020	27.065	7.367
Russian Federation	2020	6.917	-23.417
Turkey	2020	26.025	49.149
Ukraine	2020	1.406	-50.615
USA	2020	11.623	-8.242
Vietnam	2020	-7.400	-24.118
South Africa	2020	2.704	-6.887
Argentina	2021	-19.045	81.935
Australia	2021	2.845	-13.526
Brazil	2021	3.359	115.045
Canada	2021	11.660	42.613
Switzerland	2021	48.664	-7.578
Chile	2021	2.847	17.209
China	2021	16.228	9.693
Colombia	2021	7.754	-23.696
Costa Rica	2021	3.828	-33.018

Continued on following page

Table 14. Continued

Location	Year	Value	Growth rates (%)
EU	2021	17.566	-8.758
United Kingdom	2021	23.378	23.549
Indonesia	2021	17.189	7.183
India	2021	-8.089	-7.098
Iceland	2021	57.993	1.353
Israel	2021	14.846	-3.321
Japan	2021	37.517	-13.746
Kazakhstan	2021	5.715	-0.349
Korea, Rep.	2021	48.674	1.588
Mexico	2021	8.095	-14.951
Norway	2021	49.557	-7.198
New Zealand	2021	0.639	-34.663
OECD	2021	15.880	-12.877
Philippines	2021	28.909	6.813
Russian Federation	2021	3.696	-46.566
Turkey	2021	15.129	-41.867
Ukraine	2021	1.140	-18.919
USA	2021	10.548	-9.249
Vietnam	2021	-5.919	-20.014
South Africa	2021	2.318	-14.275

In percentage of the GDP, the TSE, on average across world countries, decreased -42% (standard deviation of 234%) in 2019, increased by 19% (standard deviation of 87%) in 2020 and decreased by -6% (standard deviation of 31%) in 2021 (tables 11 and 12). The producer support estimate (% of gross farm receipts) increased, on average, by 9% (standard deviation of 35%) in 2019, decreased by -1% (standard deviation of 28%) in 2020 and -1% (standard deviation of 34%) in 2021 (tables 13 and 14). The consumer support estimate (% of agricultural consumption) has the results presented in tables 15 and 16. On average, the general services support estimate decreased by -3% in 2019 (standard deviation of 26%) and increased by 2% (standard deviation of 21%) and 5% (standard deviation of 40%) in 2020 and 2021, respectively (tables 17 and 18). The ratio between the farm prices and the border reference price (tables 19 and 20) highlights the price distortions in, for example, Iceland.

Table 15. Consumer support estimate (% of agricultural consumption), for 2018 and 2019

Location	Year	Value	Growth rates (%)
Argentina	2018	27.038	
Australia	2018	0	
Brazil	2018	1.069	
Canada	2018	-7.568	
Switzerland	2018	-28.515	
Chile	2018	-0.28	
China	2018	-9.427	
Colombia	2018	-18.423	
Costa Rica	2018	-8.604	
EU	2018	-4.411	
United Kingdom	2018	-7.239	
Indonesia	2018	-24.983	
India	2018	22.169	
Iceland	2018	-45.137	
Israel	2018	-13.261	
Japan	2018	-36.424	
Kazakhstan	2018	7.268	
Korea, Rep.	2018	-42.772	
Mexico	2018	-2.701	
Norway	2018	-38.371	
New Zealand	2018	-0.736	
OECD	2018	-7.247	
Philippines	2018	-24.895	
Russian Federation	2018	-9.003	
Turkey	2018	-10.733	
Ukraine	2018	-1.282	
USA	2018	13.676	
Vietnam	2018	4.002	
South Africa	2018	-2.669	
Argentina	2019	31.185	15.338
Australia	2019	0	0.000
Brazil	2019	0.99	-7.390
Canada	2019	-8.023	-6.012
Switzerland	2019	-28.6	-0.298
Chile	2019	-0.267	4.643
China	2019	-10.409	-10.417
Colombia	2019	-19.11	-3.729
Costa Rica	2019	-12.545	-45.804

Continued on following page

Table 15. Continued

Location	Year	Value	Growth rates (%)
EU	2019	-4.256	3.514
United Kingdom	2019	-6.31	12.833
Indonesia	2019	-25.272	-1.157
India	2019	21.418	-3.388
Iceland	2019	-40.862	9.471
Israel	2019	-12.086	8.861
Japan	2019	-36.146	0.763
Kazakhstan	2019	3.544	-51.238
Korea, Rep.	2019	-40.423	5.492
Mexico	2019	-2.991	-10.737
Norway	2019	-33.234	13.388
New Zealand	2019	-1.067	-44.973
OECD	2019	-7.033	2.953
Philippines	2019	-24.841	0.217
Russian Federation	2019	-5.663	37.099
Turkey	2019	-12.98	-20.935
Ukraine	2019	-2.752	-114.665
USA	2019	13.198	-3.495
Vietnam	2019	0.249	-93.778
South Africa	2019	-2.475	7.269

Table 16. Consumer support estimate (% of agricultural consumption), for 2020 and 2021

Location	Year	Value	Growth rates (%)
Argentina	2020	12.944	-58.493
Australia	2020	0	0.000
Brazil	2020	-0.005	-100.505
Canada	2020	-8.611	-7.329
Switzerland	2020	-35.382	-23.713
Chile	2020	-0.228	14.607
China	2020	-13.571	-30.378
Colombia	2020	-15.86	17.007
Costa Rica	2020	-13.705	-9.247
EU	2020	-3.298	22.509
United Kingdom	2020	-4.632	26.593
Indonesia	2020	-17.534	30.619
India	2020	51.99	142.740
Iceland	2020	-42.1	-3.030
Israel	2020	-13.538	-12.014
Japan	2020	-36.248	-0.282
Kazakhstan	2020	6.576	85.553
Korea, Rep.	2020	-41.166	-1.838
Mexico	2020	-3.329	-11.301
Norway	2020	-31.281	5.877
New Zealand	2020	-1.466	-37.395
OECD	2020	-5.668	19.409
Philippines	2020	-26.225	-5.571
Russian Federation	2020	-4.117	27.300
Turkey	2020	-13.326	-2.666
Ukraine	2020	-0.797	71.039
USA	2020	17.716	34.232
Vietnam	2020	-0.091	-136.546
South Africa	2020	-1.83	26.061
Argentina	2021	21.422	65.498
Australia	2021	0	0.000
Brazil	2021	-1.671	-33320.000
Canada	2021	-9.74	-13.111
Switzerland	2021	-28.482	19.501
Chile	2021	-1.086	-376.316
China	2021	-14.076	-3.721
Colombia	2021	-10.382	34.540
Costa Rica	2021	-9.447	31.069

Continued on following page

Table 16. Continued

Location	Year	Value	Growth rates (%)
EU	2021	-2.619	20.588
United Kingdom	2021	-12.962	-179.836
Indonesia	2021	-16.913	3.542
India	2021	33.421	-35.716
Iceland	2021	-44.023	-4.568
Israel	2021	-14.283	-5.503
Japan	2021	-32.866	9.330
Kazakhstan	2021	7.607	15.678
Korea, Rep.	2021	-41.626	-1.117
Mexico	2021	-2.544	23.581
Norway	2021	-24.653	21.189
New Zealand	2021	-0.674	54.025
OECD	2021	-3.567	37.068
Philippines	2021	-27.508	-4.892
Russian Federation	2021	-0.671	83.702
Turkey	2021	-12.376	7.129
Ukraine	2021	-0.676	15.182
USA	2021	21.111	19.163
Vietnam	2021	-1.132	-1143.956
South Africa	2021	-1.549	15.355

Table 17. General services support estimate (% of total agricultural support), for 2018 and 2019

Location	Year	Value	Growth rates (%)
Australia	2018	56.315	
Brazil	2018	36.590	
Canada	2018	30.882	
Switzerland	2018	11.332	
Chile	2018	54.484	
China	2018	16.163	
Colombia	2018	12.543	
Costa Rica	2018	26.216	
EU	2018	10.494	
United Kingdom	2018	10.517	
Indonesia	2018	3.989	
Iceland	2018	4.398	
Israel	2018	19.480	

Continued on following page

Table 17. Continued

Location	Year	Value	Growth rates (%)
Japan	2018	19.857	
Kazakhstan	2018	24.461	
Korea, Rep.	2018	14.827	
Mexico	2018	7.280	
Norway	2018	4.754	
New Zealand	2018	81.085	
OECD	2018	13.616	
Philippines	2018	17.909	
Russian Federation	2018	17.170	
Turkey	2018	20.767	
Ukraine	2018	24.865	
USA	2018	11.072	
South Africa	2018	24.460	
Australia	2019	41.680	-25.988
Brazil	2019	37.469	2.402
Canada	2019	27.206	-11.903
Switzerland	2019	11.003	-2.903
Chile	2019	55.150	1.222
China	2019	12.900	-20.188
Colombia	2019	10.731	-14.446
Costa Rica	2019	18.881	-27.979
EU	2019	10.816	3.068
United Kingdom	2019	12.123	15.271
Indonesia	2019	7.434	86.362
Iceland	2019	4.432	0.773
Israel	2019	18.546	-4.795
Japan	2019	21.841	9.991
Kazakhstan	2019	14.062	-42.513
Korea, Rep.	2019	16.685	12.531
Mexico	2019	6.751	-7.266
Norway	2019	4.924	3.576
New Zealand	2019	78.334	-3.393
OECD	2019	13.295	-2.358
Philippines	2019	15.794	-11.810
Russian Federation	2019	19.652	14.455
Turkey	2019	10.009	-51.803
Ukraine	2019	17.897	-28.023
USA	2019	10.562	-4.606
South Africa	2019	34.046	39.191

Table 18. General services support estimate (% of total agricultural support), for 2020 and 2021

Location	Year	Value	Growth rates (%)
Australia	2020	46.638	11.895
Brazil	2020	31.733	-15.309
Canada	2020	27.734	1.941
Switzerland	2020	10.144	-7.807
Chile	2020	55.696	0.990
China	2020	12.359	-4.194
Colombia	2020	9.846	-8.247
Costa Rica	2020	18.216	-3.522
EU	2020	12.317	13.878
United Kingdom	2020	11.759	-3.003
Indonesia	2020	8.328	12.026
Iceland	2020	4.384	-1.083
Israel	2020	17.725	-4.427
Japan	2020	23.791	8.928
Kazakhstan	2020	18.061	28.438
Korea, Rep.	2020	12.937	-22.463
Mexico	2020	8.714	29.077
Norway	2020	4.881	-0.873
New Zealand	2020	64.921	-17.123
OECD	2020	13.214	-0.609
Philippines	2020	16.978	7.497
Russian Federation	2020	21.498	9.393
Turkey	2020	6.107	-38.985
Ukraine	2020	31.317	74.985
USA	2020	8.818	-16.512
South Africa	2020	35.201	3.392
Australia	2021	46.638	0.000
Brazil	2021	15.994	-49.598
Canada	2021	17.507	-36.875
Switzerland	2021	11.280	11.199
Chile	2021	53.396	-4.130
China	2021	11.577	-6.327
Colombia	2021	18.058	83.404
Costa Rica	2021	26.724	46.706
EU	2021	12.850	4.327
United Kingdom	2021	7.939	-32.486
Indonesia	2021	7.022	-15.682
Iceland	2021	3.875	-11.610

Continued on following page

Table 18. Continued

Location	Year	Value	Growth rates (%)
Israel	2021	15.442	-12.880
Japan	2021	20.623	-13.316
Kazakhstan	2021	13.389	-25.868
Korea, Rep.	2021	13.318	2.945
Mexico	2021	8.775	0.700
Norway	2021	5.590	14.526
New Zealand	2021	74.339	14.507
OECD	2021	12.383	-6.289
Philippines	2021	14.180	-16.480
Russian Federation	2021	27.556	28.179
Turkey	2021	15.603	155.494
Ukraine	2021	31.219	-0.313
USA	2021	8.234	-6.623
South Africa	2021	35.371	0.483

Table 19. Producer protection (total, ratio), for 2018 and 2019

Location	Year	Value	Growth rates (%)
Argentina	2018	0.788	
Australia	2018	1.000	
Brazil	2018	1.003	
Canada	2018	1.043	
Switzerland	2018	1.434	
Chile	2018	1.001	
China	2018	1.108	
Colombia	2018	1.145	
Costa Rica	2018	1.033	
EU	2018	1.050	
United Kingdom	2018	1.074	
Indonesia	2018	1.262	
India	2018	0.873	
Iceland	2018	2.000	
Israel	2018	1.097	
Japan	2018	1.621	
Kazakhstan	2018	0.970	
Korea, Rep.	2018	1.742	
Mexico	2018	1.062	

Continued on following page

Table 19. Continued

Location	Year	Value	Growth rates (%)
Norway	2018	1.753	
New Zealand	2018	1.004	
OECD	2018	1.100	
Philippines	2018	1.364	
Russian Federation	2018	1.069	
Turkey	2018	1.116	
Ukraine	2018	1.011	
USA	2018	1.054	
Vietnam	2018	0.901	
South Africa	2018	1.038	
Argentina	2019	0.768	-2.538
Australia	2019	1.000	0.000
Brazil	2019	1.000	-0.299
Canada	2019	1.046	0.288
Switzerland	2019	1.429	-0.349
Chile	2019	1.001	0.000
China	2019	1.114	0.542
Colombia	2019	1.128	-1.485
Costa Rica	2019	1.051	1.742
EU	2019	1.048	-0.190
United Kingdom	2019	1.065	-0.838
Indonesia	2019	1.252	-0.792
India	2019	0.876	0.344
Iceland	2019	1.879	-6.050
Israel	2019	1.117	1.823
Japan	2019	1.635	0.864
Kazakhstan	2019	1.010	4.124
Korea, Rep.	2019	1.659	-4.765
Mexico	2019	1.065	0.282
Norway	2019	1.684	-3.936
New Zealand	2019	1.006	0.199
OECD	2019	1.092	-0.727
Philippines	2019	1.356	-0.587

Continued on following page

Table 19. Continued

Location	Year	Value	Growth rates (%)
Russian Federation	2019	1.051	-1.684
Turkey	2019	1.146	2.688
Ukraine	2019	1.020	0.890
USA	2019	1.029	-2.372
Vietnam	2019	0.910	0.999
South Africa	2019	1.021	-1.638

Table 20. Producer protection (total, ratio), for 2020 and 2021

Location	Year	Value	Growth rates (%)
Argentina	2020	0.899	17.057
Australia	2020	1.000	0.000
Brazil	2020	1.007	0.700
Canada	2020	1.051	0.478
Switzerland	2020	1.595	11.617
Chile	2020	1.001	0.000
China	2020	1.146	2.873
Colombia	2020	1.104	-2.128
Costa Rica	2020	1.055	0.381
EU	2020	1.039	-0.859
United Kingdom	2020	1.052	-1.221
Indonesia	2020	1.178	-5.911
India	2020	0.828	-5.479
Iceland	2020	1.909	1.597
Israel	2020	1.156	3.491
Japan	2020	1.663	1.713
Kazakhstan	2020	0.994	-1.584
Korea, Rep.	2020	1.732	4.400
Mexico	2020	1.057	-0.751
Norway	2020	1.607	-4.572
New Zealand	2020	1.008	0.199
OECD	2020	1.088	-0.366
Philippines	2020	1.393	2.729
Russian Federation	2020	1.026	-2.379
Turkey	2020	1.160	1.222
Ukraine	2020	1.005	-1.471
USA	2020	1.023	-0.583

Continued on following page

Table 20. Continued

Location	Year	Value	Growth rates (%)
Vietnam	2020	0.931	2.308
South Africa	2020	1.020	-0.098
Argentina	2021	0.828	-7.898
Australia	2021	1.000	0.000
Brazil	2021	1.039	3.178
Canada	2021	1.055	0.381
Switzerland	2021	1.451	-9.028
Chile	2021	1.004	0.300
China	2021	1.171	2.182
Colombia	2021	1.072	-2.899
Costa Rica	2021	1.034	-1.991
EU	2021	1.031	-0.770
United Kingdom	2021	1.126	7.034
Indonesia	2021	1.165	-1.104
India	2021	0.825	-0.362
Iceland	2021	1.957	2.514
Israel	2021	1.151	-0.433
Japan	2021	1.515	-8.900
Kazakhstan	2021	0.993	-0.101
Korea, Rep.	2021	1.747	0.866
Mexico	2021	1.033	-2.271
Norway	2021	1.455	-9.459
New Zealand	2021	1.004	-0.397
OECD	2021	1.078	-0.919
Philippines	2021	1.440	3.374
Russian Federation	2021	0.992	-3.314
Turkey	2021	1.133	-2.328
Ukraine	2021	1.005	0.000
USA	2021	1.040	1.662
Vietnam	2021	0.942	1.182
South Africa	2021	1.017	-0.294

CONCLUSION

This chapter proposed to analyse the consequences from, namely, the Covid-19 pandemic on agricultural support worldwide, considering data from the OECD database for the years 2018-2021. The indicators considered were the following: TSE, total support estimate; PSE, producer support estimate; CSE, consumer support estimate; GSSE, general services support estimate. For a deeper analysis, growth rates (%) were calculated for each one of these variables.

From the literature survey, it is worth highlighting that the pandemic opened opportunities for the policymakers, for example, to rethink the agricultural systems and the food chains. In fact, the new realities require changes in the current paradigms, particularly to deal with the usual constraints, including the institutional ones. One of the most relevant challenges for the several stakeholders is to avoid setbacks in the advances achieved. In any case, designing policy measures for a great diversity of realities, such as what happens in the European Union, creates additional difficulties in times of crisis. The design process of the policies is crucial for effective implementation and here the availability of information is fundamental, as well as the scientific research and the involvement of the farmers. The current frameworks promoted economic support in some cases (OECD countries) and market strategies in other circumstances (emerging countries). In the future, it will be important to consider together the health and social dimensions, as well as the demand and supply market sides.

The findings obtained from the data assessment show some signs of decreases in agricultural support (million US dollars) in 2021 for the OECD countries. Maybe, this is an impact of the pandemic with some delay. Nonetheless, this context deserves special attention in future research to better understand the cause behind these decreases in times of crisis. For the data disaggregated by country, the average results reveal that the TSE (millions of US dollars) increase by 10% in 2020 and 4% in 2021. Despite the influence in these findings of extreme values verified in some countries, 2021 has lower growth rate, compared to 2020, confirming, in some way, the conclusions presented before about the trends for the year 2021. The growth rates for the TSE in the percentage of the GDP present a similar tendency for 2021.

In terms of practical implications, this contribution highlights that the new context created by the pandemic needs deeper assessments and more adjusted policy instruments to deal with the new realities and paradigms. For policy recommendations, it is suggested to address in the process of policy design other sustainability dimensions and the demand and supply market sides. For future research, it could be interesting to understand the causes behind the reductions in the levels/growth rates of agricultural support for the year 2021.

ACKNOWLEDGMENT

This work is funded by National Funds through the FCT - Foundation for Science and Technology, I.P., within the scope of the project Refª UIDB/00681/2020. Furthermore we would like to thank the CERNAS Research Centre and the Polytechnic Institute of Viseu for their support.

REFERENCES

Adhikari, J., Timsina, J., Khadka, S. R., Ghale, Y., & Ojha, H. (2021). COVID-19 impacts on agriculture and food systems in Nepal: Implications for SDGs. *Agricultural Systems, 186*. Scopus. doi:10.1016/j.agsy.2020.102990

Alfaro-Navarro, J.-L., & Andrés-Martínez, M.-E. (2021). A longitudinal and cross-sectional analysis of the distribution of Common Agricultural Policy aids in European countries. *Agricultural Economics (Czech Republic), 67*(9), 351–362. Scopus. doi:10.17221/87/2021-AGRICECON

Ali, A., & Imran, M. (2021). National spatial data infrastructure vs. Cadastre system for economic development: Evidence from Pakistan. *Land, 10*(2), 1–17. Scopus. doi:10.3390/land10020188

Ali, M. S., Siddik, M. S., Nur-E-Alam, S. M., Khanam, T., & Khatun, M. M. (2020). Analysis of the problems and minimization initiatives addressed by bangladeshi farmers during the covid-19 pandemic. *Asian Journal of Agriculture and Rural Development, 10*(4), 773–781. Scopus. doi:10.18488/journal.ajard.2020.104.773.781

Altieri, M. A., & Nicholls, C. I. (2020). Agroecology: Challenges and opportunities for farming in the Anthropocene. *International Journal of Agriculture and Natural Resources, 47*(3), 204–215. Scopus. doi:10.7764/ijanr.v47i3.2281

Anderson, K. (2022). Agriculture in a more uncertain global trade environment. *Agricultural Economics (United Kingdom), 53*(4), 563–579. Scopus. doi:10.1111/agec.12726

Baležentis, T., Morkūnas, M., Žičkienė, A., Volkov, A., Ribašauskienė, E., & Štreimikienė, D. (2021). Policies for rapid mitigation of the crisis' effects on agricultural supply chains: A multi-criteria decision support system with monte carlo simulation. *Sustainability (Switzerland), 13*(21). Scopus. doi:10.3390/su132111899

Belasco, E. J., & Smith, V. (2022). The impact of policy design on payment concentration in Ad-hoc disaster Relief: Lessons from the Market Facilitation and Coronavirus food Assistance programs. *Food Policy, 106*. Scopus. doi:10.1016/j.foodpol.2021.102189

Bezáková, M., & Bezák, P. (2022). Which sustainability objectives are difficult to achieve? The mid-term evaluation of predicted scenarios in remote mountain agricultural landscapes in Slovakia. *Land Use Policy, 115*. Scopus. doi:10.1016/j.landusepol.2022.106020

Bisht, I. S., Rana, J. C., Yadav, R., & Ahlawat, S. P. (2020). Mainstreaming agricultural biodiversity in traditional production landscapes for sustainable development: The Indian scenario. *Sustainability (Switzerland), 12*(24), 1–25. Scopus. doi:10.3390/su122410690

Bitsoumanou Nkounkou, J., & Temple, L. (2021). Resilience of the food sector in the face of the Covid-19 crisis and prospects for agricultural policies in the Republic of Congo. *Cahiers Agricultures, 30*. Scopus. doi:10.1051/cagri/2021025

Cao, L., Li, T., Wang, R., & Zhu, J. (2021). Impact of COVID-19 on China's agricultural trade. *China Agricultural Economic Review, 13*(1), 1–21. Scopus. doi:10.1108/CAER-05-2020-0079

Cheng, Y., Nadeem, M., Haq, S. U., Prisca, K., Aziz, B., Imran, M., & Shahbaz, P. (2022). Maintaining Quality of Life during the Pandemic: Managing Economic, Social, and Health Well-Being Amid the COVID-19 Crisis of Agricultural Entrepreneurs. *Sustainability (Switzerland), 14*(23). Scopus. doi:10.3390/su142315597

Détang-Dessendre, C., Guyomard, H., Réquillart, V., & Soler, L.-G. (2020). Changing agricultural systems and food diets to prevent and mitigate global health shocks. *Sustainability (Switzerland), 12*(16). Scopus. doi:10.3390/su12166462

Espinoza Lastra, Ó. R., Lluglla Luna, L. A., & Padilla Buñay, A. (2021). Post-covid-19 rural entrepreneurship: Considerations for a sustainable economic recovery. [Scopus.]. *Universidad y Sociedad, 13*(S1), 443–451.

Ewert, B., & Loer, K. (2022). *COVID-19 as a Catalyst for Policy Change: The Role of Trigger Points and Spillover Effects. German Politics.* Scopus., doi:10.1080/09644008.2022.2088733

Gascón, J. (2020). COVID-19, state of emergency and family farming in Spain: Rural markets in Barcelona and self-consumption orchards in Alcaine (Teruel). *Ager, 2020*(30), 177–206. Scopus. doi:10.4422/ager.2020.13

Gruère, G., & Brooks, J. (2021). Viewpoint: Characterising early agricultural and food policy responses to the outbreak of COVID-19. *Food Policy, 100*. Scopus. doi:10.1016/j.foodpol.2020.102017

Gurbuz, I. B., & Ozkan, G. (2021). Will agriculture beat the odds against covid-19? The covid-19 outbreak and its effect on agricultural supply in Turkey. *New Medit, 20*(2), 15–25. Scopus. doi:10.30682/nm2102b

Harris, J., Depenbusch, L., Pal, A. A., Nair, R. M., & Ramasamy, S. (2020). Food system disruption: Initial livelihood and dietary effects of COVID-19 on vegetable producers in India. *Food Security, 12*(4), 841–851. Scopus. doi:10.1007/s12571-020-01064-5

Hoanga, H. T. K., Thompson, W., & Kwon, S. (2021). Low-income household food consumption consequences of rice policy and pandemic impacts on income and price in Thailand. [Scopus.]. *Journal of Food Distribution Research, 52*(2), 76–100.

Jiménez, D. E., & Saldarriaga-Isaza, A. (2022). Economic impacts of agricultural policy responses to the outbreak of COVID-19. *Estudios Gerenciales, 38*(163), 200–210. Scopus. doi:10.18046/j.estger.2022.163.4924

Jouili, M., & Elloumi, M. (2022). Extraversion versus développement agricole autocentré: Le cas des pays du Maghreb. *Canadian Journal of Development Studies*. Scopus. doi:10.1080/02255189.2022.2092459

Karagöl, E. T., Akyüz, M., & Karul, C. (2021). *Is health management of the COVID-19 pandemic a cause of agricultural commodity prices? New evidences from bootstrap fourier causality test. Duzce Medical Journal, 23(Special Issue 1), 86–93.* Scopus., doi:10.18678/dtfd.879572

Ker, A. P. (2020). Risk management in Canada's agricultural sector in light of COVID-19. *Canadian Journal of Agricultural Economics, 68*(2), 251–258. Scopus. doi:10.1111/cjag.12232

Kondratieva, N. (2021). *EU agricultural digitalization decalogue. Istoriya, 12, 16–23.* Scopus., doi:10.18254/S207987840015224-5

Kumar, P., Singh, S. S., Pandey, A. K., Singh, R. K., Srivastava, P. K., Kumar, M., Dubey, S. K., Sah, U., Nandan, R., Singh, S. K., Agrawal, P., Kushwaha, A., Rani, M., Biswas, J. K., & Drews, M. (2021). Multi-level impacts of the COVID-19 lockdown on agricultural systems in India: The case of Uttar Pradesh. *Agricultural Systems, 187*. Scopus. doi:10.1016/j.agsy.2020.103027

Kurczewska, U. (2021). European solidarity and its limitations during the covid-19 pandemic. [Scopus.]. *European Journal of Science and Theology, 17*(5), 1–13.

Leakey, R. R. B., Mabhaudhi, T., & Gurib-Fakim, A. (2021). African lives matter: Wild food plants matter for livelihoods, justice, and the environment—A policy brief for agricultural reform and new crops. *Sustainability (Switzerland), 13*(13). Scopus. doi:10.3390/su13137252

Munonye, J., Osuji, E., Olaolu, M., Okoisu, A., Obi, J., Eze, G., Ibrahim-Olesin, S., Njoku, L., Amadi, M., Izuogu, C., & Azuamairo, G. (2022). Perceived Effects of COVID-19 Pandemic on Food Security in Southeast Nigeria. *Frontiers in Sustainable Food Systems, 6*. Scopus. doi:10.3389/fsufs.2022.936157

Neganova, V. P., & Chistyakov, Yu. F. (2020). Development of agricultural foreign trade in the post-soviet countries. *Economy of Region, 16*(2), 597–611. Scopus. doi:10.17059/2020-2-20

OECD. (2023). *Agriculture—OECD Data*. TheOECD. https://data.oecd.org/agriculture.htm

Patyka, N., Khodakivska, O., Mohylnyi, O., & Pugachov, M. (2021). Ukraine's Agrarian Sector in the Conditions of COVID-19 Distribution and Restrictive Quarantine Measures: Methodological Principles of Empirical Evaluation. *Scientific Horizons, 24*(12), 55–69. Scopus. doi:10.48077/scihor.24(12).2021.55-69

Popescu, G. C., & Popescu, M. (2022). COVID-19 pandemic and agriculture in Romania: Effects on agricultural systems, compliance with restrictions and relations with authorities. *Food Security, 14*(2), 557–567. Scopus. doi:10.1007/s12571-021-01239-8

Sama-Berrocal, C., & Martínez-Azúa, B. C. (2022). How Has the COVID-19 Pandemic Affected the Different Branches of the Agri-Food Industry in Extremadura (Spain)? *Land, 11*(6). Scopus. doi:10.3390/land11060938

Sánchez, M. V., Cicowiez, M., & Ortega, A. (2022). Prioritizing public investment in agriculture for post-COVID-19 recovery: A sectoral ranking for Mexico. *Food Policy, 109*. Scopus. doi:10.1016/j.foodpol.2022.102251

Scopus. (2023). *Scopus Database*. Scopus. https://www.scopus.com/search/form.uri?display=basic#basic

Scown, M. W., Brady, M. V., & Nicholas, K. A. (2020). Billions in Misspent EU Agricultural Subsidies Could Support the Sustainable Development Goals. *One Earth, 3*(2), 237–250. Scopus. doi:10.1016/j.oneear.2020.07.011

Srairi, M. T. (2021). Rethinking the agricultural development model in Morocco in the post Covid-19 era. *Cahiers Agricultures, 30*(6). Scopus. doi:10.1051/cagri/2021003

Stata. (2023). *Statistical software for data science*. Stata. https://www.stata.com/

StataCorp. (2017a). *Stata 15 Base Reference Manual*. Stata Press.

StataCorp. (2017b). *Stata Statistical Software: Release 15*. StataCorp LLC.

Varshney, D., Kumar, A., Mishra, A. K., Rashid, S., & Joshi, P. K. (2021). COVID-19, Government Transfer Payments, and Investment Decisions in Farming Business: Evidence from Northern India. *Applied Economic Perspectives and Policy, 43*(1), 248–269. Scopus. doi:10.1002/aepp.13144

Vik, J. (2020). The agricultural policy trilemma: On the wicked nature of agricultural policy making. *Land Use Policy, 99*. Scopus. doi:10.1016/j.landusepol.2020.105059

Wang, Y., Peng, D., Yu, L., Zhang, Y., Yin, J., Zhou, L., Zheng, S., Wang, F., & Li, C. (2020). Monitoring Crop Growth during the Period of the Rapid Spread of COVID-19 in China by Remote Sensing. *IEEE Journal of Selected Topics in Applied Earth Observations and Remote Sensing, 13*, 6195–6205. Scopus. doi:10.1109/JSTARS.2020.3029434

Yovchevska, P. (2021). Bulgarian agriculture in the focus of representative statistical surveys. [Scopus.]. *Bulgarian Journal of Agricultural Science*, *27*(5), 859–864.

Zysk, E., Mroczkowski, T., & Dawidowicz, A. (2022). Small Farms as "Data Producers" for the Needs of Agricultural Management Information System. *Geomatics and Environmental Engineering, 16*(3), 79–109. Scopus. doi:10.7494/geom.2022.16.3.79

Chapter 6
The Reality of Sustainable Agriculture in Recent Years

ABSTRACT

One of the greatest current challenges for the farming sector is to find a compromise among agricultural production to deal with the increased demand worldwide and the environmental impacts for more sustainable development. This is, in fact, a concern for the policymakers and international organisations and led FAO (Food and Agriculture Organization of the United Nations) to propose the CSA (climate-smart agriculture) approach. This chapter intends to analyse the impacts of recent events, namely the Covid-19 pandemic, on the sustainability of agriculture worldwide. Data from OECD (Organisation for Economic Co-operation and Development) were considered for the agricultural land and nutrient balance over the period 2019-2021. There is a long way to run to improve the sustainability of agriculture worldwide and the expectation about the contributions of the pandemic to open opportunities to mitigate the environmental impacts from the farming sector are not yet visible.

INTRODUCTION

The recent implications from international exogenous disturbances spread worldwide (Ali et al., 2020), over diverse markets (Dokić et al., 2020), household groups (Lian et al., 2023) and human security (Tansuchat et al., 2022). Bringing difficulties for the whole human dimensions, including research activities (Spies et al., 2021). They spread over the different dimensions of sustainability with negative externalities, nonetheless opened/reinforced new opportunities to improve frameworks of more sustainable development, including in agricultural practices (Adnan & Nordin, 2021).

These instabilities also (re)open the discussion, namely in the more vulnerable countries (Yang & Liu, 2022), about the need to promote more sustainable agriculture (Al-Saidi et al., 2022). This debate highlights the interrelationship of farming sustainability with the following domains: food sovereignty (Azenzem, 2022); food security (Wudil et al., 2022); rural revitalisation (Khan et al., 2022); financial risk management (Polukhin & Panarina, 2022); urban agriculture (Sridhar et al., 2022); safety of agricultural products (Teng et al., 2022); energy production (Vo, 2020); and agroecological transitions (Nelles, 2023).

DOI: 10.4018/978-1-6684-8923-9.ch006

In addition, it shows the need for more resilient (Hafez & Attia, 2020) and stable (Li et al., 2023) chains. Alternative productions and supplementary food have been identified in the current times (Kumar et al., 2021). These discussions are relevant to better address the post-pandemic scenario (Martey et al., 2022). Nonetheless, in some circumstances, there is evidence of no significant changes during the pandemic (Gómez Serna & Bernal Rivas, 2020). In other cases, there are signs of mixed effects (Satama et al., 2022).

The new technologies and the digital transition, within the context of Agriculture 4.0, may bring important contributions to more sustainable development in the agricultural sector (Qin et al., 2022), including unmanned aerial vehicles (Ahmad et al., 2020) and smart approaches for food security (Rasul, 2021). The same happens, in some conditions, about the added value for agriculture from nanotechnology, adjusted agronomic practices (Koç & Karayiğit, 2022) and more sustainable decisions (Pathirana & Carimi, 2022).

The diversification of agricultural incomes and the interlinkage of farming activities with other sectors (tourism and small industry) are referred often to as solutions to improve the sustainability of regions and villages (Auliah et al., 2022). The programmes of education, namely from the higher education institutions play here a fundamental role (Baptista et al., 2021), as well as the insights from the scientific community (Ikehi et al., 2022), the agricultural policies (Ling et al., 2023), the public regulations (Manian et al., 2022), infrastructure facilities (Paria et al., 2022) and extension services (Pampori, 2021).

Sustainability in agriculture is interrelated with many factors (from supply and demand sides) and some of which may constrain the strategies designed to promote more eco-friendly practices (Arroyo-Lambaer et al., 2022). From the demand side, consumers are currently informed about the benefits of healthy food (Guiné et al., 2022) and this influences the agricultural sustainability framework, namely when the customers prefer organic products (Zámková et al., 2022). In these contexts, adjusted water management, for example, is crucial (Barbosa, 2022).

This research proposes to bring more insights into sustainable agriculture worldwide, assessing statistical information from the OECD (2023) database for the period 2019-2021. Some of these data were analysed through the Stata (StataCorp, 2017a, 2017b; Stata, 2023) software. The Scopus (2023) database was taken into account for the literature analysis. For deeper analysis, normalised values ($(x_i - x_{minimum})/(x_{maximum} - x_{minimum})$), growth rates and logarithms results were obtained.

AGRICULTURAL LAND

The total agricultural land (thousand hectares) in 2019 has the greatest values in the following countries (from the list presented in table 1): China; USA; Australia; Brazil; Russian Federation; Kazakhstan; European Union; India; Argentina; South Africa; Mexico; Canada; Indonesia; Colombia; Ukraine; Turkey; France; Spain; and the United Kingdom. For the data available in the database used, the total agricultural land (1000 hectares) decreased, on average, -0.5% (standard deviation of 1.8%) in 2020 and -0.6% (standard deviation of 1.3%) in 2021 (tables 2 and 3).

Table 1. Total agricultural land (thousand hectares), for 2019

Location	Value	Logarithm	Growth rates (%)
Argentina	108382	11.593	
Australia	383801	12.858	
Austria	2652	7.883	
Belgium	1359	7.215	
Bulgaria	5037	8.525	
Brazil	236879	12.375	
Canada	63335	11.056	
Switzerland	1512	7.321	
Chile	15671	9.660	
China	528509	13.178	
Colombia	49609	10.812	
Costa Rica	1776	7.482	
Cyprus	125	4.828	
Czech Republic	3524	8.167	
Germany	16666	9.721	
Denmark	2626	7.873	
Spain	24372	10.101	
Estonia	988	6.896	
EU27_2020	162926	12.001	
EU28	180455	12.103	
Finland	2274	7.729	
France	29024	10.276	
United Kingdom	17529	9.772	
Greece	5153	8.547	
Croatia	1504	7.316	
Hungary	5310	8.577	
Indonesia	62300	11.040	
India	179045	12.095	
Ireland	4524	8.417	
Iceland	1591	7.372	
Israel	289	5.666	
Italy	13150	9.484	
Japan	4397	8.389	
Kazakhstan	213974	12.274	
Korea, Rep.	1676	7.424	
Lithuania	2975	7.998	
Luxembourg	132	4.883	
Latvia	1959	7.580	

Continued on following page

Table 1. Continued

Location	Value	Logarithm	Growth rates (%)
Mexico	96106	11.473	
Malta	12	2.485	
Netherlands	1816	7.504	
Norway	982	6.890	
New Zealand	10428	9.252	
Philippines	12643	9.445	
Poland	14550	9.585	
Portugal	3967	8.286	
Romania	13826	9.534	
Russian Federation	215494	12.281	
Slovak Republic	1916	7.558	
Slovenia	480	6.174	
Sweden	3005	8.008	
Turkey	37797	10.540	
Ukraine	41892	10.643	
USA	418375	12.944	
Vietnam	12388	9.424	
South Africa	96341	11.476	

Table 2. Total agricultural land (thousand hectares), for 2020

Location	Value	Logarithm	Growth rates (%)
Argentina	108382	11.593	0.000
Australia	377002	12.840	-1.771
Austria	2647	7.881	-0.189
Belgium	1367	7.220	0.589
Bulgaria	5047	8.527	0.199
Brazil	236879	12.375	0.000
Canada	62950	11.050	-0.608
Switzerland	1510	7.320	-0.132
Chile	15710	9.662	0.249
China	528508	13.178	0.000
Colombia	48243	10.784	-2.754
Costa Rica	1762	7.474	-0.788
Cyprus	128	4.852	2.400
Czech Republic	3524	8.167	0.000
Germany	16595	9.717	-0.426
Denmark	2620	7.871	-0.228
Spain	24435	10.104	0.258
Estonia	985	6.893	-0.304
EU27_2020	161964	11.995	-0.590
Finland	2270	7.728	-0.176
France	28898	10.272	-0.434
Greece	5268	8.569	2.232
Croatia	1506	7.317	0.133
Hungary	4998	8.517	-5.876
Indonesia	62300	11.040	0.000
India	179045	12.095	0.000
Ireland	4511	8.414	-0.287
Iceland	1591	7.372	0.000
Italy	13122	9.482	-0.213
Japan	4372	8.383	-0.569
Kazakhstan	214003	12.274	0.014
Korea, Rep.	1656	7.412	-1.193
Lithuania	2943	7.987	-1.076
Luxembourg	132	4.883	0.000
Latvia	1969	7.585	0.510
Mexico	97138	11.484	1.074
Malta	11	2.398	-8.333
Netherlands	1814	7.503	-0.110

Continued on following page

Table 2. Continued

Location	Value	Logarithm	Growth rates (%)
Norway	986	6.894	0.407
New Zealand	10234	9.233	-1.860
Philippines	12675	9.447	0.253
Poland	14755	9.599	1.409
Portugal	3970	8.287	0.076
Romania	13049	9.476	-5.620
Russian Federation	215494	12.281	0.000
Slovak Republic	1910	7.555	-0.313
Slovenia	484	6.182	0.833
Sweden	3006	8.008	0.033
Turkey	37762	10.539	-0.093
Ukraine	41892	10.643	0.000
Vietnam	12360	9.422	-0.226
South Africa	96341	11.476	0.000

Table 3. Total agricultural land (thousand hectares), for 2021

Location	Value	Logarithm	Growth rates (%)
Austria	2602	7.864	-1.700
Belgium	1368	7.221	0.073
Bulgaria	5047	8.527	0.000
Canada	62180	11.038	-1.223
Cyprus	123	4.812	-3.906
Czech Republic	3530	8.169	0.170
Germany	16592	9.717	-0.018
Denmark	2618	7.870	-0.076
Spain	24420	10.103	-0.061
Estonia	987	6.895	0.203
Finland	2268	7.727	-0.088
Greece	5137	8.544	-2.487
Croatia	1476	7.297	-1.992
Hungary	5049	8.527	1.020
Ireland	4337	8.375	-3.857
Iceland	1591	7.372	0.000
Italy	12987	9.472	-1.029
Japan	4349	8.378	-0.526
Lithuania	2938	7.985	-0.170
Luxembourg	133	4.890	0.758
Latvia	1970	7.586	0.051
Malta	11	2.398	0.000
Netherlands	1812	7.502	-0.110
New Zealand	10175	9.228	-0.577
Poland	14522	9.583	-1.579
Portugal	3980	8.289	0.252
Romania	13079	9.479	0.230
Slovak Republic	1856	7.526	-2.827
Slovenia	479	6.172	-1.033
Sweden	3003	8.007	-0.100
Turkey	38089	10.548	0.866

The arable and permanent cropland (1000 hectares) had the highest values in 2019 in India, the USA, China, Russian Federation, European Union, Australia, Brazil, Indonesia, Canada, Ukraine, Argentina, Kazakhstan, Turkey, Mexico, France, Spain, South Africa, Germany, Vietnam and Poland (table 4).

On average and for the information available in the database, this land also decreased by -0.4% (standard deviation of 2.7%) in 2020 and -0.6% (standard deviation of 1.9%) in 2022 (tables 5 and 6).

Table 4. Arable and permanent cropland (thousand hectares), for 2019

Location	Value	Logarithm	Growth rates (%)
Argentina	33701	10.425	
Australia	64350	11.072	
Austria	1393	7.239	
Belgium	883	6.783	
Bulgaria	3629	8.197	
Brazil	63518	11.059	
Canada	46177	10.740	
Switzerland	429	6.061	
Chile	1656	7.412	
China	135675	11.818	
Colombia	9872	9.197	
Costa Rica	576	6.356	
Cyprus	124	4.820	
Czech Republic	2532	7.837	
Germany	11915	9.386	
Denmark	2419	7.791	
Spain	17115	9.748	
Estonia	699	6.550	
EU27_2020	111585	11.623	
EU28	117713	11.676	
Finland	2250	7.719	
France	19414	9.874	
United Kingdom	6128	8.721	
Greece	3022	8.014	
Croatia	898	6.800	
Hungary	4519	8.416	
Indonesia	51300	10.845	
India	168669	12.036	
Ireland	444	6.096	
Iceland	123	4.812	
Italy	9358	9.144	
Japan	3800	8.243	
Kazakhstan	29510	10.292	
Korea, Rep.	1643	7.404	
Lithuania	2248	7.718	
Luxembourg	64	4.159	
Latvia	1328	7.191	
Mexico	22129	10.005	

Continued on following page

Table 4. Continued

Location	Value	Logarithm	Growth rates (%)
Malta	12	2.485	
Netherlands	1049	6.956	
Norway	803	6.688	
New Zealand	703	6.555	
Philippines	11143	9.319	
Poland	11422	9.343	
Portugal	1876	7.537	
Romania	9378	9.146	
Russian Federation	123442	11.724	
Slovak Republic	1397	7.242	
Slovenia	202	5.308	
Sweden	2544	7.841	
Turkey	23180	10.051	
Ukraine	33777	10.428	
USA	153108	11.939	
Vietnam	11746	9.371	
South Africa	12413	9.426	

Table 5. Arable and permanent cropland (thousand hectares), for 2020

Location	Value	Logarithm	Growth rates (%)
Argentina	33701	10.425	0.000
Australia	65286	11.087	1.455
Austria	1391	7.238	-0.144
Belgium	891	6.792	0.906
Bulgaria	3643	8.201	0.386
Brazil	63518	11.059	0.000
Canada	45325	10.722	-1.845
Switzerland	431	6.066	0.466
Chile	1695	7.435	2.355
China	135674	11.818	-0.001
Colombia	8739	9.076	-11.477
Costa Rica	562	6.332	-2.431
Cyprus	126	4.836	1.613
Czech Republic	2534	7.838	0.079
Germany	11865	9.381	-0.420
Denmark	2398	7.782	-0.868
Spain	17019	9.742	-0.561
Estonia	701	6.553	0.286
EU27_2020	110827	11.616	-0.679
Finland	2248	7.718	-0.089
France	19251	9.865	-0.840
Greece	3016	8.012	-0.199
Croatia	970	6.877	8.018
Hungary	4265	8.358	-5.621
Indonesia	51300	10.845	0.000
India	168669	12.036	0.000
Ireland	445	6.098	0.225
Iceland	123	4.812	0.000
Italy	9383	9.147	0.267
Japan	3777	8.237	-0.605
Kazakhstan	29685	10.298	0.593
Korea, Rep.	1624	7.393	-1.156
Lithuania	2286	7.735	1.690
Luxembourg	64	4.159	0.000
Latvia	1343	7.203	1.130
Mexico	22869	10.038	3.344
Malta	11	2.398	-8.333
Netherlands	1042	6.949	-0.667

Continued on following page

Table 5. Continued

Location	Value	Logarithm	Growth rates (%)
Norway	808	6.695	0.623
New Zealand	681	6.524	-3.129
Philippines	11175	9.321	0.287
Poland	11552	9.355	1.138
Portugal	1857	7.527	-1.013
Romania	8886	9.092	-5.246
Russian Federation	123442	11.724	0.000
Slovak Republic	1391	7.238	-0.429
Slovenia	204	5.318	0.990
Sweden	2542	7.841	-0.079
Turkey	23145	10.050	-0.151
Ukraine	33777	10.428	0.000
Vietnam	11718	9.369	-0.238
South Africa	12413	9.426	0.000

Table 6. Arable and permanent cropland (thousand hectares), for 2021

Location	Value	Logarithm	Growth rates (%)
Austria	1393	7.239	0.144
Belgium	892	6.793	0.112
Bulgaria	3650	8.202	0.192
Canada	43621	10.683	-3.760
Cyprus	121	4.796	-3.968
Czech Republic	2524	7.834	-0.395
Germany	11862	9.381	-0.025
Denmark	2384	7.777	-0.584
Spain	16924	9.736	-0.558
Estonia	706	6.560	0.713
Finland	2247	7.717	-0.044
Greece	2807	7.940	-6.930
Croatia	937	6.843	-3.402
Hungary	4295	8.365	0.703
Ireland	437	6.080	-1.798
Iceland	123	4.812	0.000
Italy	9358	9.144	-0.266
Japan	3755	8.231	-0.582
Lithuania	2316	7.748	1.312
Luxembourg	64	4.159	0.000
Latvia	1372	7.224	2.159
Malta	11	2.398	0.000
Netherlands	1041	6.948	-0.096
New Zealand	690	6.537	1.322
Poland	11481	9.348	-0.615
Portugal	1850	7.523	-0.377
Romania	8989	9.104	1.159
Slovak Republic	1344	7.203	-3.379
Slovenia	204	5.318	0.000
Sweden	2539	7.840	-0.118
Turkey	23472	10.064	1.413

For the permanent pasture (1000 hectares), in 2019, the countries with the biggest values are the following (table 7): China; Australia; USA; Kazakhstan; Brazil; Russian Federation; South Africa; Argentina; Mexico; European Union; Colombia; Canada; Turkey; Chile; and the United Kingdom.

The permanent pasture decreased, on average, -0.6% (3.1% for the standard deviation) and -0.5% (2.7% for the standard deviation), respectively in the years 2020 and 2021 (tables 8 and 9).

Table 7. Permanent pasture (thousand hectares), for 2019

Location	Value	Logarithm	Growth rates (%)
Argentina	74681	11.221	
Australia	319452	12.674	
Austria	1259	7.138	
Belgium	475	6.163	
Bulgaria	1408	7.250	
Brazil	173361	12.063	
Canada	17159	9.750	
Switzerland	1083	6.987	
Chile	14015	9.548	
China	392834	12.881	
Colombia	39737	10.590	
Costa Rica	1200	7.090	
Cyprus	2	0.693	
Czech Republic	992	6.900	
Germany	4751	8.466	
Denmark	207	5.333	
Spain	7256	8.890	
Estonia	289	5.666	
EU27_2020	50792	10.835	
EU28	62192	11.038	
Finland	24	3.178	
France	9611	9.171	
United Kingdom	11400	9.341	
Greece	2132	7.665	
Croatia	606	6.407	
Hungary	790	6.672	
Indonesia	11000	9.306	
India	10376	9.247	
Ireland	4080	8.314	
Iceland	1468	7.292	
Italy	3792	8.241	
Japan	597	6.392	

Continued on following page

Table 7. Continued

Location	Value	Logarithm	Growth rates (%)
Kazakhstan	184464	12.125	
Korea, Rep.	33	3.497	
Lithuania	727	6.589	
Luxembourg	68	4.220	
Latvia	632	6.449	
Mexico	73977	11.212	
Malta	0		
Netherlands	768	6.644	
Norway	179	5.187	
New Zealand	9725	9.182	
Philippines	1500	7.313	
Poland	3128	8.048	
Portugal	2091	7.645	
Romania	4448	8.400	
Russian Federation	92052	11.430	
Slovak Republic	518	6.250	
Slovenia	278	5.628	
Sweden	461	6.133	
Turkey	14617	9.590	
Ukraine	7534	8.927	
USA	265266	12.488	
Vietnam	642	6.465	
South Africa	83928	11.338	

Table 8. Arable and permanent cropland (thousand hectares), for 2020

Location	Value	Logarithm	Growth rates (%)
Argentina	74681	11.221	0.000
Australia	311716	12.650	-2.422
Austria	1256	7.136	-0.238
Belgium	476	6.165	0.211
Bulgaria	1404	7.247	-0.284
Brazil	173361	12.063	0.000
Canada	17626	9.777	2.722
Switzerland	1079	6.984	-0.369
Chile	14015	9.548	0.000
China	392834	12.881	0.000
Colombia	39504	10.584	-0.586
Costa Rica	1200	7.090	0.000
Cyprus	2	0.693	0.000
Czech Republic	990	6.898	-0.202
Germany	4730	8.462	-0.442
Denmark	222	5.403	7.246
Spain	7416	8.911	2.205
Estonia	285	5.652	-1.384
EU27_2020	50642	10.833	-0.295
Finland	22	3.091	-8.333
France	9647	9.174	0.375
Greece	2252	7.720	5.629
Croatia	537	6.286	-11.386
Hungary	733	6.597	-7.215
Indonesia	11000	9.306	0.000
India	10376	9.247	0.000
Ireland	4067	8.311	-0.319
Iceland	1468	7.292	0.000
Italy	3739	8.227	-1.398
Japan	595	6.389	-0.335
Kazakhstan	184318	12.124	-0.079
Korea, Rep.	33	3.497	0.000
Lithuania	657	6.488	-9.629

Continued on following page

Table 8. Continued

Location	Value	Logarithm	Growth rates (%)
Luxembourg	68	4.220	0.000
Latvia	626	6.439	-0.949
Mexico	74269	11.215	0.395
Malta	0		0.000
Netherlands	772	6.649	0.521
Norway	178	5.182	-0.559
New Zealand	9553	9.165	-1.769
Philippines	1500	7.313	0.000
Poland	3203	8.072	2.398
Portugal	2113	7.656	1.052
Romania	4163	8.334	-6.407
Russian Federation	92052	11.430	0.000
Slovak Republic	519	6.252	0.193
Slovenia	280	5.635	0.719
Sweden	464	6.140	0.651
Turkey	14617	9.590	0.000
Ukraine	7534	8.927	0.000
Vietnam	642	6.465	0.000
South Africa	83928	11.338	0.000

Table 9. Arable and permanent cropland (thousand hectares), for 2021

Location	Value	Logarithm	Growth rates (%)
Austria	1210	7.098	-3.662
Belgium	476	6.165	0.000
Bulgaria	1397	7.242	-0.499
Canada	18560	9.829	5.299
Cyprus	2	0.693	0.000
Czech Republic	1006	6.914	1.616
Germany	4730	8.462	0.000
Denmark	234	5.455	5.405
Spain	7496	8.922	1.079
Estonia	281	5.638	-1.404
Finland	21	3.045	-4.545
Greece	2330	7.754	3.464
Croatia	540	6.292	0.559
Hungary	754	6.625	2.865
Ireland	3901	8.269	-4.082
Iceland	1468	7.292	0.000
Italy	3629	8.197	-2.942
Japan	593	6.385	-0.336
Lithuania	622	6.433	-5.327
Luxembourg	69	4.234	1.471
Latvia	599	6.395	-4.313
Malta	0		0.000
Netherlands	771	6.648	-0.130
New Zealand	9485	9.157	-0.712
Poland	3041	8.020	-5.058
Portugal	2130	7.664	0.805
Romania	4090	8.316	-1.754
Slovak Republic	512	6.238	-1.349
Slovenia	276	5.620	-1.429
Sweden	464	6.140	0.000
Turkey	14617	9.590	0.000

NUTRIENT BALANCE

The nutrient balance for nitrogen (kg/ha) presents the higher values for 2019 in the following countries (table 10): Korea, Rep.; Netherlands; India; Vietnam; Czech Republic; New Zealand; Slovak Republic; Switzerland; Germany; and Croatia. The balance is negative for South Africa, Kazakhstan, Romania and Indonesia.

For phosphorus, the greatest values (table 11) appear in the Republic of Korea, Vietnam, India, Brazil, New Zealand, Turkey, China, Costa Rica, Philippines, Portugal, Colombia, Netherlands, Poland and Switzerland.

Table 10. Nutrient balance (Kg/ha), for nitrogen (2019)

Location	Value	Logarithm	Normalised value
Korea, Rep.	228	5.429	1.000
Netherlands	166	5.112	0.786
India	134	4.898	0.676
Vietnam	127	4.844	0.652
Czech Republic	76	4.331	0.476
New Zealand	63	4.143	0.431
Slovak Republic	63	4.143	0.431
Switzerland	59	4.078	0.417
Germany	53	3.970	0.397
Croatia	53	3.970	0.397
Philippines	48	3.871	0.379
Poland	47	3.850	0.376
Portugal	45	3.807	0.369
Finland	44	3.784	0.366
Slovenia	43	3.761	0.362
Lithuania	41	3.714	0.355
Costa Rica	41	3.714	0.355
Austria	34	3.526	0.331
Turkey	32	3.466	0.324
Mexico	31	3.434	0.321
USA	31	3.434	0.321
China	31	3.434	0.321
Brazil	21	3.045	0.286
Australia	19	2.944	0.279
Ukraine	15	2.708	0.266
Latvia	12	2.485	0.255
Iceland	7	1.946	0.238
Colombia	7	1.946	0.238
Russian Federation	7	1.946	0.238
Argentina	3	1.099	0.224
South Africa	-7		0.190
Kazakhstan	-14		0.166
Romania	-25		0.128
Indonesia	-62		0.000

Table 11. Nutrient balance (Kg/ha), for phosphorus (2019)

Location	Value	Logarithm	Normalised value
Korea, Rep.	44	3.784	1.000
Vietnam	32	3.466	0.760
India	28	3.332	0.680
Brazil	17	2.833	0.460
New Zealand	10	2.303	0.320
Turkey	9	2.197	0.300
China	9	2.197	0.300
Costa Rica	9	2.197	0.300
Philippines	8	2.079	0.280
Portugal	6	1.792	0.240
Colombia	6	1.792	0.240
Netherlands	4	1.386	0.200
Poland	4	1.386	0.200
Switzerland	4	1.386	0.200
Finland	3	1.099	0.180
USA	3	1.099	0.180
Mexico	2	0.693	0.160
Croatia	2	0.693	0.160
Argentina	2	0.693	0.160
Indonesia	2	0.693	0.160
Australia	1	0.000	0.140
Austria	1	0.000	0.140
Iceland	1	0.000	0.140
Slovenia	1	0.000	0.140
Latvia	1	0.000	0.140
Czech Republic	0		0.120
Russian Federation	0		0.120
South Africa	0		0.120
Lithuania	-1		0.100
Kazakhstan	-2		0.080
Ukraine	-2		0.080
Slovak Republic	-3		0.060
Germany	-5		0.020
Romania	-6		0.000

The nutrient balance in tonnes, for nitrogen and phosphorus, presented, in 2019, the highest values for the big countries (tables 12 and 13), such as India, China, USA, Brazil, Australia and Russian Federation.

In any case, there are some differences in the distribution worldwide of these two nutrients balance in tonnes. The logarithm and normalized values highlight great differences among the countries.

Table 12. Nutrient balance (tonnes), for nitrogen (2019)

Location	Value	Logarithm	Normalised value
India	23997610	16.993	1.000
China	16559496	16.622	0.733
USA	12774796	16.363	0.598
Australia	7433113	15.821	0.406
Brazil	5054018	15.436	0.321
Mexico	2933168	14.892	0.245
Vietnam	1567530	14.265	0.196
Russian Federation	1466515	14.198	0.192
Turkey	1215702	14.011	0.183
Germany	889610	13.699	0.171
Poland	688952	13.443	0.164
New Zealand	658016	13.397	0.163
Ukraine	604952	13.313	0.161
Philippines	598825	13.303	0.161
Korea, Rep.	381790	12.853	0.153
Colombia	351893	12.771	0.152
Argentina	334909	12.722	0.151
Netherlands	301100	12.615	0.150
Czech Republic	266358	12.493	0.149
Portugal	179035	12.095	0.146
Lithuania	121238	11.706	0.144
Slovak Republic	121187	11.705	0.144
Finland	99385	11.507	0.143
Austria	90474	11.413	0.143
Switzerland	89580	11.403	0.143
Croatia	80397	11.295	0.142
Costa Rica	72412	11.190	0.142
Latvia	24356	10.101	0.140
Slovenia	20736	9.940	0.140
Iceland	10577	9.266	0.140
Romania	-344357		0.127
South Africa	-698592		0.114
Kazakhstan	-3002979		0.032
Indonesia	-3889878		0.000

Table 13. Nutrient balance (tonnes), for phosphorus (2019)

Location	Value	Logarithm	Normalised value
China	5008742	15.427	1.000
India	4963184	15.418	0.992
Brazil	3943379	15.188	0.806
USA	1451493	14.188	0.353
Vietnam	395611	12.888	0.161
Australia	360447	12.795	0.155
Turkey	329812	12.706	0.149
Colombia	297613	12.604	0.143
Argentina	235012	12.367	0.132
Mexico	181846	12.111	0.122
Indonesia	103163	11.544	0.108
New Zealand	100188	11.515	0.108
Philippines	98320	11.496	0.107
Korea, Rep.	74332	11.216	0.103
Russian Federation	57943	10.967	0.100
Poland	52758	10.873	0.099
South Africa	25987	10.165	0.094
Portugal	23052	10.046	0.093
Costa Rica	16073	9.685	0.092
Finland	7460	8.917	0.091
Netherlands	7234	8.887	0.091
Switzerland	5301	8.576	0.090
Austria	3524	8.167	0.090
Croatia	2468	7.811	0.090
Iceland	2137	7.667	0.090
Latvia	1951	7.576	0.090
Slovenia	315	5.753	0.089
Czech Republic	-110		0.089
Lithuania	-3954		0.089
Slovak Republic	-5944		0.088
Germany	-79284		0.075
Romania	-84745		0.074
Ukraine	-89569		0.073
Kazakhstan	-491052		0.000

CONCLUSION

This study sought to investigate the implications of the Covid-19 pandemic on sustainable agriculture, considering statistical information from the OECD database for the years 2019-2021 related to agricultural land (total, arable and permanent cropland, and permanent pasture) and nutrient balance (nitrogen and phosphorus). To better explore the data, normalised values, growth rates and logarithms results were found.

The scientific literature highlights the relevant negative impacts of the pandemic on the different dimensions of sustainability, nonetheless brought new perspectives to change the paradigms, namely in agricultural practices. These disruptions also promoted the discussion about agricultural sustainable development, specifically in terms of food sovereignty, rural revitalisation, urban agriculture and agro-ecology transition. However, despite this debate and great expectations for the future, the reality shows that the old trends remain. The smart approach, in the context of Agriculture 4.0, and the nanotechnologies may support improvements in farming efficiency with positive impacts on the environment. The diversification and the multifunctionality of agriculture will be crucial to increase the resilience of the sector. The concerns of the consumers with sustainability suggest good perspectives from the demand side for more eco-friendly productions and agri-food chains.

The statistical information assessment reveals that big countries (or groups of countries), such as China, the USA, Australia, Brazil, Russia Federation, the European Union and India are those with the highest total agricultural land. On the other hand, there is a trend of some stability in this land for the years 2020 and 2021 (growth rates of -0.5% and -0.6%, respectively). A similar pattern was found for arable and permanent cropland and permanent pasture. In general, there were no relevant changes in agricultural land after the pandemic. The nutrient balance for nitrogen and phosphorus shows the distinct reality worldwide for these two nutrients.

This chapter highlights that the expectations created with the pandemic for improvements in the sustainability of agriculture seem to have been overestimated. In any case, the discussion is (re) open and this is already a great advance. In terms of policy recommendations, it would be important to take advantage of the debate created to promote transitions in agricultural practices for more eco-friendly approaches. For future research, it could be relevant to investigate, with more recent data, the real impacts of the pandemic on the nutrient balance and other environmental indicators.

ACKNOWLEDGMENT

This work is funded by National Funds through the FCT - Foundation for Science and Technology, I.P., within the scope of the project Refª UIDB/00681/2020. Furthermore we would like to thank the CERNAS Research Centre and the Polytechnic Institute of Viseu for their support.

REFERENCES

Adnan, N., & Nordin, S. M. (2021). How COVID 19 effect Malaysian paddy industry? Adoption of green fertilizer a potential resolution. *Environment, Development and Sustainability, 23*(6), 8089–8129. Scopus. doi:10.1007/s10668-020-00978-6

Ahmad, A., Ordoñez, J., Cartujo, P., & Martos, V. (2020). Remotely piloted aircraft (RPA) in agriculture: A pursuit of sustainability. *Agronomy, 11*(1). Scopus. doi:10.3390/agronomy11010007

Al-Saidi, M., Saad, S. A. G., & Elagib, N. A. (2022). *From scenario to mounting risks: COVID-19's perils for development and supply security in the Sahel. Environment, Development and Sustainability.* Scopus., doi:10.100710668-022-02303-9

Ali, M. S., Siddik, M. S., Nur-E-Alam, S. M., Khanam, T., & Khatun, M. M. (2020). Analysis of the problems and minimization initiatives addressed by bangladeshi farmers during the covid-19 pandemic. *Asian Journal of Agriculture and Rural Development, 10*(4), 773–781. Scopus. doi:10.18488/journal.ajard.2020.104.773.781

Arroyo-Lambaer, D., Zambrano, L., Rivas, M. I., Vázquez-Mendoza, D. L., Figueroa, F., Puente-Uribe, M. B., Espinosa-García, A. C., Tapia-Palacios, M. A., Mazari-Hiriart, M., Revollo-Fernández, D., Jiménez-Serna, A., Covarrubias, M., & Sumano, C. (2022). Identifying Urban Agriculture Needs and Challenges for the Implementation of Green Labeling in Xochimilco, Mexico. *Frontiers in Sustainable Cities, 4.* Scopus. doi:10.3389/frsc.2022.892341

Auliah, A., Prayitno, G., Ari, I. R. D., Rahmawati, Wardani, L. E., & Meidiana, C. (2022). The Role of Social Capital Facing Pandemic COVID-19 in Tourism Village to Support Sustainable Agriculture (Empirical Evidence from Two Tourism Villages in Indonesia). *Economies, 10*(12). Scopus. doi:10.3390/economies10120320

Azenzem, R. (2022). Oilseeds development in Morocco in the current international context. *OCL - Oilseeds and Fats, Crops and Lipids, 29.* Scopus. doi:10.1051/ocl/2022032

Baptista, F., Lourenço, P., Fitas da Cruz, V., Silva, L. L., Silva, J. R., Correia, M., Picuno, P., Dimitriou, E., & Papadakis, G. (2021). Which are the best practices for MSc programmes in sustainable agriculture? *Journal of Cleaner Production, 303.* Scopus. doi:10.1016/j.jclepro.2021.126914

Barbosa, F. A. R. (2022). Limnology And The Sustainable Use Of Water In Brazil: Visions And Challenges. *Oecologia Australis, 26*(2), 112–127. Scopus. doi:10.4257/oeco.2022.2602.03

Dokić, D., Gavran, M., Gregić, M., & Gantner, V. (2020). The Impact of Trade Balance of Agri-Food Products on the State's Ability to Withstand the Crisis. *HighTech and Innovation Journal, 1*(3), 107–111. Scopus. doi:10.28991/HIJ-2020-01-03-02

Gómez Serna, L. C., & Bernal Rivas, J. (2020). Food production, access and dietary diversity in agroecological farming families in times of COVID-19. [Scopus.]. *Agroalimentaria, 26*(51), 39–52.

Guiné, R. P. F., Florença, S. G., Costa, D. T. V. A., Çelik, S., Ferreira, M., Cardoso, A. P., Çetin, S., & Costa, C. A. (2022). Comparative Study about the Consumption of Organic Food Products on Samples of Portuguese and Turkish Consumers under the COVID-19 Pandemic Context. *Agronomy, 12*(6). Scopus. doi:10.3390/agronomy12061385

Hafez, H. M., & Attia, Y. A. (2020). Challenges to the Poultry Industry: Current Perspectives and Strategic Future After the COVID-19 Outbreak. *Frontiers in Veterinary Science, 7.* Scopus. doi:10.3389/fvets.2020.00516

Ikehi, M. E., Ifeanyieze, F. O., Onu, F. M., Ejiofor, T. E., & Nwankwo, C. U. (2022). *Assessing climate change mitigation and adaptation strategies and agricultural innovation systems in the Niger Delta. GeoJournal.* Scopus., doi:10.100710708-022-10596-6

Khan, N., Ma, J., Kassem, H. S., Kazim, R., Ray, R. L., Ihtisham, M., & Zhang, S. (2022). Rural Farmers' Cognition and Climate Change Adaptation Impact on Cash Crop Productivity: Evidence from a Recent Study. *International Journal of Environmental Research and Public Health, 19*(19). Scopus. doi:10.3390/ijerph191912556

Koç, E., & Karayiğit, B. (2022). Assessment of Biofortification Approaches Used to Improve Micronutrient-Dense Plants That Are a Sustainable Solution to Combat Hidden Hunger. *Journal of Soil Science and Plant Nutrition, 22*(1), 475–500. Scopus. doi:10.1007/s42729-021-00663-1

Kumar, A., Anju, T., Kumar, S., Chhapekar, S. S., Sreedharan, S., Singh, S., Choi, S. R., Ramchiary, N., & Lim, Y. P. (2021). Integrating omics and gene editing tools for rapid improvement of traditional food plants for diversified and sustainable food security. *International Journal of Molecular Sciences, 22*(15). Scopus. doi:10.3390/ijms22158093

Li, N., Chen, M., & Huang, D. (2023). How Do Logistics Disruptions Affect Rural Households? Evidence from COVID-19 in China. *Sustainability (Switzerland), 15*(1). Scopus. doi:10.3390/su15010465

Lian, T., Haq, S. U., Shahbaz, P., Zhao, L., Nadeem, M., & Aziz, B. (2023). Changing Food Patterns during the Pandemic: Exploring the Role of Household Dynamics and Income Stabilization Strategies. *Sustainability (Switzerland), 15*(1). Scopus. doi:10.3390/su15010123

Ling, J., Xue, Y., Yang, C., & Zhang, Y. (2023). Effect of Farmers' Awareness of Climate Change on Their Willingness to Adopt Low-Carbon Production: Based on the TAM-SOR Model. *International Journal of Environmental Research and Public Health, 20*(1). Scopus. doi:10.3390/ijerph20010619

Manian, M. A., Khoshbakht, K., Mahmoudi, H., & Liaghati, H. (2022). Dynamic Conservation in Risk Society: A Case Study of COVID-19 Pandemic Risk in Kashan Qanat Irrigated Agriculture. *Frontiers in Public Health, 10*. Scopus. doi:10.3389/fpubh.2022.882943

Martey, E., Etwire, P. M., Adzawla, W., Atakora, W., & Bindraban, P. S. (2022). Perceptions of COVID-19 shocks and adoption of sustainable agricultural practices in Ghana. *Journal of Environmental Management, 320*. Scopus. doi:10.1016/j.jenvman.2022.115810

Nelles, W. (2023). Academic network-building for agroecology and sustainable agri-food systems in Southeast Asia: Critical reflections on a regional initiative. *Agroecology and Sustainable Food Systems, 47*(1), 126–155. Scopus. doi:10.1080/21683565.2022.2134955

OECD. (2023). *Agriculture—OECD Data*. TheOECD. https://data.oecd.org/agriculture.htm

Pampori, Z. A. (2021). Saga of Hunger in India- Challenges, Chances and Resolutions: A Review. *Asian Journal of Dairy and Food Research, 40*(4), 398–407. Scopus. doi:10.18805/ajdfr.R-2183

Paria, B., Mishra, P., & Behera, B. (2022). Climate change and transition in cropping patterns: District level evidence from West Bengal, India. *Environmental Challenges, 7*. Scopus. doi:10.1016/j.envc.2022.100499

Pathirana, R., & Carimi, F. (2022). Management and Utilization of Plant Genetic Resources for a Sustainable Agriculture. *Plants, 11*(15). Scopus. doi:10.3390/plants11152038

Polukhin, A. A., & Panarina, V. I. (2022). *Financial Risk Management for Sustainable Agricultural Development Based on Corporate Social Responsibility in the Interests of Food Security. Risks, 10(1).* Scopus., doi:10.3390/risks10010017

Qin, T., Wang, L., Zhou, Y., Guo, L., Jiang, G., & Zhang, L. (2022). Digital Technology-and-Services-Driven Sustainable Transformation of Agriculture: Cases of China and the EU. *Agriculture (Switzerland), 12*(2). Scopus. doi:10.3390/agriculture12020297

Rasul, G. (2021). A Framework for Addressing the Twin Challenges of COVID-19 and Climate Change for Sustainable Agriculture and Food Security in South Asia. *Frontiers in Sustainable Food Systems, 5*. Scopus. doi:10.3389/fsufs.2021.679037

Satama, M., Singaña Tapia, D. A., & Paul, C. (2022). Adoption of sustainable agricultural practices during the pandemic: The case of Ecuadorian family farming systems. *Journal of Agribusiness in Developing and Emerging Economies*. Scopus. doi:10.1108/JADEE-04-2022-0067

Scopus. (2023). *Scopus Database*. Scopus. https://www.scopus.com/search/form.uri?display=basic#basic

Spies, M., Schick, A., Karomatov, S., Bakokhoja, B., Zikriyokhon, K., Jobirov, S., Bloch, R., & Ibisch, P. L. (2021). Adapting a Participatory and Ecosystem-Based Assessment Impacted by the Pandemic: Lessons Learned With Farmers in Tajikistan. *Frontiers in Sustainable Food Systems, 5*. Scopus. doi:10.3389/fsufs.2021.750252

Sridhar, A., Balakrishnan, A., Jacob, M. M., Sillanpää, M., & Dayanandan, N. (2022). *Global impact of COVID-19 on agriculture: Role of sustainable agriculture and digital farming. Environmental Science and Pollution Research.* Scopus., doi:10.100711356-022-19358-w

Stata. (2023). *Statistical software for data science.* Stata. https://www.stata.com/

StataCorp. (2017a). *Stata 15 Base Reference Manual.* Stata Press.

StataCorp. (2017b). *Stata Statistical Software: Release 15.* StataCorp LLC.

Tansuchat, R., Suriyankietkaew, S., Petison, P., Punjaisri, K., & Nimsai, S. (2022). Impacts of COVID-19 on Sustainable Agriculture Value Chain Development in Thailand and ASEAN. *Sustainability (Switzerland), 14*(20). Scopus. doi:10.3390/su142012985

Teng, Y., Pang, B., Zhang, M., & Guo, X. (2022). Driving mechanism of farmers' green production behavior under normalization of COVID-19 prevention and control: A case study in China. *Frontiers in Public Health, 10*. Scopus. doi:10.3389/fpubh.2022.826846

Vo, D. H. (2020). Sustainable agriculture & energy in the U.S.: A link between ethanol production and the acreage for corn. *Economics and Sociology, 13*(3), 259–268. Scopus. doi:10.14254/2071-789X.2020/13-3/16

Wudil, A. H., Usman, M., Rosak-Szyrocka, J., Pilař, L., & Boye, M. (2022). Reversing Years for Global Food Security: A Review of the Food Security Situation in Sub-Saharan Africa (SSA). *International Journal of Environmental Research and Public Health, 19*(22). Scopus. doi:10.3390/ijerph192214836

Yang, X., & Liu, W. (2022). Agricultural Production Networks and Upgrading from a Global–Local Perspective: A Review. *Land, 11*(10). Scopus. doi:10.3390/land11101864

Zámková, M., Rojík, S., Prokop, M., Činčalová, S., & Stolín, R. (2022). Czech Consumers' Preference for Organic Products in Online Grocery Stores during the COVID-19 Pandemic. *International Journal of Environmental Research and Public Health, 19*(20). Scopus. doi:10.3390/ijerph192013316

Chapter 7
Implications of the Pandemic and Recent Conflicts in the European Union Dairy Sector

ABSTRACT

It is important to create a picture of the main implications of the Covid-19 pandemic and the Russia-Ukraine crisis on the evolution of this sector. This research has as its principal objective to highlight implications from the most recent events in the European Union dairy sector. Statistical information from the Eurostat database was considered over the period January 2019-October 2022 for variables related to milk production and dairy products: raw cows' milk delivered to dairies; raw cream delivered to dairies (in milk equivalent); drinking milk; cream for direct consumption; milk and cream powders, excluding skimmed milk powders; skimmed milk powder; concentrated milk; acidified milk (yoghurts and other); butter (including dehydrated butter and ghee, and other fats and oils derived from milk; dairy spreads); and cheese from cows' milk (pure). These variables were not significantly changed with the pandemic and recent crises.

INTRODUCTION

All economic sectors were impacted by the recent international events (Chen, 2021), as well as health conditions (Miller et al., 2022). The negative consequences from the Covid-19 in the supply circumstances of the dairy sector (Abhijit et al., 2021) and on the demand side (Aiswarya & Bhagya, 2021) were also visible, particularly in the beginning (Hambardzumyan & Gevorgyan, 2022).

The effects of the pandemic on the different countries, regions, producers and groups of the population were not equal (Mumena, 2020), including for the dairy framework (Nikooyeh et al., 2022). In some cases, such as the French organic milk cattle farms, the impacts were null or moderate (Perrin & Martin, 2021). In any case, the disturbances in the dairy businesses promoted alternative approaches (Mishra & Shukla, 2022).

DOI: 10.4018/978-1-6684-8923-9.ch007

The higher costs of milk production, lower milk prices (Bhandari et al., 2021) and lack of operating capital (Qingbin et al., 2020) were identified as some of the factors that contributed to the problems in the sector during the pandemic, with changes in the competitiveness (Bórawski et al., 2022). The levels of organisation and resilience of the sector influence the adjustments in production (Dugué et al., 2021) and income availability affects the consumption of milk and dairy products (Nohra et al., 2022). The environmental impacts from the sector continue to be a concern (Kumar et al., 2021).

In turn, the pandemic highlights the relevance of milk products for the human diet (Guzel-Seydim et al., 2021) and health (Nili et al., 2022), as well as for food security (Prasad & Kothari, 2022), namely in the children (Sodri et al., 2021). In fact, dairy products are important to improve oral well-being (Farias da Cruz et al., 2022) and to prevent the severity of the pandemic (Ali et al., 2021). These products also contribute to reduce the odds of Covid-19 (Darand et al., 2022) and prevent viruses (Salama et al., 2022), as healthy functional foods (Flis et al., 2022), with health benefits (Yilmaz et al., 2022).

In some circumstances, the consumption of dairy products increased during the Covid-19 constraints (Almeida & Almeida, 2021), including between the students (Aydin & Demir, 2022) and in other cases decreased (Hoteit et al., 2022). In any case, the pandemic changed the consumer daily habits (de Melo Souza et al., 2022), nutritional conditions (Quilliot et al., 2021) and behaviour towards food consumption (Ben Hassen et al., 2021) and management (Bogevska et al., 2022). These contexts were verified also among children and adolescents (Pourghazi et al., 2022).

These impacts created diverse difficulties in the agricultural and food chains, including in dairy value channels (Biswal et al., 2020) and specific markets (Panta et al., 2023). These challenges required solutions adjusted to each case and permanent monitoring to assess strategies that in many cases were innovative and intended to deal with the new contexts generated by the restrictions of the pandemic (Benedek et al., 2021). Nonetheless, the constraints in the agri-food chains are not new and were present in other previous shocks, such as the African swine fever (Chatellier et al., 2022).

Taking into account the perspectives presented before, this research proposes to present more insights about the implications of the Covid-19 pandemic and the Russia-Ukraine conflict on the European Union dairy sector. In this way, data from the Eurostat (2023) were considered for the period 2019-2022 and the following variables: raw cows' milk delivered to dairies; raw cream delivered to dairies (in milk equivalent); drinking milk; cream for direct consumption; milk and cream powders (excluding skimmed milk powders); skimmed milk powder; concentrated milk; acidified milk (yoghurts and other); butter (including dehydrated butter and ghee, and other fats and oils derived from milk); cheese from cows' milk (pure). The Stata (StataCorp, 2017a, 2017b; Stata, 2023) software procedures were followed to obtain some results. The Scopus (2023) database was consulted. The normalised values were calculated through the following equation: $(x_i-x_{minimum})/(x_{maximum}-x_{minimum})$.

RAW COWS' MILK DELIVERED

The raw cows' milk delivered to dairies in 1000 tonnes (table 1), on average over the period January 2019-October 2022, presents the highest values for the following European Union countries: Germany; France; Netherlands; Poland; Italy; Ireland; Spain; Denmark; Belgium; and Austria. On the other hand, this indicator, on average, increased by 0.1% (standard deviation of 6.4%) in 2019, decreased by -0.1% in 2020 and 2021 (standard deviation of 5.0% and 6.9%, respectively) and increased by 0.2% (standard deviation of 6.6%) in 2022. It is not visible that the Covid-19 pandemic and the Russia-Ukraine conflict have had impacts on this variable. In fact, the lowest and biggest growth rates were verified at the beginning of 2021 and 2022 (table 2).

Table 1. Average values over the period January 2019 – October 2022 (thousand tonnes)

Country	Value	Normalised value
Germany	2688.613	1.000
France	2031.564	0.755
Netherlands	1149.385	0.427
Poland	1040.987	0.386
Italy	1035.924	0.385
Ireland	741.090	0.275
Spain	615.176	0.228
Denmark	470.679	0.174
Belgium	366.381	0.135
Austria	264.352	0.097
Czechia	262.017	0.096
Sweden	229.802	0.084
Finland	191.821	0.070
Portugal	158.324	0.058
Hungary	138.166	0.050
Lithuania	113.572	0.041
Romania	94.175	0.034
Slovakia	68.779	0.024
Latvia	66.821	0.024
Estonia	65.628	0.023
Bulgaria	56.352	0.020
Greece	54.105	0.019
Slovenia	48.206	0.017
Croatia	35.680	0.012
Cyprus	22.737	0.007
Malta	3.395	0.000
Luxembourg		

Table 2. Average values across the European Union countries (thousand tonnes)

Period	Value	Growth rates (%)
2019-01	445.822	
2019-02	418.546	-6.118
2019-03	482.325	15.238
2019-04	486.107	0.784
2019-05	503.996	3.680
2019-06	473.165	-6.117
2019-07	472.119	-0.221
2019-08	458.991	-2.781
2019-09	434.653	-5.303
2019-10	440.193	1.275
2019-11	420.495	-4.475
2019-12	441.770	5.060
2020-01	453.327	
2020-02	439.052	-3.149
2020-03	488.662	11.299
2020-04	487.924	-0.151
2020-05	503.487	3.190
2020-06	478.506	-4.962
2020-07	480.632	0.444
2020-08	460.697	-4.148
2020-09	440.016	-4.489
2020-10	444.195	0.950
2020-11	424.805	-4.365
2020-12	442.843	4.246
2021-01	450.945	
2021-02	422.987	-6.200
2021-03	495.298	17.095
2021-04	495.865	0.114
2021-05	516.672	4.196
2021-06	482.423	-6.629
2021-07	481.083	-0.278
2021-08	467.811	-2.759
2021-09	438.930	-6.174
2021-10	442.613	0.839
2021-11	422.174	-4.618
2021-12	437.257	3.573
2022-01	448.585	
2022-02	424.267	-5.421

Continued on following page

Table 2. Continued

Period	Value	Growth rates (%)
2022-03	489.058	15.271
2022-04	488.271	-0.161
2022-05	507.447	3.927
2022-06	480.285	-5.353
2022-07	481.870	0.330
2022-08	466.101	-3.272
2022-09	442.759	-5.008
2022-10	450.032	1.643

RAW CREAM DELIVERED

Italy, Belgium and Spain are the countries that present values in thousand tonnes for the raw cream delivered to dairies (in milk equivalent), on average over the monthly period from the beginning of 2019 until October 2022 (table 3). This variable increased, on average, by 2.1% (standard deviation of 7.3%) in 2019, 1.0% (standard deviation of 8.1%) in 2020, 0.4% (standard deviation of 6.6%) in 2021 and 3.8% (standard deviation of 9.3%) in 2022 (table 4). There are some fluctuations, but they seem more cyclical than promoted by external shocks. In any case, if there were impacts from the pandemic and conflict they were residual.

Table 3. Average values over the period January 2019 – October 2022 (thousand tonnes)

Country	Value	Normalised value
Italy	66.698	1.000
Belgium	0.077	0.001
Spain	0.021	0.000
Austria	0.000	0.000
Bulgaria	0.000	0.000
Cyprus	0.000	0.000
Czechia	0.000	0.000
Germany	0.000	0.000
Denmark	0.000	0.000
Estonia	0.000	0.000
Greece	0.000	0.000
Finland	0.000	0.000
France	0.000	0.000
Croatia	0.000	0.000
Hungary	0.000	0.000
Ireland	0.000	0.000
Lithuania	0.000	0.000
Luxembourg	0.000	0.000
Latvia	0.000	0.000
Malta	0.000	0.000
Netherlands	0.000	0.000
Poland	0.000	0.000
Portugal	0.000	0.000
Romania	0.000	0.000
Sweden	0.000	0.000
Slovenia	0.000	0.000
Slovakia	0.000	0.000

Table 4. Average values across the European Union countries (thousand tonnes)

Period	Value	Growth rates (%)
2019-01	2.574	
2019-02	2.265	-12.005
2019-03	2.119	-6.446
2019-04	2.410	13.733
2019-05	2.630	9.129
2019-06	2.738	4.106
2019-07	2.702	-1.315
2019-08	2.957	9.437
2019-09	2.989	1.082
2019-10	3.000	0.368
2019-11	3.043	1.433
2019-12	3.161	3.878
2020-01	2.714	
2020-02	2.778	2.358
2020-03	3.129	12.635
2020-04	3.058	-2.269
2020-05	2.764	-9.614
2020-06	3.161	14.363
2020-07	3.198	1.171
2020-08	3.504	9.568
2020-09	3.353	-4.309
2020-10	3.080	-8.142
2020-11	3.017	-2.045
2020-12	2.921	-3.182
2021-01	2.732	
2021-02	2.520	-7.760
2021-03	2.412	-4.286
2021-04	2.689	11.484
2021-05	2.595	-3.496
2021-06	2.787	7.399
2021-07	2.741	-1.651
2021-08	2.603	-5.035
2021-09	2.575	-1.076
2021-10	2.739	6.369
2021-11	2.595	-5.257
2021-12	2.794	7.669
2022-01	2.447	
2022-02	2.533	3.515

Continued on following page

Table 4. Continued

Period	Value	Growth rates (%)
2022-03	2.355	-7.027
2022-04	2.419	2.718
2022-05	2.448	1.199
2022-06	3.056	24.837
2022-07	3.218	5.301
2022-08	3.055	-5.065
2022-09	3.057	0.065
2022-10	3.321	8.636

DRINKING MILK

Drinking milk has the greatest value, over the period considered, in Germany, Spain, France, Italy, Poland, Austria, Sweden, Portugal, Belgium, Czechia, Finland, Hungary and Ireland (table 5). The drinking milk, on average, increased by 0.01% (standard deviation of 5.66%) in 2019, 0.42% (standard deviation of 6.55%) in 2020, 0.18% (standard deviation of 4.67%) in 2021 and decreased -0.44% (standard deviation of 5.20%) in 2022 (table 6).

Table 5. Average values over the period January 2019 – October 2022 (thousand tonnes)

Country	Value	Normalised value
Germany	366.863	1.000
Spain	309.158	0.841
France	241.305	0.655
Italy	199.587	0.540
Poland	162.641	0.438
Austria	65.276	0.171
Sweden	58.198	0.151
Portugal	56.673	0.147
Belgium	55.595	0.144
Czechia	52.664	0.136
Finland	47.160	0.121
Hungary	44.788	0.114
Ireland	44.523	0.114
Denmark	41.385	0.105
Greece	31.850	0.079
Romania	30.116	0.074
Slovakia	19.964	0.046
Croatia	18.783	0.043
Slovenia	13.189	0.028
Estonia	8.948	0.016
Lithuania	6.836	0.010
Bulgaria	6.226	0.008
Cyprus	4.645	0.004
Latvia	3.176	0.000
Luxembourg		
Malta		
Netherlands		

Table 6. Average values across the European Union countries (thousand tonnes)

Period	Value	Growth rates (%)
2019-01	81.110	
2019-02	75.054	-7.466
2019-03	80.490	7.243
2019-04	79.619	-1.082
2019-05	79.631	0.015
2019-06	71.575	-10.117
2019-07	74.560	4.170
2019-08	74.643	0.111
2019-09	73.878	-1.025
2019-10	79.723	7.912
2019-11	76.924	-3.511
2019-12	79.933	3.912
2020-01	82.139	
2020-02	75.745	-7.784
2020-03	87.363	15.338
2020-04	84.891	-2.830
2020-05	79.007	-6.931
2020-06	76.681	-2.944
2020-07	76.493	-0.245
2020-08	75.102	-1.818
2020-09	77.050	2.594
2020-10	80.975	5.094
2020-11	79.727	-1.541
2020-12	84.270	5.698
2021-01	82.184	
2021-02	78.031	-5.053
2021-03	85.674	9.795
2021-04	80.222	-6.364
2021-05	80.094	-0.160
2021-06	76.268	-4.777
2021-07	74.976	-1.694
2021-08	76.731	2.341
2021-09	76.501	-0.300
2021-10	78.313	2.369
2021-11	79.745	1.829
2021-12	82.923	3.985
2022-01	81.838	
2022-02	77.290	-5.557

Continued on following page

Table 6. Continued

Period	Value	Growth rates (%)
2022-03	85.456	10.565
2022-04	80.721	-5.541
2022-05	78.709	-2.493
2022-06	75.328	-4.296
2022-07	73.880	-1.922
2022-08	76.262	3.224
2022-09	76.052	-0.275
2022-10	77.861	2.379

CREAM FOR DIRECT CONSUMPTION

Germany, France, Poland, Belgium, Spain, Italy, Sweden, Austria, Romania, Finland, Czechia, Denmark, Croatia, and Slovakia are the countries with more cream for direct consumption (1000 tonnes) on average over the period considered (table 7). On average, this variable increased by around 1-2% (table 8) in the years taken into account (2019, 2020, 2021 and 2022). The highest growth rate was in 2021 (1.8%) and the lowest was in 2020 (0.5%). The standard deviations were, respectively, 5.6%, 7.7%, 7.3% and 7.7, for the years considered.

Table 7. Average values over the period January 2019 – October 2022 (thousand tonnes)

Country	Value	Normalised value
Germany	44.882	1.000
France	44.648	0.995
Poland	21.674	0.481
Belgium	21.051	0.467
Spain	13.367	0.295
Italy	11.810	0.260
Sweden	6.314	0.137
Austria	5.906	0.128
Romania	5.602	0.121
Finland	5.448	0.117
Czechia	5.239	0.113
Denmark	3.943	0.084
Croatia	2.618	0.054
Slovakia	2.530	0.052
Estonia	2.457	0.050
Latvia	2.270	0.046
Portugal	2.062	0.042
Ireland	2.007	0.040
Greece	1.023	0.018
Slovenia	0.975	0.017
Hungary	0.816	0.014
Lithuania	0.548	0.008
Bulgaria	0.266	0.001
Cyprus	0.202	0.000
Luxembourg		
Malta		
Netherlands		

Table 8. Average values across the European Union countries (thousand tonnes)

Period	Value	Growth rates (%)
2019-01	8.245	
2019-02	8.087	-1.916
2019-03	9.100	12.526
2019-04	9.011	-0.978
2019-05	9.250	2.652
2019-06	8.344	-9.795
2019-07	8.273	-0.851
2019-08	8.381	1.305
2019-09	8.342	-0.465
2019-10	8.895	6.629
2019-11	8.664	-2.597
2019-12	8.837	1.997
2020-01	8.654	
2020-02	8.053	-6.945
2020-03	9.075	12.691
2020-04	8.207	-9.565
2020-05	8.154	-0.646
2020-06	8.840	8.413
2020-07	9.370	5.995
2020-08	8.709	-7.054
2020-09	9.160	5.179
2020-10	9.125	-0.382
2020-11	8.350	-8.493
2020-12	8.858	6.084
2021-01	8.140	
2021-02	7.947	-2.371
2021-03	9.509	19.655
2021-04	8.673	-8.792
2021-05	9.055	4.404
2021-06	8.940	-1.270
2021-07	8.532	-4.564
2021-08	8.600	0.797
2021-09	9.049	5.221
2021-10	9.156	1.182
2021-11	9.170	0.153
2021-12	9.662	5.365
2022-01	8.671	
2022-02	8.169	-5.789

Continued on following page

Table 8. Continued

Period	Value	Growth rates (%)
2022-03	9.790	19.843
2022-04	9.573	-2.217
2022-05	9.483	-0.940
2022-06	9.051	-4.556
2022-07	8.699	-3.889
2022-08	8.917	2.506
2022-09	9.015	1.099
2022-10	9.267	2.795

MILK AND CREAM POWDERS

The milk and cream powders, excluding skimmed milk powders, have the values presented in decreasing order in table 9 for the European Union countries, on average over the period January 2019-October 2022 (thousand tonnes). Netherlands, Germany, France, Denmark and Belgium are the top 5 countries. This variable decreased, on average, in 2020 and 2022 (table 10), respectively -0.2% (standard deviation of 10.5%) and -0.01% (standard deviation of 11.3%). In 2021 this indicator increased by 1.4% (standard deviation of 18.3%) and in 2019 increased by 0.3% (standard deviation of 8.3%).

Table 9. Average values over the period January 2019 – October 2022 (thousand tonnes)

Country	Value	Normalised value
Netherlands	11.935	1.000
Germany	10.513	0.881
France	10.220	0.856
Denmark	5.993	0.502
Belgium	4.130	0.346
Poland	2.573	0.216
Sweden	2.562	0.215
Spain	1.296	0.109
Czechia	0.865	0.072
Portugal	0.793	0.066
Slovakia	0.087	0.007
Austria	0.079	0.007
Romania	0.071	0.006
Lithuania	0.028	0.002
Bulgaria	0.000	0.000
Cyprus	0.000	0.000
Estonia	0.000	0.000
Greece	0.000	0.000
Croatia	0.000	0.000
Hungary	0.000	0.000
Ireland	0.000	0.000
Italy	0.000	0.000
Luxembourg	0.000	0.000
Malta	0.000	0.000
Slovenia	0.000	0.000
Finland		
Latvia		

Table 10. Average values across the European Union countries (thousand tonnes)

Period	Value	Growth rates (%)
2019-01	2.379	
2019-02	2.120	-10.887
2019-03	2.160	1.887
2019-04	2.320	7.407
2019-05	2.629	13.319
2019-06	2.268	-13.731
2019-07	2.385	5.159
2019-08	2.293	-3.857
2019-09	2.373	3.489
2019-10	2.307	-2.781
2019-11	2.188	-5.158
2019-12	2.366	8.135
2020-01	2.656	
2020-02	2.314	-12.877
2020-03	2.572	11.150
2020-04	2.798	8.787
2020-05	2.610	-6.719
2020-06	2.215	-15.134
2020-07	2.185	-1.354
2020-08	2.065	-5.492
2020-09	2.247	8.814
2020-10	2.136	-4.940
2020-11	2.093	-2.013
2020-12	2.470	18.012
2021-01	2.270	
2021-02	1.972	-13.128
2021-03	2.347	19.016
2021-04	2.464	4.985
2021-05	2.056	-16.558
2021-06	1.888	-8.171
2021-07	2.246	18.962
2021-08	1.872	-16.652
2021-09	1.826	-2.457
2021-10	1.827	0.055
2021-11	1.618	-11.440
2021-12	2.286	41.286
2022-01	2.132	
2022-02	1.799	-15.619

Continued on following page

Table 10. Continued

Period	Value	Growth rates (%)
2022-03	2.039	13.341
2022-04	2.150	5.444
2022-05	2.232	3.814
2022-06	2.044	-8.423
2022-07	1.910	-6.556
2022-08	1.891	-0.995
2022-09	1.705	-9.836
2022-10	2.025	18.768

SKIMMED MILK POWDER

Table 11 shows that France, Germany, Ireland, Belgium and Poland are the leaders countries for skimmed milk powder (thousand tonnes, on average over 2019-2022). Portugal, for example, appears in this ranking with 1.919 thousand tonnes and France with 33.674 thousand tonnes. This variable decreased on average in 2021 and 2022 to around -1%, as presented in table 12, and increase by close to 1% in 2019 and 2020. The standard deviation for 2019, 2020, 2021 and 2022 was, respectively, 15.0%, 15.4%, 13.6% and 10.6%.

Table 11. Average values over the period January 2019 – October 2022 (thousand tonnes)

Country	Value	Normalised value
France	33.674	1.000
Germany	31.545	0.937
Ireland	14.784	0.439
Belgium	13.831	0.411
Poland	13.217	0.392
Netherlands	7.022	0.209
Finland	4.274	0.127
Denmark	3.889	0.115
Sweden	3.180	0.094
Portugal	1.919	0.057
Lithuania	1.734	0.051
Czechia	1.523	0.045
Spain	1.128	0.033
Estonia	0.967	0.029
Austria	0.485	0.014
Slovakia	0.256	0.008
Romania	0.089	0.003
Bulgaria	0.000	0.000
Cyprus	0.000	0.000
Greece	0.000	0.000
Croatia	0.000	0.000
Hungary	0.000	0.000
Italy	0.000	0.000
Luxembourg	0.000	0.000
Malta	0.000	0.000
Slovenia	0.000	0.000
Latvia		

Table 12. Average values across the European Union countries (thousand tonnes)

Period	Value	Growth rates (%)
2019-01	5.963	
2019-02	5.503	-7.714
2019-03	6.033	9.631
2019-04	6.846	13.476
2019-05	7.241	5.770
2019-06	6.410	-11.476
2019-07	5.507	-14.087
2019-08	5.270	-4.304
2019-09	4.803	-8.861
2019-10	5.090	5.975
2019-11	4.435	-12.868
2019-12	6.012	35.558
2020-01	6.416	
2020-02	5.728	-10.723
2020-03	6.247	9.061
2020-04	7.139	14.279
2020-05	7.046	-1.303
2020-06	6.601	-6.316
2020-07	6.773	2.606
2020-08	5.399	-20.286
2020-09	5.181	-4.038
2020-10	5.207	0.502
2020-11	4.474	-14.077
2020-12	6.084	35.986
2021-01	6.341	
2021-02	5.413	-14.635
2021-03	6.689	23.573
2021-04	7.026	5.038
2021-05	7.899	12.425
2021-06	6.791	-14.027
2021-07	6.010	-11.501
2021-08	5.324	-11.414
2021-09	4.942	-7.175
2021-10	4.471	-9.531
2021-11	4.555	1.879
2021-12	5.360	17.673
2022-01	6.130	
2022-02	5.329	-13.067

Continued on following page

Table 12. Continued

Period	Value	Growth rates (%)
2022-03	6.472	21.449
2022-04	6.794	4.975
2022-05	7.296	7.389
2022-06	6.825	-6.456
2022-07	6.510	-4.615
2022-08	5.780	-11.214
2022-09	5.732	-0.830
2022-10	5.638	-1.640

CONCENTRATED MILK

Netherlands, Germany, France, Belgium, Spain, Poland, Lithuania, Czechia and Greece are the countries that present values for concentrated milk (table 13). This variable presents, on average, negative growth rates for 2019, 2020 and 2021, around -1%, such as may be observed in table 14, with standard deviations of 5.8%, 9.2% and 5.4%, respectively. In 2022, this indicator increased, on average, by 0.36% (standard deviation of 7.6%).

Table 13. Average values over the period January 2019 – October 2022 (thousand tonnes)

Country	Value	Normalised value
Netherlands	32.467	1.000
Germany	26.048	0.802
France	9.032	0.278
Belgium	4.863	0.150
Spain	4.133	0.127
Poland	2.675	0.082
Lithuania	1.860	0.057
Czechia	0.823	0.025
Greece	0.387	0.012
Bulgaria	0.000	0.000
Cyprus	0.000	0.000
Denmark	0.000	0.000
Estonia	0.000	0.000
Finland	0.000	0.000
Croatia	0.000	0.000
Hungary	0.000	0.000
Ireland	0.000	0.000
Italy	0.000	0.000
Luxembourg	0.000	0.000
Malta	0.000	0.000
Portugal	0.000	0.000
Romania	0.000	0.000
Sweden	0.000	0.000
Slovenia	0.000	0.000
Slovakia	0.000	0.000
Austria		
Latvia		

Table 14. Average values across the European Union countries (thousand tonnes)

Period	Value	Growth rates (%)
2019-01	3.437	
2019-02	3.211	-6.576
2019-03	3.472	8.128
2019-04	3.370	-2.938
2019-05	3.354	-0.475
2019-06	3.078	-8.229
2019-07	3.261	5.945
2019-08	3.097	-5.029
2019-09	3.336	7.717
2019-10	3.370	1.019
2019-11	3.217	-4.540
2019-12	3.069	-4.601
2020-01	3.927	
2020-02	3.446	-12.249
2020-03	4.033	17.034
2020-04	3.809	-5.554
2020-05	3.565	-6.406
2020-06	3.956	10.968
2020-07	4.064	2.730
2020-08	3.703	-8.883
2020-09	3.685	-0.486
2020-10	3.767	2.225
2020-11	3.650	-3.106
2020-12	3.221	-11.753
2021-01	4.109	
2021-02	3.879	-5.597
2021-03	4.207	8.456
2021-04	4.009	-4.706
2021-05	3.968	-1.023
2021-06	3.788	-4.536
2021-07	3.808	0.528
2021-08	4.012	5.357
2021-09	3.750	-6.530
2021-10	3.619	-3.493
2021-11	3.835	5.968
2021-12	3.594	-6.284
2022-01	3.736	
2022-02	3.556	-4.818

Continued on following page

Table 14. Continued

Period	Value	Growth rates (%)
2022-03	4.164	17.098
2022-04	3.690	-11.383
2022-05	3.677	-0.352
2022-06	3.769	2.502
2022-07	3.816	1.247
2022-08	3.822	0.157
2022-09	3.740	-2.145
2022-10	3.774	0.909

ACIDIFIED MILK

The acidified milk (yoghurts and other) has the average values (1000 tonnes over 2019-2022) presented in table 15, where Germany, France, Spain, Poland, Belgium and the Netherlands are the leaders countries. On average, the acidified milk decreased in 2019 (-1.5%), 2020 (-0.9%) and 2021 (-0.7%) and increased by 0.4% in 2022 (table 16). The standard deviations were, respectively, 6.5%, 5.6%, 5.9% and 5.3% for the years considered.

Table 15. Average values over the period January 2019 – October 2022 (thousand tonnes)

Country	Value	Normalised value
Germany	149.174	1.000
France	113.855	0.763
Spain	79.850	0.535
Poland	45.931	0.308
Belgium	26.728	0.179
Netherlands	24.854	0.167
Italy	23.014	0.154
Austria	21.974	0.147
Sweden	19.123	0.128
Romania	18.547	0.124
Czechia	15.372	0.103
Bulgaria	14.134	0.095
Finland	13.813	0.093
Greece	13.368	0.090
Denmark	10.083	0.068
Portugal	9.943	0.067
Hungary	9.762	0.065
Croatia	7.494	0.050
Lithuania	6.241	0.042
Slovakia	4.288	0.029
Slovenia	3.834	0.026
Estonia	3.315	0.022
Latvia	3.007	0.020
Cyprus	0.455	0.003
Ireland	0.000	0.000
Luxembourg		
Malta		

Table 16. Average values across the European Union countries (thousand tonnes)

Period	Value	Growth rates (%)
2019-01	27.800	
2019-02	25.402	-8.626
2019-03	27.405	7.885
2019-04	27.650	0.894
2019-05	28.414	2.763
2019-06	26.488	-6.778
2019-07	28.325	6.935
2019-08	26.616	-6.034
2019-09	26.393	-0.838
2019-10	27.689	4.910
2019-11	24.898	-10.080
2019-12	23.137	-7.073
2020-01	27.123	
2020-02	25.648	-5.438
2020-03	29.133	13.588
2020-04	27.668	-5.029
2020-05	27.931	0.951
2020-06	27.019	-3.265
2020-07	27.414	1.462
2020-08	26.584	-3.028
2020-09	27.285	2.637
2020-10	26.676	-2.232
2020-11	25.255	-5.327
2020-12	24.285	-3.841
2021-01	26.113	
2021-02	25.065	-4.013
2021-03	28.945	15.480
2021-04	26.933	-6.951
2021-05	26.978	0.167
2021-06	27.032	0.200
2021-07	26.747	-1.054
2021-08	26.218	-1.978
2021-09	26.465	0.942
2021-10	26.013	-1.708
2021-11	25.211	-3.083
2021-12	23.748	-5.803
2022-01	24.602	
2022-02	24.657	0.224

Continued on following page

Table 16. Continued

Period	Value	Growth rates (%)
2022-03	27.681	12.264
2022-04	26.025	-5.982
2022-05	26.550	2.017
2022-06	26.167	-1.443
2022-07	25.697	-1.796
2022-08	26.562	3.366
2022-09	25.510	-3.961
2022-10	25.295	-0.843

BUTTER

The butter (incl. dehydrated butter and ghee, and other fats and oils derived from milk; dairy spreads) appears with the highest values in Germany, France and Ireland (table 17), with 40.527, 34.132 and 22.347 thousand tonnes (on average over the period considered), respectively. The butter increased by less than 1% in the four years taken into account in this research (table 18). The standard deviations were, respectively, 9.0%, 7.7%, 9.1% and 6.8% for the period 2019-2022.

Table 17. Average values over the period January 2019 – October 2022 (thousand tonnes)

Country	Value	Normalised value
Germany	40.527	1.000
France	34.132	0.842
Ireland	22.347	0.551
Poland	19.524	0.482
Netherlands	11.528	0.284
Belgium	9.427	0.233
Italy	8.147	0.201
Denmark	6.535	0.161
Finland	5.008	0.124
Spain	3.120	0.077
Austria	3.042	0.075
Portugal	2.522	0.062
Czechia	2.165	0.053
Sweden	1.503	0.037
Lithuania	1.066	0.026
Romania	0.923	0.023
Slovakia	0.872	0.022
Hungary	0.845	0.021
Croatia	0.393	0.010
Estonia	0.385	0.009
Latvia	0.259	0.006
Slovenia	0.205	0.005
Greece	0.131	0.003
Bulgaria	0.090	0.002
Cyprus	0.000	0.000
Malta	0.000	0.000
Luxembourg		

Table 18. Average values across the European Union countries (thousand tonnes)

Period	Value	Growth rates (%)
2019-01	6.937	
2019-02	6.292	-9.298
2019-03	7.188	14.240
2019-04	7.465	3.854
2019-05	7.508	0.576
2019-06	6.568	-12.520
2019-07	6.690	1.857
2019-08	6.168	-7.803
2019-09	5.957	-3.421
2019-10	6.556	10.055
2019-11	6.074	-7.352
2019-12	6.749	11.113
2020-01	6.876	
2020-02	6.566	-4.508
2020-03	7.440	13.311
2020-04	7.846	5.457
2020-05	7.361	-6.181
2020-06	6.773	-7.988
2020-07	6.849	1.122
2020-08	6.128	-10.527
2020-09	6.239	1.811
2020-10	6.792	8.864
2020-11	6.452	-5.006
2020-12	6.923	7.300
2021-01	6.727	
2021-02	6.169	-8.295
2021-03	7.578	22.840
2021-04	7.696	1.557
2021-05	7.432	-3.430
2021-06	6.836	-8.019
2021-07	6.515	-4.696
2021-08	6.226	-4.436
2021-09	5.985	-3.871
2021-10	6.175	3.175
2021-11	5.977	-3.206
2021-12	6.513	8.968
2022-01	6.202	
2022-02	6.310	1.741

Continued on following page

Table 18. Continued

Period	Value	Growth rates (%)
2022-03	7.274	15.277
2022-04	7.345	0.976
2022-05	7.463	1.607
2022-06	6.811	-8.736
2022-07	6.527	-4.170
2022-08	6.427	-1.532
2022-09	6.123	-4.730
2022-10	6.370	4.034

CHEESE

The results for cheese from cows' milk (pure) are not so different from those presented before in the previous section for the other dairy products (tables 19 and 20). In this case, Germany and France are the top countries and the average growth rates for each one of the years considered, when positive (2020 and 2021) are modest (0.05% and 0.1%, respectively). In 2019 this indicator decreased by -0.2% and in 2022 decreased by -0.01.

Table 19. Average values over the period January 2019 – October 2022 (thousand tonnes)

Country	Value	Normalised value
Germany	203.152	1.000
France	141.492	0.696
Italy	96.627	0.476
Netherlands	78.607	0.387
Poland	76.396	0.376
Denmark	38.387	0.189
Ireland	23.851	0.117
Austria	17.619	0.087
Spain	16.164	0.080
Czechia	12.263	0.060
Belgium	8.863	0.044
Lithuania	8.206	0.040
Hungary	7.705	0.038
Bulgaria	7.422	0.036
Romania	7.385	0.036
Finland	6.980	0.034
Sweden	6.835	0.034
Portugal	5.376	0.026
Latvia	4.312	0.021
Estonia	3.896	0.019
Slovakia	3.200	0.016
Croatia	2.696	0.013
Greece	1.865	0.009
Slovenia	1.328	0.006
Cyprus	0.014	0.000
Luxembourg		
Malta		

Table 20. Average values across the European Union countries (thousand tonnes)

Period	Value	Growth rates (%)
2019-01	31.258	
2019-02	28.770	-7.960
2019-03	31.548	9.656
2019-04	31.734	0.590
2019-05	32.425	2.177
2019-06	30.104	-7.158
2019-07	31.534	4.750
2019-08	31.048	-1.541
2019-09	29.841	-3.888
2019-10	31.295	4.872
2019-11	30.186	-3.544
2019-12	30.137	-0.162
2020-01	30.738	
2020-02	29.141	-5.196
2020-03	32.685	12.162
2020-04	31.161	-4.663
2020-05	31.988	2.654
2020-06	31.871	-0.366
2020-07	31.930	0.185
2020-08	30.997	-2.922
2020-09	31.034	0.119
2020-10	31.627	1.911
2020-11	29.994	-5.163
2020-12	30.534	1.800
2021-01	30.775	
2021-02	29.345	-4.647
2021-03	34.536	17.690
2021-04	32.603	-5.597
2021-05	33.349	2.288
2021-06	32.018	-3.991
2021-07	31.310	-2.211
2021-08	31.664	1.131
2021-09	31.706	0.133
2021-10	31.574	-0.416
2021-11	30.929	-2.043
2021-12	30.670	-0.837
2022-01	32.006	
2022-02	30.109	-5.927

Continued on following page

Table 20. Continued

Period	Value	Growth rates (%)
2022-03	33.540	11.395
2022-04	31.952	-4.735
2022-05	32.893	2.945
2022-06	31.858	-3.147
2022-07	31.281	-1.811
2022-08	32.032	2.401
2022-09	31.565	-1.458
2022-10	31.642	0.244

CONCLUSION

This research proposed to analyse how the recent shocks from the pandemic and Russia-Ukraine crisis affected the European Union dairy sector. For that, data from the Eurostat were considered from January 2019 until October 2022 for variables related to raw cows' milk delivered to dairies, raw cream delivered to dairies (in milk equivalent), drinking milk, cream for direct consumption, milk and cream powders (excluding skimmed milk powders), skimmed milk powder, concentrated milk, acidified milk (yoghurts and other), butter (including dehydrated butter and ghee, and other fats and oils derived from milk; dairy spreads) and cheese from cows' milk (pure).

The literature highlights that the pandemic impacted, in some circumstances, the production of dairy products, namely in the beginning. In other cases, the consequences were residual. The main problems for the dairy sector, during the pandemic, came from the higher costs, lower milk prices and constraints in the operating capital. On the demand side, the lack of income by the consumers, in some contexts, was another difficulty. This is particularly worrying when is known the importance of dairy products for the human diet and to prevent of diseases. Nonetheless, these implications did not spread in a similar way worldwide. On the other hand, some of the challenges created by the pandemic are not new and come from other exogenous shocks, such as climate change and African swine fever, for example.

The results obtained from the data assessment show that in the context of the European Union, Germany, France, Spain, Italy, Netherlands, Denmark, Poland and Ireland, for example, are important countries in the production of dairy products. In turn, some fluctuations were identified for the variables analysed over the months of the period considered, but they seem more cyclical than promoted by the international shocks. In any case, if there were impacts from the international events on the dairy production they were residual and distinct for the different dairy products.

In terms of practical implications, some problems were identified in the supply and demand of dairy products, as a consequence of the pandemic and Russia-Ukraine crisis, with implications in the respective chains and the human diet. For policy recommendations, it is suggested awareness campaigns, informing the population about the importance of dairy products for human health. In future research, it could be interesting to analyse the impacts of international events on the European production of other agri-food products.

ACKNOWLEDGMENT

This work is funded by National Funds through the FCT - Foundation for Science and Technology, I.P., within the scope of the project Refª UIDB/00681/2020. Furthermore we would like to thank the CERNAS Research Centre and the Polytechnic Institute of Viseu for their support.

REFERENCES

Abhijit, D. A. S., Sivaram, M., & Thejesh, S. (2021). Economic impact of COVID-19 pandemic on dairy sector: A meta-analysis. [Scopus.]. *The Indian Journal of Animal Sciences, 91*(7), 582–594.

Aiswarya, A., & Bhagya, D. (2021). Effect of Covid 19 lockdown on the lifestyle and dietary diversity of women handloom workers. *Clinical Epidemiology and Global Health, 12*. Scopus. doi:10.1016/j.cegh.2021.100856

Ali, A. M., Kunugi, H., Abdelmageed, H. A., Mandour, A. S., Ahmed, M. E., Ahmad, S., & Hendawy, A. O. (2021). Vitamin k in covid-19—Potential anti-covid-19 properties of fermented milk fortified with bee honey as a natural source of vitamin k and probiotics. *Fermentation, 7*(4). Scopus. doi:10.3390/fermentation7040202

Almeida, E. M. T. C., & Almeida, M. T. C. (2021). Attitude and behavioral changes of dairy consumers during the new coronavirus pandemic in Brazil. *International Journal of Dairy Science, 16*(2), 67–74. Scopus. doi:10.3923/ijds.2021.67.74

Aydin, E., & Demir, P. A. (2022). The Effect Of The Pandemic On The Consumption Of Animal Products: The Case Of Kafkas University Of Turkey. *Online Journal of Animal and Feed Research, 12*(1), 37–45. Scopus. doi:10.51227/ojafr.2022.6

Ben Hassen, T., El Bilali, H., Allahyari, M. S., Berjan, S., & Fotina, O. (2021). Food purchase and eating behavior during the COVID-19 pandemic: A cross-sectional survey of Russian adults. *Appetite, 165*. Scopus. doi:10.1016/j.appet.2021.105309

Benedek, Z., Fertö, I., Marreiros, C. G., De Aguiar, P. M., Pocol, C. B., Čechura, L., Poder, A., Pääso, P., & Bakucs, Z. (2021). Farm diversification as a potential success factor for small-scale farmers constrained by COVID-related lockdown. Contributions from a survey conducted in four European countries during the first wave of COVID-19. *PLoS ONE, 16*(5 May). Scopus. doi:10.1371/journal.pone.0251715

Bhandari, G., Lal, P., Chaudhary, U., Haritha, K., Malhotra, R., & Chandel, B. S. (2021). Assessing snowball effect of COVID-19 pandemic on Indian dairy sector. [Scopus.]. *The Indian Journal of Animal Sciences, 91*(12), 1011–1017.

Biswal, J., Vijayalakshmy, K., & Rahman, H. (2020). Impact of COVID-19 and associated lockdown on livestock and poultry sectors in India. *Veterinary World, 13*(9), 1928–1933. Scopus. doi:10.14202/vetworld.2020.1928-1933

Bogevska, Z., Berjan, S., El Bilali, H., Sadegh Allahyari, M., Radosavac, A., & Davitkovska, M. (2022). Exploring food shopping, consumption and waste habits in North Macedonia during the COVID-19 pandemic. *Socio-Economic Planning Sciences, 82*. Scopus. doi:10.1016/j.seps.2021.101150

Bórawski, P., Bełdycka-Bórawska, A., Parzonko, A., Rokicki, T., & Holden, L. (2022). Changes in the comparative advantage of Polish dairy products. *Agricultural Economics (Czech Republic), 68*(12), 464–475. Scopus. doi:10.17221/322/2022-AGRICECON

Chatellier, V., Chaumet, J.-M., & Pouch, T. (2022). Covid-19, internationalagriculturaleconomy and-animalsectors: The case ofChina, the UnitedStates andthe EU. *Inra Productions Animales, 35*(1), 1–20. Scopus. doi:10.20870/productions-animales.2022.35.1.5376

Chen, Q. (2021). Prospects for chinese-russian cooperation in the dairy sector. *R-Economy, 7*(1), 52–60. Scopus. doi:10.15826/recon.2021.7.1.005

Darand, M., Hassanizadeh, S., Marzban, A., Mirzaei, M., & Hosseinzadeh, M. (2022). The association between dairy products and the risk of COVID-19. *European Journal of Clinical Nutrition, 76*(11), 1583–1589. Scopus. doi:10.1038/s41430-022-01149-8

de Melo Souza, T. C., Liboredo, J. C., Ferreira, L. G., Daniel, M. M., Di Renzo, L., Pivari, F., & Anastácio, L. R. (2022). Daily Habits of Brazilians at Different Moments of the COVID-19 Pandemic. *Nutrients, 14*(23). Scopus. doi:10.3390/nu14235136

Dugué, P., Sraïri, M. T., Jamin, J.-Y., Vall, E., Mburu, J., Ndambi, A., Sall, C., Camara, A. D., Sow, A., Ba, K., Corniaux, C., Diaw, A., Seck, D., Vigne, M., Audouin, S., Rakotomalala, L. J. E., Rakotonoely, L. N., Ferreira, F. D., Véromalalanirina, E., … Duteurtre, G. (2021). Early effects of the COVID-19 outbreak on the African dairy industry: Cases of Burkina Faso, Kenya, Madagascar, and Senegal. *Cahiers Agricultures, 30*. Scopus. doi:10.1051/cagri/2020047

Eurostat. (2023). *Several Statistics and Information*. https://ec.europa.eu/eurostat

Farias da Cruz, M., Baraúna Magno, M., Alves Jural, L., Pimentel, T. C., Masterson Tavares Pereira Ferreira, D., Almeida Esmerino, E., Luis Paiva Anciens Ramos, G., Vicente Gomila, J., Cristina Silva, M., Cruz, A. G. D., da Silva Fidalgo, T. K., & Cople Maia, L. (2022). Probiotics and dairy products in dentistry: A bibliometric and critical review of randomized clinical trials. *Food Research International, 157*. Scopus. doi:10.1016/j.foodres.2022.111228

Flis, Z., Szczecina, J., & Molik, E. (2022). The role of sheep's milk bioactive substances in the prevention of metabolic and viral diseases. *Journal of Animal and Feed Sciences, 31*(3), 211–216. Scopus. doi:10.22358/jafs/151020/2022

Guzel-Seydim, Z. B., Gökırmaklı, Ç., & Greene, A. K. (2021). A comparison of milk kefir and water kefir: Physical, chemical, microbiological and functional properties. *Trends in Food Science and Technology, 113*, 42–53. Scopus. doi:10.1016/j.tifs.2021.04.041

Hambardzumyan, G., & Gevorgyan, S. (2022). *The impact of COVID-19 on the small and medium dairy farms and comparative analysis of customers' behavior in Armenia. Future Foods, 5*. Scopus., doi:10.1016/j.fufo.2021.100110

HoteitM.MortadaH.Al-JawaldehA.IbrahimC.MansourR. (2022). COVID-19 home isolation and food consumption patterns: Investigating the correlates of poor dietary diversity in Lebanon: A cross-sectional study. *F1000Research, 11*. Scopus. doi:10.12688/f1000research.75761.1

Kumar, M., Kumar Choubey, V., Deepak, A., Gedam, V. V., & Raut, R. D. (2021). Life cycle assessment (LCA) of dairy processing industry: A case study of North India. *Journal of Cleaner Production, 326*. Scopus. doi:10.1016/j.jclepro.2021.129331

Miller, L. C., Neupane, S., Joshi, N., Lohani, M., Sah, K., & Shrestha, B. (2022). Dairy Animal Ownership and Household Milk Production Associated with Better Child and Family Diet in Rural Nepal during the COVID-19 Pandemic. *Nutrients, 14*(10). Scopus. doi:10.3390/nu14102074

Mishra, A., & Shukla, A. (2022). *Gyan Fresh: Digital Transformation of Dairy Business with Resilience and Technology Innovation. FIIB Business Review*. Scopus., doi:10.1177/23197145221113374

Mumena, W. A. (2020). Impact of COVID-19 curfew on eating habits, food intake, and weight according to food security status in Saudi Arabia: A retrospective study. *Progress in Nutrition, 22*(4). Scopus. doi:10.23751/pn.v22i4.9883

Nikooyeh, B., Rabiei, S., Amini, M., Ghodsi, D., Rasekhi, H., Doustmohammadian, A., Abdollahi, Z., Minaie, M., Sadeghi, F., & Neyestani, T. R. (2022). COVID-19 epidemic lockdown-induced remarkable decrease in dairy products consumption of Iran population: Does it really matter? National Food and Nutrition Surveillance. *BMC Nutrition, 8*(1). Scopus. doi:10.1186/s40795-022-00612-w

Nili, H., Bouzari, M., Attaran, H. R., Ghalegolab, N., Rabani, M., & Mahmoudian, A. (2022). Hyper-Immune Bovine Milk as an Immunological and Nutritional Supplement for COVID-19. *Frontiers in Nutrition, 9*. Scopus. doi:10.3389/fnut.2022.868964

Nohra, R. G., Naim, E., Chaaban, T., & Rothan-Tondeur, M. (2022). Nurses' eating habits in Lebanon during the economic and health crises: A cross-sectional study. *Archives of Public Health, 80*(1). Scopus. doi:10.1186/s13690-021-00775-1

Panta, R., Paswan, V. K., Kanetkar, P., Bunkar, D. S., Rose, H., & Bakshi, S. (2023). Exploring trade prospects of Chhurpi and the present status of Chhurpi producers and exporters of Nepal. *Journal of Ethnic Foods, 10*(1). Scopus. doi:10.1186/s42779-022-00165-0

Perrin, A., & Martin, G. (2021). Resilience of French organic dairy cattle farms and supply chains to the Covid-19 pandemic. *Agricultural Systems, 190*. Scopus. doi:10.1016/j.agsy.2021.103082

Pourghazi, F., Eslami, M., Ehsani, A., Ejtahed, H.-S., & Qorbani, M. (2022). Eating habits of children and adolescents during the COVID-19 era: A systematic review. *Frontiers in Nutrition, 9*. Scopus. doi:10.3389/fnut.2022.1004953

Prasad, A., & Kothari, N. (2022). Cow products: Boon to human health and food security. *Tropical Animal Health and Production, 54*(1). Scopus. doi:10.1007/s11250-021-03014-5

Qingbin, W., Liu, C.-Q., Zhao, Y.-F., Kitsos, A., Cannella, M., Wang, S.-K., & Han, L. (2020). Impacts of the COVID-19 pandemic on the dairy industry: Lessons from China and the United States and policy implications. *Journal of Integrative Agriculture, 19*(12), 2903–2915. Scopus. doi:10.1016/S2095-3119(20)63443-8

Quilliot, D., Gérard, M., Bonsack, O., Malgras, A., Vaillant, M.-F., Di Patrizio, P., Jaussaud, R., Ziegler, O., & Nguyen-Thi, P.-L. (2021). Impact of severe SARS-CoV-2 infection on nutritional status and subjective functional loss in a prospective cohort of COVID-19 survivors. *BMJ Open, 11*(7). Scopus. doi:10.1136/bmjopen-2021-048948

Salama, H. H., Trif, M., Rusu, A. V., & Bhattacharya, S. (2022). Application of Functional and Edible Coatings and Films as Promising Strategies for Developing Dairy Functional Products—A Review on Yoghurt Case. *Coatings, 12*(6). Scopus. doi:10.3390/coatings12060838

Scopus. (2023). *Scopus Database*. https://www.scopus.com/search/form.uri?display=basic#basic

Sodri, N. I., Mohamed-Yassin, M.-S., Nor, N. S. M., & Ismail, I. A. (2021). Rickets due to severe vitamin d and calcium deficiency during the covid-19 pandemic in Malaysia. *American Journal of Case Reports, 22*(1). Scopus. doi:10.12659/AJCR.934216

Stata. (2023). *Statistical software for data science*. Stata. https://www.stata.com/

StataCorp. (2017a). *Stata 15 Base Reference Manual*. Stata Press.

StataCorp. (2017b). *Stata Statistical Software: Release 15*. StataCorp LLC.

Yilmaz, B., Sharma, H., Melekoglu, E., & Ozogul, F. (2022). Recent developments in dairy kefir-derived lactic acid bacteria and their health benefits. *Food Bioscience, 46*. Scopus. doi:10.1016/j.fbio.2022.101592

Chapter 8
The Evolution of Poultry Farming in the European Union:
Our Days

ABSTRACT

Poultry meat has a relevant weight in the diet of the European Union citizens, considering their properties for human health and the characteristics of the associated sectors. The questions here are to understand the implications of the most recent occurrences worldwide on the poultry sector, including in the international markets. In this context, this study aims to assess the consequences of the pandemic and recent conflicts in the European Union poultry farming. Statistical information from Eurostat was considered for the period January 2019 – September 2022 for the following variables related to the chicks of laying hen breeds (laying): chicks hatched, eggs placed in incubation, exports intra-EU, exports – total, exports extra-EU, imports intra-EU, imports – total, imports extra-EU. The implications from the most recent events seem modest in this sector, including in the international markets.

INTRODUCTION

The human diet, where poultry meat is included, influences the spread of diseases, such as Covid-19 (Abdel-Wahab et al., 2020). A balanced consumption of poultry meat has benefits for health (Mafruchati, 2020), as a source of protein (Zargaraan et al., 2021). The pandemic (re)open the discussion about poultry management (Hafez & Attia, 2020), alternative productions (Mnisi et al., 2021) and the use of different feed sources (Attia et al., 2021), because the antimicrobial residues (Hassan et al., 2021).

Another debate was about the transference of viral diseases during the several steps of the meat food chains, including poultry meat processing (Bailey et al., 2022). Nonetheless, the role of the poultry sector in the viruses maintenance (Suarez et al., 2020), susceptibility (Hobbs & Reid, 2021) and replication (Berhane et al., 2021) were not confirmed. This is true namely for chicken (Hafez et al., 2021) and ducks (Shi et al., 2020), at least for great probabilities (Kumar et al., 2021). These results contradict the rumours of Covid-19 transmission through meat poultry (Das & Samanta, 2021) and eggs (Kolluri et al., 2021).

DOI: 10.4018/978-1-6684-8923-9.ch008

In any case, the signs of survival capacity of the viruses in some foods require special attention. Specifically, the susceptibility of the poultry sector to diverse diseases (Schat, 2022), such as different coronaviruses (Tizard, 2020), frequent outbreaks (Uwishema et al., 2021) and potential transmission for humans (Yang et al., 2022), as a source of zoonotic diseases (Zmrhal et al., 2022), suggest safe practices (Dhakal et al., 2021). Suggest also adjusted management of the risks in live poultry markets (Guo et al., 2022). The same care are suggested for the entire poultry industry (Keep et al., 2022) and backyard chickens (Larsen et al., 2022), where scientific research may bring important contributions (Li et al., 2021).

Covid-19 created challenges in diverse dimensions of society (Fulber & Kamen, 2022), human lives (Pourghazi et al., 2022) and public health (Pramod et al., 2021). This event created also disturbances in poultry educational programmes (Karcher et al., 2022) and operators' training (Violet et al., 2022). The pandemics are also phases of threats for the animals (Leung & Hon, 2023).

Poultry farming is vulnerable to the pandemic shocks (Biswal et al., 2020), namely small farmers and in countries with traditional problems related to poverty in some regions and social groups (Bamidele & Amole, 2021). These impacts were particularly problematic at the beginning of the Covid-19 disease (Brosch & Cartanza, 2021). The pandemic significantly impacted the egg and chicken output (Fang et al., 2021), partnerships (Nurlaelah et al., 2022) and the whole poultry sector (Maples et al., 2021). In any case, the variations in the poultry markets are not new, as a consequence of the recent shocks (Mubeena et al., 2021). Nonetheless, the impacts seem to be visible in the Greek imports of poultry meat (Stamatopoulou & Tzimitra-Kalogianni, 2022).

In this perspective, the study here presented aims to analyse the implications of the events that occurred since 2019 in poultry farming worldwide, namely in the European Union, considering data from the Eurostat (2023) database for the following indicators: chicks hatched; eggs placed in incubation; exports intra-EU; exports – total; exports extra-EU; imports intra-EU; imports – total; imports extra-EU. Some assessments of the statistical information were carried out following the Stata (StataCorp, 2017a, 2017b; Stata, 2023) software procedures. Information from Scopus (2023) was considered. The normalised values were obtained using the following ratio: $(x_i\text{-}x_{minimum})/(x_{maximum}\text{-}x_{minimum})$.

CHICKS HATCHED

The chicks hatched (chicks of laying hen breeds (laying), live animals, 1000 units) present the highest values, on average over the period January 2019-September 2022, in France, Netherlands, Spain, Poland, Germany, Czechia, Italy, Austria, Hungary, Belgium and Bulgaria (Table 1). On average, this variable decreased in 2019 (-1.8%), 2020 (-1.5%) and 2021 (-0.5%), and increased by 0.4% in 2022 (Table 2). The standard deviations were, respectively, 12.3%, 11.2%, 10.9% and 16.7%.

Table 1. Average values over the period January 2019 – September 2022 (chicks of laying hen breeds (laying), live animals, 1000 units)

Country	Value	Normalised Value
France	4131.097	1.000
Netherlands	3593.085	0.870
Spain	3184.511	0.771
Poland	3069.204	0.743
Germany	2832.935	0.686
Czechia	1074.530	0.260
Italy	809.689	0.196
Austria	793.642	0.192
Hungary	699.155	0.169
Belgium	519.526	0.126
Bulgaria	313.175	0.076
Finland	307.733	0.074
Denmark	292.822	0.071
Greece	226.818	0.055
Croatia	52.080	0.013
Cyprus	38.697	0.009
Romania	32.939	0.008
Estonia	0.000	0.000
Ireland	0.000	0.000
Lithuania	0.000	0.000
Luxembourg	0.000	0.000
Latvia	0.000	0.000
Malta	0.000	0.000
Portugal		
Sweden		
Slovenia		
Slovakia		

Table 2. Average values across the European Union countries (chicks of laying hen breeds (laying), live animals, 1000 units)

Period	Value	Growth Rates (%)
2019-01	1222.382	
2019-02	1168.516	-4.407
2019-03	1206.345	3.237
2019-04	1219.077	1.055
2019-05	1327.368	8.883
2019-06	1142.138	-13.955
2019-07	1176.289	2.990
2019-08	947.270	-19.470
2019-09	1002.113	5.790
2019-10	1194.935	19.242
2019-11	938.502	-21.460
2019-12	921.490	-1.813
2020-01	1280.064	
2020-02	1035.642	-19.095
2020-03	1017.257	-1.775
2020-04	1007.346	-0.974
2020-05	1167.989	15.947
2020-06	1140.454	-2.357
2020-07	1091.430	-4.299
2020-08	947.211	-13.214
2020-09	938.497	-0.920
2020-10	1113.661	18.664
2020-11	1001.839	-10.041
2020-12	1018.179	1.631
2021-01	1168.869	
2021-02	1034.697	-11.479
2021-03	1197.042	15.690
2021-04	1258.326	5.120
2021-05	1075.126	-14.559
2021-06	1181.279	9.874
2021-07	1167.678	-1.151
2021-08	936.651	-19.785
2021-09	1007.163	7.528
2021-10	1034.950	2.759
2021-11	1012.042	-2.213
2021-12	1039.647	2.728
2022-01	1032.588	

Continued on following page

Table 2. Continued

Period	Value	Growth Rates (%)
2022-02	857.340	-16.972
2022-03	1038.772	21.162
2022-04	977.849	-5.865
2022-05	1030.537	5.388
2022-06	987.596	-4.167
2022-07	828.958	-16.063
2022-08	756.184	-8.779
2022-09	971.445	28.467

EGGS PLACED IN INCUBATION

Netherlands, France, Poland, Spain, Belgium, Germany, Czechia, Italy and Austria are the leaders' countries for the average values of eggs placed in incubation, as highlighted in Table 3. On the other hand, this indicator increased around 3% (standard deviation of 18.6%), on average, in 2022, after negative growth rates in 2019 (-0.9%), 2020 (-0.8%) and 2021 (-1.1%). The standard deviations for 2019, 2020 and 2021 were, respectively, 10.3%, 6.8% and 8.6% (Table 4).

Table 3. Average values over the period January 2019 – September 2022 (chicks of laying hen breeds (laying), live animals, 1000 units)

Country	Value	Normalised Value
Netherlands	11982.889	1.000
France	11429.824	0.954
Poland	7998.707	0.668
Spain	7777.599	0.649
Belgium	6897.272	0.576
Germany	6680.365	0.557
Czechia	3053.085	0.255
Italy	2070.844	0.173
Austria	1989.174	0.166
Hungary	1862.212	0.155
Bulgaria	851.499	0.071
Finland	766.622	0.064
Denmark	723.172	0.060
Greece	579.532	0.048
Croatia	126.000	0.011
Cyprus	57.709	0.005
Romania	41.928	0.003
Estonia	0.000	0.000
Ireland	0.000	0.000
Lithuania	0.000	0.000
Luxembourg	0.000	0.000
Latvia	0.000	0.000
Malta	0.000	0.000
Portugal		
Sweden		
Slovenia		
Slovakia		

Table 4. Average values across the European Union countries (chicks of laying hen breeds (laying), live animals, 1000 units)

Period	Value	Growth Rates (%)
2019-01	3558.346	
2019-02	3497.136	-1.720
2019-03	3497.493	0.010
2019-04	3356.070	-4.044
2019-05	3758.350	11.987
2019-06	3130.030	-16.718
2019-07	2919.009	-6.742
2019-08	2950.286	1.071
2019-09	3047.395	3.292
2019-10	2925.015	-4.016
2019-11	2551.972	-12.754
2019-12	3051.000	19.555
2020-01	3610.483	
2020-02	3105.314	-13.992
2020-03	3071.620	-1.085
2020-04	2983.326	-2.875
2020-05	3148.555	5.538
2020-06	2901.962	-7.832
2020-07	2872.090	-1.029
2020-08	2839.640	-1.130
2020-09	2828.223	-0.402
2020-10	3127.274	10.574
2020-11	3342.987	6.898
2020-12	3232.160	-3.315
2021-01	3790.950	
2021-02	3260.819	-13.984
2021-03	3738.131	14.638
2021-04	3537.000	-5.381
2021-05	3748.743	5.987
2021-06	3516.455	-6.196
2021-07	3288.791	-6.474
2021-08	3181.101	-3.274
2021-09	2972.115	-6.570
2021-10	2921.390	-1.707
2021-11	2892.325	-0.995
2021-12	3234.585	11.833
2022-01	3047.666	

Continued on following page

Table 4. Continued

Period	Value	Growth Rates (%)
2022-02	2691.447	-11.688
2022-03	3089.350	14.784
2022-04	2831.316	-8.352
2022-05	3266.313	15.364
2022-06	3091.760	-5.344
2022-07	2240.656	-27.528
2022-08	2597.641	15.932
2022-09	3317.766	27.722

EXPORTS INTRA-EU

The great exporters (intra-EU) of chicks of laying hen breeds (laying) live animals (on average over the period considered) are Belgium, Netherlands, Hungary, Czechia, Germany, France, Austria, Portugal and Spain (Table 5). In turn, the European Union exports (intra-UE) increased by around 6% (standard deviation of 30.3%) in 2021 and 4% (standard deviation of 26.5%) in 2022 (Table 6). In 2019 and 2020, this indicator decreased by -1.5% (standard deviation of 23.9%) and -2.6% (standard deviation of 21.3%), respectively.

Table 5. Average values over the period January 2019 – September 2022 (chicks of laying hen breeds (laying), live animals, 1000 units)

Country	Value	Normalised Value
Belgium	1241.230	1.000
Netherlands	1233.289	0.994
Hungary	739.641	0.596
Czechia	691.935	0.557
Germany	535.182	0.431
France	252.602	0.204
Austria	185.722	0.150
Portugal	115.311	0.093
Spain	91.280	0.074
Italy	56.022	0.045
Finland	23.844	0.019
Ireland	18.911	0.015
Poland	9.213	0.007
Greece	6.243	0.005
Slovenia	5.107	0.004
Croatia	0.843	0.001
Bulgaria	0.474	0.000
Luxembourg	0.283	0.000
Cyprus	0.000	0.000
Denmark	0.000	0.000
Estonia	0.000	0.000
Lithuania	0.000	0.000
Latvia	0.000	0.000
Malta	0.000	0.000
Romania	0.000	0.000
Slovakia	0.000	0.000
Sweden		

Table 6. Average values across the European Union countries (chicks of laying hen breeds (laying), live animals, 1000 units)

Period	Value	Growth Rates (%)
2019-01	197.191	
2019-02	232.580	17.947
2019-03	241.380	3.784
2019-04	313.918	30.051
2019-05	262.680	-16.322
2019-06	235.557	-10.325
2019-07	195.682	-16.928
2019-08	236.304	20.759
2019-09	156.519	-33.764
2019-10	209.664	33.954
2019-11	149.873	-28.518
2019-12	123.453	-17.628
2020-01	260.781	
2020-02	262.338	0.597
2020-03	294.384	12.216
2020-04	248.576	-15.561
2020-05	231.295	-6.952
2020-06	275.291	19.022
2020-07	209.755	-23.806
2020-08	160.666	-23.403
2020-09	185.140	15.233
2020-10	247.152	33.495
2020-11	157.254	-36.374
2020-12	152.188	-3.222
2021-01	168.894	
2021-02	193.476	14.555
2021-03	312.035	61.278
2021-04	301.483	-3.382
2021-05	222.080	-26.337
2021-06	252.941	13.896
2021-07	182.149	-27.988
2021-08	143.538	-21.197
2021-09	211.715	47.498
2021-10	197.707	-6.616
2021-11	163.382	-17.362
2021-12	210.670	28.943
2022-01	196.129	

Continued on following page

Table 6. Continued

Period	Value	Growth Rates (%)
2022-02	167.735	-14.477
2022-03	259.997	55.005
2022-04	226.017	-13.069
2022-05	222.183	-1.696
2022-06	248.071	11.652
2022-07	165.735	-33.190
2022-08	195.007	17.662
2022-09	212.910	9.181

EXPORTS: TOTAL

The greatest total exports of chicks of laying hen breeds (laying) live animals, on average between 2019 and 2022, were carried out by the same leading countries (Belgium, Netherlands, Hungary, Czechia, Germany, France, Austria, Portugal and Spain) presented in the previous section for the intra-EU exports (Table 7). The total exports (Table 8) increased by 3% (standard deviation of 24.9%) in 2021 and 4% in 2022 (standard deviation of 13.1%). The average growth rates for 2019 and 2020 were, respectively -1.4% and -0.7%.

Table 7. Average values over the period January 2019 – September 2022 (chicks of laying hen breeds (laying), live animals, 1000 units)

Country	Value	Normalised Value
Belgium	2247.808	1.000
Netherlands	1826.156	0.812
Hungary	1004.908	0.447
Czechia	800.378	0.356
Germany	538.760	0.240
France	500.419	0.223
Austria	218.985	0.097
Portugal	135.462	0.060
Spain	111.483	0.050
Bulgaria	65.188	0.029
Italy	57.156	0.025
Greece	40.095	0.018
Finland	23.844	0.011
Poland	20.524	0.009
Ireland	18.911	0.008
Slovenia	5.107	0.002
Luxembourg	3.144	0.001
Croatia	0.843	0.000
Cyprus	0.000	0.000
Denmark	0.000	0.000
Estonia	0.000	0.000
Lithuania	0.000	0.000
Latvia	0.000	0.000
Malta	0.000	0.000
Romania	0.000	0.000
Slovakia	0.000	0.000
Sweden		

Table 8. Average values across the European Union countries (chicks of laying hen breeds (laying), live animals, 1000 units)

Period	Value	Growth Rates (%)
2019-01	267.460	
2019-02	324.618	21.371
2019-03	379.114	16.788
2019-04	469.956	23.962
2019-05	376.652	-19.854
2019-06	327.147	-13.143
2019-07	327.836	0.211
2019-08	333.782	1.814
2019-09	247.170	-25.949
2019-10	324.995	31.486
2019-11	247.224	-23.930
2019-12	177.798	-28.082
2020-01	361.143	
2020-02	379.136	4.982
2020-03	387.732	2.267
2020-04	293.089	-24.409
2020-05	268.902	-8.252
2020-06	343.413	27.709
2020-07	305.585	-11.015
2020-08	255.236	-16.476
2020-09	272.131	6.619
2020-10	408.458	50.096
2020-11	261.638	-35.945
2020-12	253.137	-3.249
2021-01	281.132	
2021-02	293.474	4.390
2021-03	484.593	65.123
2021-04	416.155	-14.123
2021-05	342.396	-17.724
2021-06	361.296	5.520
2021-07	329.219	-8.878
2021-08	252.200	-23.394
2021-09	317.434	25.866
2021-10	295.465	-6.921
2021-11	274.838	-6.981
2021-12	303.878	10.566
2022-01	255.198	

Continued on following page

Table 8. Continued

Period	Value	Growth Rates (%)
2022-02	261.134	2.326
2022-03	332.839	27.459
2022-04	309.216	-7.097
2022-05	315.927	2.170
2022-06	312.651	-1.037
2022-07	264.382	-15.439
2022-08	304.334	15.111
2022-09	321.766	5.728

EXPORTS EXTRA-EU

The top 10 countries with the biggest exports extra-EU of chicks of laying hen breeds (laying) live animals are the following (Table 9): Belgium; Netherlands; Hungary; France; Czechia; Bulgaria; Greece; Austria; Spain; and Portugal. These exports increased by around 3% (standard deviation of 32.6%) in 2019, 8% (standard deviation of 44.5%) in 2020, 2% (standard deviation of 30.3%) in 2021 and 14% (standard deviation of 36.3%) in 2022 (Table 10).

Table 9. Average values over the period January 2019 – September 2022 (chicks of laying hen breeds (laying), live animals, 1000 units)

Country	Value	Normalised Value
Belgium	1006.579	1.000
Netherlands	592.867	0.589
Hungary	272.040	0.270
France	247.816	0.246
Czechia	108.443	0.108
Bulgaria	64.713	0.064
Greece	33.851	0.034
Austria	33.263	0.033
Spain	20.203	0.020
Portugal	20.151	0.020
Poland	11.311	0.011
Germany	3.578	0.004
Luxembourg	2.861	0.003
Italy	1.133	0.001
Cyprus	0.000	0.000
Denmark	0.000	0.000
Estonia	0.000	0.000
Finland	0.000	0.000
Croatia	0.000	0.000
Ireland	0.000	0.000
Lithuania	0.000	0.000
Latvia	0.000	0.000
Malta	0.000	0.000
Romania	0.000	0.000
Slovenia	0.000	0.000
Slovakia	0.000	0.000
Sweden		

Table 10. Average values across the European Union countries (chicks of laying hen breeds (laying), live animals, 1000 units)

Period	Value	Growth Rates (%)
2019-01	70.269	
2019-02	92.037	30.978
2019-03	137.735	49.652
2019-04	156.039	13.289
2019-05	109.224	-30.002
2019-06	91.590	-16.145
2019-07	132.154	44.289
2019-08	87.205	-34.013
2019-09	87.024	-0.208
2019-10	110.525	27.005
2019-11	97.353	-11.918
2019-12	54.344	-44.178
2020-01	100.362	
2020-02	116.798	16.377
2020-03	93.348	-20.077
2020-04	44.513	-52.315
2020-05	37.607	-15.515
2020-06	68.123	81.144
2020-07	95.830	40.672
2020-08	94.571	-1.314
2020-09	86.991	-8.015
2020-10	161.305	85.427
2020-11	104.385	-35.287
2020-12	100.949	-3.292
2021-01	107.748	
2021-02	95.998	-10.905
2021-03	165.656	72.562
2021-04	110.086	-33.545
2021-05	120.316	9.293
2021-06	104.021	-13.544
2021-07	141.188	35.730
2021-08	104.316	-26.116
2021-09	97.586	-6.452
2021-10	93.998	-3.677
2021-11	106.998	13.830
2021-12	89.624	-16.238
2022-01	54.525	

Continued on following page

Table 10. Continued

Period	Value	Growth Rates (%)
2022-02	89.663	64.444
2022-03	73.375	-18.166
2022-04	94.544	28.850
2022-05	105.908	12.020
2022-06	61.997	-41.461
2022-07	98.648	59.117
2022-08	104.121	5.548
2022-09	103.413	-0.680

IMPORTS INTRA-EU

The great importers (intra-EU) of chicks of laying hen breeds (laying) are France, Netherlands, Slovakia, Portugal, Spain, Latvia, Germany, Czechia, Greece, Ireland, Bulgaria and Belgium (Table 11). These imports increased by 5% (standard deviation of 37.4%) in 2020, 4% (standard deviation of 31.2%) in 2021 and 5% (standard deviation of 31.0%) in 2022, on average, as presented in Table 12.

Table 11. Average values over the period January 2019 – September 2022 (chicks of laying hen breeds (laying), live animals, 1000 units)

Country	Value	Normalised Value
France	268.626	1.000
Netherlands	242.222	0.902
Slovakia	212.243	0.790
Portugal	196.166	0.730
Spain	182.727	0.680
Latvia	146.255	0.544
Germany	138.953	0.517
Czechia	128.892	0.480
Greece	119.965	0.447
Ireland	113.756	0.423
Bulgaria	94.431	0.352
Belgium	73.768	0.275
Croatia	61.137	0.228
Poland	57.757	0.215
Italy	55.000	0.205
Lithuania	32.639	0.122
Hungary	19.908	0.074
Cyprus	11.288	0.042
Austria	10.590	0.039
Luxembourg	0.056	0.000
Denmark	0.000	0.000
Estonia	0.000	0.000
Finland	0.000	0.000
Romania	0.000	0.000
Slovenia	0.000	0.000
Malta		
Sweden		

Table 12. Average values across the European Union countries (chicks of laying hen breeds (laying), live animals, 1000 units)

Period	Value	Growth Rates (%)
2019-01	96.462	
2019-02	84.348	-12.558
2019-03	87.860	4.164
2019-04	94.866	7.974
2019-05	88.465	-6.747
2019-06	85.342	-3.530
2019-07	74.595	-12.593
2019-08	87.143	16.822
2019-09	70.730	-18.835
2019-10	95.379	34.849
2019-11	82.475	-13.529
2019-12	50.992	-38.173
2020-01	100.262	
2020-02	79.160	-21.047
2020-03	96.321	21.679
2020-04	95.744	-0.599
2020-05	79.560	-16.903
2020-06	131.864	65.742
2020-07	69.266	-47.472
2020-08	70.930	2.402
2020-09	71.026	0.135
2020-10	119.864	68.761
2020-11	74.054	-38.218
2020-12	85.965	16.084
2021-01	71.619	
2021-02	81.611	13.952
2021-03	102.442	25.525
2021-04	76.614	-25.212
2021-05	104.846	36.850
2021-06	84.982	-18.946
2021-07	50.368	-40.731
2021-08	79.724	58.283
2021-09	92.701	16.277
2021-10	105.301	13.592
2021-11	68.018	-35.406
2021-12	69.946	2.835
2022-01	72.438	

Continued on following page

Table 12. Continued

Period	Value	Growth Rates (%)
2022-02	70.567	-2.583
2022-03	83.762	18.699
2022-04	81.999	-2.105
2022-05	99.693	21.578
2022-06	136.644	37.065
2022-07	60.293	-55.876
2022-08	51.149	-15.166
2022-09	69.895	36.650

IMPORTS: TOTAL

The results for the total imports are very similar to those presented previously (France, Netherlands, Slovakia, Portugal, Spain, Latvia, Germany, Czechia, Greece, Ireland, Bulgaria and Belgium) for the imports intra-EU (Tables 13-14), considering that the imports extra-EU are very low, such as presented in the next section.

Table 13. Average values over the period January 2019 – September 2022 (chicks of laying hen breeds (laying), live animals, 1000 units)

Country	Value	Normalised Value
France	270.296	1.000
Netherlands	242.222	0.896
Slovakia	208.151	0.770
Portugal	196.166	0.726
Spain	182.727	0.676
Latvia	146.255	0.541
Germany	139.148	0.515
Czechia	128.892	0.477
Greece	120.857	0.447
Ireland	113.756	0.421
Bulgaria	94.431	0.349
Belgium	73.768	0.273
Croatia	61.137	0.226
Poland	57.757	0.214
Italy	55.000	0.203
Lithuania	32.639	0.121
Hungary	19.908	0.074
Cyprus	11.288	0.042
Austria	10.590	0.039
Luxembourg	0.056	0.000
Denmark	0.000	0.000
Estonia	0.000	0.000
Finland	0.000	0.000
Romania	0.000	0.000
Slovenia	0.000	0.000
Malta		
Sweden		

Table 14. Average values across the European Union countries (chicks of laying hen breeds (laying), live animals, 1000 units)

Period	Value	Growth Rates (%)
2019-01	96.462	
2019-02	84.348	-12.558
2019-03	87.954	4.275
2019-04	91.411	3.930
2019-05	88.708	-2.957
2019-06	85.342	-3.794
2019-07	74.840	-12.306
2019-08	87.143	16.439
2019-09	71.199	-18.296
2019-10	95.690	34.398
2019-11	82.475	-13.810
2019-12	50.992	-38.173
2020-01	100.262	
2020-02	79.343	-20.864
2020-03	96.473	21.590
2020-04	95.744	-0.756
2020-05	79.560	-16.903
2020-06	125.063	57.193
2020-07	69.385	-44.520
2020-08	70.930	2.227
2020-09	71.026	0.135
2020-10	119.864	68.761
2020-11	74.148	-38.140
2020-12	86.000	15.984
2021-01	71.664	
2021-02	81.747	14.070
2021-03	102.442	25.316
2021-04	76.614	-25.212
2021-05	104.846	36.850
2021-06	84.982	-18.946
2021-07	50.368	-40.731
2021-08	79.724	58.283
2021-09	92.701	16.277
2021-10	105.301	13.592
2021-11	68.018	-35.406
2021-12	69.946	2.835
2022-01	65.463	

Continued on following page

Table 14. Continued

Period	Value	Growth Rates (%)
2022-02	70.567	7.797
2022-03	83.762	18.699
2022-04	81.999	-2.105
2022-05	99.693	21.578
2022-06	137.964	38.389
2022-07	60.867	-55.882
2022-08	51.666	-15.117
2022-09	69.895	35.282

IMPORTS EXTRA-EU

Only France, Greece and Germany present value for the imports extra-EU (Table 15). These findings together with those presented in Table 16 suggest that the imports of chicks of laying hen breeds (laying) live animals outside the European Union are very reduced and null in some cases.

Table 15. Average values over the period January 2019 – September 2022 (chicks of laying hen breeds (laying), live animals, 1000 units)

Country	Value	Normalised Value
France	1.670	1.000
Greece	0.892	0.534
Germany	0.195	0.117
Austria	0.000	0.000
Belgium	0.000	0.000
Bulgaria	0.000	0.000
Cyprus	0.000	0.000
Czechia	0.000	0.000
Denmark	0.000	0.000
Estonia	0.000	0.000
Spain	0.000	0.000
Finland	0.000	0.000
Croatia	0.000	0.000
Hungary	0.000	0.000
Ireland	0.000	0.000
Italy	0.000	0.000
Lithuania	0.000	0.000
Luxembourg	0.000	0.000
Latvia	0.000	0.000
Malta	0.000	0.000
Netherlands	0.000	0.000
Poland	0.000	0.000
Portugal	0.000	0.000
Romania	0.000	0.000
Slovenia	0.000	0.000
Slovakia	0.000	0.000
Sweden		

Table 16. Average values across the European Union countries (chicks of laying hen breeds (laying), live animals, 1000 units)

Period	Value
2019-01	0.000
2019-02	0.000
2019-03	0.090
2019-04	0.393
2019-05	0.243
2019-06	0.000
2019-07	0.235
2019-08	0.000
2019-09	0.450
2019-10	0.298
2019-11	0.000
2019-12	0.000
2020-01	0.000
2020-02	0.176
2020-03	0.146
2020-04	0.000
2020-05	0.000
2020-06	0.000
2020-07	0.110
2020-08	0.000
2020-09	0.000
2020-10	0.000
2020-11	0.087
2020-12	0.032
2021-01	0.044
2021-02	0.130
2021-03	0.000
2021-04	0.000
2021-05	0.000
2021-06	0.000
2021-07	0.000
2021-08	0.000
2021-09	0.000
2021-10	0.000
2021-11	0.000
2021-12	0.000
2022-01	0.000

Continued on following page

Table 16. Continued

Period	Value
2022-02	0.000
2022-03	0.000
2022-04	0.000
2022-05	0.000
2022-06	1.269
2022-07	0.530
2022-08	0.470
2022-09	0.000

CONCLUSION

This chapter intended to assess the impacts of the Covid-19 pandemic and Russia-Ukraine crisis on the European Union farming poultry. To achieve these objectives, statistical information from January 2019 to September 2022 (Euostat) was considered for the indicators associated with the chicks of laying hen breeds (laying), namely chicks hatched, eggs placed in incubation, exports intra-EU, exports – total; exports extra-EU, imports intra-EU, imports – total and imports extra-EU.

The survey of the scientific literature shows the importance of poultry meat for the human diet, particularly as a source of protein. Nonetheless, this sector is under diverse threats, specifically those related to the conditions of production, feed sources and diseases treatments. These are old problems to which the pandemic and the Russia-Ukraine conflict added new difficulties, in certain circumstances. Another debate was about the risks of viral disease transmission by the poultry sector, including through the meat, because of some outbreaks in the poultry industry. Generally, these rumours were not confirmed. In any case, the vulnerability of the poultry sector to several diseases requires special attention. In turn, this sector has a small dimension and reduced dynamics in some regions and countries, becoming it more susceptible to external shocks.

The data assessment reveals the importance of some countries for the European Union framework related to poultry farming (chicks of laying hen breeds (laying)), such as the following: France, Netherlands, Spain, Poland, Germany, Czechia, Italy, Austria, Hungary, Belgium and Bulgaria. In general, the consequences of the pandemic and the Russia-Ukraine crisis on poultry farming are not visible. There are some signs of increases in 2022, but this needs confirmation with more data by the end of the year. The great exporters for this indicator are, for example, Belgium, Netherlands, Hungary, Czechia, Germany, France, Austria, Portugal and Spain. The top importers are France, Netherlands, Slovakia, Portugal, Spain, Latvia, Germany, Czechia, Greece, Ireland, Bulgaria and Belgium. The impacts on the international markets for this indicator seem null or modest. There is some evidence of increase in international trade after the pandemic for this variable, however, this needs deeper analysis to understand if this is a consequence of cyclical evolutions, or an implication of changes in consumer behaviour, for example.

This contribution revealed the vulnerability of the poultry sector to diseases and international shocks, which requires special attention by the governments and international organisations, nonetheless the impacts on the European Union sector from the most recent events seem null, or modest. For policymakers are required adjusted policy instruments to promote a more sustainable management of the sector.

For future research, it is suggested to better understand the causes that promote average increases in the international markets after the pandemic.

ACKNOWLEDGMENT

This work is funded by National Funds through the FCT - Foundation for Science and Technology, I.P., within the scope of the project Refª UIDB/00681/2020. Furthermore we would like to thank the CERNAS Research Centre and the Polytechnic Institute of Viseu for their support.

REFERENCES

Abdel-Wahab, T. I., Abdel-Wahab, S. I., Abdelwahab, E. I., & Adel, M. M. (2020). A step forward to control of COVID-19. *International Journal of Pharmaceutical Research, 12*, 1167–1190. doi:10.31838/ijpr/2020.SP1.191

Attia, A. I., Reda, F. M., Patra, A. K., Elnesr, S. S., Attia, Y. A., & Alagawany, M. (2021). Date (Phoenix dactylifera l.) by-products: Chemical composition, nutritive value and applications in poultry nutrition, an updating review. *Animals, 11*(4). doi:10.3390/ani11041133

Bailey, E. S., Curcic, M., & Sobsey, M. D. (2022). Persistence of Coronavirus Surrogates on Meat and Fish Products during Long-Term Storage. *Applied and Environmental Microbiology, 88*(12). doi:10.1128/aem.00504-22

Bamidele, O., & Amole, T. A. (2021). Impact of covid-19 on smallholder poultry farmers in nigeria. *Sustainability (Switzerland), 13*(20). doi:10.3390/su132011475

Berhane, Y., Suderman, M., Babiuk, S., & Pickering, B. (2021). Susceptibility of turkeys, chickens and chicken embryos to SARS-CoV-2. *Transboundary and Emerging Diseases, 68*(6), 3038–3042. doi:10.1111/tbed.13970

Biswal, J., Vijayalakshmy, K., & Rahman, H. (2020). Impact of COVID-19 and associated lockdown on livestock and poultry sectors in India. *Veterinary World, 13*(9), 1928–1933. doi:10.14202/vetworld.2020.1928-1933

Brosch, C., & Cartanza, G. (2021). Covid-19 acutely impacted the delmarva poultry industry in early 2020. *Delaware Journal of Public Health, 7*(1). doi:10.32481/djph.2021.01.008

Das, P. K., & Samanta, I. (2021). Role of backyard poultry in South-East Asian countries: Post COVID-19 perspective. *World's Poultry Science Journal, 77*(2), 415–426. doi:10.1080/00439339.2021.1893620

Dhakal, J., Jia, M., Joyce, J. D., Moore, G. A., Ovissipour, R., & Bertke, A. S. (2021). Survival of severe acute respiratory syndrome coronavirus 2 (Sars-cov-2) and herpes simplex virus 1 (hsv-1) on foods stored at refrigerated temperature. *Foods, 10*(5). doi:10.3390/foods10051005

Eurostat. (2023). *Several Statistics and Information*. https://ec.europa.eu/eurostat

Fang, P., Belton, B., Zhang, X., & Ei Win, H. (2021). Impacts of COVID-19 on Myanmar's chicken and egg sector, with implications for the sustainable development goals. *Agricultural Systems, 190*. doi:10.1016/j.agsy.2021.103094

Fulber, J. P. C., & Kamen, A. A. (2022). Development and Scalable Production of Newcastle Disease Virus-Vectored Vaccines for Human and Veterinary Use. *Viruses, 14*(5). doi:10.3390/v14050975

Guo, J., Song, W., Ni, X., Zhou, K., Wu, J., Liu, W., Xia, W., He, F., Tong, X., Fan, G., Deng, Z., Liao, Z., Chen, H., & Chen, S. (2022). The Impact of the Closure of the Live Poultry Market due to COVID-19 on the Avian Influenza Virus in Nanchang, Jiangxi Province, China. *American Journal of Tropical Medicine and Hygiene, 106*(1), 127–131. doi:10.4269/ajtmh.21-0732

Hafez, H. M., & Attia, Y. A. (2020). Challenges to the Poultry Industry: Current Perspectives and Strategic Future After the COVID-19 Outbreak. *Frontiers in Veterinary Science, 7*. doi:10.3389/fvets.2020.00516

Hafez, H. M., Attia, Y. A., Bovera, F., Abd El-Hack, M. E., Khafaga, A. F., & de Oliveira, M. C. (2021). Influence of COVID-19 on the poultry production and environment. *Environmental Science and Pollution Research, 28*(33), 44833–44844. doi:10.1007/s11356-021-15052-5

Hassan, M. M., El Zowalaty, M. E., Lundkvist, Å., Järhult, J. D., Khan Nayem, M. R., Tanzin, A. Z., Badsha, M. R., Khan, S. A., & Ashour, H. M. (2021). Residual antimicrobial agents in food originating from animals. *Trends in Food Science and Technology, 111*, 141–150. doi:10.1016/j.tifs.2021.01.075

Hobbs, E. C., & Reid, T. J. (2021). Animals and SARS-CoV-2: Species susceptibility and viral transmission in experimental and natural conditions, and the potential implications for community transmission. *Transboundary and Emerging Diseases, 68*(4), 1850–1867. doi:10.1111/tbed.13885

Karcher, E. L., Koltes, D., Wenner, B., & Wells, J. (2022). Sparking curiosity and engagement through online curriculum. *Poultry Science, 101*(2). doi:10.1016/j.psj.2021.101577

Keep, S., Stevenson-Leggett, P., Dowgier, G., Foldes, K., Webb, I., Fones, A., Littolff, K., Everest, H., Britton, P., & Bickerton, E. (2022). A Temperature-Sensitive Recombinant of Avian Coronavirus Infectious Bronchitis Virus Provides Complete Protection against Homologous Challenge. *Journal of Virology, 96*(17). doi:10.1128/jvi.01100-22

Kolluri, G., Tyagi, J. S., & Sasidhar, P. V. K. (2021). Research Note: Indian poultry industry vis-à-vis coronavirus disease 2019: A situation analysis report. *Poultry Science, 100*(3). doi:10.1016/j.psj.2020.11.011

Kumar, A., Panwar, A., Batra, K., De, S., & Maan, S. (2021). In-silico analysis of angiotensin-converting enzyme 2 (ACE2) of livestock, pet and poultry animals to determine its susceptibility to SARS–CoV-2 infection. *Combinatorial Chemistry and High Throughput Screening, 24*(10), 1769–1783. doi:10.2174/1386207323666201110144542

Larsen, K. M., DeCicco, M., Hood, K., & Etter, A. J. (2022). Salmonella enterica frequency in backyard chickens in Vermont and biosecurity knowledge and practices of owners. *Frontiers in Veterinary Science, 9*. doi:10.3389/fvets.2022.979548

Leung, K. K. Y., & Hon, K. L. (2023). Lessons from Animal Culling during Human Pandemics: Is Vaccination a Viable Option for Animals? *Current Pediatric Reviews, 19*(1), 2–4. doi:10.2174/1573396318666220316124155

Li, Y., Zhuang, Q., Jiang, L., Jiang, W., Peng, C., Jiang, N., Zhang, F., Yu, X., Wang, S., Yuan, L., Hou, G., Liu, S., Wang, J., Yu, J., Li, J., Zhao, C., Huang, B., Liu, H., & Wang, K. (2021). Traceable surveillance and genetic diversity analysis of coronaviruses in poultry from China in 2019. *Virus Research, 306*. doi:10.1016/j.virusres.2021.198566

Mafruchati, M. (2020). Broiler Chicken vs. Turkey Meat; which One Has the Least Bad Fat to Avoid Positive Case of COVID-19? *Systematic Reviews in Pharmacy, 11*(10), 799–802. doi:10.31838/srp.2020.10.119

Maples, J. G., Thompson, J. M., Anderson, J. D., & Anderson, D. P. (2021). Estimating COVID-19 Impacts on the Broiler Industry. *Applied Economic Perspectives and Policy, 43*(1), 315–328. doi:10.1002/aepp.13089

Mnisi, C. M., Marareni, M., Manyeula, F., & Madibana, M. J. (2021). A way forward for the South African quail sector as a potential contributor to food and nutrition security following the aftermath of COVID-19: A review. *Agriculture and Food Security, 10*(1). doi:10.1186/s40066-021-00331-8

Mubeena, M., Muthukumar, M., & Barbuddhe, S. B. (2021). Trends and seasonal variations in broilers prices. *Veterinary Practitioner, 22*(2), 165–171.

Nurlaelah, S., Harnianti, Asnawi, A., & Sirajuddin, S. N. (2022). Income analysis on broiler chicken farming in partnership system during the covid-19 pandemic in Tanralili district, Indonesia. *Online Journal of Animal and Feed Research, 12*(4), 210–216. doi:10.51227/ojafr.2022.28

Pourghazi, F., Eslami, M., Ehsani, A., Ejtahed, H.-S., & Qorbani, M. (2022). Eating habits of children and adolescents during the COVID-19 era: A systematic review. *Frontiers in Nutrition, 9*. doi:10.3389/fnut.2022.1004953

Pramod, R. K., Nair, A. V., Tambare, P. K., Chauhan, K., Kumar, T. V., Rajan, R. A., Mani, B. M., Asaf, M., & Pandey, A. K. (2021). Reverse zoonosis of coronavirus disease-19: Present status and the control by one health approach. *Veterinary World, 14*(10), 2817–2826. doi:10.14202/vetworld.2021.2817-2826

Schat, K. A. (2022). Northeastern Conference on Avian Diseases from 1928-2021: 93 Years of Contributions to Organized Avian Medicine. *Avian Diseases, 66*(1), 7–19. 10.1637/aviandiseases-21-00074

Scopus. (2023). *Scopus Database*. https://www.scopus.com/search/form.uri?display=basic#basic

Shi, J., Wen, Z., Zhong, G., Yang, H., Wang, C., Huang, B., Liu, R., He, X., Shuai, L., Sun, Z., Zhao, Y., Liu, P., Liang, L., Cui, P., Wang, J., Zhang, X., Guan, Y., Tan, W., Wu, G., … Bu, Z. (2020). Susceptibility of ferrets, cats, dogs, and other domesticated animals to SARS-coronavirus 2. *Science, 368*(6494), 1016–1020. doi:10.1126/science.abb7015

Stamatopoulou, M., & Tzimitra-Kalogianni, I. (2022). Consumers' profile analysis for chicken meat, during the first wave of COVID-19 pandemic: Case of northern Greece. *Agraarteadus, 33*(1), 162–176. doi:10.15159/jas.22.01

Stata. (2023). *Statistical software for data science*. https://www.stata.com/

StataCorp. (2017a). *Stata 15 Base Reference Manual*. Stata Press.

StataCorp. (2017b). *Stata Statistical Software: Release 15*. StataCorp LLC.

Suarez, D. L., Pantin-Jackwood, M. J., Swayne, D. E., Lee, S. A., DeBlois, S. M., & Spackman, E. (2020). Lack of Susceptibility to SARS-CoV-2 and MERS-CoV in Poultry. *Emerging Infectious Diseases, 26*(12), 3074–3076. doi:10.3201/eid2612.202989 PMID:33219803

Tizard, I. R. (2020). Vaccination against coronaviruses in domestic animals. *Vaccine, 38*(33), 5123–5130. doi:10.1016/j.vaccine.2020.06.026

Uwishema, O., Adriano, L. F., Chalhoub, E., Onyeaka, H., Mhanna, M., David, S. C., Nasrallah, Y., Ribeiro, L. L. P. A., & Berjaoui, C. (2021). Bird flu outbreak amidst COVID-19 pandemic in South Africa: Efforts and challenges at hand. *Journal of Medical Virology, 93*(10), 5676–5679. doi:10.1002/jmv.27124

Violet, A. O., Elizabeth, A. O., & Cynthia, E. N. (2022). Enterprise Training Needs of Poultry Farmers During Covid-19 Pandemic in Enugu State, Nigeria. *Journal of Agricultural Extension, 26*(2), 129–137. doi:10.4314/jae.v26i1.16S

Yang, R., Sun, H., Gao, F., Luo, K., Huang, Z., Tong, Q., Song, H., Han, Q., Liu, J., Lan, Y., Qi, J., Li, H., Chen, S., Xu, M., Qiu, J., Zeng, G., Zhang, X., Huang, C., Pei, R., … Liu, J. (2022). Human infection of avian influenza A H3N8 virus and the viral origins: A descriptive study. *The Lancet Microbe, 3*(11), e824–e834. doi:10.1016/S2666-5247(22)00192-6

Zargaraan, A., Nasrabadi, F. M., Mehr, A. R., Eskandari, S., Salmani, Y., & Esfarjani, F. (2021). A Mixed-Method Study of Chicken Meat Safety in Iran during the COVID-19 Pandemic: SWOT Analysis. *Poultry Science Journal, 9*(2), 233–243. doi:10.22069/psj.2021.19049.1690

Zmrhal, V., Svoradova, A., Batik, A., & Slama, P. (2022). Three-Dimensional Avian Hematopoietic Stem Cell Cultures as a Model for Studying Disease Pathogenesis. *Frontiers in Cell and Developmental Biology, 9*. doi:10.3389/fcell.2021.730804

Chapter 9
Characterisation of the European Union Meat Sector Before and After the Pandemic

ABSTRACT

For the food security and food sovereignty of each country, the meat sector plays a determinant role. Nonetheless, on the other hand, to mitigate the carbon impacts, some national and international institutions suggest reductions in meat production and consumption. This chapter proposes to highlight the evolution of this sector after the COVID-19 pandemic. Data from Eurostat were considered over the period January 2017 – September 2022 for the following variables related to slaughtering in slaughterhouses in the European Union countries: bovine meat, pig meat, meat of sheep and goats, and poultry meat. The greatest average values for the European Union context over the period considered were found for the pig meat, followed by the poultry meat, bovine meat, and meat of sheep and goats, respectively. The pandemic and the Russia-Ukraine conflict seem to have reduced implications here.

INTRODUCTION

Since 2019/2020 the world assists to, at least, two great international shocks with impacts that spread over the diverse domains of human life (Mignogna et al., 2022). The pandemic is one of the most impactful international occurrence (Chatellier et al., 2022). The implications on the several dimensions of the markets were effective (Vall Castelló & Lopez Casasnovas, 2021), namely in the prices, with consequences for the supply and demand sides, including in the international trade (Das & Samanta, 2021).

The behaviour of the consumers and households was affected by the lockdowns (Morávková et al., 2022) and this had implications on the levels of food consumption (Górska et al., 2021b), including meat (Górska et al., 2021a). The pandemic impacted food consumption behaviour (Skotnicka et al., 2021) and health conditions (Yazlcl et al., 2022), but the diet also had implications on infections (Schaalan et al., 2022). With or without Covid-19 the behaviour of consumers among the different social groups is not

DOI: 10.4018/978-1-6684-8923-9.ch009

equal. In general, young consumers are more concerned with their health and this is reflected in their consumption decisions (Wojciechowska-Solis et al., 2022).

The prices of some food products (such as meat, fish, seafood and vegetables) were more affected (Akter, 2020) than others. The structure, organisation, diversification and dimension of the food supply chains were decisive for the success of the farmers, specifically the smaller ones, during the lockdown (Benedek et al., 2021). The resilience of the agri-food sector played also an important role, including that of the meat chains (Grinberga-Zalite et al., 2021). The food chain coordination and cooperation among the partners make the difference to take advantage of opportunities (Fleming et al., 2021). The African Swine Fever brought additional disruptions to the meat sector and markets (Duflot, 2021).

On the other hand, the conjuncture created by the pandemic and the associated risks for human health, for example, boosted the solution of old problems (Ban et al., 2022). The food sector is always associated with risks of contamination with micro-organisms and respective diseases, particularly incorrectly prepared meat (Mihai et al., 2022). This is a context in which the scientific community brings relevant knowledge in terms of analysis and evaluation (Bastin et al., 2022). In turn, the food processing facilities in certain regions and countries have experienced outbreaks (Dasgupta & Finch, 2020). The workplaces are often environments of outbreaks, increasing in this way the risks of transmission (Ingram et al., 2021). In any case, the ways the workers are contaminated are not always clear (Mallet et al., 2021). The pandemic does not affect the whole population in a similar way (McClure et al., 2020).

Considering this framework and the discussion about the production and consumption of meat (Dagevos & Verbeke, 2022), including during the pandemic (Losso et al., 2021), it is intended here to assess the European Union meat sector before and during the pandemic and the Russia-Ukraine crisis. For that, statistical information from the Eurostat (2023) database was considered and the Stata (StataCorp, 2017a, 2017b; Stata, 2023) software procedures were considered to obtain some results. The information taken into account is relative to the following dimensions: bovine meat; pigmeat; meat of sheep and goats; poultry meat. Scopus (2023) data were taken into account. To better understand the data, normalised values $((x_i-x_{minimum})/(x_{maximum}-x_{minimum}))$ were found.

BOVINE MEAT

France, Germany, Italy, Spain, Ireland, Poland, Netherlands, Belgium, Austria, Sweden, Denmark, Portugal and Finland are the top countries with the highest bovine meat production (slaughtering in slaughterhouses), on average, over the period January 2017-September 2022 (thousand tonnes), such as highlighted in Table 1. This indicator increased in 2017 (average of 0.6% and standard deviation of 10.2%), 2020 (average of 0.2% and standard deviation of 8.0%), 2021 (average of 1.4% and standard deviation of 7.8%) and 2022 (average of 1.8% and standard deviation of 9.1%) (Tables 2-3). In 2022, this variable, on average, decreased by -5% (standard deviation of 9.3%) in July, increased by 10% (standard deviation of 13.0%) in August and by 9% (standard deviation of 21.8%) in September (Tables 4-5). The average decreases in 2018 and 2019 were -0.4% and -0.2%, respectively.

Table 1. Average values from January 2017 to September 2022 (thousand tonnes)

Country	Value	Normalised Value
France	118.982	1.000
Germany	89.994	0.756
Italy	63.536	0.534
Spain	57.424	0.482
Ireland	51.472	0.432
Poland	46.502	0.390
Netherlands	36.186	0.304
Belgium	21.757	0.182
Austria	18.474	0.155
Sweden	11.360	0.095
Denmark	10.272	0.086
Portugal	8.097	0.067
Finland	7.138	0.059
Czechia	5.908	0.049
Croatia	3.601	0.030
Romania	3.584	0.029
Lithuania	3.488	0.029
Greece	3.062	0.025
Slovenia	2.981	0.024
Hungary	2.357	0.019
Latvia	1.279	0.010
Luxembourg	0.834	0.006
Estonia	0.724	0.005
Slovakia	0.683	0.005
Bulgaria	0.557	0.004
Cyprus	0.492	0.003
Malta	0.089	0.000

Table 2. Average values across the European Union countries (thousand tonnes), for 2017-2019

Period	Value	Growth Rates (%)
2017-01	20.597	
2017-02	19.441	-5.612
2017-03	22.602	16.259
2017-04	19.094	-15.521
2017-05	22.003	15.235
2017-06	21.078	-4.204
2017-07	19.924	-5.475
2017-08	21.708	8.954
2017-09	21.743	0.161
2017-10	22.734	4.558
2017-11	23.559	3.629
2017-12	21.006	-10.837
2018-01	21.962	
2018-02	19.486	-11.274
2018-03	22.370	14.800
2018-04	20.511	-8.310
2018-05	21.610	5.358
2018-06	20.926	-3.165
2018-07	21.201	1.314
2018-08	22.198	4.703
2018-09	21.385	-3.662
2018-10	24.283	13.552
2018-11	23.477	-3.319
2018-12	20.189	-14.005
2019-01	22.327	
2019-02	19.406	-13.083
2019-03	20.947	7.941
2019-04	21.379	2.062
2019-05	21.863	2.264
2019-06	19.066	-12.793
2019-07	21.475	12.635
2019-08	20.805	-3.120
2019-09	21.091	1.375
2019-10	23.827	12.972
2019-11	22.603	-5.137
2019-12	21.068	-6.791

Table 3. Average values across the European Union countries (thousand tonnes), for 2020-2022

Period	Value	Growth Rates (%)
2020-01	22.193	
2020-02	19.868	-10.476
2020-03	21.209	6.750
2020-04	18.882	-10.972
2020-05	19.047	0.874
2020-06	21.598	13.393
2020-07	21.636	0.176
2020-08	20.451	-5.477
2020-09	22.050	7.819
2020-10	22.781	3.315
2020-11	21.029	-7.691
2020-12	21.939	4.327
2021-01	19.189	
2021-02	19.059	-0.677
2021-03	22.797	19.613
2021-04	20.383	-10.589
2021-05	20.080	-1.487
2021-06	20.967	4.417
2021-07	20.567	-1.908
2021-08	21.018	2.193
2021-09	21.882	4.111
2021-10	21.369	-2.344
2021-11	22.892	7.127
2021-12	21.722	-5.111
2022-01	19.207	
2022-02	19.049	-0.823
2022-03	22.301	17.072
2022-04	20.135	-9.713
2022-05	21.003	4.311
2022-06	20.035	-4.609
2022-07	18.869	-5.820
2022-08	21.093	11.787
2022-09	21.551	2.171

Table 4. Values for June-July 2022

Location	Period	Value	Logarithm	Growth Rates (%)
Austria	2022-06	15.760	2.757	
Belgium	2022-06	19.530	2.972	
Bulgaria	2022-06	0.420	-0.868	
Cyprus	2022-06	0.440	-0.821	
Czechia	2022-06	5.580	1.719	
Germany	2022-06	73.440	4.296	
Denmark	2022-06	10.630	2.364	
Estonia	2022-06	0.630	-0.462	
Greece	2022-06	2.910	1.068	
Spain	2022-06	65.240	4.178	
Finland	2022-06	6.790	1.915	
France	2022-06	114.710	4.742	
Croatia	2022-06	3.490	1.250	
Hungary	2022-06	2.170	0.775	
Ireland	2022-06	48.250	3.876	
Italy	2022-06	56.380	4.032	
Lithuania	2022-06	3.230	1.172	
Luxembourg	2022-06	0.800	-0.223	
Latvia	2022-06	1.000	0.000	
Malta	2022-06	0.070	-2.659	
Netherlands	2022-06	35.250	3.562	
Poland	2022-06	47.730	3.866	
Portugal	2022-06	9.600	2.262	
Romania	2022-06	2.800	1.030	
Sweden	2022-06	10.900	2.389	
Slovenia	2022-06	2.580	0.948	
Slovakia	2022-06	0.610	-0.494	
Austria	2022-07	13.900	2.632	-11.802
Belgium	2022-07	16.860	2.825	-13.671
Bulgaria	2022-07	0.410	-0.892	-2.381
Cyprus	2022-07	0.360	-1.022	-18.182
Czechia	2022-07	4.800	1.569	-13.978
Germany	2022-07	71.910	4.275	-2.083
Denmark	2022-07	8.000	2.079	-24.741
Estonia	2022-07	0.620	-0.478	-1.587
Greece	2022-07	3.090	1.128	6.186
Spain	2022-07	62.160	4.130	-4.721
Finland	2022-07	6.150	1.816	-9.426

Continued on following page

Table 4. Continued

Location	Period	Value	Logarithm	Growth Rates (%)
France	2022-07	103.380	4.638	-9.877
Croatia	2022-07	3.630	1.289	4.011
Hungary	2022-07	2.120	0.751	-2.304
Ireland	2022-07	49.330	3.899	2.238
Italy	2022-07	57.830	4.058	2.572
Lithuania	2022-07	3.500	1.253	8.359
Luxembourg	2022-07	0.740	-0.301	-7.500
Latvia	2022-07	1.180	0.166	18.000
Malta	2022-07	0.070	-2.659	0.000
Netherlands	2022-07	31.490	3.450	-10.667
Poland	2022-07	43.200	3.766	-9.491
Portugal	2022-07	9.320	2.232	-2.917
Romania	2022-07	2.840	1.044	1.429
Sweden	2022-07	9.760	2.278	-10.459
Slovenia	2022-07	2.160	0.770	-16.279
Slovakia	2022-07	0.640	-0.446	4.918

Table 5. Values for August-September 2022

Location	Period	Value	Logarithm	Growth Rates (%)
Austria	2022-08	17.660	2.871	27.050
Belgium	2022-08	20.090	3.000	19.158
Bulgaria	2022-08	0.360	-1.022	-12.195
Cyprus	2022-08	0.350	-1.050	-2.778
Czechia	2022-08	5.500	1.705	14.583
Germany	2022-08	81.850	4.405	13.823
Denmark	2022-08	11.340	2.428	41.750
Estonia	2022-08	0.710	-0.342	14.516
Greece	2022-08	2.610	0.959	-15.534
Spain	2022-08	65.680	4.185	5.663
Finland	2022-08	7.980	2.077	29.756
France	2022-08	117.490	4.766	13.649
Croatia	2022-08	4.110	1.413	13.223
Hungary	2022-08	2.210	0.793	4.245
Ireland	2022-08	55.990	4.025	13.501
Italy	2022-08	60.180	4.097	4.064
Lithuania	2022-08	3.310	1.197	-5.429
Luxembourg	2022-08	0.770	-0.261	4.054
Latvia	2022-08	1.170	0.157	-0.847
Malta	2022-08	0.080	-2.526	14.286
Netherlands	2022-08	35.870	3.580	13.909
Poland	2022-08	45.630	3.821	5.625
Portugal	2022-08	9.890	2.292	6.116
Romania	2022-08	3.170	1.154	11.620
Sweden	2022-08	12.180	2.500	24.795
Slovenia	2022-08	2.730	1.004	26.389
Slovakia	2022-08	0.610	-0.494	-4.688
Austria	2022-09	17.140	2.841	-2.945
Belgium	2022-09	20.760	3.033	3.335
Bulgaria	2022-09	0.740	-0.301	105.556
Cyprus	2022-09	0.390	-0.942	11.429
Czechia	2022-09	5.850	1.766	6.364
Germany	2022-09	87.170	4.468	6.500
Denmark	2022-09	10.770	2.377	-5.026
Estonia	2022-09	0.810	-0.211	14.085
Greece	2022-09	3.220	1.169	23.372
Spain	2022-09	64.410	4.165	-1.934
Finland	2022-09	7.000	1.946	-12.281

Continued on following page

Table 5. Continued

Location	Period	Value	Logarithm	Growth Rates (%)
France	2022-09	116.960	4.762	-0.451
Croatia	2022-09	3.760	1.324	-8.516
Hungary	2022-09	2.230	0.802	0.905
Ireland	2022-09	57.600	4.054	2.876
Italy	2022-09	65.260	4.178	8.441
Lithuania	2022-09	4.100	1.411	23.867
Luxembourg	2022-09	0.840	-0.174	9.091
Latvia	2022-09	1.280	0.247	9.402
Malta	2022-09	0.080	-2.526	0.000
Netherlands	2022-09	38.000	3.638	5.938
Poland	2022-09	45.290	3.813	-0.745
Portugal	2022-09	8.370	2.125	-15.369
Romania	2022-09	3.310	1.197	4.416
Sweden	2022-09	12.540	2.529	2.956
Slovenia	2022-09	3.240	1.176	18.681
Slovakia	2022-09	0.760	-0.274	24.590

PIG MEAT

Germany, Spain, France, Poland and the Netherlands are the leaders' countries for pigmeat (thousand tonnes), on average from January 2017 until September 2022, as highlighted in Table 6. Pig meat production increased by 0.3% (standard deviation of 8.4%) in 2017, 0.01% (standard deviation of 6.3%) in 2020 and 0.2% (standard deviation of 6.8%) in 2021. Decreased -1.0% (standard deviation of 8.0%) in 2018, -0.8% (standard deviation of 6.5%) in 2019 and -1.2% (standard deviation of 8.3%) in 2022 (Tables 7-8). In 2022, this indicator decreased in July (average of -4.5% and standard deviation of 5.1%) and September (average of -3.3% and standard deviation of 8.4%) and increased by 6.9% (standard deviation of 8.1%) in August (Tables 9-10).

Table 6. Average values from January 2017 to September 2022 (thousand tonnes)

Country	Value	Normalised Value
Germany	427.006	1.000
Spain	397.240	0.930
France	182.310	0.426
Poland	164.293	0.384
Netherlands	134.389	0.314
Denmark	132.569	0.310
Italy	115.060	0.269
Belgium	89.349	0.209
Austria	41.719	0.097
Hungary	36.725	0.085
Portugal	29.862	0.069
Romania	27.129	0.063
Ireland	26.227	0.061
Sweden	20.578	0.047
Czechia	17.634	0.040
Finland	14.427	0.033
Croatia	6.542	0.014
Greece	6.410	0.014
Bulgaria	6.274	0.014
Lithuania	5.738	0.013
Slovakia	4.792	0.010
Estonia	3.658	0.008
Cyprus	3.545	0.007
Latvia	2.976	0.006
Slovenia	1.870	0.003
Luxembourg	1.052	0.002
Malta	0.378	0.000

Table 7. Average values across the European Union countries (thousand tonnes), for 2017-2019

Period	Value	Growth Rates (%)
2017-01	71.769	
2017-02	66.496	-7.347
2017-03	74.165	11.533
2017-04	62.771	-15.363
2017-05	71.249	13.506
2017-06	67.443	-5.342
2017-07	64.210	-4.794
2017-08	67.156	4.588
2017-09	70.063	4.329
2017-10	72.131	2.952
2017-11	72.908	1.077
2017-12	71.481	-1.957
2018-01	78.221	
2018-02	68.128	-12.903
2018-03	73.963	8.565
2018-04	69.103	-6.571
2018-05	71.947	4.116
2018-06	66.668	-7.337
2018-07	68.175	2.260
2018-08	67.379	-1.168
2018-09	66.841	-0.798
2018-10	76.219	14.030
2018-11	74.651	-2.057
2018-12	67.584	-9.467
2019-01	78.141	
2019-02	70.114	-10.272
2019-03	70.660	0.779
2019-04	70.404	-0.362
2019-05	70.389	-0.021
2019-06	63.135	-10.306
2019-07	68.509	8.512
2019-08	65.777	-3.988
2019-09	68.240	3.744
2019-10	75.057	9.990
2019-11	72.534	-3.361
2019-12	70.318	-3.055

Table 8. Average values across the European Union countries (thousand tonnes), for 2020-2022

Period	Value	Growth Rates (%)
2020-01	76.009	
2020-02	68.525	-9.846
2020-03	73.947	7.912
2020-04	65.599	-11.289
2020-05	65.983	0.585
2020-06	69.211	4.892
2020-07	69.935	1.046
2020-08	67.561	-3.395
2020-09	72.390	7.148
2020-10	73.380	1.368
2020-11	75.929	3.474
2020-12	74.544	-1.824
2021-01	74.731	
2021-02	72.055	-3.581
2021-03	80.410	11.595
2021-04	70.076	-12.852
2021-05	69.408	-0.953
2021-06	69.912	0.726
2021-07	66.925	-4.273
2021-08	69.400	3.698
2021-09	71.583	3.146
2021-10	70.493	-1.523
2021-11	77.144	9.435
2021-12	74.298	-3.689
2022-01	73.082	
2022-02	69.377	-5.070
2022-03	74.754	7.750
2022-04	65.053	-12.977
2022-05	70.249	7.987
2022-06	65.896	-6.197
2022-07	61.291	-6.988
2022-08	66.845	9.062
2022-09	64.935	-2.857

Table 9. Values for June-July 2022

Location	Period	Value	Logarithm	Growth Rates (%)
Austria	2022-06	38.220	3.643	
Belgium	2022-06	82.870	4.417	
Bulgaria	2022-06	7.110	1.962	
Cyprus	2022-06	3.260	1.182	
Czechia	2022-06	18.020	2.891	
Germany	2022-06	374.110	5.925	
Denmark	2022-06	124.960	4.828	
Estonia	2022-06	3.440	1.235	
Greece	2022-06	5.470	1.699	
Spain	2022-06	392.500	5.973	
Finland	2022-06	14.350	2.664	
France	2022-06	180.750	5.197	
Croatia	2022-06	6.600	1.887	
Hungary	2022-06	35.060	3.557	
Ireland	2022-06	27.220	3.304	
Italy	2022-06	100.230	4.607	
Lithuania	2022-06	5.360	1.679	
Luxembourg	2022-06	1.080	0.077	
Latvia	2022-06	2.980	1.092	
Malta	2022-06	0.340	-1.079	
Netherlands	2022-06	140.910	4.948	
Poland	2022-06	140.060	4.942	
Portugal	2022-06	26.870	3.291	
Romania	2022-06	20.680	3.029	
Sweden	2022-06	20.950	3.042	
Slovenia	2022-06	1.740	0.554	
Slovakia	2022-06	4.050	1.399	
Austria	2022-07	35.770	3.577	-6.410
Belgium	2022-07	77.990	4.357	-5.889
Bulgaria	2022-07	7.280	1.985	2.391
Cyprus	2022-07	3.190	1.160	-2.147
Czechia	2022-07	15.830	2.762	-12.153
Germany	2022-07	352.480	5.865	-5.782
Denmark	2022-07	122.200	4.806	-2.209
Estonia	2022-07	3.540	1.264	2.907
Greece	2022-07	5.480	1.701	0.183
Spain	2022-07	350.870	5.860	-10.606
Finland	2022-07	13.830	2.627	-3.624

Continued on following page

Table 9. Continued

Location	Period	Value	Logarithm	Growth Rates (%)
France	2022-07	165.070	5.106	-8.675
Croatia	2022-07	5.940	1.782	-10.000
Hungary	2022-07	32.900	3.493	-6.161
Ireland	2022-07	25.460	3.237	-6.466
Italy	2022-07	89.610	4.495	-10.596
Lithuania	2022-07	5.380	1.683	0.373
Luxembourg	2022-07	1.050	0.049	-2.778
Latvia	2022-07	2.970	1.089	-0.336
Malta	2022-07	0.350	-1.050	2.941
Netherlands	2022-07	131.910	4.882	-6.387
Poland	2022-07	131.930	4.882	-5.805
Portugal	2022-07	26.720	3.285	-0.558
Romania	2022-07	21.020	3.045	1.644
Sweden	2022-07	21.120	3.050	0.811
Slovenia	2022-07	1.520	0.419	-12.644
Slovakia	2022-07	3.450	1.238	-14.815

Table 10. Values for August-September 2022

Location	Period	Value	Logarithm	Growth Rates (%)
Austria	2022-08	39.880	3.686	11.490
Belgium	2022-08	82.180	4.409	5.372
Bulgaria	2022-08	6.860	1.926	-5.769
Cyprus	2022-08	3.410	1.227	6.897
Czechia	2022-08	18.180	2.900	14.845
Germany	2022-08	371.130	5.917	5.291
Denmark	2022-08	154.580	5.041	26.498
Estonia	2022-08	3.910	1.364	10.452
Greece	2022-08	5.040	1.617	-8.029
Spain	2022-08	375.540	5.928	7.031
Finland	2022-08	15.060	2.712	8.894
France	2022-08	180.190	5.194	9.160
Croatia	2022-08	6.150	1.816	3.535
Hungary	2022-08	36.020	3.584	9.483
Ireland	2022-08	26.320	3.270	3.378
Italy	2022-08	102.840	4.633	14.764
Lithuania	2022-08	5.410	1.688	0.558
Luxembourg	2022-08	0.990	-0.010	-5.714
Latvia	2022-08	2.970	1.089	0.000
Malta	2022-08	0.330	-1.109	-5.714
Netherlands	2022-08	143.830	4.969	9.036
Poland	2022-08	143.740	4.968	8.952
Portugal	2022-08	30.650	3.423	14.708
Romania	2022-08	21.600	3.073	2.759
Sweden	2022-08	22.140	3.097	4.830
Slovenia	2022-08	1.710	0.536	12.500
Slovakia	2022-08	4.160	1.426	20.580
Austria	2022-09	39.570	3.678	-0.777
Belgium	2022-09	85.230	4.445	3.711
Bulgaria	2022-09	6.230	1.829	-9.184
Cyprus	2022-09	3.460	1.241	1.466
Czechia	2022-09	16.510	2.804	-9.186
Germany	2022-09	366.830	5.905	-1.159
Denmark	2022-09	99.740	4.603	-35.477
Estonia	2022-09	3.430	1.233	-12.276
Greece	2022-09	4.970	1.603	-1.389
Spain	2022-09	391.730	5.971	4.311
Finland	2022-09	13.500	2.603	-10.359

Continued on following page

Table 10. Continued

Location	Period	Value	Logarithm	Growth Rates (%)
France	2022-09	177.710	5.180	-1.376
Croatia	2022-09	6.350	1.848	3.252
Hungary	2022-09	34.970	3.554	-2.915
Ireland	2022-09	27.140	3.301	3.116
Italy	2022-09	101.910	4.624	-0.904
Lithuania	2022-09	5.110	1.631	-5.545
Luxembourg	2022-09	1.060	0.058	7.071
Latvia	2022-09	3.020	1.105	1.684
Malta	2022-09	0.350	-1.050	6.061
Netherlands	2022-09	141.710	4.954	-1.474
Poland	2022-09	147.260	4.992	2.449
Portugal	2022-09	28.100	3.336	-8.320
Romania	2022-09	21.010	3.045	-2.731
Sweden	2022-09	20.740	3.032	-6.323
Slovenia	2022-09	1.540	0.432	-9.942
Slovakia	2022-09	4.070	1.404	-2.163

MEAT OF SHEEP AND GOATS

The meat of sheep and goats presents the highest values (thousand tonnes, on average) in the following countries (Table 11): Spain; France; Greece; Ireland; Italy; Germany; Netherlands; Portugal; Romania; and Austria. The production of this meat increased around 4-7% in the years considered: 2017 (average of 7% and standard deviation of 23%), 2018 (average of 7% and standard deviation of 26%), 2019 (average of 7% and standard deviation of 27%), 2020 (average of 6% and standard deviation of 24%), 2021 (average of 6% and standard deviation of 27%) and 2022 (average of 4% and standard deviation of 26%), such as presented in Tables 12-13. This variable increased by 10% in July 2022, 1% in August and 2% in September (Tables 14-15).

Table 11. Average values from January 2017 to September 2022 (thousand tonnes)

Country	Value	Normalised Value
Spain	10.766	1.000
France	7.315	0.679
Greece	5.787	0.538
Ireland	5.534	0.514
Italy	2.669	0.248
Germany	2.009	0.187
Netherlands	1.374	0.128
Portugal	0.858	0.080
Romania	0.793	0.074
Austria	0.620	0.058
Cyprus	0.439	0.041
Sweden	0.415	0.039
Belgium	0.249	0.023
Bulgaria	0.212	0.020
Finland	0.115	0.011
Denmark	0.094	0.009
Poland	0.088	0.008
Hungary	0.061	0.006
Latvia	0.041	0.004
Slovakia	0.021	0.002
Malta	0.014	0.001
Estonia	0.014	0.001
Lithuania	0.013	0.001
Slovenia	0.013	0.001
Czechia	0.008	0.001
Luxembourg	0.004	0.000

Table 12. Average values across the European Union countries (thousand tonnes), for 2017-2019

Period	Value	Growth Rates (%)
2017-01	1.230	
2017-02	1.260	2.439
2017-03	1.712	35.873
2017-04	2.256	31.776
2017-05	1.630	-27.748
2017-06	1.483	-9.018
2017-07	1.394	-6.001
2017-08	1.652	18.508
2017-09	1.526	-7.627
2017-10	1.457	-4.522
2017-11	1.417	-2.745
2017-12	2.061	45.448
2018-01	1.346	
2018-02	1.235	-8.247
2018-03	1.917	55.223
2018-04	1.954	1.930
2018-05	1.738	-11.054
2018-06	1.560	-10.242
2018-07	1.594	2.179
2018-08	1.893	18.758
2018-09	1.403	-25.885
2018-10	1.518	8.197
2018-11	1.408	-7.246
2018-12	2.157	53.196
2019-01	1.384	
2019-02	1.327	-4.118
2019-03	1.583	19.292
2019-04	2.378	50.221
2019-05	1.680	-29.352
2019-06	1.530	-8.929
2019-07	1.630	6.536
2019-08	1.737	6.564
2019-09	1.448	-16.638
2019-10	1.589	9.738
2019-11	1.358	-14.537
2019-12	2.127	56.627

Table 13. Average values across the European Union countries (thousand tonnes), for 2020-2022

Period	Value	Growth Rates (%)
2020-01	1.384	
2020-02	1.248	-9.827
2020-03	1.630	30.609
2020-04	2.246	37.791
2020-05	1.553	-30.855
2020-06	1.452	-6.504
2020-07	1.743	20.041
2020-08	1.492	-14.400
2020-09	1.406	-5.764
2020-10	1.476	4.979
2020-11	1.348	-8.672
2020-12	1.970	46.142
2021-01	1.307	
2021-02	1.244	-4.820
2021-03	2.154	73.151
2021-04	2.056	-4.550
2021-05	1.773	-13.765
2021-06	1.534	-13.480
2021-07	1.769	15.319
2021-08	1.516	-14.302
2021-09	1.511	-0.330
2021-10	1.449	-4.103
2021-11	1.438	-0.759
2021-12	1.992	38.526
2022-01	1.344	
2022-02	1.345	0.074
2022-03	1.878	39.628
2022-04	2.584	37.593
2022-05	1.648	-36.223
2022-06	1.559	-5.400
2022-07	1.88	20.590
2022-08	1.601	-14.840
2022-09	1.495	-6.621

Table 14. Values for June-July 2022

Location	Period	Value	Logarithm	Growth Rates (%)
Austria	2022-06	0.630	-0.462	
Belgium	2022-06	0.220	-1.514	
Bulgaria	2022-06	0.070	-2.659	
Cyprus	2022-06	0.510	-0.673	
Czechia	2022-06	0.000		
Germany	2022-06	1.780	0.577	
Denmark	2022-06	0.050	-2.996	
Estonia	2022-06	0.010	-4.605	
Greece	2022-06	3.690	1.306	
Spain	2022-06	11.430	2.436	
Finland	2022-06			
France	2022-06	7.410	2.003	
Croatia	2022-06			
Hungary	2022-06	0.050	-2.996	
Ireland	2022-06	5.270	1.662	
Italy	2022-06	1.740	0.554	
Lithuania	2022-06			
Luxembourg	2022-06	0.000		
Latvia	2022-06	0.040	-3.219	
Malta	2022-06	0.010	-4.605	
Netherlands	2022-06	1.290	0.255	
Poland	2022-06	0.110	-2.207	
Portugal	2022-06	0.950	-0.051	
Romania	2022-06	0.210	-1.561	
Sweden	2022-06	0.370	-0.994	
Slovenia	2022-06	0.010	-4.605	
Slovakia	2022-06			
Austria	2022-07	0.640	-0.446	1.587
Belgium	2022-07	0.220	-1.514	0.000
Bulgaria	2022-07	0.060	-2.813	-14.286
Cyprus	2022-07	0.530	-0.635	3.922
Czechia	2022-07	0.000		
Germany	2022-07	2.520	0.924	41.573
Denmark	2022-07	0.070	-2.659	40.000
Estonia	2022-07	0.010	-4.605	0.000
Greece	2022-07	4.240	1.445	14.905
Spain	2022-07	12.040	2.488	5.337
Finland	2022-07			

Continued on following page

Table 14. Continued

Location	Period	Value	Logarithm	Growth Rates (%)
France	2022-07	8.250	2.110	11.336
Croatia	2022-07			
Hungary	2022-07			
Ireland	2022-07	5.730	1.746	8.729
Italy	2022-07	2.000	0.693	14.943
Lithuania	2022-07			
Luxembourg	2022-07	0.010	-4.605	
Latvia	2022-07	0.050	-2.996	25.000
Malta	2022-07	0.010	-4.605	0.000
Netherlands	2022-07	1.940	0.663	50.388
Poland	2022-07	0.150	-1.897	36.364
Portugal	2022-07	0.730	-0.315	-23.158
Romania	2022-07			
Sweden	2022-07	0.280	-1.273	-24.324
Slovenia	2022-07	0.010	-4.605	0.000
Slovakia	2022-07			

Table 15. Values for August-September 2022

Location	Period	Value	Logarithm	Growth Rates (%)
Austria	2022-08	0.520	-0.654	-18.750
Belgium	2022-08	0.220	-1.514	0.000
Bulgaria	2022-08	0.100	-2.303	66.667
Cyprus	2022-08	0.680	-0.386	28.302
Czechia	2022-08	0.000		
Germany	2022-08	1.670	0.513	-33.730
Denmark	2022-08	0.060	-2.813	-14.286
Estonia	2022-08	0.010	-4.605	0.000
Greece	2022-08	4.160	1.426	-1.887
Spain	2022-08	11.040	2.402	-8.306
Finland	2022-08			
France	2022-08	7.260	1.982	-12.000
Croatia	2022-08			
Hungary	2022-08			
Ireland	2022-08	6.170	1.820	7.679
Italy	2022-08	1.700	0.531	-15.000
Lithuania	2022-08	0.020	-3.912	
Luxembourg	2022-08	0.010	-4.605	0.000
Latvia	2022-08	0.050	-2.996	0.000
Malta	2022-08	0.010	-4.605	0.000
Netherlands	2022-08	1.350	0.300	-30.412
Poland	2022-08	0.120	-2.120	-20.000
Portugal	2022-08	0.860	-0.151	17.808
Romania	2022-08	0.380	-0.968	
Sweden	2022-08	0.420	-0.868	50.000
Slovenia	2022-08	0.010	-4.605	0.000
Slovakia	2022-08			
Austria	2022-09	0.590	-0.528	13.462
Belgium	2022-09	0.210	-1.561	-4.545
Bulgaria	2022-09	0.110	-2.207	10.000
Cyprus	2022-09	0.390	-0.942	-42.647
Czechia	2022-09	0.000		
Germany	2022-09	1.830	0.604	9.581
Denmark	2022-09	0.070	-2.659	16.667
Estonia	2022-09	0.020	-3.912	
Greece	2022-09	4.030	1.394	-3.125
Spain	2022-09	9.330	2.233	-15.489
Finland	2022-09			

Continued on following page

Table 15. Continued

Location	Period	Value	Logarithm	Growth Rates (%)
France	2022-09	6.260	1.834	-13.774
Croatia	2022-09			
Hungary	2022-09			
Ireland	2022-09	6.540	1.878	5.997
Italy	2022-09	1.590	0.464	-6.471
Lithuania	2022-09	0.020	-3.912	0.000
Luxembourg	2022-09	0.010	-4.605	0.000
Latvia	2022-09	0.050	-2.996	0.000
Malta	2022-09	0.010	-4.605	0.000
Netherlands	2022-09	1.300	0.262	-3.704
Poland	2022-09	0.120	-2.120	0.000
Portugal	2022-09	0.710	-0.342	-17.442
Romania	2022-09	0.590	-0.528	55.263
Sweden	2022-09	0.600	-0.511	42.857
Slovenia	2022-09	0.010	-4.605	0.000
Slovakia	2022-09			

POULTRY MEAT

Poland, France, Spain, Germany and Italy are the top 5 countries with the highest averages (thousand tonnes) for poultry meat production (Table 16). The average growth rates for the years considered, when positive (in 2017, 2020, 2021 and 2022) are modest, with growth rates of less than 1% (Tables 17-18). Over the months considered for 2022, this indicator only presents positive average growth rates for August (5% and standard deviation of 9.3%), such as highlighted in Tables 19-20.

Table 16. Average values from January 2017 to September 2022 (thousand tonnes)

Country	Value	Normalised Value
Poland	213.419	1.000
France	137.724	0.645
Spain	136.738	0.641
Germany	130.884	0.613
Italy	110.683	0.519
Hungary	42.918	0.201
Romania	38.149	0.179
Belgium	38.027	0.178
Portugal	29.409	0.138
Greece	19.235	0.090
Sweden	14.186	0.066
Czechia	14.013	0.066
Ireland	13.778	0.065
Denmark	13.015	0.061
Finland	11.711	0.055
Austria	11.701	0.055
Bulgaria	9.415	0.044
Lithuania	8.258	0.039
Slovakia	6.182	0.029
Slovenia	5.909	0.028
Croatia	5.693	0.027
Latvia	2.888	0.014
Cyprus	2.188	0.010
Malta	0.331	0.002
Luxembourg	0.000	0.000
Estonia		
Netherlands		

Table 17. Average values across the European Union countries (thousand tonnes), for 2017-2019

Period	Value	Growth Rates (%)
2017-01	41.060	
2017-02	37.287	-9.189
2017-03	43.144	15.708
2017-04	38.360	-11.088
2017-05	43.797	14.174
2017-06	41.650	-4.902
2017-07	39.269	-5.717
2017-08	42.839	9.091
2017-09	39.199	-8.497
2017-10	41.716	6.421
2017-11	41.361	-0.851
2017-12	39.752	-3.890
2018-01	42.373	
2018-02	38.045	-10.214
2018-03	42.312	11.216
2018-04	41.405	-2.144
2018-05	43.077	4.038
2018-06	41.427	-3.830
2018-07	42.120	1.673
2018-08	41.741	-0.900
2018-09	39.575	-5.189
2018-10	46.272	16.922
2018-11	43.121	-6.810
2018-12	40.208	-6.755
2019-01	43.967	
2019-02	39.400	-10.387
2019-03	41.648	5.706
2019-04	42.960	3.150
2019-05	44.581	3.773
2019-06	40.305	-9.592
2019-07	43.526	7.992
2019-08	41.533	-4.579
2019-09	42.319	1.892
2019-10	46.260	9.313
2019-11	42.514	-8.098
2019-12	42.529	0.035

Table 18. Average values across the European Union countries (thousand tonnes), for 2020-2022

Period	Value	Growth Rates (%)
2020-01	44.087	
2020-02	39.739	-9.862
2020-03	43.497	9.457
2020-04	43.159	-0.777
2020-05	42.570	-1.365
2020-06	43.028	1.076
2020-07	43.613	1.360
2020-08	41.273	-5.365
2020-09	43.771	6.052
2020-10	44.719	2.166
2020-11	43.657	-2.375
2020-12	44.890	2.824
2021-01	40.527	
2021-02	37.716	-6.936
2021-03	44.343	17.571
2021-04	41.254	-6.966
2021-05	40.655	-1.452
2021-06	41.715	2.607
2021-07	42.412	1.671
2021-08	43.847	3.383
2021-09	44.711	1.970
2021-10	42.451	-5.055
2021-11	43.395	2.224
2021-12	43.653	0.595
2022-01	39.607	
2022-02	35.846	-9.496
2022-03	41.504	15.784
2022-04	38.604	-6.987
2022-05	40.672	5.357
2022-06	41.024	0.865
2022-07	39.009	-4.912
2022-08	41.724	6.960
2022-09	41.662	-0.149

Table 19. Values for June-July 2022

Location	Period	Value	Logarithm	Growth Rates (%)
Austria	2022-06	12.120	2.495	
Belgium	2022-06	39.780	3.683	
Bulgaria	2022-06	9.970	2.300	
Cyprus	2022-06	2.110	0.747	
Czechia	2022-06	14.750	2.691	
Germany	2022-06	133.000	4.890	
Denmark	2022-06	13.060	2.570	
Estonia	2022-06			
Greece	2022-06	20.660	3.028	
Spain	2022-06	139.020	4.935	
Finland	2022-06	12.890	2.556	
France	2022-06	110.320	4.703	
Croatia	2022-06	7.690	2.040	
Hungary	2022-06	38.740	3.657	
Ireland	2022-06	12.290	2.509	
Italy	2022-06	103.130	4.636	
Lithuania	2022-06	7.430	2.006	
Luxembourg	2022-06	0.000		
Latvia	2022-06	2.870	1.054	
Malta	2022-06	0.370	-0.994	
Netherlands	2022-06			
Poland	2022-06	244.160	5.498	
Portugal	2022-06	31.490	3.450	
Romania	2022-06	42.080	3.740	
Sweden	2022-06	14.740	2.691	
Slovenia	2022-06	6.010	1.793	
Slovakia	2022-06	6.930	1.936	
Austria	2022-07	12.380	2.516	2.145
Belgium	2022-07	34.630	3.545	-12.946
Bulgaria	2022-07	9.540	2.255	-4.313
Cyprus	2022-07	2.290	0.829	8.531
Czechia	2022-07	14.010	2.640	-5.017
Germany	2022-07	127.000	4.844	-4.511
Denmark	2022-07	12.320	2.511	-5.666
Estonia	2022-07			
Greece	2022-07	22.260	3.103	7.744
Spain	2022-07	131.860	4.882	-5.150
Finland	2022-07	12.080	2.492	-6.284

Continued on following page

Table 19. Continued

Location	Period	Value	Logarithm	Growth Rates (%)
France	2022-07	109.000	4.691	-1.197
Croatia	2022-07	6.550	1.879	-14.824
Hungary	2022-07	36.430	3.595	-5.963
Ireland	2022-07	15.700	2.754	27.746
Italy	2022-07	102.590	4.631	-0.524
Lithuania	2022-07	7.090	1.959	-4.576
Luxembourg	2022-07	0.000		
Latvia	2022-07	2.950	1.082	2.787
Malta	2022-07	0.370	-0.994	0.000
Netherlands	2022-07			
Poland	2022-07	218.190	5.385	-10.636
Portugal	2022-07	30.990	3.434	-1.588
Romania	2022-07	40.960	3.713	-2.662
Sweden	2022-07	13.660	2.614	-7.327
Slovenia	2022-07	6.000	1.792	-0.166
Slovakia	2022-07	6.380	1.853	-7.937

Table 20. Values for August-September 2022

Location	Period	Value	Logarithm	Growth Rates (%)
Austria	2022-08	11.020	2.400	-10.985
Belgium	2022-08	39.110	3.666	12.937
Bulgaria	2022-08	10.500	2.351	10.063
Cyprus	2022-08	2.080	0.732	-9.170
Czechia	2022-08	15.400	2.734	9.921
Germany	2022-08	132.000	4.883	3.937
Denmark	2022-08	13.500	2.603	9.578
Estonia	2022-08			
Greece	2022-08	23.750	3.168	6.694
Spain	2022-08	140.680	4.946	6.689
Finland	2022-08	13.150	2.576	8.858
France	2022-08	126.290	4.839	15.862
Croatia	2022-08	6.340	1.847	-3.206
Hungary	2022-08	42.440	3.748	16.497
Ireland	2022-08	12.370	2.515	-21.210
Italy	2022-08	106.210	4.665	3.529
Lithuania	2022-08	7.480	2.012	5.501
Luxembourg	2022-08	0.000		
Latvia	2022-08	3.150	1.147	6.780
Malta	2022-08	0.340	-1.079	-8.108
Netherlands	2022-08			
Poland	2022-08	230.180	5.439	5.495
Portugal	2022-08	34.110	3.530	10.068
Romania	2022-08	43.510	3.773	6.226
Sweden	2022-08	16.020	2.774	17.277
Slovenia	2022-08	6.580	1.884	9.667
Slovakia	2022-08	6.880	1.929	7.837
Austria	2022-09	11.730	2.462	6.443
Belgium	2022-09	38.610	3.654	-1.278
Bulgaria	2022-09	10.200	2.322	-2.857
Cyprus	2022-09	2.180	0.779	4.808
Czechia	2022-09	14.170	2.651	-7.987
Germany	2022-09	129.000	4.860	-2.273
Denmark	2022-09	12.940	2.560	-4.148
Estonia	2022-09			
Greece	2022-09	20.510	3.021	-13.642
Spain	2022-09	142.990	4.963	1.642
Finland	2022-09	12.440	2.521	-5.399

Continued on following page

Table 20. Continued

Location	Period	Value	Logarithm	Growth Rates (%)
France	2022-09	129.520	4.864	2.558
Croatia	2022-09	6.050	1.800	-4.574
Hungary	2022-09	45.160	3.810	6.409
Ireland	2022-09	16.880	2.826	36.459
Italy	2022-09	110.090	4.701	3.653
Lithuania	2022-09	6.830	1.921	-8.690
Luxembourg	2022-09	0.000		
Latvia	2022-09	3.060	1.118	-2.857
Malta	2022-09	0.340	-1.079	0.000
Netherlands	2022-09			
Poland	2022-09	226.560	5.423	-1.573
Portugal	2022-09	31.010	3.434	-9.088
Romania	2022-09	42.630	3.753	-2.023
Sweden	2022-09	15.770	2.758	-1.561
Slovenia	2022-09	6.260	1.834	-4.863
Slovakia	2022-09	6.630	1.892	-3.634

CONCLUSION

This research sought to understand the meat sector before and during the pandemic in the European Union member-states, with data from the Eurostat database for the months between January 2017 and September 2022. The variables considered are related to slaughtering in slaughterhouses. The indicators taken into account were bovine meat, pigmeat, the meat of sheep and goats and poultry meat.

The insights from the literature reveal that the most recent international shocks had consequences on the markets, namely in the prices and trade. Particularly, the lockdowns, during the pandemic, impacted food consumption with implications for human well-being. Nonetheless, these impacts were not uniform over the diverse social groups, depending on their age and social condition. In turn, some food prices were more affected, promoting changes in the consumers' behaviour. In any case, other shocks, such as the African Swine Fever, also promoted disturbances in the meat markets. In these contexts, workplaces, particularly those associated with meat processing, in certain circumstances, are vulnerable to outbreaks. In addition, the meat sector is associated with activities where the risks of contamination, during meat processing and from animal diseases, are real. This calls always for special care in the supply chains.

In the European Union context, the bovine meat, related to slaughtering in slaughterhouses, increased, on average, in the years during the pandemic (0.2% in 2020, 1.4% in 2021 and 1.8% in 2022) after decreases in 2018 and 2019. The pigmeat had average growth rates of 0.01% and 0.2% in 2020 and 2021, respectively, and negative growth rates of around -1% for 2018, 2019 and 2022. The meat of sheep and goats increased, on average, over the years considered, however, with slightly lower growth rates after the beginning of the pandemic. Poultry meat also presented positive (however modest) during the pandemic.

This chapter highlights that there are some signs of substitution effects in meat consumption that promoted changes in the respective markets, specifically in the supply. However, this needs further assessment in future research. On the other hand, it could be relevant that the policymakers create policy instruments that promote a more balanced consumption among the different meat alternatives.

ACKNOWLEDGMENT

This work is funded by National Funds through the FCT - Foundation for Science and Technology, I.P., within the scope of the project Refª UIDB/00681/2020. Furthermore we would like to thank the CERNAS Research Centre and the Polytechnic Institute of Viseu for their support.

REFERENCES

Akter, S. (2020). The impact of COVID-19 related 'stay-at-home' restrictions on food prices in Europe: Findings from a preliminary analysis. *Food Security, 12*(4), 719–725. doi:10.1007/s12571-020-01082-3

Ban, C., Bohle, D., & Naczyk, M. (2022). A perfect storm: COVID-19 and the reorganisation of the German meat industry. *Transfer, 28*(1), 101–118. doi:10.1177/10242589221081943

Bastin, B., Thompson, W., Benzinger, M. J., Crowley, E. S., Leonte, A.-M., Vandoros, E. J., Thomas, D., Hughes, A., Crabtree, D., Evans, K., & Sohier, D. (2022). Evaluation of the Thermo Scientific SureTect Salmonella Species PCR Assay in a Broad Range of Foods and Select Environmental Surfaces: Pre-Collaborative and Collaborative Study: First Action 2021.02. *Journal of AOAC International, 105*(1), 167–190. doi:10.1093/jaoacint/qsab122

Benedek, Z., Fertö, I., Marreiros, C. G., De Aguiar, P. M., Pocol, C. B., Čechura, L., Poder, A., Pääso, P., & Bakucs, Z. (2021). Farm diversification as a potential success factor for small-scale farmers constrained by COVID-related lockdown. Contributions from a survey conducted in four European countries during the first wave of COVID-19. *PLoS ONE, 16*(5). doi:10.1371/journal.pone.0251715

Chatellier, V., Chaumet, J.-M., & Pouch, T. (2022). Covid-19, international agricultural economy and animal sectors: The case of China, the United States and the EU. *Inra Productions Animales, 35*(1), 1–20. doi:10.20870/productions-animales.2022.35.1.5376

Dagevos, H., & Verbeke, W. (2022). Meat consumption and flexitarianism in the Low Countries. *Meat Science, 192*. doi:10.1016/j.meatsci.2022.108894

Das, P. K., & Samanta, I. (2021). Role of backyard poultry in South-East Asian countries: Post COVID-19 perspective. *World's Poultry Science Journal, 77*(2), 415–426. doi:10.1080/00439339.2021.1893620

Dasgupta, K., & Finch, S. E. (2020). A Case of Pediatric Multisystem Inflammatory Syndrome Temporally Associated with COVID-19 in South Dakota. *South Dakota Medicine : The Journal of the South Dakota State Medical Association*, *73*(6), 246–251. PMID:32580256

Duflot, B. (2021). Which consequences of african swine fever on the global swine industry and other animal industries? *Bulletin de l'Academie Veterinaire de France, 174*. doi:10.3406/bavf.2021.70947

Eurostat. (2023). *Several Statistics and Information*. https://ec.europa.eu/eurostat

Fleming, E., Griffith, G., Mounter, S., Hartmann, M., & Simons, J. (2021). Food Value Chain Coordination in Practice: European and Australian Case Studies of the Creation of Chain Good Innovations*. *International Journal on Food System Dynamics, 12*(3), 196–205. doi:10.18461/ijfsd.v12i2.85

Górska, P., Górna, I., Miechowicz, I., & Przysławski, J. (2021a). Changes in eating behaviour during sars-cov-2 pandemic among the inhabitants of five European countries. *Foods, 10*(7). doi:10.3390/foods10071624

Górska, P., Górna, I., Miechowicz, I., & Przysławski, J. (2021b). Changes in life situations during the sars-cov-2 virus pandemic and their impact on eating behaviors for residents of europe, australia as well as North and South America. *Nutrients, 13*(10). doi:10.3390/nu13103570

Grinberga-Zalite, G., Pilvere, I., Muska, A., & Kruzmetra, Z. (2021). Resilience of meat supply chains during and after covid-19 crisis. *Emerging Science Journal, 5*(1), 57–66. doi:10.28991/esj-2021-01257

Ingram, C., Downey, V., Roe, M., Chen, Y., Archibald, M., Kallas, K.-A., Kumar, J., Naughton, P., Uteh, C. O., Rojas-Chaves, A., Shrestha, S., Syed, S., Cléirigh Büttner, F., Buggy, C., & Perrotta, C. (2021). COVID-19 Prevention and Control Measures in Workplace Settings: A Rapid Review and Meta-Analysis. *International Journal of Environmental Research and Public Health, 18*(15), 7847. doi:10.3390/ijerph18157847 PMID:34360142

Losso, J. N., Losso, M. J. N., Toc, M., Inungu, J. N., & Finley, J. W. (2021). The Young Age and Plant-Based Diet Hypothesis for Low SARS-CoV-2 Infection and COVID-19 Pandemic in Sub-Saharan Africa. *Plant Foods for Human Nutrition, 76*(3), 270–280. doi:10.1007/s11130-021-00907-6

Mallet, Y., Pivette, M., Revest, M., Angot, E., Valence, M., Dupin, C., Picard, N., Brelivet, G., Seyler, T., Ballet, S., Le Tertre, A., & Guillois, Y. (2021). Identification of Workers at Increased Risk of Infection During a COVID-19 Outbreak in a Meat Processing Plant, France, May 2020. *Food and Environmental Virology, 13*(4), 535–543. doi:10.1007/s12560-021-09500-1

McClure, E. S., Vasudevan, P., Bailey, Z., Patel, S., & Robinson, W. R. (2020). Racial capitalism within public health-how occupational settings drive covid-19 disparities. *American Journal of Epidemiology, 189*(11), 1244–1253. doi:10.1093/aje/kwaa126

Mignogna, C., Costanzo, S., Ghulam, A., Cerletti, C., Donati, M. B., De Gaetano, G., Iacoviello, L., & Bonaccio, M. (2022). Impact of Nationwide Lockdowns Resulting from the First Wave of the COVID-19 Pandemic on Food Intake, Eating Behaviors, and Diet Quality: A Systematic Review. *Advances in Nutrition, 13*(2), 388–423. doi:10.1093/advances/nmab130

Mihai, I. F., Anita, D., Dorneanu, O. S., Luca, C. M., Manciuc, C. D., Budacu, C. C., Roşu, F. M., Savuta, G., Anita, A., & Vâţă, A. (2022). Seroprevalence of Anti-Hepatitis E Virus Antibodies among Patients from a Tertiary Hospital from Northeast Romania. *Medicina (Lithuania), 58*(8). doi:10.3390/medicina58081020

Morávková, M., Veselá, L., & Kubíčková, L. (2022). Changes in households' stances on wasting food during the COVID-19 pandemic. *Waste Forum, 3*, 161–178.

Schaalan, M., Abou Warda, A. E., Osman, S. M., Fathy, S., Sarhan, R. M., Boshra, M. S., Sarhan, N., Gaber, S., & Ali, A. M. A. (2022). The Impact of Sociodemographic, Nutritional, and Health Factors on the Incidence and Complications of COVID-19 in Egypt: A Cross-Sectional Study. *Viruses, 14*(3). doi:10.3390/v14030448

Scopus. (2023). *Scopus Database*. https://www.scopus.com/search/form.uri?display=basic#basic

Skotnicka, M., Karwowska, K., Kłobukowski, F., Wasilewska, E., & Małgorzewicz, S. (2021). Dietary habits before and during the covid-19 epidemic in selected European countries. *Nutrients, 13*(5). doi:10.3390/nu13051690

Stata. (2023). *Statistical software for data science*. https://www.stata.com/

StataCorp. (2017a). *Stata 15 Base Reference Manual*. Stata Press.

StataCorp. (2017b). *Stata Statistical Software: Release 15*. StataCorp LLC.

Vall Castelló, J., & Lopez Casasnovas, G. (2021). The effect of lockdowns and infection rates on supermarket sales. *Economics and Human Biology, 40*. doi:10.1016/j.ehb.2020.100947

Wojciechowska-Solis, J., Kowalska, A., Bieniek, M., Ratajczyk, M., & Manning, L. (2022). Comparison of the Purchasing Behaviour of Polish and United Kingdom Consumers in the Organic Food Market during the COVID-19 Pandemic. *International Journal of Environmental Research and Public Health, 19*(3). doi:10.3390/ijerph19031137

Yazlcl, D., Fersahoğlu, M. M., Fersahoğlu, T., Bulut, N. E., Çiğiltepe, H., Çeler, Ö., Sancak, S., Sulu, C., Durcan, E., Şahin, S., Polat Korkmaz, Ö., Bozdoğan Polat, S. H., Taşkln, E., İşeri, C., Gürsoy, E., Küçük Yetgin, M., Kaya, T., Özdemir, F., Mestanoğlu İstanbullu, N., … Yumuk, V. D. (2022). Status of Weight Change, Lifestyle Behaviors, Depression, Anxiety, and Diabetes Mellitus in a Cohort with Obesity during the COVID-19 Lockdown: Turk-Com Study Group. *Obesity Facts, 15*(4), 528–539. doi:10.1159/000522658

Chapter 10
Evolution of the Agricultural Producer Markets in the European Union

ABSTRACT

It is important to understand if the increases verified by the consumers in the agricultural and food prices have their sources in the agricultural productions increased costs or in the food chains costs/ speculation. In this way, it is proposed to assess the European Union agricultural producer markets before and after the pandemic. Data from the Eurostat were considered over the period 2016-2021 for the following producer prices (euros): soft wheat – prices per 100 kg; durum wheat – prices per 100 kg; maize – prices per 100 kg; rice – prices per 100 kg; soya – prices per 100 kg; tomatoes in the open: all qualities – prices per 100 kg; lettuce in the open: all qualities – prices per 100 kg; carrots: all qualities – prices per 100 kg; onions: all qualities – prices per 100 kg; and dessert apples: all varieties – prices per 100 kg. There are visible impacts from the pandemic on the producer prices, but they are not equal across the different production.

INTRODUCTION

The Covid-19 pandemic has had impacts on our lives (Ronaghi, 2022) in diverse ways (Cariappa et al., 2022) and may get worse old problems (Ibiyemi & Oldewage-Theron, 2022), including those from global warming (Koo et al., 2021) and China–USA trade crisis (Wu & Turvey, 2020). The efforts to improve food security were annulled by the restrictions measures to control the infection (Merchant et al., 2022).

The contexts created by these international and exogenous shocks have been influenced by several factors, such as social media (Bansal et al., 2020) and public opinion (Liu et al., 2022). On the other hand, the perceptions about the effects of the pandemic differed among countries, regions (Helfenstein et al., 2022) and locations (Hammond et al., 2022).

Agriculture is fundamental for the economic development of any country and this was visible during the pandemic (Srairi, 2021). The agricultural and food prices when rising affect the nutritional conditions

DOI: 10.4018/978-1-6684-8923-9.ch010

(Ivascu et al., 2021), namely of the populations with fewer incomes (Boughton et al., 2021). The reductions in agricultural prices have implications for the farmers' income (Harris et al., 2020) and sometimes are accompanied by higher production costs (Uddin et al., 2021), because of rises in input prices (Vatta et al., 2022). The pandemic had impacts on agricultural prices (Ceballos et al., 2021) volatility (Yan et al., 2021) and sales volume (Pilvere et al., 2021), because of disruptions (Ejeromedoghene et al., 2020) and reluctance in the supply chains (Selim & Eltarabily, 2022).

These implications call for adjusted policy interventions (Nugroho, 2021), particularly in more vulnerable communities (Ghosh-Jerath et al., 2022), to minimise the effects on the farmers (Pieter et al., 2022) and to deal with the connectivity of the markets (Tanaka & Guo, 2020). The new technologies and artificial intelligence may produce innovative insights to better support the farmers (Peng et al., 2021), including the internet (Zheng et al., 2022), to achieve adjusted prices (Kumar, 2022), as well as resilience and organisation of the farmers (Lopez-Ridaura et al., 2021).

The resilience and dimension of the supply chains have also their importance (McBurney et al., 2021), particularly to deal with lockdowns. In medium and long supply chains there are more operators to add value to the prices, beyond the producers (Mila et al., 2022). The Russia-Ukraine conflict created additional challenges for these markets (Feng et al., 2023).

However, in some circumstances were not identified relevant impacts from the pandemic on the agricultural markets (Hambardzumyan & Gevorgyan, 2022). In other cases, these consequences were identified as temporary (Wang et al., 2020). In addition, agritourism, for example, seems to have attracted more customers during the infection outbreak (Zawadka et al., 2022). The management of the pandemic's impacts on several economic sectors, including agriculture, allowed to avoid more dramatic consequences, nonetheless the concerns with the environment seem to persist (Edwards et al., 2022).

Following these highlights from the literature, this study aims to analyse the impacts on the European Union agricultural prices, in the producer markets, specifically from the Covid-19 pandemic. Data from the Eurostat (2023) database were considered for the period 2016-2021 and the following agricultural products (euros/100 kg): soft wheat; durum wheat; maize; rice; soya; tomatoes in the open (all qualities); lettuce in the open (all qualities); carrots (all qualities); onions (all qualities); dessert apples (all varieties). Some results were found with the Stata (StataCorp, 2017a, 2017b; Stata, 2023) software. Information from Scopus (2023) was considered for the literature review. Normalised values $((x_i-x_{minimum})/(x_{maximum}-x_{minimum}))$ were also found for the data assessment.

SOFT WHEAT

Tables 1-3 show that the highest soft wheat producer prices (per 100 kg) appeared in 2021 and the top countries are the following: Netherlands (24.35 euros); Spain (23.64 euros); Belgium (23.3 euros); Portugal (23.28 euros); Greece (23.27 euros); Germany (21.96 euros); Finland (21.55 euros); Poland (21.2 euros); and Estonia (21.1 euros). For example, the prices in Netherlands were 14.9 euros in 2016, 15.25 euros in 2017, 19.4 euros in 2018, 16.2 euros in 2019 and 18.4 euros in 2020. On average these prices were 14.6 euros (2.7 for the logarithm and 0.2 for the normalized value) in 2016, 15.4 euros (2.7 for the logarithm and 0.3 for the normalized value) in 2017, 17.0 euros (2.8 for the logarithm and 0.4 for the normalized value) in 2018, 16.7 euros (2.8 for the logarithm and 0.4 for the normalized value) in 2019, 16.8 euros (2.8 for the logarithm and 0.4 for the normalized value) in 2020 and 20.7 euros (3.0 for the logarithm and 0.7 for the normalized value) in 2021. It seems that the pandemic impacts on soft wheat prices started in 2021, however, this needs deeper assessment.

Table 1. The highest producer prices (per 100 kg) for the period 2016-2021

Location	Year	Value	Logarithm	Normalised Value For The Whole Period And Countries
Netherlands	2021	24.350	3.193	1.000
Spain	2021	23.640	3.163	0.945
Belgium	2021	23.300	3.148	0.918
Portugal	2021	23.280	3.148	0.916
Greece	2021	23.270	3.147	0.916
Germany	2021	21.960	3.089	0.813
Finland	2021	21.550	3.070	0.781
Poland	2021	21.200	3.054	0.754
Estonia	2021	21.100	3.049	0.746
Portugal	2019	20.720	3.031	0.716
Bulgaria	2021	20.450	3.018	0.695
Lithuania	2021	20.160	3.004	0.673
France	2016	20.100	3.001	0.668
Sweden	2021	20.090	3.000	0.667
Portugal	2020	20.040	2.998	0.663
Hungary	2021	19.990	2.995	0.659
Finland	2018	19.980	2.995	0.659
Italy	2019	19.940	2.993	0.655
Italy	2018	19.820	2.987	0.646
Italy	2020	19.810	2.986	0.645
Greece	2020	19.770	2.984	0.642
Latvia	2021	19.750	2.983	0.641
Czechia	2021	19.580	2.975	0.627
Romania	2021	19.510	2.971	0.622
Denmark	2021	19.480	2.969	0.620
Netherlands	2018	19.400	2.965	0.613
Greece	2019	19.080	2.949	0.588
Greece	2017	18.840	2.936	0.570
Slovenia	2021	18.780	2.933	0.565
Italy	2017	18.770	2.932	0.564
Spain	2019	18.660	2.926	0.555
Spain	2020	18.570	2.922	0.548

Continued on following page

Table 1. Continued

Location	Year	Value	Logarithm	Normalised Value For The Whole Period And Countries
Greece	2016	18.520	2.919	0.545
Slovakia	2021	18.520	2.919	0.545
Greece	2018	18.450	2.915	0.539
Portugal	2018	18.450	2.915	0.539
Sweden	2018	18.430	2.914	0.538
Finland	2019	18.400	2.912	0.535
Netherlands	2020	18.400	2.912	0.535
Spain	2018	18.290	2.906	0.527
Croatia	2021	18.150	2.899	0.516
Denmark	2019	17.970	2.889	0.502
Luxembourg	2018	17.930	2.886	0.498
Italy	2016	17.820	2.880	0.490
Portugal	2017	17.820	2.880	0.490
Spain	2017	17.640	2.870	0.476
Austria	2021	17.450	2.859	0.461

Table 2. The intermediate producer prices (per 100 kg) for the period 2016-2021

Location	Year	Value	Logarithm	Normalised Value For The Whole Period And Countries
Belgium	2020	17.450	2.859	0.461
Germany	2020	17.210	2.845	0.442
Luxembourg	2020	17.180	2.844	0.440
Latvia	2018	17.170	2.843	0.439
Lithuania	2018	17.160	2.843	0.438
Portugal	2016	17.060	2.837	0.430
Finland	2020	17.050	2.836	0.430
Latvia	2020	17.040	2.836	0.429
Poland	2018	17.040	2.836	0.429
Czechia	2019	16.950	2.830	0.422
Denmark	2018	16.850	2.824	0.414
Poland	2020	16.850	2.824	0.414
Germany	2018	16.840	2.824	0.413
Poland	2019	16.810	2.822	0.411
Estonia	2019	16.790	2.821	0.409
Estonia	2020	16.780	2.820	0.409
Bulgaria	2020	16.760	2.819	0.407
Lithuania	2020	16.730	2.817	0.405
Germany	2019	16.690	2.815	0.402
Belgium	2018	16.660	2.813	0.399
Denmark	2020	16.640	2.812	0.398
Estonia	2018	16.410	2.798	0.380
Romania	2020	16.330	2.793	0.373
Slovenia	2019	16.320	2.792	0.373
Latvia	2019	16.280	2.790	0.370
Lithuania	2019	16.250	2.788	0.367
Netherlands	2019	16.200	2.785	0.363
Czechia	2018	15.990	2.772	0.347
Slovenia	2018	15.940	2.769	0.343
Finland	2017	15.910	2.767	0.341
Luxembourg	2019	15.820	2.761	0.334
Czechia	2020	15.760	2.757	0.329
Spain	2016	15.730	2.756	0.327
Poland	2017	15.620	2.749	0.318
Belgium	2019	15.590	2.747	0.316
Estonia	2017	15.500	2.741	0.309

Continued on following page

Table 2. Continued

Location	Year	Value	Logarithm	Normalised Value For The Whole Period And Countries
Lithuania	2017	15.490	2.740	0.308
Bulgaria	2017	15.470	2.739	0.306
Hungary	2019	15.470	2.739	0.306
Austria	2018	15.410	2.735	0.302
Bulgaria	2018	15.400	2.734	0.301
Slovenia	2017	15.400	2.734	0.301
Romania	2019	15.380	2.733	0.299
Hungary	2020	15.350	2.731	0.297
Slovakia	2019	15.310	2.729	0.294
Hungary	2018	15.270	2.726	0.291
Netherlands	2017	15.250	2.725	0.289
Bulgaria	2019	15.210	2.722	0.286
Finland	2016	15.140	2.717	0.280
Germany	2017	15.130	2.717	0.280
Slovakia	2018	15.120	2.716	0.279
Luxembourg	2017	15.000	2.708	0.270
Croatia	2019	14.990	2.707	0.269
Denmark	2017	14.970	2.706	0.267
Netherlands	2016	14.900	2.701	0.262
Austria	2020	14.880	2.700	0.260
Latvia	2017	14.820	2.696	0.255

Table 3. The lowest producer prices (per 100 kg) for the period 2016-2021

Location	Year	Value	Logarithm	Normalised Value For The Whole Period And Countries
Slovenia	2020	14.820	2.696	0.255
Sweden	2020	14.780	2.693	0.252
Slovakia	2020	14.780	2.693	0.252
Estonia	2016	14.640	2.684	0.241
Austria	2019	14.630	2.683	0.241
Croatia	2018	14.610	2.682	0.239
Romania	2018	14.610	2.682	0.239
Czechia	2017	14.510	2.675	0.231
Lithuania	2016	14.500	2.674	0.230
Austria	2017	14.460	2.671	0.227
Luxembourg	2016	14.350	2.664	0.219
Poland	2016	14.230	2.655	0.209
Romania	2017	14.230	2.655	0.209
Belgium	2017	14.140	2.649	0.202
Denmark	2016	14.100	2.646	0.199
Bulgaria	2016	14.060	2.643	0.196
Germany	2016	14.050	2.643	0.195
Romania	2016	14.030	2.641	0.194
Hungary	2017	14.020	2.640	0.193
Slovakia	2017	13.980	2.638	0.190
Croatia	2020	13.940	2.635	0.187
Sweden	2019	13.910	2.633	0.184
Croatia	2017	13.890	2.631	0.183
Sweden	2017	13.880	2.630	0.182
Czechia	2016	13.690	2.617	0.167
Sweden	2016	13.540	2.606	0.155
Latvia	2016	13.460	2.600	0.149
Slovenia	2016	13.380	2.594	0.143
Belgium	2016	13.340	2.591	0.140
Hungary	2016	12.770	2.547	0.095
Slovakia	2016	12.290	2.509	0.058
Croatia	2016	12.190	2.501	0.050
Austria	2016	11.550	2.447	0.000
Luxembourg	2021			

DURUM WHEAT

For the durum wheat the biggest producer prices, for the period 2016-2021, were verified in 2021 for Spain, Greece, Hungary, Austria, Cyprus and Portugal (Table 4). In these countries, the prices increased significantly in 2021. On average, the prices were 22.1 euros (logarithm of 3.1 and normalized value of 0.3) in 2016, 19.8 euros (logarithm of 3.0 and normalized value of 0.2) in 2017, 20.2 euros (logarithm of 3.0 and normalized value of 0.3) in 2018, 20.9 euros (logarithm of 3.0 and normalized value of 0.3) in 2019, 23.0 euros (logarithm of 3.1 and normalized value of 0.4) in 2020 and 27.7 euros (logarithm of 3.3 and normalized value of 0.6) in 2021.

Table 4. Producer prices (per 100 kg) for the period 2016-2021

Location	Year	Value	Logarithm	Normalised Value
France	2016	37.700	3.630	1.000
Spain	2021	35.380	3.566	0.902
Greece	2021	33.140	3.501	0.807
Cyprus	2016	31.200	3.440	0.724
Hungary	2021	31.090	3.437	0.720
Austria	2021	28.530	3.351	0.611
Cyprus	2021	28.280	3.342	0.601
Italy	2020	27.430	3.312	0.564
Portugal	2021	27.360	3.309	0.561
Cyprus	2018	26.030	3.259	0.505
Hungary	2020	24.790	3.210	0.453
Cyprus	2020	24.750	3.209	0.451
Spain	2020	24.360	3.193	0.434
Italy	2019	23.520	3.158	0.399
Portugal	2016	23.500	3.157	0.398
Austria	2020	23.100	3.140	0.381
Greece	2020	23.010	3.136	0.377
Portugal	2017	23.000	3.135	0.377
Portugal	2020	23.000	3.135	0.377
Portugal	2019	22.970	3.134	0.375
Cyprus	2019	22.730	3.124	0.365
Italy	2016	22.410	3.110	0.352
Spain	2016	22.280	3.104	0.346
Spain	2017	22.180	3.099	0.342
Croatia	2021	22.180	3.099	0.342
Croatia	2017	22.010	3.091	0.335
Germany	2020	22.000	3.091	0.334
Spain	2019	21.870	3.085	0.329

Continued on following page

Table 4. Continued

Location	Year	Value	Logarithm	Normalised Value
Cyprus	2017	21.820	3.083	0.327
Portugal	2018	21.810	3.082	0.326
Italy	2018	21.790	3.081	0.325
Italy	2017	21.780	3.081	0.325
Bulgaria	2021	21.700	3.077	0.321
Germany	2019	21.500	3.068	0.313
Germany	2021	21.400	3.063	0.309
Croatia	2018	21.010	3.045	0.292
Germany	2018	21.000	3.045	0.292
Greece	2019	20.240	3.008	0.260
Austria	2019	20.030	2.997	0.251
Hungary	2019	19.820	2.987	0.242
Croatia	2020	19.720	2.982	0.237
Croatia	2019	19.480	2.969	0.227
Greece	2017	19.430	2.967	0.225
Spain	2018	19.100	2.950	0.211
Greece	2016	18.860	2.937	0.201
Greece	2018	18.830	2.935	0.200
Hungary	2018	18.830	2.935	0.200
Austria	2018	18.270	2.905	0.176
Austria	2017	18.240	2.904	0.175
Bulgaria	2020	18.240	2.904	0.175
Austria	2016	17.880	2.884	0.159
Hungary	2016	16.750	2.818	0.112
Bulgaria	2019	16.580	2.808	0.104
Bulgaria	2016	16.330	2.793	0.094
Bulgaria	2018	15.580	2.746	0.062
Bulgaria	2017	15.290	2.727	0.050
Croatia	2016	14.390	2.667	0.011
Hungary	2017	14.120	2.648	0.000

MAIZE

Greece, Spain, Portugal and Germany (Tables 5-6) are the top countries for maize producer prices in 2021. For this crop and these countries, a great increase in producer prices occurred in 2021. The average prices for 2016-2021 were, respectively, the following: 15.4 euros (2.7 for logarithm and 0.2 for normalized value); 15.3 euros (2.7 for logarithm and 0.2 for normalized value); 15.8 euros (2.8 for logarithm and 0.2 for normalized value); 15.3 euros (2.7 for logarithm and 0.2 for normalized value);

15.7 euros (2.7 for logarithm and 0.2 for normalized value); and 21.3 euros (3.1 for logarithm and 0.6 for normalized value).

Table 5. The highest producer prices (per 100 kg) for the period 2016-2021

Location	Year	Value	Logarithm	Normalised Value For The Whole Period And Countries
Greece	2021	27.870	3.328	1.000
Spain	2021	26.280	3.269	0.899
Portugal	2021	24.200	3.186	0.767
Germany	2021	22.260	3.103	0.644
Greece	2016	21.510	3.069	0.596
Greece	2018	21.310	3.059	0.583
Greece	2020	21.270	3.057	0.581
Greece	2019	21.180	3.053	0.575
Slovenia	2021	20.790	3.034	0.550
Bulgaria	2021	20.590	3.025	0.537
Greece	2017	20.550	3.023	0.535
Romania	2021	20.520	3.021	0.533
Hungary	2021	20.310	3.011	0.520
Croatia	2021	20.270	3.009	0.517
France	2016	20.100	3.001	0.506
Austria	2021	19.770	2.984	0.485
Spain	2020	19.500	2.970	0.468
Lithuania	2021	19.410	2.966	0.463
Slovakia	2021	19.240	2.957	0.452
Czechia	2021	19.150	2.952	0.446
Italy	2018	18.980	2.943	0.435
Italy	2017	18.940	2.941	0.433
Italy	2020	18.490	2.917	0.404
Italy	2016	18.480	2.917	0.403
Italy	2019	18.170	2.900	0.384
Luxembourg	2016	17.750	2.876	0.357
Spain	2019	17.710	2.874	0.355
Spain	2018	17.640	2.870	0.350
Poland	2021	17.580	2.867	0.346
Portugal	2020	17.430	2.858	0.337
Portugal	2018	17.320	2.852	0.330
Spain	2017	17.180	2.844	0.321

Continued on following page

Table 5. Continued

Location	Year	Value	Logarithm	Normalised Value For The Whole Period And Countries
Spain	2016	16.880	2.826	0.302
Portugal	2019	16.840	2.824	0.299
Portugal	2016	16.740	2.818	0.293
Romania	2020	16.740	2.818	0.293
Germany	2020	16.650	2.812	0.287
Portugal	2017	16.650	2.812	0.287
Germany	2019	16.520	2.805	0.279
Germany	2018	16.500	2.803	0.278
Luxembourg	2019	16.500	2.803	0.278
Romania	2016	16.480	2.802	0.276
Luxembourg	2017	16.000	2.773	0.246
Czechia	2019	15.970	2.771	0.244
Lithuania	2020	15.940	2.769	0.242
Czechia	2018	15.670	2.752	0.225
Germany	2017	15.660	2.751	0.224
Luxembourg	2018	15.500	2.741	0.214

Table 6. The lowest producer prices (per 100 kg) for the period 2016-2021

Location	Year	Value	Logarithm	Normalised Value For The Whole Period And Countries
Bulgaria	2020	15.450	2.738	0.211
Lithuania	2018	15.450	2.738	0.211
Romania	2018	15.260	2.725	0.199
Germany	2016	15.180	2.720	0.194
Romania	2019	15.170	2.719	0.193
Luxembourg	2020	15.000	2.708	0.182
Romania	2017	14.880	2.700	0.175
Czechia	2020	14.740	2.691	0.166
Slovenia	2018	14.690	2.687	0.163
Bulgaria	2017	14.580	2.680	0.156
Poland	2018	14.530	2.676	0.152
Hungary	2018	14.460	2.671	0.148
Czechia	2016	14.430	2.669	0.146
Slovakia	2018	14.390	2.667	0.144
Czechia	2017	14.360	2.664	0.142
Bulgaria	2018	14.310	2.661	0.139
Bulgaria	2016	14.240	2.656	0.134
Hungary	2020	14.180	2.652	0.130
Lithuania	2017	14.150	2.650	0.128
Slovenia	2017	14.130	2.648	0.127
Hungary	2017	14.120	2.648	0.126
Bulgaria	2019	14.040	2.642	0.121
Austria	2017	14.020	2.640	0.120
Austria	2018	14.010	2.640	0.119
Croatia	2017	13.940	2.635	0.115
Lithuania	2019	13.920	2.633	0.114
Poland	2019	13.920	2.633	0.114
Slovakia	2019	13.710	2.618	0.100
Croatia	2016	13.570	2.608	0.091
Slovakia	2017	13.500	2.603	0.087
Poland	2020	13.480	2.601	0.086
Hungary	2019	13.470	2.600	0.085
Slovakia	2020	13.450	2.599	0.084
Hungary	2016	13.380	2.594	0.079
Austria	2020	13.060	2.570	0.059
Slovenia	2016	13.060	2.570	0.059

Continued on following page

Table 6. Continued

Location	Year	Value	Logarithm	Normalised Value For The Whole Period And Countries
Croatia	2018	12.890	2.556	0.048
Croatia	2020	12.840	2.553	0.045
Slovakia	2016	12.830	2.552	0.044
Poland	2017	12.800	2.549	0.043
Austria	2019	12.790	2.549	0.042
Austria	2016	12.650	2.538	0.033
Slovenia	2020	12.490	2.525	0.023
Slovenia	2019	12.370	2.515	0.015
Lithuania	2016	12.320	2.511	0.012
Poland	2016	12.180	2.500	0.003
Croatia	2019	12.130	2.496	0.000
Luxembourg	2021			

RICE

Italy has, in general, the highest prices for rice (Table 7) and for 2021 the greatest prices were found in Spain and Portugal. In Spain, these producer prices increased by around 20% in 2021. On average the prices were 31.3 euros (3.4 for logarithm and 0.4 for normalized values) in 2016, 27.4 euros (3.3 for logarithm and 0.2 for normalized values) in 2017, 28.8 euros (3.4 for logarithm and 0.3 for normalized values) in 2018, 30.9 euros (3.4 for logarithm and 0.4 for normalized values) in 2019, 34.3 euros (3.5 for logarithm and 0.5 for normalized values) in 2020 and 33.5 euros (3.5 for logarithm and 0.5 for normalized values) in 2021.

Table 7. Producer prices (per 100 kg) for the period 2016-2021

Location	Year	Value	Logarithm	Normalised Value
Italy	2016	45.670	3.821	1.000
Italy	2019	42.290	3.745	0.857
Italy	2020	42.180	3.742	0.853
Spain	2021	37.880	3.634	0.671
Portugal	2020	37.720	3.630	0.664
Portugal	2021	35.490	3.569	0.570
Bulgaria	2020	34.810	3.550	0.542
Italy	2017	34.570	3.543	0.531
Bulgaria	2021	34.270	3.534	0.519
Italy	2018	34.200	3.532	0.516
Portugal	2019	33.300	3.506	0.478
Spain	2020	31.700	3.456	0.410
Portugal	2018	31.220	3.441	0.390
Bulgaria	2016	30.910	3.431	0.377
Spain	2019	30.510	3.418	0.360
Bulgaria	2019	30.300	3.411	0.351
Spain	2016	30.080	3.404	0.342
Spain	2018	29.190	3.374	0.304
Spain	2017	29.010	3.368	0.297
Portugal	2017	27.650	3.320	0.239
Portugal	2016	27.400	3.311	0.229
Bulgaria	2018	26.440	3.275	0.188
Greece	2021	26.310	3.270	0.183
Greece	2020	25.060	3.221	0.130
Greece	2019	24.530	3.200	0.108
Hungary	2019	24.510	3.199	0.107
Bulgaria	2017	24.030	3.179	0.087
Greece	2018	23.030	3.137	0.044
Greece	2016	22.600	3.118	0.026
Greece	2017	21.980	3.090	0.000
Hungary	2016			
Hungary	2017			
Hungary	2018			
Hungary	2020			
Hungary	2021			

SOYA

For soya producer prices the greatest values were found in 2021 for Croatia, Germany, Slovenia and Austria (Table 8). In Slovenia, for example, these producer prices increased by around 60% in 2021. The average prices for 2016-2021 were the following (respectively): 34.0 euros (logarithm was 3.5 and normalized value 0.2); 35.0 euros (logarithm was 3.5 and normalized value 0.3); 34.1 euros (logarithm was 3.5 and normalized value 0.2); 32.3 euros (logarithm was 3.5 and normalized value 0.2); 36.7 euros (logarithm was 3.6 and normalized value 0.3); and 49.3 euros (logarithm was 3.9 and normalized value 0.8).

Table 8. Producer prices (per 100 kg) for the period 2016-2021

Location	Year	Value	Logarithm	Normalised Value
Croatia	2021	55.780	4.021	1.000
Germany	2021	53.600	3.982	0.923
Slovenia	2021	50.710	3.926	0.821
Austria	2021	49.400	3.900	0.775
Spain	2020	48.440	3.880	0.741
Hungary	2021	48.040	3.872	0.727
Spain	2018	47.950	3.870	0.724
Spain	2019	47.540	3.862	0.710
Bulgaria	2021	46.630	3.842	0.678
Spain	2016	45.900	3.826	0.652
Spain	2017	45.510	3.818	0.638
Germany	2020	44.900	3.804	0.617
Romania	2021	41.250	3.720	0.488
Hungary	2020	38.090	3.640	0.377
Bulgaria	2016	36.750	3.604	0.329
Hungary	2017	36.030	3.584	0.304
Croatia	2017	35.840	3.579	0.297
Austria	2020	35.090	3.558	0.271
Austria	2017	34.480	3.540	0.249
Austria	2018	34.260	3.534	0.242
Bulgaria	2020	34.240	3.533	0.241
Slovakia	2017	34.160	3.531	0.238
Hungary	2016	33.810	3.521	0.226
Hungary	2018	33.810	3.521	0.226
Slovakia	2018	33.750	3.519	0.224
Croatia	2020	33.740	3.519	0.223
Germany	2018	33.700	3.517	0.222
Austria	2016	33.300	3.506	0.208

Continued on following page

Table 8. Continued

Location	Year	Value	Logarithm	Normalised Value
Bulgaria	2018	33.270	3.505	0.207
Germany	2019	33.000	3.497	0.197
Slovakia	2020	32.650	3.486	0.185
Slovakia	2016	32.570	3.483	0.182
Bulgaria	2017	32.390	3.478	0.176
Slovenia	2020	32.150	3.470	0.167
Slovenia	2017	31.940	3.464	0.160
Hungary	2019	31.860	3.461	0.157
Slovenia	2016	31.500	3.450	0.144
Austria	2019	31.380	3.446	0.140
Slovenia	2018	31.110	3.438	0.131
Romania	2016	30.950	3.432	0.125
Croatia	2016	30.890	3.430	0.123
Slovakia	2019	30.680	3.424	0.116
Romania	2020	30.590	3.421	0.112
France	2016	30.400	3.414	0.106
Croatia	2018	30.340	3.412	0.104
Bulgaria	2019	29.700	3.391	0.081
Slovenia	2019	29.380	3.380	0.070
Croatia	2019	29.330	3.379	0.068
Romania	2017	29.330	3.379	0.068
Romania	2018	29.010	3.368	0.057
Romania	2019	27.400	3.311	0.000

TOMATOES

Czechia, Germany, Portugal and Malta presented the highest producer prices in 2021 for tomatoes in the open (all qualities), as presented in Tables 9-10. In some of these countries, 2020 was a year with decreases in the prices for this crop. The prices were (on average) 64.6 euros in 2016 (4.1 for logarithm and 0.2 for normalised value), 76.7 euros in 2017 (4.2 for logarithm and 0.3 for normalised value), 71.6 euros in 2018 (4.1 for logarithm and 0.3 for normalised value), 76.4 euros in 2019 (4.3 for logarithm and 0.3 for normalised value), 77.0 euros in 2020 (4.3 for logarithm and 0.3 for normalised value) and 84.6 euros in 2021 (4.4 for logarithm and 0.3 for normalised value).

Table 9. The highest producer prices (per 100 kg) for the period 2016-2021

Location	Year	Value	Logarithm	Normalised Value For The Whole Period And Countries
Germany	2017	205.630	5.326	1.000
Germany	2018	203.030	5.313	0.986
Czechia	2020	164.050	5.100	0.769
Czechia	2021	151.480	5.020	0.699
Czechia	2019	118.930	4.779	0.517
Germany	2021	104.680	4.651	0.438
Portugal	2021	99.250	4.598	0.408
Malta	2017	98.150	4.586	0.402
Malta	2021	97.860	4.584	0.400
Portugal	2019	96.080	4.565	0.390
Italy	2019	91.510	4.516	0.365
Czechia	2016	91.070	4.512	0.362
Malta	2019	91.050	4.511	0.362
Slovakia	2021	89.800	4.498	0.355
Portugal	2016	88.770	4.486	0.349
Czechia	2018	88.630	4.484	0.349
Portugal	2020	87.860	4.476	0.344
Czechia	2017	86.230	4.457	0.335
Slovakia	2020	86.100	4.456	0.335
Romania	2020	85.980	4.454	0.334
Italy	2020	85.040	4.443	0.329
Romania	2021	84.730	4.439	0.327
Malta	2020	83.010	4.419	0.317
Germany	2019	81.820	4.405	0.311
Hungary	2021	81.740	4.404	0.310
Italy	2018	80.040	4.383	0.301
Germany	2020	80.010	4.382	0.301
Italy	2017	79.250	4.373	0.296
Portugal	2018	79.130	4.371	0.296
Hungary	2019	78.910	4.368	0.295
Italy	2016	78.700	4.366	0.293
Malta	2018	78.470	4.363	0.292
Malta	2016	78.220	4.360	0.291
Romania	2019	77.760	4.354	0.288
Spain	2021	77.240	4.347	0.285
Croatia	2021	75.820	4.328	0.277

Continued on following page

Table 9. Continued

Location	Year	Value	Logarithm	Normalised Value For The Whole Period And Countries
Slovakia	2019	75.690	4.327	0.277
Portugal	2017	74.070	4.305	0.268
Slovakia	2018	72.810	4.288	0.261
Spain	2017	72.630	4.285	0.260
Greece	2019	71.700	4.272	0.254
Slovakia	2017	71.310	4.267	0.252
Greece	2021	71.020	4.263	0.251
Romania	2017	69.820	4.246	0.244
Greece	2018	69.020	4.234	0.239
Croatia	2019	69.010	4.234	0.239

Table 10. The lowest producer prices (per 100 kg) for the period 2016-2021

Location	Year	Value	Logarithm	Normalised Value For The Whole Period And Countries
Slovakia	2016	68.780	4.231	0.238
Germany	2016	68.630	4.229	0.237
Romania	2016	68.370	4.225	0.236
Greece	2020	68.270	4.223	0.235
Romania	2018	66.390	4.196	0.225
Spain	2018	64.860	4.172	0.216
Spain	2020	64.820	4.172	0.216
Spain	2019	64.740	4.170	0.216
Hungary	2016	62.700	4.138	0.204
Hungary	2018	59.820	4.091	0.188
Croatia	2020	59.550	4.087	0.187
Croatia	2016	58.450	4.068	0.181
Greece	2017	58.160	4.063	0.179
Greece	2016	56.500	4.034	0.170
Croatia	2017	56.440	4.033	0.169
Hungary	2020	55.840	4.022	0.166
Spain	2016	54.320	3.995	0.158
Hungary	2017	53.330	3.976	0.152
Poland	2021	49.430	3.901	0.130
Poland	2020	48.420	3.880	0.125
Croatia	2018	46.580	3.841	0.115
Poland	2019	44.760	3.801	0.104
Poland	2018	41.030	3.714	0.084
Poland	2017	38.370	3.647	0.069
Bulgaria	2017	33.320	3.506	0.041
Poland	2016	32.630	3.485	0.037
Bulgaria	2021	32.300	3.475	0.035
Bulgaria	2016	32.270	3.474	0.035
Bulgaria	2020	31.520	3.451	0.031
Bulgaria	2019	30.750	3.426	0.026
Austria	2018	27.170	3.302	0.006
Bulgaria	2018	26.010	3.258	0.000

Table 11. The highest producer prices (per 100 kg) for the period 2016-2021

Location	Year	Value	Logarithm	Normalised Value For The Whole Period And Countries
Czechia	2021	163.650	5.098	1.000
Latvia	2021	156.440	5.053	0.951
Latvia	2020	154.720	5.042	0.939
Latvia	2019	149.340	5.006	0.903
Romania	2016	148.540	5.001	0.897
Czechia	2020	147.500	4.994	0.890
Romania	2020	145.920	4.983	0.880
Romania	2017	137.670	4.925	0.824
Romania	2021	137.360	4.923	0.821
Romania	2019	135.710	4.911	0.810
Denmark	2018	134.630	4.903	0.803
Lithuania	2021	134.620	4.902	0.803
Latvia	2018	133.050	4.891	0.792
Czechia	2019	130.810	4.874	0.777
Latvia	2017	128.890	4.859	0.764
Denmark	2021	128.810	4.858	0.763
Romania	2018	121.400	4.799	0.713
Denmark	2020	120.900	4.795	0.710
Czechia	2018	120.250	4.790	0.705
Czechia	2016	117.190	4.764	0.684
Germany	2021	116.570	4.758	0.680
Czechia	2017	116.120	4.755	0.677
Slovakia	2016	113.410	4.731	0.659
Lithuania	2020	113.130	4.729	0.657
Denmark	2017	113.010	4.727	0.656
Slovakia	2018	112.920	4.727	0.655
Lithuania	2018	112.660	4.724	0.654
Slovakia	2019	112.570	4.724	0.653
Lithuania	2019	111.030	4.710	0.643
Denmark	2019	109.400	4.695	0.632
Slovenia	2021	108.910	4.691	0.628
Hungary	2018	107.380	4.676	0.618
Slovenia	2019	105.490	4.659	0.605
Slovakia	2017	105.100	4.655	0.602
Denmark	2016	103.150	4.636	0.589
Cyprus	2018	102.440	4.629	0.584

Continued on following page

Table 11. Continued

Location	Year	Value	Logarithm	Normalised Value For The Whole Period And Countries
Slovenia	2020	102.240	4.627	0.583
Slovakia	2021	96.930	4.574	0.547
Slovakia	2020	95.010	4.554	0.534
Germany	2020	93.510	4.538	0.524
Slovenia	2018	92.240	4.524	0.515
Germany	2018	90.730	4.508	0.505
Germany	2016	83.090	4.420	0.453
Lithuania	2017	80.880	4.393	0.438
Germany	2017	79.290	4.373	0.427

LETTUCE

Czechia, Latvia, Romania and Lithuania are the top countries with the greatest lettuce in the open (all qualities) producer prices in 2021. For this crop and these countries the growth rates obtained for 2021 are relatively different from those identified for previous years (Tables 11-12). On average, the prices changed from 73.0 euros in 2016 to 85.2 euros in 2019, 86.3 euros in 2020 and 93.1 euros in 2021.

Table 12. The lowest producer prices (per 100 kg) for the period 2016-2021

Location	Year	Value	Logarithm	Normalised Value For The Whole Period And Countries
Hungary	2016	78.940	4.369	0.425
Cyprus	2017	77.920	4.356	0.418
Germany	2019	77.460	4.350	0.415
Lithuania	2016	76.990	4.344	0.411
Croatia	2021	76.620	4.339	0.409
Greece	2017	76.340	4.335	0.407
Hungary	2017	76.310	4.335	0.407
Cyprus	2021	76.160	4.333	0.406
Slovenia	2017	76.030	4.331	0.405
Cyprus	2019	74.160	4.306	0.392
Greece	2020	73.890	4.303	0.390
Slovenia	2016	73.790	4.301	0.390
Cyprus	2020	73.360	4.295	0.387
Greece	2021	72.990	4.290	0.384
Greece	2019	72.670	4.286	0.382
Hungary	2020	72.440	4.283	0.381
Croatia	2019	72.400	4.282	0.380
Hungary	2019	72.180	4.279	0.379
Croatia	2018	69.290	4.238	0.359
Cyprus	2016	69.200	4.237	0.359
Croatia	2017	68.980	4.234	0.357
Austria	2018	67.690	4.215	0.348
Greece	2018	66.860	4.203	0.343
Croatia	2020	64.720	4.170	0.328
Austria	2017	63.860	4.157	0.322
Austria	2016	62.500	4.135	0.313
Greece	2016	60.510	4.103	0.300
Hungary	2021	57.290	4.048	0.278
Portugal	2018	56.930	4.042	0.275
Portugal	2016	56.720	4.038	0.274
Portugal	2020	53.440	3.979	0.252
Portugal	2021	53.160	3.973	0.250
Belgium	2021	50.910	3.930	0.234
Croatia	2016	50.190	3.916	0.229
Portugal	2019	47.580	3.862	0.212
Portugal	2017	40.910	3.711	0.166

Continued on following page

Table 12. Continued

Location	Year	Value	Logarithm	Normalised Value For The Whole Period And Countries
Belgium	2019	40.360	3.698	0.163
Spain	2017	39.430	3.675	0.156
Belgium	2018	32.120	3.469	0.107
Spain	2016	32.020	3.466	0.106
Malta	2021	31.280	3.443	0.101
Spain	2019	27.870	3.328	0.078
Spain	2021	27.490	3.314	0.075
Malta	2020	26.350	3.271	0.068
Malta	2016	25.510	3.239	0.062
Malta	2017	25.190	3.226	0.060
Malta	2018	25.020	3.220	0.059
Spain	2020	24.730	3.208	0.057
Malta	2019	24.650	3.205	0.056
Spain	2018	22.550	3.116	0.042
Belgium	2017	19.830	2.987	0.023
Belgium	2020	18.520	2.919	0.014
Belgium	2016	16.400	2.797	0.000
Luxembourg	2016			
Luxembourg	2017			
Luxembourg	2018			
Luxembourg	2019			
Luxembourg	2020			
Luxembourg	2021			

CARROTS

For the carrots (all qualities) the countries with the biggest producer prices in 2021 were Finland, Romania, Ireland, Malta, Denmark and Greece (Tables 13-15). The average prices for the years 2016, 2017, 2018, 2019, 2020 and 2021 were respectively the following: 32.7 euros; 30.7 euros; 38.7 euros; 41.0 euros; 35.2 euros; and 36.7 euros. The pandemic impacts here seemed to be more modest.

Table 13. The highest producer prices (per 100 kg) for the period 2016-2021

Location	Year	Value	Logarithm	Normalised Value For The Whole Period And Countries
Finland	2019	93.850	4.542	1.000
Finland	2018	88.320	4.481	0.936
Finland	2021	78.710	4.366	0.825
Finland	2020	72.770	4.287	0.756
Romania	2019	72.700	4.286	0.755
Finland	2017	68.610	4.228	0.708
Romania	2020	67.790	4.216	0.698
Romania	2021	66.040	4.190	0.678
Romania	2018	65.750	4.186	0.675
Finland	2016	65.070	4.175	0.667
Ireland	2021	55.000	4.007	0.550
Greece	2019	54.630	4.001	0.546
Malta	2021	53.470	3.979	0.533
Romania	2016	52.560	3.962	0.522
Denmark	2021	52.210	3.955	0.518
Slovenia	2019	51.840	3.948	0.514
Denmark	2020	50.440	3.921	0.498
Malta	2019	49.670	3.905	0.489
Greece	2021	49.430	3.901	0.486
Malta	2020	49.150	3.895	0.483
Denmark	2018	48.180	3.875	0.471
Ireland	2019	48.000	3.871	0.469
Ireland	2020	48.000	3.871	0.469
Cyprus	2019	47.720	3.865	0.466
Romania	2017	47.710	3.865	0.466
Denmark	2019	47.500	3.861	0.464
Greece	2018	46.460	3.839	0.452
Greece	2020	46.200	3.833	0.449
Belgium	2019	46.150	3.832	0.448
Slovenia	2020	46.100	3.831	0.447
Sweden	2018	44.990	3.806	0.435
Slovenia	2016	44.360	3.792	0.427
Belgium	2018	43.910	3.782	0.422
Cyprus	2020	43.900	3.782	0.422
Austria	2016	43.040	3.762	0.412
Croatia	2019	42.180	3.742	0.402

Continued on following page

Table 13. Continued

Location	Year	Value	Logarithm	Normalised Value For The Whole Period And Countries
Malta	2017	41.840	3.734	0.398
Greece	2017	41.280	3.720	0.392
Malta	2018	40.960	3.713	0.388
Slovenia	2021	40.700	3.706	0.385
Austria	2018	40.400	3.699	0.381
Slovakia	2021	40.310	3.697	0.380
Ireland	2016	40.000	3.689	0.377
Ireland	2017	40.000	3.689	0.377
Slovakia	2017	39.890	3.686	0.376
Sweden	2019	39.850	3.685	0.375
Cyprus	2018	39.620	3.679	0.372

Table 14. The intermediate producer prices (per 100 kg) for the period 2016-2021

Location	Year	Value	Logarithm	Normalised Value For The Whole Period And Countries
Greece	2016	39.590	3.679	0.372
Slovakia	2016	39.560	3.678	0.372
Hungary	2018	38.970	3.663	0.365
Denmark	2017	38.430	3.649	0.359
Croatia	2018	38.360	3.647	0.358
Ireland	2018	38.000	3.638	0.354
Belgium	2016	37.910	3.635	0.353
Croatia	2021	37.760	3.631	0.351
Cyprus	2016	37.270	3.618	0.345
Slovakia	2019	37.270	3.618	0.345
Croatia	2020	37.100	3.614	0.343
Denmark	2016	36.710	3.603	0.339
Austria	2017	36.470	3.596	0.336
Czechia	2018	36.460	3.596	0.336
Sweden	2021	36.160	3.588	0.332
Slovakia	2020	36.080	3.586	0.331
Spain	2019	36.070	3.585	0.331
Cyprus	2021	35.930	3.582	0.330
Sweden	2016	35.660	3.574	0.327
Hungary	2021	35.180	3.560	0.321
Cyprus	2017	34.390	3.538	0.312
Slovakia	2018	34.350	3.537	0.311
Malta	2016	34.060	3.528	0.308
Germany	2018	33.610	3.515	0.303
Sweden	2017	33.350	3.507	0.300
Czechia	2019	32.960	3.495	0.295
Belgium	2017	32.670	3.486	0.292
Lithuania	2019	32.440	3.479	0.289
Spain	2018	32.260	3.474	0.287
Sweden	2020	31.590	3.453	0.279
Czechia	2021	31.240	3.442	0.275
Portugal	2021	31.050	3.436	0.273
Hungary	2019	30.780	3.427	0.270
Portugal	2018	30.260	3.410	0.264
Spain	2016	29.640	3.389	0.257
Hungary	2016	29.330	3.379	0.253

Continued on following page

Table 14. Continued

Location	Year	Value	Logarithm	Normalised Value For The Whole Period And Countries
Czechia	2020	29.220	3.375	0.252
Lithuania	2018	29.030	3.368	0.250
Germany	2021	28.730	3.358	0.246
Spain	2017	28.260	3.341	0.241
Czechia	2016	28.150	3.338	0.240
Germany	2016	27.910	3.329	0.237
Germany	2019	27.910	3.329	0.237
Spain	2020	27.520	3.315	0.232
Germany	2020	27.110	3.300	0.228
Lithuania	2021	26.950	3.294	0.226
Latvia	2019	26.100	3.262	0.216
Bulgaria	2019	25.650	3.245	0.211
Hungary	2020	25.510	3.239	0.209
Czechia	2017	25.450	3.237	0.208
Portugal	2019	25.310	3.231	0.207
Latvia	2018	25.170	3.226	0.205
Bulgaria	2021	24.910	3.215	0.202
Portugal	2020	24.800	3.211	0.201
Croatia	2017	24.680	3.206	0.200
Bulgaria	2020	24.510	3.199	0.198

Table 15. The lowest producer prices (per 100 kg) for the period 2016-2021

Location	Year	Value	Logarithm	Normalised Value For The Whole Period And Countries
Croatia	2016	24.060	3.181	0.192
Hungary	2017	23.880	3.173	0.190
Latvia	2021	23.360	3.151	0.184
Spain	2021	23.300	3.148	0.184
Lithuania	2020	23.110	3.140	0.181
Germany	2017	23.070	3.139	0.181
Lithuania	2016	22.820	3.128	0.178
Belgium	2020	22.350	3.107	0.173
Netherlands	2018	21.940	3.088	0.168
Portugal	2016	21.910	3.087	0.167
Latvia	2016	21.740	3.079	0.165
Bulgaria	2017	21.560	3.071	0.163
Bulgaria	2018	20.640	3.027	0.153
Netherlands	2019	20.180	3.005	0.147
Latvia	2017	19.250	2.958	0.137
Portugal	2017	19.000	2.944	0.134
Lithuania	2017	18.960	2.942	0.133
Bulgaria	2016	18.230	2.903	0.125
Netherlands	2021	16.930	2.829	0.110
Latvia	2020	16.610	2.810	0.106
Netherlands	2020	15.490	2.740	0.093
Netherlands	2016	14.460	2.671	0.081
Poland	2018	12.700	2.542	0.061
Poland	2019	12.420	2.519	0.058
Poland	2021	10.700	2.370	0.038
Belgium	2021	9.860	2.288	0.028
Poland	2020	9.780	2.280	0.027
Poland	2016	8.900	2.186	0.017
Poland	2017	8.400	2.128	0.011
Netherlands	2017	7.440	2.007	0.000
Luxembourg	2016			
Luxembourg	2017			
Luxembourg	2018			
Luxembourg	2019			
Luxembourg	2020			
Luxembourg	2021			
Slovenia	2017			
Slovenia	2018			

ONIONS

For the onions (all qualities) producer prices, Finland, Malta, Romania and Portugal presented the highest values in 2021, but without visible significant changes in the trend verified for the years 2016-2010 (Tables 16-18). In fact, the prices were 29.3 euros in 2016, 26.0 euros in 2017, 31.9 euros in 2018, 41.0 euros in 2019, 33.6 euros in 2020 and 34.8 euros in 2021.

Table 16. The highest producer prices (per 100 kg) for the period 2016-2021

Location	Year	Value	Logarithm	Normalised Value For The Whole Period And Countries
Malta	2019	85.160	4.445	1.000
Finland	2019	81.930	4.406	0.959
Finland	2018	81.750	4.404	0.957
Finland	2021	78.430	4.362	0.915
Malta	2016	75.400	4.323	0.876
Finland	2020	75.340	4.322	0.875
Malta	2020	74.260	4.308	0.862
Romania	2019	72.910	4.289	0.845
Finland	2017	71.550	4.270	0.827
Malta	2021	71.410	4.268	0.826
Romania	2020	65.930	4.189	0.756
Finland	2016	63.480	4.151	0.725
Romania	2021	59.130	4.080	0.670
Romania	2018	56.510	4.034	0.637
Malta	2017	53.440	3.979	0.598
Slovenia	2018	51.580	3.943	0.574
Portugal	2021	48.470	3.881	0.535
Denmark	2019	47.590	3.863	0.524
Romania	2016	47.430	3.859	0.522
Greece	2019	46.660	3.843	0.512
Portugal	2019	45.090	3.809	0.492
Czechia	2019	43.710	3.778	0.474
Romania	2017	43.560	3.774	0.472
Slovenia	2017	42.650	3.753	0.461
Ireland	2021	42.000	3.738	0.453
Slovenia	2021	41.670	3.730	0.449
Portugal	2018	41.400	3.723	0.445
Slovenia	2019	41.260	3.720	0.443

Continued on following page

Table 16. Continued

Location	Year	Value	Logarithm	Normalised Value For The Whole Period And Countries
Slovenia	2016	40.930	3.712	0.439
Portugal	2020	40.820	3.709	0.438
Greece	2018	40.670	3.705	0.436
Greece	2016	40.030	3.690	0.428
Hungary	2019	39.820	3.684	0.425
Ireland	2019	37.500	3.624	0.396
Poland	2019	37.290	3.619	0.393
Greece	2020	36.780	3.605	0.387

Table 17. The intermediate producer prices (per 100 kg) for the period 2016-2021

Location	Year	Value	Logarithm	Normalised Value For The Whole Period And Countries
Portugal	2016	36.720	3.603	0.386
Slovenia	2020	36.060	3.585	0.377
Slovakia	2019	35.590	3.572	0.371
Belgium	2020	34.850	3.551	0.362
Slovakia	2020	34.660	3.546	0.360
Malta	2018	34.370	3.537	0.356
Lithuania	2019	34.020	3.527	0.352
Cyprus	2021	33.960	3.525	0.351
Greece	2021	33.830	3.521	0.349
Slovakia	2021	33.690	3.517	0.347
Croatia	2019	32.770	3.490	0.336
Belgium	2021	32.320	3.476	0.330
Sweden	2019	32.280	3.474	0.329
Greece	2017	31.810	3.460	0.323
Denmark	2020	31.350	3.445	0.318
Slovakia	2016	31.320	3.444	0.317
Cyprus	2019	30.840	3.429	0.311
Germany	2019	30.830	3.428	0.311
Denmark	2021	30.220	3.409	0.303
Portugal	2017	30.180	3.407	0.303
Croatia	2021	30.000	3.401	0.301
Cyprus	2018	29.970	3.400	0.300
Czechia	2018	29.910	3.398	0.299
Slovakia	2018	29.410	3.381	0.293
Cyprus	2020	29.280	3.377	0.291
Czechia	2016	28.850	3.362	0.286
Czechia	2020	28.310	3.343	0.279
Hungary	2018	28.070	3.335	0.276
Ireland	2016	28.000	3.332	0.275
Ireland	2017	28.000	3.332	0.275
Ireland	2018	28.000	3.332	0.275
Ireland	2020	28.000	3.332	0.275
Sweden	2018	27.980	3.331	0.275
Hungary	2021	27.850	3.327	0.273
Hungary	2020	27.780	3.324	0.272
Latvia	2019	27.770	3.324	0.272

Continued on following page

Table 17. Continued

Location	Year	Value	Logarithm	Normalised Value For The Whole Period And Countries
Germany	2021	27.740	3.323	0.272
Czechia	2021	27.690	3.321	0.271
Slovakia	2017	26.950	3.294	0.262
Spain	2019	26.920	3.293	0.261
Netherlands	2019	26.860	3.291	0.261
Latvia	2018	26.240	3.267	0.253
Latvia	2021	26.070	3.261	0.251
Denmark	2018	26.010	3.258	0.250
Denmark	2016	25.630	3.244	0.245
Cyprus	2017	25.610	3.243	0.245
Sweden	2021	25.490	3.238	0.243
Belgium	2019	25.370	3.234	0.242
Sweden	2020	25.260	3.229	0.240
Netherlands	2018	25.250	3.229	0.240
Austria	2018	25.230	3.228	0.240
Poland	2018	25.160	3.225	0.239

Table 18. The lowest producer prices (per 100 kg) for the period 2016-2021

Location	Year	Value	Logarithm	Normalised Value For The Whole Period And Countries
Lithuania	2020	24.690	3.206	0.233
Bulgaria	2018	24.330	3.192	0.229
Denmark	2017	24.310	3.191	0.228
Lithuania	2021	23.800	3.170	0.222
Germany	2018	23.280	3.148	0.215
Bulgaria	2016	23.240	3.146	0.215
Latvia	2020	23.140	3.142	0.214
Spain	2018	23.040	3.137	0.212
Sweden	2017	22.910	3.132	0.211
Sweden	2016	22.770	3.125	0.209
Austria	2016	22.730	3.124	0.208
Cyprus	2016	22.700	3.122	0.208
Poland	2020	22.440	3.111	0.205
Germany	2020	22.410	3.110	0.204
Lithuania	2018	22.410	3.110	0.204
Croatia	2020	22.370	3.108	0.204
Poland	2021	21.600	3.073	0.194
Czechia	2017	21.460	3.066	0.192
Spain	2020	21.330	3.060	0.191
Bulgaria	2017	21.170	3.053	0.189
Germany	2016	21.060	3.047	0.187
Spain	2021	20.450	3.018	0.179
Hungary	2016	20.190	3.005	0.176
Hungary	2017	19.780	2.985	0.171
Latvia	2016	19.720	2.982	0.170
Spain	2017	19.680	2.980	0.170
Poland	2016	19.580	2.975	0.168
Spain	2016	19.400	2.965	0.166
Bulgaria	2019	19.320	2.961	0.165
Croatia	2016	19.140	2.952	0.163
Lithuania	2016	19.140	2.952	0.163
Latvia	2017	18.680	2.927	0.157
Bulgaria	2020	17.950	2.888	0.148
Croatia	2017	17.890	2.884	0.147
Bulgaria	2021	17.120	2.840	0.137
Croatia	2018	16.600	2.809	0.131

Continued on following page

Table 18. Continued

Location	Year	Value	Logarithm	Normalised Value For The Whole Period And Countries
Netherlands	2020	16.200	2.785	0.126
Belgium	2018	15.560	2.745	0.117
Lithuania	2017	15.170	2.719	0.112
Poland	2017	15.130	2.717	0.112
Netherlands	2016	14.430	2.669	0.103
Austria	2017	14.050	2.643	0.098
Germany	2017	13.900	2.632	0.096
Belgium	2017	13.830	2.627	0.095
Netherlands	2021	13.580	2.609	0.092
Belgium	2016	13.090	2.572	0.086
Netherlands	2017	6.300	1.841	0.000

DESSERT APPLES: ALL VARIETIES - PRICES PER 100 KG

The producer prices for the dessert apples (all varieties) presented the biggest values in 2021 for Finland, Luxembourg, Sweden and Austria (Tables 19-21). These prices seem to follow a different trend in 2020 and 2021 relative to the previous years. On average, the prices increased from 51.7 euros in 2016 to 57.7 euros in 2019, 68.3 euros in 2020 and 69.7 euros in 2021.

Table 19. The highest producer prices (per 100 kg) for the period 2016-2021

Location	Year	Value	Logarithm	Normalised Value For The Whole Period And Countries
Finland	2020	166.440	5.115	1.000
Finland	2021	164.640	5.104	0.988
Finland	2017	164.030	5.100	0.984
Finland	2019	158.920	5.068	0.951
Finland	2018	147.780	4.996	0.878
Finland	2016	146.710	4.988	0.871
Luxembourg	2021	145.000	4.977	0.860
Luxembourg	2017	143.000	4.963	0.847
Luxembourg	2020	143.000	4.963	0.847
Luxembourg	2018	140.000	4.942	0.827
Luxembourg	2019	140.000	4.942	0.827
Luxembourg	2016	138.000	4.927	0.814
Sweden	2020	115.750	4.751	0.669
Sweden	2021	96.430	4.569	0.542
Sweden	2017	93.520	4.538	0.523
Sweden	2019	91.380	4.515	0.509
Austria	2021	90.080	4.501	0.501
Cyprus	2017	89.780	4.497	0.499
Cyprus	2020	85.880	4.453	0.473
Sweden	2018	85.560	4.449	0.471
Cyprus	2018	85.540	4.449	0.471
Netherlands	2018	85.470	4.448	0.471
Cyprus	2019	85.180	4.445	0.469
Netherlands	2021	83.400	4.424	0.457
Austria	2020	80.730	4.391	0.440
Portugal	2021	80.070	4.383	0.435
Cyprus	2021	79.350	4.374	0.431
Netherlands	2020	72.660	4.286	0.387
Latvia	2020	71.560	4.271	0.380
Malta	2021	71.280	4.267	0.378
Portugal	2020	70.810	4.260	0.375
Portugal	2017	69.880	4.247	0.369
Belgium	2018	69.710	4.244	0.368
Cyprus	2016	69.300	4.238	0.365
Portugal	2018	69.150	4.236	0.364
Netherlands	2017	68.890	4.233	0.362

Continued on following page

Table 19. Continued

Location	Year	Value	Logarithm	Normalised Value For The Whole Period And Countries
Germany	2018	67.510	4.212	0.353
Italy	2018	67.000	4.205	0.350
Portugal	2019	67.000	4.205	0.350
Latvia	2021	66.970	4.204	0.350
Greece	2021	66.520	4.198	0.347
Romania	2020	66.350	4.195	0.346
Slovenia	2021	65.760	4.186	0.342
Greece	2018	64.870	4.172	0.336
Greece	2020	64.610	4.168	0.334
Denmark	2021	64.410	4.165	0.333

Table 20. The intermediate producer prices (per 100 kg) for the period 2016-2021

Location	Year	Value	Logarithm	Normalised Value For The Whole Period And Countries
Portugal	2016	63.850	4.157	0.329
Malta	2020	63.460	4.150	0.327
Denmark	2020	63.440	4.150	0.327
Germany	2021	62.920	4.142	0.323
Latvia	2019	62.900	4.142	0.323
Slovenia	2017	62.850	4.141	0.323
Belgium	2021	62.800	4.140	0.323
Germany	2020	62.000	4.127	0.317
Romania	2021	61.770	4.123	0.316
Greece	2019	61.350	4.117	0.313
Netherlands	2019	61.100	4.113	0.312
Romania	2017	60.630	4.105	0.308
Sweden	2016	60.590	4.104	0.308
Austria	2018	59.880	4.092	0.304
Latvia	2018	59.700	4.089	0.302
Belgium	2020	59.580	4.087	0.302
Romania	2016	59.460	4.085	0.301
Greece	2017	58.300	4.066	0.293
Romania	2018	57.800	4.057	0.290
Czechia	2021	57.610	4.054	0.289
Germany	2017	57.470	4.051	0.288
Malta	2019	56.990	4.043	0.285
Austria	2017	56.440	4.033	0.281
Slovenia	2020	56.090	4.027	0.279
Greece	2016	55.760	4.021	0.277
Spain	2021	55.320	4.013	0.274
Czechia	2020	54.960	4.007	0.271
Spain	2018	54.070	3.990	0.266
Italy	2016	54.000	3.989	0.265
Italy	2017	53.990	3.989	0.265
Slovakia	2021	53.270	3.975	0.260
Romania	2019	51.840	3.948	0.251
Lithuania	2020	51.740	3.946	0.250
Italy	2019	51.710	3.946	0.250
Croatia	2020	50.990	3.932	0.245
Italy	2020	50.360	3.919	0.241

Continued on following page

Table 20. Continued

Location	Year	Value	Logarithm	Normalised Value For The Whole Period And Countries
Croatia	2021	49.940	3.911	0.239
Spain	2020	49.900	3.910	0.238
Belgium	2017	49.620	3.904	0.236
Slovenia	2016	49.490	3.902	0.236
Latvia	2017	49.470	3.901	0.235
Slovenia	2019	49.020	3.892	0.233
Netherlands	2016	48.680	3.885	0.230
Austria	2019	48.220	3.876	0.227
Malta	2018	47.010	3.850	0.219
Lithuania	2021	46.590	3.841	0.217
Lithuania	2018	46.270	3.834	0.215
Germany	2016	46.100	3.831	0.213
Malta	2016	45.990	3.828	0.213
Czechia	2017	45.580	3.819	0.210
Czechia	2019	45.580	3.819	0.210
Denmark	2018	45.570	3.819	0.210
Slovakia	2020	45.410	3.816	0.209
Germany	2019	44.990	3.806	0.206

Table 21. The lowest producer prices (per 100 kg) for the period 2016-2021

Location	Year	Value	Logarithm	Normalised Value For The Whole Period And Countries
Latvia	2016	44.460	3.795	0.203
Hungary	2020	44.130	3.787	0.201
Spain	2017	43.820	3.780	0.199
Denmark	2019	43.610	3.775	0.197
Spain	2019	43.380	3.770	0.196
Spain	2016	43.020	3.762	0.193
Czechia	2018	42.700	3.754	0.191
Malta	2017	42.580	3.751	0.190
Slovakia	2017	41.750	3.732	0.185
Denmark	2017	40.910	3.711	0.180
Croatia	2017	39.750	3.683	0.172
Lithuania	2017	39.190	3.668	0.168
Slovenia	2018	39.140	3.667	0.168
Denmark	2016	38.840	3.659	0.166
Slovakia	2018	38.390	3.648	0.163
Croatia	2019	37.720	3.630	0.159
Lithuania	2019	37.590	3.627	0.158
Slovakia	2019	37.400	3.622	0.157
Croatia	2016	37.240	3.617	0.156
Belgium	2019	37.230	3.617	0.155
Croatia	2018	36.750	3.604	0.152
Lithuania	2016	36.690	3.603	0.152
Czechia	2016	36.250	3.590	0.149
Hungary	2021	35.610	3.573	0.145
Belgium	2016	35.360	3.566	0.143
Austria	2016	34.940	3.554	0.141
Hungary	2017	34.720	3.547	0.139
Slovakia	2016	34.720	3.547	0.139
Hungary	2019	29.470	3.383	0.105
Bulgaria	2020	28.770	3.359	0.100
Bulgaria	2021	28.490	3.350	0.098
Bulgaria	2018	28.150	3.338	0.096
Hungary	2016	26.340	3.271	0.084
Hungary	2018	26.160	3.264	0.083
Bulgaria	2019	26.120	3.263	0.083
Bulgaria	2017	24.880	3.214	0.075

Continued on following page

Table 21. Continued

Location	Year	Value	Logarithm	Normalised Value For The Whole Period And Countries
Poland	2017	24.380	3.194	0.072
Bulgaria	2016	22.680	3.121	0.060
Poland	2020	20.850	3.037	0.048
Poland	2019	15.350	2.731	0.012
Poland	2018	14.530	2.676	0.007
Poland	2021	13.830	2.627	0.003
Poland	2016	13.440	2.598	0.000

CONCLUSION

This contribution intended to show the impacts of the pandemic on the European Union agricultural producer prices. To achieve these aims, data from the Eurostat were considered for the years from 2016 to 2021 and for the following crops: soft wheat; durum wheat; maize; rice; soya; tomatoes in the open: all qualities; lettuce in the open: all qualities; carrots: all qualities; onions: all qualities; and dessert apples: all varieties.

The pandemic frameworks, namely those related to the lockdowns worsened old problems related, particularly with food security. On the other hand, social media influenced public opinion and perceptions about conjunctures. The agricultural sector plays a determinant role in economic and social stability. The changes in agricultural prices affect the human diet, namely in the groups with greater difficulties. The pandemic impacted the agricultural markets with consequences on the respective prices. Shorter agricultural supply chains seem to be a solution to deal with the challenges created by the pandemic. In any case, the consequences of the Covid-19 context were not equal for the different agricultural chains.

In the European Union, the average prices for soft wheat remained similar in 2020 to those verified in 2019 and increased by about 23% in 2021. The average prices of durum wheat increased by around 10% in 2020 and 20% in 2021. For maize, the average prices remained stable in 2020 and increased by about 36% in 2021. The average rice prices increased by 11% in 2020 and decreased in 2021. For the soya, the average prices increased by 14% in 2020 and 34% in 2021. For the tomatoes, the increases in 2021 were around 10%, lettuce 8%, carrots 4% (with decreases in 2020), onions 4% (with decreases in 2020) and dessert apples 2%.

This study shows that the greatest average prices increases were verified for wheat, maize and soya. These impacts have a delay of one year and were verified in 2021. On the other hand, the prices of tomatoes, lettuce, carrots, onions and dessert apples seem to be less affected (in some cases with decreases in 2020 relative to 2019). In future research, it could be important to understand the causes of these rises, specifically in 2021, and conclude if the roots of these changes are related to the lockdowns and the respective restrictions, or if there are here speculative effects. For policy recommendations, it is suggested to create policy measures that increase the resilience of the agricultural markets and chains and become them less vulnerable to external shocks. This is important to reduce effects as those verified for the cereal markets, for example.

ACKNOWLEDGMENT

This work is funded by National Funds through the FCT - Foundation for Science and Technology, I.P., within the scope of the project Ref[a] UIDB/00681/2020. Furthermore we would like to thank the CERNAS Research Centre and the Polytechnic Institute of Viseu for their support.

REFERENCES

Bansal, E., Chaturvedi, K. R., Saroha, K., Kaur, J., Mehra, P., Ahmad, F., & Vynnychenko, N. (2020). Covid-19: Social media impact on agricultural prices and consumer's buying behavior. *International Journal of Agricultural and Statistics Sciences*, *16*, 953–957.

Boughton, D., Goeb, J., Lambrecht, I., Headey, D., Takeshima, H., Mahrt, K., Masias, I., Goudet, S., Ragasa, C., Maredia, M. K., Minten, B., & Diao, X. (2021). Impacts of COVID-19 on agricultural production and food systems in late transforming Southeast Asia: The case of Myanmar. *Agricultural Systems, 188*. doi:10.1016/j.agsy.2020.103026

Cariappa, A. A., Acharya, K. K., Adhav, C. A., R, S., Ramasundaram, P., Kumar, A., Singh, S., & Singh, G. P. (2022). COVID-19 induced lockdown effect on wheat supply chain and prices in India – Insights from state interventions led resilience. *Socio-Economic Planning Sciences, 84*. doi:10.1016/j.seps.2022.101366

Ceballos, F., Kannan, S., & Kramer, B. (2021). Crop prices, farm incomes, and food security during the COVID-19 pandemic in India: Phone-based producer survey evidence from Haryana State. *Agricultural Economics (United Kingdom), 52*(3), 525–542. doi:10.1111/agec.12633

Edwards, R., Halimatussadiah, A., Moeis, F. R., & Maulia, R. F. (2022). Agriculture, Development and Sustainability in the Covid-19 Era. *Bulletin of Indonesian Economic Studies, 58*(1), 1–30. doi:10.1080/00074918.2022.2056935

Ejeromedoghene, O., Tesi, J. N., Uyanga, V. A., Adebayo, A. O., Nwosisi, M. C., Tesi, G. O., & Akinyeye, R. O. (2020). Food security and safety concerns in animal production and public health issues in Africa: A perspective of COVID-19 pandemic era. *Ethics, Medicine and Public Health, 15*. doi:10.1016/j.jemep.2020.100600

Eurostat. (2023). *Several Statistics and Information*. https://ec.europa.eu/eurostat

Feng, F., Jia, N., & Lin, F. (2023). *Quantifying the impact of Russia–Ukraine crisis on food security and trade pattern: Evidence from a structural general equilibrium trade model. China Agricultural Economic Review*. doi:10.1108/CAER-07-2022-0156

Ghosh-Jerath, S., Kapoor, R., Dhasmana, A., Singh, A., Downs, S., & Ahmed, S. (2022). Effect of COVID-19 Pandemic on Food Systems and Determinants of Resilience in Indigenous Communities of Jharkhand State, India: A Serial Cross-Sectional Study. *Frontiers in Sustainable Food Systems, 6*. doi:10.3389/fsufs.2022.724321

Hambardzumyan, G., & Gevorgyan, S. (2022). *The impact of COVID-19 on the small and medium dairy farms and comparative analysis of customers' behavior in Armenia. Future Foods, 5*. doi:10.1016/j.fufo.2021.100110

Hammond, J., Siegal, K., Milner, D., Elimu, E., Vail, T., Cathala, P., Gatera, A., Karim, A., Lee, J.-E., Douxchamps, S., Tu, M. T., Ouma, E., Lukuyu, B., Lutakome, P., Leitner, S., Wanyama, I., Thi, T. P., Phuc, P. T. H., Herrero, M., & van Wijk, M. (2022). Perceived effects of COVID-19 restrictions on smallholder farmers: Evidence from seven lower- and middle-income countries. *Agricultural Systems, 198*. doi:10.1016/j.agsy.2022.103367

Harris, J., Depenbusch, L., Pal, A. A., Nair, R. M., & Ramasamy, S. (2020). Food system disruption: Initial livelihood and dietary effects of COVID-19 on vegetable producers in India. *Food Security, 12*(4), 841–851. doi:10.1007/s12571-020-01064-5

Helfenstein, J., Bürgi, M., Debonne, N., Dimopoulos, T., Diogo, V., Dramstad, W., Edlinger, A., Garcia-Martin, M., Hernik, J., Kizos, T., Lausch, A., Levers, C., Mohr, F., Moreno, G., Pazur, R., Siegrist, M., Swart, R., Thenail, C., Verburg, P. H., … Herzog, F. (2022). Farmer surveys in Europe suggest that specialized, intensive farms were more likely to perceive negative impacts from COVID-19. *Agronomy for Sustainable Development, 42*(5). doi:10.1007/s13593-022-00820-5

Ibiyemi, T. E., & Oldewage-Theron, W. (2022). A review of food and nutrition security in Nigeria. *African Journal of Food, Agriculture, Nutrition and Development, 22*(1).[REMOVED HYPERLINK FIELD]

Ivascu, L., Frank Ahimaz, D., Arulanandam, B. V., & Tirian, G.-O. (2021). The perception and degree of adoption by urbanites towards urban farming. *Sustainability (Switzerland), 13*(21). doi:10.3390/su132112151

Koo, J. H., Lee, S. W., Lee, G. E., & Pyeon, M. W. (2021). A Study on Smart Farmer Service Using Community Mapping. *Journal of the Korean Society of Surveying, Geodesy, Photogrammetry and Cartography, 39*(6), 419–427. doi:10.7848/ksgpc.2021.39.6.419

Kumar, N. P. (2022). *Modeling the impact of covid-19 on the farm produce availability and pricing in India. Interdisciplinary Journal of Information, Knowledge, and Management, 17, 35–63*. doi:10.28945/4897

Liu, Y., Liu, S., Ye, D., Tang, H., & Wang, F. (2022). Dynamic impact of negative public sentiment on agricultural product prices during COVID-19. *Journal of Retailing and Consumer Services, 64*. doi:10.1016/j.jretconser.2021.102790

Lopez-Ridaura, S., Sanders, A., Barba-Escoto, L., Wiegel, J., Mayorga-Cortes, M., Gonzalez-Esquivel, C., Lopez-Ramirez, M. A., Escoto-Masis, R. M., Morales-Galindo, E., & García-Barcena, T. S. (2021). Immediate impact of COVID-19 pandemic on farming systems in Central America and Mexico. *Agricultural Systems, 192*. doi:10.1016/j.agsy.2021.103178

McBurney, M., Tuaza, L. A., Ayol, C., & Johnson, C. A. (2021). Land and livelihood in the age of COVID-19: Implications for indigenous food producers in Ecuador. *Journal of Agrarian Change, 21*(3), 620–628. doi:10.1111/joac.12417

Merchant, E. V., Fatima, T., Fatima, A., Maiyo, N., Mutuku, V., Keino, S., Simon, J. E., Hoffman, D. J., & Downs, S. M. (2022). The Influence of Food Environments on Food Security Resilience during the COVID-19 Pandemic: An Examination of Urban and Rural Difference in Kenya. *Nutrients, 14*(14). doi:10.3390/nu14142939

Mila, F. A., Nahar, A., Amin, M. R., Culas, R. J., & Ahmed, A. (2022). Empirical assessment of onion supply chain constraints in Bangladesh: A pre-covid to covid situation. *Journal of Agriculture and Food Research, 10*. doi:10.1016/j.jafr.2022.100418

Nugroho, A. D. (2021). Does covid-19 have effects on the indonesian horticultural subsector? *Bulgarian Journal of Agricultural Science*, *27*(5), 865–874.

Peng, L., Lu, G., Pang, K., & Yao, Q. (2021). Optimal farmer's income from farm products sales on live streaming with random rewards: Case from China's rural revitalisation strategy. *Computers and Electronics in Agriculture, 189*. doi:10.1016/j.compag.2021.106403

Pieter, L. A. G., Utomo, M. M. B., Suhartono, S., Sudomo, A., Sanudin, S., Fauziyah, E., Widyaningsih, T. S., Palmolina, M., Hani, A., & Siagian, C. M. (2022). The Nexus of COVID-19 Pandemic and Rural Agroforestry Farmers' Livelihoods in Tasikmalaya Regency, East Priangan, Indonesia. *Forest and Society, 6*(1), 335–354. doi:10.24259/fs.v6i1.18773

Pilvere, I., Upite, I., Muska, A., Zdanovskis, K., Nipers, A., & Janmere, L. (2021). Resilience of milk supply chains during and after the CoviD-19 crisis in Latvia. *Rural Sustainability Research, 45*(340), 53–64. doi:10.2478/plua-2021-0007

Ronaghi, M. (2022). Effects of COVID-19 on Iran's Livestock and Meat Market. *Journal of Agricultural Science and Technology*, *24*(5), 1017–1028.

Scopus. (2023). *Scopus Database*. https://www.scopus.com/search/form.uri?display=basic#basic

Selim, T., & Eltarabily, M. G. (2022). Impact of COVID-19 lockdown on small-scale farming in Northeastern Nile Delta of Egypt and learned lessons for water conservation potentials. *Ain Shams Engineering Journal, 13*(4). doi:10.1016/j.asej.2021.11.018

Srairi, M. T. (2021). Rethinking the agricultural development model in Morocco in the post Covid-19 era. *Cahiers Agricultures, 30*(6). doi:10.1051/cagri/2021003

Stata. (2023). *Statistical software for data science*. https://www.stata.com/

StataCorp. (2017a). *Stata 15 Base Reference Manual*. Stata Press.

StataCorp. (2017b). *Stata Statistical Software: Release 15*. StataCorp LLC.

Tanaka, T., & Guo, J. (2020). *International price volatility transmission and structural change: A market connectivity analysis in the beef sector. Humanities and Social Sciences Communications, 7(1)*. doi:10.105741599-020-00657-x

Uddin, M. M., Akter, A., Khaleduzzaman, A. B. M., Sultana, M. N., & Hemme, T. (2021). Application of the Farm Simulation Model approach on economic loss estimation due to Coronavirus (COVID-19) in Bangladesh dairy farms—Strategies, options, and way forward. *Tropical Animal Health and Production, 53*(1). doi:10.1007/s11250-020-02471-8

Vatta, K., Bhogal, S., Green, A. S., Sharma, H., Petrie, C. A., & Dixit, S. (2022). COVID-19 Pandemic-Induced Disruptions and Implications for National Food Security and Farm Incomes: Farm-Level Evidence from Indian Punjab. *Sustainability (Switzerland), 14*(8). doi:10.3390/su14084452

Wang, Y., Wang, J., & Wang, X. (2020). COVID-19, supply chain disruption and China's hog market: A dynamic analysis. *China Agricultural Economic Review, 12*(3), 427–443. doi:10.1108/CAER-04-2020-0053

Wu, Y., & Turvey, C. G. (2020). The impact of the China–USA trade war on USA Chapter 12 farm bankruptcies. Agricultural Finance Review, 81(3), 386–414. doi:10.1108/AFR-05-2020-0076

Yan, W., Cai, Y., Lin, F., & Ambaw, D. T. (2021). The Impacts of Trade Restrictions on World Agricultural Price Volatility during the COVID-19 Pandemic. *China and World Economy, 29*(6), 139–158. doi:10.1111/cwe.12398

Zawadka, J., Jęczmyk, A., Wojcieszak-Zbierska, M. M., Niedbała, G., Uglis, J., & Pietrzak-Zawadka, J. (2022). Socio-Economic Factors Influencing Agritourism Farm Stays and Their Safety during the COVID-19 Pandemic: Evidence from Poland. *Sustainability (Switzerland), 14*(6). doi:10.3390/su14063526

Zheng, Y., Fan, Q., & Jia, W. (2022). Article How Much Did Internet Use Promote Grain Production? Evidence from a Survey of 1242 Farmers in 13 Provinces in China. *Foods, 11*(10). doi:10.3390/foods11101389

Chapter 11
Current Realities of Portuguese Organic Markets

ABSTRACT

Inside the European Union, the agricultural realities are different across the different countries, considering their particularities related, for example, to the socioeconomic, climate, and soil conditions. In these frameworks, Portugal has specific characteristics for several agricultural activities, including organic production. This chapter analyses producer prices in the Portuguese organic markets. Statistical information was considered from the SIMA (Sistema de Informação de Mercados Agrícolas) database for the weeks from 2017 to 2022, specifically for the producer prices of blueberries and strawberries across the different Portuguese regions and markets. The impacts of the COVID-19 pandemic and Russia-Ukraine on these Portuguese markets seem irrelevant in some cases and temporary in other cases.

INTRODUCTION

The concerns with the environment and balanced lifestyles increased the interest of several stakeholders in sustainable agricultural practices (Das et al., 2021), since the eighties (Farias et al., 2022), agroecology (Johansson et al., 2023) and agroforestry systems (Cechin et al., 2021). This context promoted also organic farming practices (Blume et al., 2021) and curiosity about their dimensions (Bonnel, 2022). Among the organic practices appear the use of organic waste as fertiliser (Cruz Nieto et al., 2022) and conservative techniques (Naveen et al., 2021). This interest encouraged the scientific community to produce insights (Hidalgo et al., 2022) and knowledge (Loebnitz et al., 2022) about these topics.

Despite the growing trend (Shahabi Ahangarkolaee & Gorton, 2021), including in East Asian countries (Wang, 2021), in some cases quickly (Zollo et al., 2021), the organic markets represent a small part of the food supply (Carmona et al., 2021), meaning that there is some way to run, where the extension services (Sarker, 2021) may have relevant contributions. The wine sector is an example where there is still potential to grow (Hauck et al., 2021).

These activities are alternative to conventional productions (Khoso et al., 2021) and may bring relevant potentialities and opportunities to the farmers (Allison et al., 2021) and the food industry, including

DOI: 10.4018/978-1-6684-8923-9.ch011

dairy producers (Badruddoza et al., 2022). Organic grain production, for example, is fundamental for the food industry (Han & Grudens-Schuck, 2022). The organic supply provides healthy and eco-friendly products (Antiushko et al., 2022) with mitigation of carbon emissions (Gualtieri et al., 2023) and soil quality preservation (Phillips & Heins, 2022). This occurs because, for example, the restrictions on the use of chemical inputs (Benzing et al., 2021).

However, the discussion about organic farming regulations is still open (Bulut et al., 2021) worldwide (Esteves et al., 2021), including in the dairy sector (Dimitri & Nehring, 2022). The same debate happens about some informality in these frameworks (Buquera & Marques, 2021), consumption motivations (Mai et al., 2021), attitudes and intentions to buy (Nautiyal & Lal, 2022), competition and coordination (Nematollahi et al., 2021), practices (Olejar et al., 2021), safety (Saleh et al., 2023) and the certification systems (Iannucci & Sacchi, 2021). Some consumers report higher trust in the certification systems of their countries (Murphy et al., 2022) and prefer to buy the national products (Rojík et al., 2022).

The prices and the quality are among the main determinants of organic food consumption (Radojević et al., 2021) in certain circumstances (Bilous et al., 2021), as well as environmental protection, ethical values and health consciousness (Diagourtas et al., 2022). Nonetheless, these markets are not immune to the current international crises (Andrusenko et al., 2022), such as the Covid-19 pandemic (Buraczyńska et al., 2022).

This contribution proposes to highlight trends in the Portuguese organic markets before and during the Covid-19 pandemic and Russia-Ukraine conflict with data from the Portuguese Agricultural Markets Information System (SIMA, 2023) for the period 2017-2022. Some of this information was explored deeper with the Stata (StataCorp, 2017a, 2017b; Stata, 2023) software. For the literature survey, the Scopus (2023) database was considered. Normalised values=$(x_i - x_{minimum})/(x_{maximum} - x_{minimum})$.

BLUEBERRIES

The Portuguese producer prices for blueberries "Organic*SE*I*Cuvete 125 g, Beira Interior" (Table 1), over 2019-2022, "Organic*SE*I*Cuvete 125 g, Beira Litoral" (Table 2), over the period 2017-2021, and "Organic*SE*II*Cuvete 125 g, Beira Litoral" (Table 3), over the years 2017-2021, do not seem to have changed significantly with the shocks from the Covid-19 pandemic and Russia-Ukraine crisis. In fact, on average, the prices for blueberries "Organic*SE*I*Cuvete 125 g, Beira Interior" were 6.1 euros in 2019 (logarithm was 1.8 and normalised value 0.6), 6.4 euros in 2020 (logarithm was 1.8 and normalised value 0.7), 5.9 euros in 2021 (logarithm was 1.8 and normalised value 0.6) and 6.2 euros in 2022 (logarithm was 1.8 and normalised value 0.6). For blueberries "Organic*SE*I*Cuvete 125 g, Beira Litoral", on average, the prices ranged from 6.0 euros in 2017, to 4.9 euros in 2018, 5.8 euros in 2019, 6.3 euros in 2020 and 5.5 euros in 2021. For blueberries "Organic*SE*II*Cuvete 125 g, Beira Litoral", the prices changed from 5.0 euros in 2017, 3.9 euros in 2018, 5.2 euros in 2020 and 4.5 euros in 2021.

*Table 1. Values for producer prices (euro/kg) of blueberries (Organic*SE*I*Cuvete 125 g, Beira Interior (region and market))*

Period	Month	Year	Value	Logarithm	Normalised Value for All Regions
24/06/2019	6	2019	6.500	1.872	0.700
01/07/2019	7	2019	6.500	1.872	0.700
08/07/2019	7	2019	6.500	1.872	0.700
15/07/2019	7	2019	6.000	1.792	0.600
22/07/2019	7	2019	5.000	1.609	0.300
15/06/2020	6	2020	7.500	2.015	1.000
22/06/2020	6	2020	7.500	2.015	1.000
29/06/2020	6	2020	6.000	1.792	0.600
06/07/2020	7	2020	6.000	1.792	0.600
13/07/2020	7	2020	6.000	1.792	0.600
20/07/2020	7	2020	6.000	1.792	0.600
27/07/2020	7	2020	6.000	1.792	0.600
03/08/2020	8	2020	6.000	1.792	0.600
31/05/2021	5	2021	6.000	1.792	0.600
07/06/2021	6	2021	6.000	1.792	0.600
14/06/2021	6	2021	6.000	1.792	0.600
21/06/2021	6	2021	6.000	1.792	0.600
28/06/2021	6	2021	6.000	1.792	0.600
05/07/2021	7	2021	6.000	1.792	0.600
12/07/2021	7	2021	6.000	1.792	0.600
19/07/2021	7	2021	6.000	1.792	0.600
26/07/2021	7	2021	6.000	1.792	0.600
02/08/2021	8	2021	5.700	1.740	0.500
09/08/2021	8	2021	5.700	1.740	0.500
30/05/2022	5	2022	6.500	1.872	0.700
06/06/2022	6	2022	6.500	1.872	0.700
13/06/2022	6	2022	6.500	1.872	0.700
20/06/2022	6	2022	6.000	1.792	0.600
27/06/2022	6	2022	6.000	1.792	0.600
04/07/2022	7	2022	6.000	1.792	0.600
11/07/2022	7	2022	6.000	1.792	0.600
18/07/2022	7	2022	6.000	1.792	0.600
25/07/2022	7	2022	6.000	1.792	0.600
01/08/2022	8	2022	6.000	1.792	0.600

Continued on following page

*Table 2. Values for producer prices (euro/kg) of blueberries (Organic*SE*I*Cuvete 125 g, Beira Litoral (region and market))*

Period	Month	Year	Value	Logarithm	Normalised Value for All Regions
03/07/2017	7	2017	6.000	1.792	0.600
10/07/2017	7	2017	6.000	1.792	0.600
17/07/2017	7	2017	6.000	1.792	0.600
24/07/2017	7	2017	6.000	1.792	0.600
31/07/2017	7	2017	6.000	1.792	0.600
07/08/2017	8	2017	6.000	1.792	0.600
14/08/2017	8	2017	6.000	1.792	0.600
25/06/2018	6	2018	6.000	1.792	0.600
02/07/2018	7	2018	6.000	1.792	0.600
09/07/2018	7	2018	6.000	1.792	0.600
16/07/2018	7	2018	4.000	1.386	0.000
23/07/2018	7	2018	4.000	1.386	0.000
30/07/2018	7	2018	4.000	1.386	0.000
06/08/2018	8	2018	4.500	1.504	0.100
27/05/2019	5	2019	5.000	1.609	0.300
03/06/2019	6	2019	5.000	1.609	0.300
10/06/2019	6	2019	5.500	1.705	0.400
17/06/2019	6	2019	6.000	1.792	0.600
24/06/2019	6	2019	6.000	1.792	0.600
01/07/2019	7	2019	6.000	1.792	0.600
08/07/2019	7	2019	6.000	1.792	0.600
15/07/2019	7	2019	6.000	1.792	0.600
22/07/2019	7	2019	6.000	1.792	0.600
29/07/2019	7	2019	6.000	1.792	0.600
01/06/2020	6	2020	7.500	2.015	1.000
08/06/2020	6	2020	7.500	2.015	1.000
15/06/2020	6	2020	7.000	1.946	0.900
22/06/2020	6	2020	6.000	1.792	0.600
29/06/2020	6	2020	6.000	1.792	0.600
06/07/2020	7	2020	6.000	1.792	0.600
13/07/2020	7	2020	6.000	1.792	0.600
20/07/2020	7	2020	6.000	1.792	0.600
27/07/2020	7	2020	6.000	1.792	0.600
03/08/2020	8	2020	6.000	1.792	0.600
10/08/2020	8	2020	6.000	1.792	0.600
17/08/2020	8	2020	6.000	1.792	0.600

Continued on following page

Table 2. Continued

Period	Month	Year	Value	Logarithm	Normalised Value for All Regions
24/08/2020	8	2020	6.000	1.792	0.600
24/05/2021	5	2021	6.000	1.792	0.600
31/05/2021	5	2021	6.000	1.792	0.600
07/06/2021	6	2021	6.000	1.792	0.600
14/06/2021	6	2021	6.000	1.792	0.600
21/06/2021	6	2021	6.000	1.792	0.600
28/06/2021	6	2021	6.000	1.792	0.600
05/07/2021	7	2021	6.000	1.792	0.600
12/07/2021	7	2021	5.000	1.609	0.300
19/07/2021	7	2021	5.000	1.609	0.300
26/07/2021	7	2021	5.000	1.609	0.300
02/08/2021	8	2021	5.000	1.609	0.300
09/08/2021	8	2021	5.000	1.609	0.300
16/08/2021	8	2021	5.000	1.609	0.300

*Table 3. Values for producer prices (euro/kg) of blueberries (Organic*SE*II*Cuvete 125 g, Beira Litoral (region and market))*

Period	Month	Year	Value	Logarithm	Normalised Value
03/07/2017	7	2017	5.000	1.609	0.700
10/07/2017	7	2017	5.000	1.609	0.700
17/07/2017	7	2017	5.000	1.609	0.700
24/07/2017	7	2017	5.000	1.609	0.700
31/07/2017	7	2017	5.000	1.609	0.700
07/08/2017	8	2017	5.000	1.609	0.700
14/08/2017	8	2017	5.000	1.609	0.700
25/06/2018	6	2018	5.000	1.609	0.700
02/07/2018	7	2018	5.000	1.609	0.700
09/07/2018	7	2018	5.000	1.609	0.700
16/07/2018	7	2018	3.000	1.099	0.000
23/07/2018	7	2018	3.000	1.099	0.000
30/07/2018	7	2018	3.000	1.099	0.000
06/08/2018	8	2018	3.000	1.099	0.000
01/06/2020	6	2020	6.000	1.792	1.000
08/06/2020	6	2020	6.000	1.792	1.000
15/06/2020	6	2020	6.000	1.792	1.000
22/06/2020	6	2020	5.000	1.609	0.700
29/06/2020	6	2020	5.000	1.609	0.700
06/07/2020	7	2020	5.000	1.609	0.700
13/07/2020	7	2020	5.000	1.609	0.700
20/07/2020	7	2020	5.000	1.609	0.700
27/07/2020	7	2020	5.000	1.609	0.700
03/08/2020	8	2020	5.000	1.609	0.700
10/08/2020	8	2020	5.000	1.609	0.700
17/08/2020	8	2020	5.000	1.609	0.700
24/08/2020	8	2020	5.000	1.609	0.700
24/05/2021	5	2021	5.000	1.609	0.700
31/05/2021	5	2021	5.000	1.609	0.700
07/06/2021	6	2021	5.000	1.609	0.700
14/06/2021	6	2021	5.000	1.609	0.700
21/06/2021	6	2021	5.000	1.609	0.700
28/06/2021	6	2021	5.000	1.609	0.700
05/07/2021	7	2021	5.000	1.609	0.700
12/07/2021	7	2021	4.000	1.386	0.300
19/07/2021	7	2021	4.000	1.386	0.300
26/07/2021	7	2021	4.000	1.386	0.300

Continued on following page

Table 3. Continued

Period	Month	Year	Value	Logarithm	Normalised Value
02/08/2021	8	2021	4.000	1.386	0.300
09/08/2021	8	2021	4.000	1.386	0.300
16/08/2021	8	2021	4.000	1.386	0.300

STRAWBERRIES

For the Portuguese producer prices of the strawberries "Organic*SE*I*Cuvete 250 g", "Organic*SE*I*Cuvete 500 g", "Organic*SE*I*Not calibrated*Box" (Tables 4-10), the findings are similar to those obtained in the previous section for the blueberries. In other words, the organic strawberry producer prices do not seem to be changed significantly with the context created by the recent events. Indeed, on average, the prices of strawberries "Organic*SE*I*Cuvete 250 g" (Beira Litoral) were 3.0 euros (logarithm was 1.1 and normalised value 0.0) in 2017 and 4.1 euros (logarithm was 1.4 and normalised value 0.4) in 2021. For strawberries "Organic*SE*I*Cuvete 500 g" (Beira Litoral), the prices changed, on average, from 4.2 euros in 2018 to 3.6 euros in 2019, 3.3 euros in 2020 and 3.5 euros in 2022. The prices of strawberries "Organic*SE*I*Not calibrated*Box" (Ribatejo and Oeste), on average, ranged from 5.1 euros in 2019, to 5.5 euros in 2020, 5.5 euros in 2021 and 5.7 euros in 2022.

*Table 4. Values for producer prices (euro/kg) of strawberries (Organic*SE*I*Cuvete 250 g)*

Region	Market	Period	Value	Logarithm	Normalised Value
Algarve	Algarve	26/02/2018	6.000	1.792	1.000
Algarve	Algarve	05/03/2018	6.000	1.792	1.000
Algarve	Algarve	12/03/2018	6.000	1.792	1.000
Algarve	Algarve	19/03/2018	6.000	1.792	1.000
Beira Litoral	Beira Litoral	03/07/2017	3.000	1.099	0.000
Beira Litoral	Beira Litoral	10/07/2017	3.000	1.099	0.000
Beira Litoral	Beira Litoral	17/07/2017	3.000	1.099	0.000
Beira Litoral	Beira Litoral	24/07/2017	3.000	1.099	0.000
Beira Litoral	Beira Litoral	31/07/2017	3.000	1.099	0.000
Beira Litoral	Beira Litoral	07/08/2017	3.000	1.099	0.000
Beira Litoral	Beira Litoral	14/08/2017	3.000	1.099	0.000
Beira Litoral	Beira Litoral	21/08/2017	3.000	1.099	0.000
Beira Litoral	Beira Litoral	28/08/2017	3.000	1.099	0.000
Beira Litoral	Beira Litoral	04/09/2017	3.000	1.099	0.000
Beira Litoral	Beira Litoral	11/09/2017	3.000	1.099	0.000
Beira Litoral	Beira Litoral	18/09/2017	3.000	1.099	0.000
Beira Litoral	Beira Litoral	25/09/2017	3.000	1.099	0.000
Beira Litoral	Beira Litoral	02/10/2017	3.000	1.099	0.000
Beira Litoral	Beira Litoral	09/10/2017	3.000	1.099	0.000
Beira Litoral	Beira Litoral	29/03/2021	5.000	1.609	0.700
Beira Litoral	Beira Litoral	05/04/2021	5.000	1.609	0.700
Beira Litoral	Beira Litoral	12/04/2021	5.000	1.609	0.700
Beira Litoral	Beira Litoral	19/04/2021	5.000	1.609	0.700
Beira Litoral	Beira Litoral	26/04/2021	5.000	1.609	0.700
Beira Litoral	Beira Litoral	03/05/2021	5.000	1.609	0.700
Beira Litoral	Beira Litoral	10/05/2021	5.000	1.609	0.700
Beira Litoral	Beira Litoral	17/05/2021	5.000	1.609	0.700
Beira Litoral	Beira Litoral	24/05/2021	4.000	1.386	0.300
Beira Litoral	Beira Litoral	31/05/2021	4.000	1.386	0.300
Beira Litoral	Beira Litoral	07/06/2021	4.000	1.386	0.300
Beira Litoral	Beira Litoral	14/06/2021	4.000	1.386	0.300
Beira Litoral	Beira Litoral	21/06/2021	4.000	1.386	0.300
Beira Litoral	Beira Litoral	28/06/2021	4.000	1.386	0.300
Beira Litoral	Beira Litoral	05/07/2021	3.500	1.253	0.200
Beira Litoral	Beira Litoral	12/07/2021	3.500	1.253	0.200
Beira Litoral	Beira Litoral	19/07/2021	3.000	1.099	0.000
Beira Litoral	Beira Litoral	26/07/2021	3.000	1.099	0.000
Beira Litoral	Beira Litoral	02/08/2021	3.000	1.099	0.000

Continued on following page

Table 4. Continued

Region	Market	Period	Value	Logarithm	Normalised Value
Beira Litoral	Beira Litoral	09/08/2021	3.000	1.099	0.000
Beira Litoral	Beira Litoral	16/08/2021	3.000	1.099	0.000

*Table 5. Values for producer prices (euro/kg) of strawberries (Organic*SE*I*Cuvete 500 g, Beira Litoral (region and market), 2018 and 2019)*

Period	Month	Year	Value	Logarithm	Normalised Value for All Periods
02/04/2018	4	2018	4.200	1.435	0.600
09/04/2018	4	2018	4.200	1.435	0.600
16/04/2018	4	2018	4.200	1.435	0.600
23/04/2018	4	2018	4.400	1.482	0.700
30/04/2018	4	2018	4.400	1.482	0.700
07/05/2018	5	2018	4.400	1.482	0.700
14/05/2018	5	2018	4.400	1.482	0.700
21/05/2018	5	2018	4.400	1.482	0.700
28/05/2018	5	2018	4.400	1.482	0.700
04/06/2018	6	2018	4.400	1.482	0.700
11/06/2018	6	2018	4.400	1.482	0.700
18/06/2018	6	2018	4.400	1.482	0.700
25/06/2018	6	2018	4.400	1.482	0.700
02/07/2018	7	2018	4.400	1.482	0.700
09/07/2018	7	2018	4.400	1.482	0.700
16/07/2018	7	2018	4.400	1.482	0.700
23/07/2018	7	2018	4.400	1.482	0.700
30/07/2018	7	2018	4.400	1.482	0.700
03/09/2018	9	2018	4.000	1.386	0.500
10/09/2018	9	2018	4.000	1.386	0.500
17/09/2018	9	2018	4.000	1.386	0.500
24/09/2018	9	2018	4.000	1.386	0.500
01/10/2018	10	2018	4.000	1.386	0.500
08/10/2018	10	2018	4.000	1.386	0.500
15/10/2018	10	2018	4.000	1.386	0.500
22/10/2018	10	2018	4.000	1.386	0.500
29/10/2018	10	2018	4.000	1.386	0.500
11/03/2019	3	2019	4.000	1.386	0.500
18/03/2019	3	2019	5.000	1.609	1.000

Continued on following page

Table 5. Continued

Period	Month	Year	Value	Logarithm	Normalised Value for All Periods
25/03/2019	3	2019	5.000	1.609	1.000
01/04/2019	4	2019	5.000	1.609	1.000
08/04/2019	4	2019	5.000	1.609	1.000
15/04/2019	4	2019	5.000	1.609	1.000
22/04/2019	4	2019	5.000	1.609	1.000
29/04/2019	4	2019	5.000	1.609	1.000
06/05/2019	5	2019	5.000	1.609	1.000
13/05/2019	5	2019	5.000	1.609	1.000
20/05/2019	5	2019	4.000	1.386	0.500
27/05/2019	5	2019	3.500	1.253	0.300
03/06/2019	6	2019	3.000	1.099	0.000
10/06/2019	6	2019	3.000	1.099	0.000
17/06/2019	6	2019	3.000	1.099	0.000
24/06/2019	6	2019	3.000	1.099	0.000
01/07/2019	7	2019	3.000	1.099	0.000
08/07/2019	7	2019	3.000	1.099	0.000
15/07/2019	7	2019	3.000	1.099	0.000
22/07/2019	7	2019	3.000	1.099	0.000
29/07/2019	7	2019	3.000	1.099	0.000
05/08/2019	8	2019	3.000	1.099	0.000
12/08/2019	8	2019	3.000	1.099	0.000
19/08/2019	8	2019	3.000	1.099	0.000
26/08/2019	8	2019	3.000	1.099	0.000
02/09/2019	9	2019	3.000	1.099	0.000
09/09/2019	9	2019	3.000	1.099	0.000
16/09/2019	9	2019	3.000	1.099	0.000
23/09/2019	9	2019	3.000	1.099	0.000
30/09/2019	9	2019	3.000	1.099	0.000
07/10/2019	10	2019	3.000	1.099	0.000
14/10/2019	10	2019	3.000	1.099	0.000
21/10/2019	10	2019	3.000	1.099	0.000

*Table 6. Values for producer prices (euro/kg) of strawberries (Organic*SE*I*Cuvete 500 g, Beira Litoral (region and market), 2020 and 2022)*

Period	Month	Year	Value	Logarithm	Normalised Value for All Periods
23/03/2020	3	2020	4.000	1.386	0.500
30/03/2020	3	2020	4.000	1.386	0.500
06/04/2020	4	2020	4.000	1.386	0.500
13/04/2020	4	2020	4.000	1.386	0.500
20/04/2020	4	2020	4.000	1.386	0.500
27/04/2020	4	2020	4.000	1.386	0.500
04/05/2020	5	2020	3.500	1.253	0.300
11/05/2020	5	2020	3.500	1.253	0.300
18/05/2020	5	2020	3.500	1.253	0.300
25/05/2020	5	2020	3.500	1.253	0.300
01/06/2020	6	2020	3.000	1.099	0.000
08/06/2020	6	2020	3.000	1.099	0.000
15/06/2020	6	2020	3.000	1.099	0.000
22/06/2020	6	2020	3.000	1.099	0.000
29/06/2020	6	2020	3.000	1.099	0.000
06/07/2020	7	2020	3.000	1.099	0.000
13/07/2020	7	2020	3.000	1.099	0.000
20/07/2020	7	2020	3.000	1.099	0.000
27/07/2020	7	2020	3.000	1.099	0.000
03/08/2020	8	2020	3.000	1.099	0.000
10/08/2020	8	2020	3.000	1.099	0.000
17/08/2020	8	2020	3.000	1.099	0.000
24/08/2020	8	2020	3.000	1.099	0.000
31/08/2020	8	2020	3.000	1.099	0.000
07/09/2020	9	2020	3.000	1.099	0.000
14/09/2020	9	2020	3.000	1.099	0.000
21/09/2020	9	2020	3.000	1.099	0.000
28/09/2020	9	2020	3.000	1.099	0.000
05/10/2020	10	2020	3.000	1.099	0.000
12/10/2020	10	2020	3.500	1.253	0.300
19/10/2020	10	2020	3.500	1.253	0.300
26/10/2020	10	2020	3.500	1.253	0.300
02/11/2020	11	2020	3.500	1.253	0.300
09/11/2020	11	2020	3.500	1.253	0.300
16/11/2020	11	2020	3.500	1.253	0.300
09/05/2022	5	2022	3.600	1.281	0.300
16/05/2022	5	2022	3.600	1.281	0.300

Continued on following page

Table 6. Continued

Period	Month	Year	Value	Logarithm	Normalised Value for All Periods
23/05/2022	5	2022	3.500	1.253	0.300
30/05/2022	5	2022	3.500	1.253	0.300
06/06/2022	6	2022	3.500	1.253	0.300
13/06/2022	6	2022	3.500	1.253	0.300
20/06/2022	6	2022	3.500	1.253	0.300
27/06/2022	6	2022	3.500	1.253	0.300
04/07/2022	7	2022	3.500	1.253	0.300

*Table 7. Values for producer prices (euro/kg) of strawberries (Organic*SE*I*Not calibrated*Box, Ribatejo e Oeste (region) - Península de Setúbal (market), 2019)*

Period	Month	Year	Value	Logarithm	Normalised Value for All Periods
06/05/2019	5	2019	5.000	1.609	0.000
13/05/2019	5	2019	5.000	1.609	0.000
20/05/2019	5	2019	5.000	1.609	0.000
27/05/2019	5	2019	5.000	1.609	0.000
03/06/2019	6	2019	5.000	1.609	0.000
10/06/2019	6	2019	5.000	1.609	0.000
24/06/2019	6	2019	5.000	1.609	0.000
01/07/2019	7	2019	5.000	1.609	0.000
08/07/2019	7	2019	5.000	1.609	0.000
15/07/2019	7	2019	5.000	1.609	0.000
22/07/2019	7	2019	5.000	1.609	0.000
29/07/2019	7	2019	5.000	1.609	0.000
05/08/2019	8	2019	5.000	1.609	0.000
12/08/2019	8	2019	5.000	1.609	0.000
19/08/2019	8	2019	5.000	1.609	0.000
26/08/2019	8	2019	5.000	1.609	0.000
02/09/2019	9	2019	5.000	1.609	0.000
09/09/2019	9	2019	5.000	1.609	0.000
16/09/2019	9	2019	5.000	1.609	0.000
23/09/2019	9	2019	5.000	1.609	0.000
30/09/2019	9	2019	5.000	1.609	0.000
07/10/2019	10	2019	5.000	1.609	0.000
14/10/2019	10	2019	5.000	1.609	0.000
21/10/2019	10	2019	5.000	1.609	0.000
28/10/2019	10	2019	5.000	1.609	0.000

Continued on following page

Table 7. Continued

Period	Month	Year	Value	Logarithm	Normalised Value for All Periods
04/11/2019	11	2019	5.000	1.609	0.000
11/11/2019	11	2019	5.000	1.609	0.000
18/11/2019	11	2019	5.000	1.609	0.000
25/11/2019	11	2019	5.000	1.609	0.000
02/12/2019	12	2019	5.000	1.609	0.000
09/12/2019	12	2019	6.000	1.792	0.700
16/12/2019	12	2019	6.000	1.792	0.700
23/12/2019	12	2019	6.000	1.792	0.700
30/12/2019	12	2019	6.000	1.792	0.700

*Table 8. Values for producer prices (euro/kg) of strawberries (Organic*SE*I*Not calibrated*Box, Ribatejo e Oeste (region) - Península de Setúbal (market), 2020)*

Period	Month	Year	Value	Logarithm	Normalised Value for All Periods
06/01/2020	1	2020	6.000	1.792	0.700
13/01/2020	1	2020	6.000	1.792	0.700
20/01/2020	1	2020	6.000	1.792	0.700
27/01/2020	1	2020	6.000	1.792	0.700
03/02/2020	2	2020	6.000	1.792	0.700
10/02/2020	2	2020	6.000	1.792	0.700
17/02/2020	2	2020	6.000	1.792	0.700
24/02/2020	2	2020	6.000	1.792	0.700
02/03/2020	3	2020	6.000	1.792	0.700
09/03/2020	3	2020	6.000	1.792	0.700
16/03/2020	3	2020	6.000	1.792	0.700
23/03/2020	3	2020	6.000	1.792	0.700
30/03/2020	3	2020	6.000	1.792	0.700
06/04/2020	4	2020	6.000	1.792	0.700
13/04/2020	4	2020	5.000	1.609	0.000
20/04/2020	4	2020	5.000	1.609	0.000
27/04/2020	4	2020	5.000	1.609	0.000
04/05/2020	5	2020	5.000	1.609	0.000
11/05/2020	5	2020	5.000	1.609	0.000
18/05/2020	5	2020	5.000	1.609	0.000
25/05/2020	5	2020	5.000	1.609	0.000
01/06/2020	6	2020	5.000	1.609	0.000
08/06/2020	6	2020	5.000	1.609	0.000

Continued on following page

Table 8. Continued

Period	Month	Year	Value	Logarithm	Normalised Value for All Periods
15/06/2020	6	2020	5.000	1.609	0.000
22/06/2020	6	2020	5.000	1.609	0.000
29/06/2020	6	2020	5.000	1.609	0.000
06/07/2020	7	2020	5.000	1.609	0.000
13/07/2020	7	2020	5.000	1.609	0.000
20/07/2020	7	2020	5.000	1.609	0.000
27/07/2020	7	2020	5.000	1.609	0.000
03/08/2020	8	2020	5.000	1.609	0.000
10/08/2020	8	2020	5.000	1.609	0.000
17/08/2020	8	2020	5.000	1.609	0.000
24/08/2020	8	2020	5.000	1.609	0.000
31/08/2020	8	2020	5.000	1.609	0.000
30/11/2020	11	2020	6.000	1.792	0.700
07/12/2020	12	2020	6.000	1.792	0.700
14/12/2020	12	2020	6.000	1.792	0.700
21/12/2020	12	2020	6.000	1.792	0.700
28/12/2020	12	2020	6.000	1.792	0.700

*Table 9. Values for producer prices (euro/kg) of strawberries (Organic*SE*I*Not calibrated*Box, Ribatejo e Oeste (region) - Península de Setúbal (market), 2021)*

Period	Month	Year	Value	Logarithm	Normalised Value for All Periods
04/01/2021	1	2021	6.000	1.792	0.700
11/01/2021	1	2021	6.000	1.792	0.700
18/01/2021	1	2021	6.000	1.792	0.700
25/01/2021	1	2021	6.000	1.792	0.700
01/02/2021	2	2021	6.000	1.792	0.700
08/02/2021	2	2021	6.000	1.792	0.700
15/02/2021	2	2021	6.000	1.792	0.700
22/02/2021	2	2021	6.000	1.792	0.700
01/03/2021	3	2021	6.000	1.792	0.700
08/03/2021	3	2021	6.000	1.792	0.700
15/03/2021	3	2021	6.000	1.792	0.700
22/03/2021	3	2021	6.000	1.792	0.700
29/03/2021	3	2021	6.000	1.792	0.700
05/04/2021	4	2021	6.000	1.792	0.700
12/04/2021	4	2021	6.000	1.792	0.700

Continued on following page

Table 9. Continued

Period	Month	Year	Value	Logarithm	Normalised Value for All Periods
19/04/2021	4	2021	6.000	1.792	0.700
26/04/2021	4	2021	6.000	1.792	0.700
03/05/2021	5	2021	6.000	1.792	0.700
10/05/2021	5	2021	5.000	1.609	0.000
17/05/2021	5	2021	5.000	1.609	0.000
24/05/2021	5	2021	5.000	1.609	0.000
31/05/2021	5	2021	5.000	1.609	0.000
07/06/2021	6	2021	5.000	1.609	0.000
14/06/2021	6	2021	5.000	1.609	0.000
21/06/2021	6	2021	5.000	1.609	0.000
28/06/2021	6	2021	5.000	1.609	0.000
05/07/2021	7	2021	5.000	1.609	0.000
12/07/2021	7	2021	5.000	1.609	0.000
19/07/2021	7	2021	5.000	1.609	0.000
26/07/2021	7	2021	5.000	1.609	0.000
02/08/2021	8	2021	5.000	1.609	0.000
09/08/2021	8	2021	5.000	1.609	0.000
16/08/2021	8	2021	5.000	1.609	0.000
23/08/2021	8	2021	5.000	1.609	0.000
30/08/2021	8	2021	5.000	1.609	0.000
06/09/2021	9	2021	5.000	1.609	0.000
13/09/2021	9	2021	5.000	1.609	0.000
20/09/2021	9	2021	5.000	1.609	0.000
27/09/2021	9	2021	5.000	1.609	0.000
04/10/2021	10	2021	5.000	1.609	0.000
11/10/2021	10	2021	5.000	1.609	0.000
18/10/2021	10	2021	5.000	1.609	0.000
25/10/2021	10	2021	5.000	1.609	0.000
01/11/2021	11	2021	5.000	1.609	0.000
08/11/2021	11	2021	6.000	1.792	0.700
15/11/2021	11	2021	6.000	1.792	0.700
22/11/2021	11	2021	6.000	1.792	0.700
29/11/2021	11	2021	6.000	1.792	0.700
06/12/2021	12	2021	6.000	1.792	0.700
13/12/2021	12	2021	6.000	1.792	0.700
20/12/2021	12	2021	6.400	1.856	1.000
27/12/2021	12	2021	6.400	1.856	1.000

*Table 10. Values for producer prices (euro/kg) of strawberries (Organic*SE*I*Not calibrated*Box, Ribatejo e Oeste (region) - Península de Setúbal (market), 2022)*

Period	Month	Year	Value	Logarithm	Normalised Value for All Periods
03/01/2022	1	2022	6.400	1.856	1.000
10/01/2022	1	2022	6.400	1.856	1.000
17/01/2022	1	2022	6.400	1.856	1.000
24/01/2022	1	2022	6.000	1.792	0.700
31/01/2022	1	2022	6.000	1.792	0.700
07/02/2022	2	2022	6.000	1.792	0.700
14/02/2022	2	2022	6.000	1.792	0.700
21/02/2022	2	2022	6.000	1.792	0.700
28/02/2022	2	2022	5.500	1.705	0.400
07/03/2022	3	2022	5.500	1.705	0.400
14/03/2022	3	2022	5.500	1.705	0.400
21/03/2022	3	2022	5.500	1.705	0.400
28/03/2022	3	2022	6.000	1.792	0.700
04/04/2022	4	2022	6.000	1.792	0.700
11/04/2022	4	2022	6.000	1.792	0.700
18/04/2022	4	2022	6.000	1.792	0.700
25/04/2022	4	2022	6.000	1.792	0.700
02/05/2022	5	2022	6.000	1.792	0.700
09/05/2022	5	2022	6.000	1.792	0.700
16/05/2022	5	2022	5.500	1.705	0.400
23/05/2022	5	2022	5.500	1.705	0.400
30/05/2022	5	2022	5.500	1.705	0.400
06/06/2022	6	2022	5.500	1.705	0.400
13/06/2022	6	2022	5.500	1.705	0.400
20/06/2022	6	2022	5.500	1.705	0.400
27/06/2022	6	2022	5.500	1.705	0.400
04/07/2022	7	2022	5.500	1.705	0.400
11/07/2022	7	2022	5.500	1.705	0.400
18/07/2022	7	2022	5.500	1.705	0.400
25/07/2022	7	2022	5.500	1.705	0.400
01/08/2022	8	2022	5.000	1.609	0.000
08/08/2022	8	2022	5.000	1.609	0.000
15/08/2022	8	2022	5.000	1.609	0.000
22/08/2022	8	2022	5.000	1.609	0.000
29/08/2022	8	2022	5.000	1.609	0.000
05/09/2022	9	2022	5.500	1.705	0.400
12/09/2022	9	2022	6.000	1.792	0.700

Continued on following page

Table 10. Continued

Period	Month	Year	Value	Logarithm	Normalised Value for All Periods
19/09/2022	9	2022	6.000	1.792	0.700
26/09/2022	9	2022	6.000	1.792	0.700
03/10/2022	10	2022	6.000	1.792	0.700

CONCLUSION

This chapter sought to bring more insights into the impacts of the Covid-19 pandemic and Russia-Ukraine conflict on the Portuguese organic markets, particularly on the producer prices of organic blueberries and strawberries (considering the information available and the objectives proposed for this research). For that, statistical information from the Portuguese agricultural markets information system (SIMA) for the years 2017 to 2022.

There is an increased concern for the environment and sustainability, specifically among the younger, and have promoted more sustainable agricultural practices (organic fertilisers and conservative approaches), where organic farming and agroecology have risen in their importance. Despite this increased interest in these eco-friendly practices worldwide, there is still a long way to run. These practices are fundamental to mitigate carbon emissions and preserving soil quality. The soil and the water quality are crucial for sustainable development and food security. In any case, the debate about the advantages of organic products and practices and about the trust in their quality is still open. Some consumers remain sceptical about the certification systems. The consumption of these products is also influenced by the relationship between quality and prices.

In the Portuguese context, the producer prices for blueberries "Organic*SE*I*Cuvete 125 g, Beira Interior", "Organic*SE*I*Cuvete 125 g, Beira Litoral" and "Organic*SE*II*Cuvete 125 g, Beira Litoral", do not seem to have been significantly affected by the frameworks from the Covid-19 pandemic and Russia-Ukraine crisis. The same was verified for the prices in the producers of the strawberries "Organic*SE*I*Cuvete 250 g" (Beira Litoral), "Organic*SE*I*Cuvete 500 g" (Beira Litoral), "Organic*SE*I*Not calibrated*Box" (Ribatejo e Oeste).

This study highlights the residual or null impact of the pandemic and Russia-Ukraine crisis on the Portuguese producer prices of the blueberries and strawberries organic markets. For future research, it would be important to compare the prices over the organic products chain, namely in the producer and the retailer and to identify the causes for the respective differences (when existing). For the policymakers, it could be important to design policy instruments that mitigate eventual differences in the trends of the prices over the chain.

ACKNOWLEDGMENT

This work is funded by National Funds through the FCT - Foundation for Science and Technology, I.P., within the scope of the project Refª UIDB/00681/2020. Furthermore we would like to thank the CERNAS Research Centre and the Polytechnic Institute of Viseu for their support.

REFERENCES

Allison, J., Jr., Burdine, K. H., Dillon, C., Smith, S. R., Butler, D. M., Bates, G. E., & Pighetti, G. M. (2021). Optimal forage and supplement balance for organic dairy farms in the Southeastern United States. *Agricultural Systems, 189*. doi:10.1016/j.agsy.2021.103048

Andrusenko, N., Martynova, L., Sharko, V., Garbazhii, K., Hyrych, S., & Vasylyshyna, O. (2022). Changes in the Organic Products Market as a Result of the 2022 Events in Eastern Europe. *WSEAS Transactions on Environment and Development, 18*, 918–929. doi:10.37394/232015.2022.18.87

Antiushko, D., Bozhko, T., Osyka, V., Shapovalova, N., Lozova, T., Fil, M., Sienohonova, L., Dudla, I., Volodavchyk, V., & Sienohonova, H. (2022). Ensuring environmental protection and economic renewal through organic production certification. *Eastern-European Journal of Enterprise Technologies, 4*(13–118), 53–60. doi:10.15587/1729-4061.2022.262328

Badruddoza, S., Carlson, A. C., & McCluskey, J. J. (2022). Long-term dynamics of US organic milk, eggs, and yogurt premiums. *Agribusiness, 38*(1), 45–72. doi:10.1002/agr.21723

Benzing, A., Piepho, H.-P., Malik, W. A., Finckh, M. R., Mittelhammer, M., Strempel, D., Jaschik, J., Neuendorff, J., Guamán, L., Mancheno, J., Melo, L., Pavón, O., Cangahuamín, R., & Ullauri, J.-C. (2021). Appropriate sampling methods and statistics can tell apart fraud from pesticide drift in organic farming. *Scientific Reports, 11*(1). doi:10.1038/s41598-021-93624-8

Bilous, S., Kyryliuk, Y., Kyryliuk, I., Proshchalykina, A., & Stachowiak, Z. (2021). Scenario forecasting of consumption of main organic livestock products in Ukraine. *Agricultural and Resource Economics, 7*(3), 22–43. doi:10.51599/are.2021.07.03.02

Blume, L., Hoischen-Taubner, S., & Sundrum, A. (2021). Effects of alfalfa leaf mass as a part of organic feeding strategies on growth and slaughtering performance of dual-purpose roosters: A preliminary study. *European Poultry Science, 85*. doi:10.1399/eps.2021.341

Bonnel, G. (2022). Étudier le travail au ras du sol. Une enquête sur le travail en agriculture biologique. *BMS Bulletin of Sociological Methodology/ Bulletin de Methodologie Sociologique, 156*(1), 92–123. doi:10.1177/07591063221128597

Bulut, E., Stout, A., Wemette, M., Llanos-Soto, S., Schell, R. C., Greiner Safi, A., Shapiro, M. A., Moroni, P., & Ivanek, R. (2021). How does public perception of antibiotic use on dairy farms contribute to self-reported purchasing of organic? *Journal of Food Science, 86*(5), 2045–2060. doi:10.1111/1750-3841.15720

Buquera, R. B., & Marques, P. E. M. (2021). Trust relationships involving organic food consumers: A case study in Sorocaba (in the state of São Paulo). *Revista de Economia e Sociologia Rural, 60*(Special Issue). Advance online publication. doi:10.1590/1806-9479.2021.251225

Buraczyńska, B. U., Żelazna, A. E., & Bojanowska, A. B. (2022). The Impact of the COVID-19 Pandemic on the Behaviours of Polish Consumers in the Organic Products Market. *Sustainability (Switzerland), 14*(19). doi:10.3390/su141911821

Carmona, I., Griffith, D. M., & Aguirre, I. (2021). Understanding the factors limiting organic consumption: The effect of marketing channel on produce price, availability, and price fairness. *Organic Agriculture, 11*(1), 89–103. doi:10.1007/s13165-020-00331-1

Cechin, A., da Silva Araújo, V., & Amand, L. (2021). Exploring the synergy between Community Supported Agriculture and agroforestry: Institutional innovation from smallholders in a brazilian rural settlement. *Journal of Rural Studies, 81*, 246–258. doi:10.1016/j.jrurstud.2020.10.031

Cruz Nieto, D. D., Rodriguez Espinoza, R. F., Legua Cárdenas, J. A., Natividad Huasupoma, D. E., Ipanaqué Roña, J. M., Abarca Rodríguez, J. J., & Ramos Pacheco, R. L. (2022). Analysis of the Yield of Radish (Raphanus Sativus L.) Fertilized with Compost Based on Organic Waste from Markets and its Relationship with its Stomatal Density. *Chemical Engineering Transactions, 94*, 1387–1392. doi:10.3303/CET2294231

Das, S., Mandal, S., & Mohanty, S. (2021). Agribusiness Opportunities of Organic Agriculture in West Bengal – An Empirical Analysis. *Indian Journal of Agricultural Economics, 76*(3), 514–532.

Diagourtas, G., Kounetas, K. E., & Simaki, V. (2022). *Consumer attitudes and sociodemographic profiles in purchasing organic food products: Evidence from a Greek and Swedish survey. British Food Journal.* doi:10.1108/BFJ-03-2022-0196

Dimitri, C., & Nehring, R. (2022). Thirty years of organic dairy in the United States: The influences of farms, the market and the organic regulation. *Renewable Agriculture and Food Systems, 37*(6), 588–602. doi:10.1017/S1742170522000254

Esteves, R. C., do Amaral Vendramini, A. L., & Accioly, F. (2021). A qualitative meta-synthesis study of the convergence between organic crop regulations in the United States, Brazil, and Europe. *Trends in Food Science and Technology, 107*, 343–357. doi:10.1016/j.tifs.2020.10.044

Farias, L. F., Guimarães Soares, J. P., & de Sousa, R. L. (2022). The organic market and the prices practiced in the main commercialization channels in the city of Goiânia, GO. *Iheringia - Serie Botanica, 77*. doi:10.21826/2446-82312022v77e2022009

Gualtieri, P., Marchetti, M., Frank, G., Cianci, R., Bigioni, G., Colica, C., Soldati, L., Moia, A., De Lorenzo, A., & Di Renzo, L. (2023). Exploring the Sustainable Benefits of Adherence to the Mediterranean Diet during the COVID-19 Pandemic in Italy. *Nutrients, 15*(1). doi:10.3390/nu15010110

Han, G., & Grudens-Schuck, N. (2022). Motivations and Challenges for Adoption of Organic Grain Production: A Qualitative Study of Iowa Organic Farmers. *Foods, 11*(21). doi:10.3390/foods11213512

Hauck, K., Szolnoki, G., & Pabst, E. (2021). Motivation factors for organic wines. An analysis from the perspective of German producers and retailers. *Wine Economics and Policy, 10*(2), 61–74. doi:10.36253/wep-9893

Hidalgo, G., Monticelli, J. M., Pedroso, J., Verschoore, J. R., & De Matos, C. A. (2022). The Influence of Formal Institution Agents on Coopetition in the Organic Food Industry. *Journal of Agricultural and Food Industrial Organization, 20*(2), 61–74. doi:10.1515/jafio-2019-0009

Iannucci, G., & Sacchi, G. (2021). The evolution of organic market between third-party certification and participatory guarantee systems. *Bio-Based and Applied Economics, 10*(3), 239–251. doi:10.36253/bae-10470

Johansson, E., Martin, R., & Mapunda, K. M. (2023). *Participatory future visions of collaborative agroecological farmer-pastoralist systems in Tanzania. Agroecology and Sustainable Food Systems*. doi:10.1080/21683565.2023.2165592

Khoso, K. K., Buarod, T., & Kaewpijit, J. (2021). Sustainable impact of organic farming: A social constructivist perspective. *Humanities, Arts and Social Sciences Studies, 21*(1), 150–161. doi:10.14456/hasss.2021.15

Loebnitz, N., Frank, P., & Otterbring, T. (2022). Stairway to organic heaven: The impact of social and temporal distance in print ads. *Journal of Business Research, 139*, 1044–1057. doi:10.1016/j.jbusres.2021.10.020

Mai, R., Hoffmann, S., & Balderjahn, I. (2021). When drivers become inhibitors of organic consumption: The need for a multistage view. *Journal of the Academy of Marketing Science, 49*(6), 1151–1174. doi:10.1007/s11747-021-00787-x

Murphy, B., Martini, M., Fedi, A., Loera, B. L., Elliott, C. T., & Dean, M. (2022). Consumer trust in organic food and organic certifications in four European countries. *Food Control, 133*. doi:10.1016/j.foodcont.2021.108484

Nautiyal, S., & Lal, C. (2022). Product knowledge as a facilitator of organic purchase intention in emerging markets: Empirical evidence from India. *Journal of Cleaner Production, 372*. doi:10.1016/j.jclepro.2022.133782

Naveen, J., Saikia, M., Borah, N., Pathak, K., & Das, R. (2021). Influence of nutrient management and moisture conservation practices on Growth and development of organic baby corn (Zea mays L.) in Assam. *Indian Journal of Agricultural Research, 55*(4), 497–500. doi:10.18805/IJARe.A-5531

Nematollahi, M., Tajbakhsh, A., & Mosadegh Sedghy, B. (2021). The reflection of competition and coordination on organic agribusiness supply chains. *Transportation Research Part E: Logistics and Transportation Review, 154*. doi:10.1016/j.tre.2021.102462

Olejar, K. J., Vasconcelos, M. C., King, P. D., Smart, R. E., Ball, K., & Field, S. K. (2021). Herbicide reduction through the use of weedmat undervine treatment and the lack of impact on the aromatic profile and volatile composition of Malbec wines. *Food Chemistry, 343*. doi:10.1016/j.foodchem.2020.128474

Phillips, H. N., & Heins, B. J. (2022). Alternative Practices in Organic Dairy Production and Effects on Animal Behavior, Health, and Welfare. *Animals, 12*(14). doi:10.3390/ani12141785

Radojević, V., Tomaš Simin, M., Glavaš Trbić, D., & Milić, D. (2021). A profile of organic food consumers—Serbia case-study. *Sustainability (Switzerland), 13*(1), 1–22. doi:10.3390/su13010131

Rojík, S., Zámková, M., Chalupová, M., Pilař, L., Prokop, M., Stolín, R., Malec, K., Appiah-Kubi, S. N. K., Maitah, M., Dziekański, P., & Prus, P. (2022). Pre-COVID-19 Organic Market in the European Union—Focus on the Czech, German, and Slovak Markets. *Agriculture (Switzerland), 12*(1). doi:10.3390/agriculture12010082

Saleh, M. M., de Carvalho, A. M., de Andrade, N. M., Fonseca, A. B. M., Keller, L. A. M., Duarte, M. C. K. H., & Franco, R. M. (2023). Bacteriological quality and antimicrobial resistance of Staphylococcus spp. And Escherichia coli isolated from organic and conventional fresh cheese. *Food Science and Technology (Brazil), 43.* Advance online publication. doi:10.1590/fst.54922

Sarker, M. A. (2021). The art and myth of organic agriculture for nature conservation and sustainable food production: Perspective of Bangladesh. *Agricultural Technology (Thailand), 17*(6), 2293–2302.

Scopus. (2023). *Scopus Database.* https://www.scopus.com/search/form.uri?display=basic#basic

Shahabi Ahangarkolaee, S., & Gorton, M. (2021). The effects of perceived regulatory efficacy, ethnocentrism and food safety concern on the demand for organic food. *International Journal of Consumer Studies, 45*(2), 273–286. doi:10.1111/ijcs.12619

SIMA. (2023). *Several Statistics and Information.* https://www.gpp.pt/index.php/sima/sima-2018

Stata. (2023). *Statistical software for data science.* https://www.stata.com/

StataCorp. (2017a). *Stata 15 Base Reference Manual.* Stata Press.

StataCorp. (2017b). *Stata Statistical Software: Release 15.* StataCorp LLC.

Wang, C.-M. (2021). Performing and counter-performing organic food markets in East Asia: The role of ahimsa, scientific knowledge and faith groups. *Geographical Journal, 187*(4), 361–372. doi:10.1111/geoj.12338

Zollo, L., Carranza, R., Faraoni, M., Díaz, E., & Martín-Consuegra, D. (2021). What influences consumers' intention to purchase organic personal care products? The role of social reassurance. *Journal of Retailing and Consumer Services, 60.* doi:10.1016/j.jretconser.2020.102432

Chapter 12
Portuguese Dairy Markets After the Pandemic

ABSTRACT

Portugal has particular conditions to produce milk and dairy products (butter and cheese, for example) in some regions (Azores, for example). This means that the dairy sector assumes special relevance for Portuguese agricultural and food production. In any case, there is a tendency for unbalances between the demand and the supply in these European Union markets, which led the European institutions to create milk quotas among 1984 and 2015. This research aims to analyse the prices of milk production (at the farm level) and dairy products (at the factory level) over the period 2017-2022 in the Portuguese mainland and Azores region. Data from the SIMA (Sistema de Informação de Mercados Agrícolas) database were considered. The consequences on the producer prices of these sectors were greater on the dairy products prices than on the milk prices and namely from the Russia-Ukraine crisis.

INTRODUCTION

The dairy context and prices always promoted interest and curiosity for diverse stakeholders, including researchers (Foltz, 2004). This is due to the volatility of the prices (Gilbert et al., 2016), particularly in some periods (Rezitis & Tremma, 2022), and the need for adjusted statistical approaches to assess the respective data (Arias & Cox, 2001). The risk of volatility transmission in the dairy markets is another topic (Rude & An, 2013) and motive of concern (Stewart & Blayney, 2011).

On the other hand, often dairy prices are vulnerable to the supply of these products (Barrett, 2009a) and this promotes regularly decreases in dairy prices (Badruddoza et al., 2022) in some cases. In other circumstances encourage increases (Barrett, 2009b) in response to the milk output (Barrett & Murray, 2013). Dairy prices are generally higher in lower-income countries, which influences the decisions of consumers (Bai et al., 2021). There may, however, in some contexts, be good prospects for the future (Philippidis et al., 2021). Dairy prices have, indeed, effects on their consumption (Sturm & Datar, 2011).

Management in the dairy sector is not easy, considering the tendencies of the evolution of the dairy prices, the contexts in the prices of the inputs (Bjornlund & Rossini, 2007), including oil price (Zmami

DOI: 10.4018/978-1-6684-8923-9.ch012

& Ben-Salha, 2019), the retailers' power (Kharin, 2015), the buyers' influence (Ordofa et al., 2021), animal diseases (Swinkels et al., 2005) and Global Dairy Trade (GDT) auctions (Olipra, 2020). The vulnerability to the international markets (Howden et al., 2016), respective dynamics (McCarthy, 2015) and predictions (Murray, 2014), also affect the dairy sector planning. Some of these dynamics are not new (Morell & Söderhäll, 2019). These challenges have created innovations in the sector to deal with the difficulties (Cadilhon et al., 2016).

Public policies also have their contribution to these frameworks (Castellari et al., 2018), particularly to support the prices (Chavas & Kim, 2005), sometimes needing adjustments (Charlebois et al., 2007) and assessment (Karelakis, 2019). The agricultural policies in the European Union also were considered to support dairy exports (Muhammad & Kilmer, 2008). It was expected that technological advances contributed to promote dairy product consumption (Dong, 2006).

The discussion about the dairy markets is in the order of the day, because the climate change (Gibson et al., 2022). The Covid-19 pandemic brought additional impacts on these markets (Liu & Rabinowitz, 2021).

From the scenario described before, in the research here presented, it is intended to assess the Portuguese producer dairy prices, using information from the Portuguese agricultural markets information (SIMA, 2023) database for the period 2017-2022. To better explore the information in these data, the Stata (StataCorp, 2017a, 2017b; Stata, 2023) software procedures were followed. For the literature survey, the Scopus (2023) database was consulted. Normalised values=$(x_i-x_{minimum})/(x_{maximum}-x_{minimum})$.

MILK PRODUCER PRICES (PORTUGAL MAINLAND)

The prices for milk purchased from individual producers in Portugal Mainland, over the period 2017-2022, seem to have increased only in 2022, more in the last months (Tables 1-2). A similar pattern was found for the monthly prices for milk purchased from reception stations and collective milking parlours in Portugal Mainland (Tables 3-4). The prices for milk purchased from individual producers, on average, were 0.3 euros in 2017 (normalized value was 0.0), 0.3 euros in 2018 (normalized value was 0.0), 0.3 euros in 2019 (normalized value was 0.0), 0.3 euros in 2020 (normalized value was 0.0), 0.3 euros in 2021 (normalized value was 0.0) and 0.4 euros in 2022 (normalized value was 0.4). The prices for milk purchased from reception stations and collective milking parlours were 0.2 euros in 2017 (normalized value was 0.1), 0.3 euros in 2018 (normalized value was 0.3), 0.3 euros in 2019 (normalized value was 0.3), 0.3 euros in 2020 (normalized value was 0.3), 0.3 euros in 2021 (normalized value was 0.3) and 0.4 euros in 2022 (normalized value was 0.6).

Table 1. Monthly prices (2017-2019) for milk purchased from individual producers (Portugal mainland)

Year	Month	EUR/Kg	Normalised Prices for All Periods	Average Fat Content (%)	Protein Content (%)
2017	1	0.3	0.0	3.9	3.3
2017	2	0.3	0.0	3.9	3.2
2017	3	0.3	0.0	3.8	3.2
2017	4	0.3	0.0	3.7	3.2
2017	5	0.3	0.0	3.7	3.2
2017	6	0.3	0.0	3.6	3.2
2017	7	0.3	0.0	3.7	3.2
2017	8	0.3	0.0	3.7	3.2
2017	9	0.3	0.0	3.8	3.3
2017	10	0.3	0.0	3.9	3.3
2017	11	0.3	0.0	3.9	3.4
2017	12	0.3	0.0	4.0	3.4
2018	1	0.3	0.0	3.9	3.3
2018	2	0.3	0.0	3.8	3.3
2018	3	0.3	0.0	3.8	3.3
2018	4	0.3	0.0	3.8	3.3
2018	5	0.3	0.0	3.7	3.2
2018	6	0.3	0.0	3.7	3.2
2018	7	0.3	0.0	3.6	3.2
2018	8	0.3	0.0	3.6	3.2
2018	9	0.3	0.0	3.7	3.2
2018	10	0.3	0.0	3.8	3.3
2018	11	0.3	0.0	4.0	3.4
2018	12	0.3	0.0	3.9	3.3
2019	1	0.3	0.0	3.9	3.3
2019	2	0.3	0.0	3.9	3.3
2019	3	0.3	0.0	3.8	3.3
2019	4	0.3	0.0	3.8	3.3
2019	5	0.3	0.0	3.7	3.3
2019	6	0.3	0.0	3.7	3.3
2019	7	0.3	0.0	3.7	3.3
2019	8	0.3	0.0	3.8	3.3
2019	9	0.3	0.0	3.8	3.3
2019	10	0.3	0.0	3.9	3.3
2019	11	0.3	0.0	3.9	3.4
2019	12	0.3	0.0	3.9	3.3

Table 2. Monthly prices (2020-2022) for milk purchased from individual producers (Portugal mainland)

Year	Month	EUR/Kg	Normalised Prices for All Periods	Average Fat Content (%)	Protein Content (%)
2020	1	0.3	0.0	3.9	3.3
2020	2	0.3	0.0	3.8	3.3
2020	3	0.3	0.0	3.8	3.3
2020	4	0.3	0.0	3.8	3.3
2020	5	0.3	0.0	3.7	3.2
2020	6	0.3	0.0	3.7	3.2
2020	7	0.3	0.0	3.7	3.2
2020	8	0.3	0.0	3.7	3.2
2020	9	0.3	0.0	3.7	3.3
2020	10	0.3	0.0	3.9	3.4
2020	11	0.3	0.0	3.9	3.4
2020	12	0.3	0.0	3.9	3.4
2021	1	0.3	0.0	3.9	3.4
2021	2	0.3	0.0	3.8	3.3
2021	3	0.3	0.0	3.7	3.3
2021	4	0.3	0.0	3.7	3.3
2021	5	0.3	0.0	3.8	3.3
2021	6	0.3	0.0	3.7	3.3
2021	7	0.3	0.0	3.7	3.2
2021	8	0.3	0.0	3.8	3.3
2021	9	0.3	0.0	3.8	3.3
2021	10	0.3	0.0	3.9	3.3
2021	11	0.3	0.0	4.0	3.4
2021	12	0.3	0.0	3.9	3.3
2022	1	0.4	0.3	3.9	3.3
2022	2	0.4	0.3	3.9	3.3
2022	3	0.4	0.3	3.8	3.3
2022	4	0.4	0.3	3.8	3.3
2022	5	0.4	0.3	3.7	3.3
2022	6	0.4	0.3	3.7	3.2
2022	7	0.4	0.3	3.7	3.2
2022	8	0.4	0.3	3.8	3.2
2022	9	0.5	0.7	3.8	3.3
2022	10	0.5	0.7	3.9	3.3
2022	11	0.6	1.0	3.8	3.3

Table 3. Monthly prices (2017-2019) for milk purchased from reception stations and collective milking parlours (Portugal mainland)

Year	Month	EUR/Kg	Normalised Prices for All Periods	Average Fat Content (%)	Protein Content (%)
2017	1	0.2	0.0	4.2	3.2
2017	2	0.2	0.0	4.2	3.2
2017	3	0.2	0.0	4.0	3.2
2017	4	0.2	0.0	3.9	3.2
2017	5	0.2	0.0	3.8	3.1
2017	6	0.2	0.0	3.8	3.1
2017	7	0.2	0.0	3.8	3.0
2017	8	0.3	0.3	3.8	3.1
2017	9	0.3	0.3	4.0	3.2
2017	10	0.3	0.3	4.1	3.2
2017	11	0.3	0.3	4.2	3.3
2017	12	0.3	0.3	4.2	3.3
2018	1	0.3	0.3	4.2	3.2
2018	2	0.3	0.3	4.1	3.2
2018	3	0.3	0.3	4.1	3.2
2018	4	0.3	0.3	3.9	3.2
2018	5	0.3	0.3	3.9	3.2
2018	6	0.3	0.3	3.9	3.1
2018	7	0.3	0.3	3.8	3.1
2018	8	0.3	0.3	3.7	3.0
2018	9	0.3	0.3	3.8	3.1
2018	10	0.3	0.3	4.0	3.2
2018	11	0.3	0.3	4.1	3.3
2018	12	0.3	0.3	4.2	3.3
2019	1	0.3	0.3	4.0	3.2
2019	2	0.3	0.3	3.9	3.2
2019	3	0.3	0.3	3.9	3.2
2019	4	0.3	0.3	3.8	3.2
2019	5	0.3	0.3	3.7	3.1
2019	6	0.3	0.3	3.9	3.1
2019	7	0.3	0.3	3.7	3.0
2019	8	0.3	0.3	3.7	3.0
2019	9	0.3	0.3	3.9	3.1
2019	10	0.3	0.3	4.0	3.1
2019	11	0.3	0.3	4.0	3.2
2019	12	0.3	0.3	4.1	3.2

Table 4. Monthly prices (2020-2022) for milk purchased from reception stations and collective milking parlours (Portugal mainland)

Year	Month	EUR/Kg	Normalised Prices for All Periods	Average Fat Content (%)	Protein Content (%)
2020	1	0.3	0.3	4.1	3.2
2020	2	0.3	0.3	4.0	3.2
2020	3	0.3	0.3	3.9	3.1
2020	4	0.3	0.3	3.9	3.1
2020	5	0.3	0.3	3.9	3.3
2020	6	0.3	0.3	3.7	3.0
2020	7	0.3	0.3	3.7	2.9
2020	8	0.3	0.3	3.7	3.0
2020	9	0.3	0.3	3.9	3.1
2020	10	0.3	0.3	4.1	3.3
2020	11	0.3	0.3	4.1	3.3
2020	12	0.3	0.3	4.1	3.3
2021	1	0.3	0.3	4.2	3.3
2021	2	0.3	0.3	4.0	3.2
2021	3	0.3	0.3	4.0	3.2
2021	4	0.3	0.3	3.9	3.2
2021	5	0.3	0.3	3.9	3.2
2021	6	0.3	0.3	3.9	3.1
2021	7	0.3	0.3	3.8	3.0
2021	8	0.3	0.3	3.8	3.0
2021	9	0.3	0.3	3.9	3.1
2021	10	0.3	0.3	4.0	3.2
2021	11	0.3	0.3	4.0	3.2
2021	12	0.3	0.3	4.1	3.2
2022	1	0.3	0.3	4.0	3.2
2022	2	0.3	0.3	4.0	3.2
2022	3	0.3	0.3	3.9	3.2
2022	4	0.4	0.7	3.9	3.1
2022	5	0.4	0.7	3.9	3.1
2022	6	0.4	0.7	3.9	3.0
2022	7	0.4	0.7	3.7	3.0
2022	8	0.4	0.7	3.7	3.0
2022	9	0.4	0.7	3.7	3.0
2022	10	0.5	1.0	3.7	3.1
2022	11	0.5	1.0	3.9	3.2

MILK PRODUCER PRICES (AZORES)

The monthly prices for milk purchased from individual producers (producers have a cooling tank on the farm; transport at the expense of the factory) in the Azores (Tables 5-6) increased only in the last months of 2022. The same for the prices for milk purchased from individual producers (producers deliver to factory reception points; transport at producer's expense) in the Azores (Tables 7-8). The prices for milk purchased from individual producers (producers have a cooling tank on the farm; transport at the expense of the factory) remained stable at 0.3 euros from 2017 to 2021 and increased in 2022 for 0.4 euros. A similar pattern was found for the prices for milk purchased from individual producers (producers deliver to factory reception points; transport at the producer's expense).

Table 5. Monthly prices (2017-2019) for milk purchased from individual producers - producers have a cooling tank on the farm; transport at the expense of the factory (Azores)

Year	Month	EUR/Kg	Normalised Prices for All Periods	Average Fat Content (%)	Protein Content (%)
2017	1	0.3	0.0	3.8	3.2
2017	2	0.3	0.0	3.7	3.2
2017	3	0.3	0.0	3.7	3.3
2017	4	0.3	0.0	3.7	3.2
2017	5	0.3	0.0	3.7	3.2
2017	6	0.3	0.0	3.7	3.2
2017	7	0.3	0.0	3.7	3.1
2017	8	0.3	0.0	3.7	3.1
2017	9	0.3	0.0	3.7	3.2
2017	10	0.3	0.0	3.8	3.2
2017	11	0.3	0.0	3.9	3.3
2017	12	0.3	0.0	3.8	3.2
2018	1	0.3	0.0	3.7	3.2
2018	2	0.3	0.0	3.7	3.2
2018	3	0.3	0.0	3.7	3.2
2018	4	0.3	0.0	3.7	3.2
2018	5	0.3	0.0	3.6	3.2
2018	6	0.3	0.0	3.6	3.1
2018	7	0.3	0.0	3.6	3.1
2018	8	0.3	0.0	3.7	3.1
2018	9	0.3	0.0	3.7	3.2
2018	10	0.3	0.0	3.8	3.3
2018	11	0.3	0.0	3.9	3.3
2018	12	0.3	0.0	3.9	3.3
2019	1	0.3	0.0	3.7	3.2
2019	2	0.3	0.0	3.7	3.2
2019	3	0.3	0.0	3.6	3.2
2019	4	0.3	0.0	3.7	3.2
2019	5	0.3	0.0	3.6	3.2
2019	6	0.3	0.0	3.7	3.2
2019	7	0.3	0.0	3.6	3.1
2019	8	0.3	0.0	3.7	3.2
2019	9	0.3	0.0	3.7	3.2
2019	10	0.3	0.0	3.8	3.3
2019	11	0.3	0.0	3.8	3.3
2019	12	0.3	0.0	3.8	3.3

Table 6. Monthly prices (2020-2022) for milk purchased from individual producers - producers have a cooling tank on the farm; transport at the expense of the factory (Azores)

Year	Month	EUR/Kg	Normalised Prices for All Periods	Average Fat Content (%)	Protein Content (%)
2020	1	0.3	0.0	3.7	3.2
2020	2	0.3	0.0	3.7	3.2
2020	3	0.3	0.0	3.6	3.2
2020	4	0.3	0.0	3.7	3.2
2020	5	0.3	0.0	3.6	3.2
2020	6	0.3	0.0	3.7	3.2
2020	7	0.3	0.0	3.7	3.1
2020	8	0.3	0.0	3.7	3.1
2020	9	0.3	0.0	3.8	3.2
2020	10	0.3	0.0	3.8	3.2
2020	11	0.3	0.0	3.9	3.3
2020	12	0.3	0.0	3.9	3.2
2021	1	0.3	0.0	3.8	3.2
2021	2	0.3	0.0	3.8	3.2
2021	3	0.3	0.0	3.7	3.1
2021	4	0.3	0.0	3.7	3.2
2021	5	0.3	0.0	3.7	3.1
2021	6	0.3	0.0	3.7	3.1
2021	7	0.3	0.0	3.7	3.1
2021	8	0.3	0.0	3.7	3.1
2021	9	0.3	0.0	3.8	3.1
2021	10	0.3	0.0	3.8	3.2
2021	11	0.3	0.0	3.9	3.2
2021	12	0.3	0.0	3.9	3.2
2022	1	0.3	0.0	3.8	3.2
2022	2	0.3	0.0	3.7	3.2
2022	3	0.3	0.0	3.7	3.2
2022	4	0.3	0.0	3.7	3.2
2022	5	0.3	0.0	3.7	3.1
2022	6	0.3	0.0	3.8	3.1
2022	7	0.4	0.5	3.8	3.1
2022	8	0.4	0.5	3.8	3.1
2022	9	0.4	0.5	3.8	3.2
2022	10	0.5	1.0	3.9	3.3
2022	11	0.5	1.0	4.1	3.3

Table 7. Monthly prices (2017-2019) for milk purchased from individual producers - producers deliver to factory reception points; transport at producer's expense (Azores)

Year	Month	EUR/Kg	Normalised Prices for All Periods	Average Fat Content (%)	Protein Content (%)
2017	1	0.3	0.0	3.8	3.2
2017	2	0.3	0.0	3.7	3.2
2017	3	0.3	0.0	3.6	3.3
2017	4	0.3	0.0	3.6	3.3
2017	5	0.3	0.0	3.6	3.2
2017	6	0.3	0.0	3.6	3.2
2017	7	0.3	0.0	3.6	3.1
2017	8	0.3	0.0	3.7	3.1
2017	9	0.3	0.0	3.7	3.2
2017	10	0.3	0.0	3.8	3.2
2017	11	0.3	0.0	3.9	3.3
2017	12	0.3	0.0	3.8	3.2
2018	1	0.3	0.0	3.7	3.2
2018	2	0.3	0.0	3.7	3.2
2018	3	0.3	0.0	3.6	3.2
2018	4	0.3	0.0	3.6	3.3
2018	5	0.3	0.0	3.6	3.2
2018	6	0.3	0.0	3.6	3.1
2018	7	0.3	0.0	3.7	3.1
2018	8	0.3	0.0	3.7	3.1
2018	9	0.3	0.0	3.8	3.2
2018	10	0.3	0.0	3.9	3.3
2018	11	0.3	0.0	3.9	3.3
2018	12	0.3	0.0	3.8	3.3
2019	1	0.3	0.0	3.7	3.2
2019	2	0.3	0.0	3.6	3.2
2019	3	0.3	0.0	3.5	3.2
2019	4	0.3	0.0	3.6	3.2
2019	5	0.3	0.0	3.6	3.2
2019	6	0.3	0.0	3.6	3.2
2019	7	0.3	0.0	3.6	3.1
2019	8	0.3	0.0	3.7	3.1
2019	9	0.3	0.0	3.8	3.2
2019	10	0.3	0.0	3.9	3.3
2019	11	0.3	0.0	3.9	3.3
2019	12	0.3	0.0	3.8	3.3

Table 8. Monthly prices (2020-2022) for milk purchased from individual producers - producers deliver to factory reception points; transport at producer's expense (Azores)

Year	Month	EUR/Kg	Normalised Prices for All Periods	Average Fat Content (%)	Protein Content (%)
2020	1	0.3	0.0	3.7	3.2
2020	2	0.3	0.0	3.6	3.2
2020	3	0.3	0.0	3.6	3.2
2020	4	0.3	0.0	3.6	3.2
2020	5	0.3	0.0	3.6	3.2
2020	6	0.3	0.0	3.7	3.2
2020	7	0.3	0.0	3.7	3.1
2020	8	0.3	0.0	3.7	3.1
2020	9	0.3	0.0	3.8	3.2
2020	10	0.3	0.0	3.9	3.2
2020	11	0.3	0.0	3.9	3.3
2020	12	0.3	0.0	3.9	3.2
2021	1	0.3	0.0	3.8	3.2
2021	2	0.3	0.0	3.7	3.1
2021	3	0.3	0.0	3.6	3.1
2021	4	0.3	0.0	3.6	3.2
2021	5	0.3	0.0	3.6	3.1
2021	6	0.3	0.0	3.6	3.1
2021	7	0.3	0.0	3.7	3.1
2021	8	0.3	0.0	3.7	3.1
2021	9	0.3	0.0	3.8	3.1
2021	10	0.3	0.0	3.8	3.2
2021	11	0.3	0.0	3.9	3.3
2021	12	0.3	0.0	3.9	3.3
2022	1	0.3	0.0	3.8	3.2
2022	2	0.3	0.0	3.7	3.2
2022	3	0.3	0.0	3.7	3.2
2022	4	0.3	0.0	3.7	3.2
2022	5	0.3	0.0	3.7	3.2
2022	6	0.3	0.0	3.7	3.1
2022	7	0.4	0.5	3.7	3.1
2022	8	0.4	0.5	3.8	3.1
2022	9	0.4	0.5	3.8	3.2
2022	10	0.5	1.0	3.9	3.3
2022	11	0.5	1.0	4.0	3.3

PACKED MILK (LEAVING THE FACTORY) PRICE INDEX

The Portuguese monthly price index for packed milk leaving the factory (full-fat, semi-skimmed and skimmed), over the period 2017-2022, only presents relevant changes in the last months of 2022 (Tables 9-10). These frameworks highlight that the shocks from the Russia-Ukraine crisis had more impacts on Portuguese dairy products prices than the events created by the Covid-19 pandemic. For full-fat packed milk leaving the factory, the price index changed, on average, from 100 in 2017 to 101 in 2018, 98 in 2019, 103 in 2020, 97 in 2021 and 117 in 2022. The semi-skimmed milk price index, on average, ranged from 75 in 2017 to 76 in 2018, 75 in 2019, 75 in 2020, 76 in 2021 and 95 in 2022. A similar pattern was found for the skimmed milk price index.

Table 9. Monthly price index (2017-2019) for packed milk (leaving the factory), 2000 = 100

Year	Month	UHT					
		Full-Fat Milk	Normalised Price Index for All Periods	Semi-Skimmed Milk	Normalised Price Index for All Periods	Skimmed Milk	Normalised Price Index for All Periods
2017	1	98	0.1	74	0.1	79	0.1
2017	2	94	0.0	71	0.0	76	0.0
2017	3	100	0.1	74	0.1	80	0.1
2017	4	100	0.1	74	0.1	80	0.1
2017	5	103	0.2	74	0.1	82	0.1
2017	6	101	0.1	76	0.1	83	0.1
2017	7	99	0.1	74	0.1	82	0.1
2017	8	100	0.1	76	0.1	83	0.1
2017	9	100	0.1	76	0.1	82	0.1
2017	10	100	0.1	76	0.1	83	0.1
2017	11	103	0.2	77	0.1	85	0.2
2017	12	101	0.1	79	0.2	85	0.2
2018	1	101	0.1	77	0.1	83	0.1
2018	2	101	0.1	76	0.1	83	0.1
2018	3	103	0.2	77	0.1	85	0.2
2018	4	99	0.1	77	0.1	83	0.1
2018	5	101	0.1	76	0.1	83	0.1
2018	6	101	0.1	77	0.1	83	0.1
2018	7	101	0.1	76	0.1	82	0.1
2018	8	101	0.1	76	0.1	81	0.1
2018	9	101	0.1	77	0.1	82	0.1
2018	10	101	0.1	74	0.1	80	0.1
2018	11	100	0.1	76	0.1	80	0.1
2018	12	100	0.1	76	0.1	80	0.1
2019	1	98	0.1	77	0.1	82	0.1
2019	2	100	0.1	77	0.1	82	0.1
2019	3	100	0.1	73	0.0	77	0.0
2019	4	95	0.0	78	0.1	82	0.1
2019	5	94	0.0	74	0.1	77	0.0
2019	6	100	0.1	75	0.1	79	0.1
2019	7	98	0.1	75	0.1	79	0.1
2019	8	100	0.1	74	0.1	77	0.0
2019	9	97	0.1	75	0.1	80	0.1
2019	10	94	0.0	74	0.1	79	0.1
2019	11	97	0.1	77	0.1	82	0.1
2019	12	97	0.1	76	0.1	77	0.0

Table 10. Monthly price index (2020-2022) for packed milk (leaving the factory), 2000 = 100

Year	Month	UHT					
		Full-Fat Milk	Normalised Price Index for All Periods	Semi-Skimmed Milk	Normalised Price Index for All Periods	Skimmed Milk	Normalised Price Index for All Periods
2020	1	95	0.0	74	0.1	79	0.1
2020	2	102	0.2	74	0.1	77	0.0
2020	3	104	0.2	74	0.1	83	0.1
2020	4	106	0.2	74	0.1	85	0.2
2020	5	105	0.2	79	0.2	88	0.3
2020	6	103	0.2	75	0.1	80	0.1
2020	7	105	0.2	76	0.1	84	0.2
2020	8	105	0.2	76	0.1	85	0.2
2020	9	104	0.2	73	0.0	85	0.2
2020	10	103	0.2	74	0.1	86	0.2
2020	11	103	0.2	74	0.1	85	0.2
2020	12	103	0.2	73	0.0	84	0.2
2021	1	98	0.1	76	0.1	80	0.1
2021	2	96	0.0	78	0.1	82	0.1
2021	3	99	0.1	76	0.1	80	0.1
2021	4	97	0.1	75	0.1	80	0.1
2021	5	95	0.0	76	0.1	81	0.1
2021	6	99	0.1	77	0.1	80	0.1
2021	7	95	0.0	75	0.1	80	0.1
2021	8	96	0.0	76	0.1	80	0.1
2021	9	97	0.1	74	0.1	80	0.1
2021	10	97	0.1	74	0.1	80	0.1
2021	11	96	0.0	77	0.1	81	0.1
2021	12	98	0.1	77	0.1	81	0.1
2022	1	98	0.1	78	0.1	82	0.1
2022	2	98	0.1	81	0.2	84	0.2
2022	3	105	0.2	82	0.2	86	0.2
2022	4	108	0.3	84	0.3	88	0.3
2022	5	107	0.3	88	0.3	90	0.3
2022	6	112	0.4	89	0.3	92	0.3
2022	7	116	0.4	92	0.4	94	0.4
2022	8	116	0.4	95	0.5	98	0.5
2022	9	121	0.5	99	0.5	102	0.5
2022	10	135	0.8	112	0.8	114	0.8
2022	11	139	0.9	117	0.9	121	0.9
2022	12	144	1.0	123	1.0	124	1.0

FLEMISH CHEESE (LEAVING THE FACTORY) MONTHLY PRICES

The monthly prices of the cheese (leaving the factory) present a similar pattern to those found previously for other dairy product prices (Table 11). In other words, only in the last months of 2022 were verified relevant changes in the prices of the Flemish cheese. The prices changed from 445 euros/100kg in 2019 (normalised value was 0.0), 471 euros/100kg in 2020 (normalised value was 0.1), 458 euros/100kg in 2021 (normalised value was 0.1) and 552 euros/100kg in 2022 (normalised value was 0.4).

Table 11. Monthly prices for Flemish cheese (leaving the factory)

Year	Month	EUR/100Kg	Normalised Prices
2019	1	440.3	0.0
2019	2	443.3	0.0
2019	3	442.5	0.0
2019	4	445.7	0.0
2019	5	443.5	0.0
2019	6	442.7	0.0
2019	7	445.0	0.0
2019	8	445.7	0.0
2019	9	441.2	0.0
2019	10	450.2	0.0
2019	11	453.4	0.0
2019	12	451.1	0.0
2020	1	449.5	0.0
2020	2	450.5	0.0
2020	3	466.5	0.1
2020	4	479.8	0.1
2020	5	477.1	0.1
2020	6	483.1	0.2
2020	7	489.6	0.2
2020	8	463.9	0.1
2020	9	468.6	0.1
2020	10	470.0	0.1
2020	11	491.5	0.2
2020	12	467.7	0.1
2021	1	455.7	0.1
2021	2	456.9	0.1
2021	3	460.1	0.1
2021	4	457.2	0.1
2021	5	460.4	0.1
2021	6	455.2	0.1

Continued on following page

Table 11. Continued

Year	Month	EUR/100Kg	Normalised Prices
2021	7	456.7	0.1
2021	8	462.8	0.1
2021	9	459.5	0.1
2021	10	455.7	0.1
2021	11	456.8	0.1
2021	12	459.4	0.1
2022	1	463.2	0.1
2022	2	481.0	0.1
2022	3	490.2	0.2
2022	4	504.7	0.2
2022	5	523.6	0.3
2022	6	528.9	0.3
2022	7	552.4	0.4
2022	8	568.5	0.5
2022	9	592.2	0.6
2022	10	652.6	0.8
2022	11	711.8	1.0

CONCLUSION

This contribution aimed to analyse how the Portuguese dairy sector was affected by the Covid-19 pandemic and the subsequent events. For that, data from the Portuguese agricultural markets information system were considered for the period 2017-2022. This statistical information is relative to the producer prices of milk, the price index of packed milk and the prices of cheese. The producer prices of milk were taken into account for Portugal Mainland and Azores (Portuguese islands).

The prices of dairy products, namely milk, present, in certain circumstances, a trend to decrease, because of unbalances between the demand and supply in the dairy markets. This was visible, for example, in the European Union which promoted the creation of milk quotas. In turn, dairy prices are often higher in lower-income nations, with consequences on the diet of the populations. Some uncertainty in the milk markets and interrelationships with other commodity markets create additional challenges for the farmers and other stakeholders present in the sector. The climate changes and the recent international shocks are additional disturbances for these markets.

In the Portuguese Mainland framework, the prices for milk purchased from individual producers have increased only in 2022, particularly in the last months. The same was found for the monthly prices for milk purchased from reception stations and collective milking parlours. In the Azores context, this is also true for the monthly prices for milk purchased from individual producers that have a cooling tank on the farm (transport at the expense of the factory) and for the prices for milk purchased from individual producers that deliver to factory reception points (transport at producer's expense). A similar pattern was found for the Portuguese monthly price index for full-fat, semi-skimmed and skimmed packed milk leaving the factory. The same was found for the cheese.

This research reveals that the implications of the pandemic and Russia-Ukraine crisis on Portuguese dairy prices were more visible in the last months of 2022, showing the greater vulnerability of this sector to the consequences of the conflict among Russia and Ukraine. For future research, it is suggested to compare these prices with those practised in the retailers' markets. In terms of policy recommendations, it could be interesting to identify the causes of these increases by the final of 2022 and try to mitigate them.

ACKNOWLEDGMENT

This work is funded by National Funds through the FCT - Foundation for Science and Technology, I.P., within the scope of the project Refª UIDB/00681/2020. Furthermore we would like to thank the CERNAS Research Centre and the Polytechnic Institute of Viseu for their support.

REFERENCES

Arias, C., & Cox, T. L. (2001). Estimation of a US dairy sector model by maximum simulated likehood. *Applied Economics, 33*(9), 1201–1211. doi:10.1080/00036840010005797

Badruddoza, S., Carlson, A. C., & McCluskey, J. J. (2022). Long-term dynamics of US organic milk, eggs, and yogurt premiums. *Agribusiness, 38*(1), 45–72. doi:10.1002/agr.21723

Bai, Y., Alemu, R., Block, S. A., Headey, D., & Masters, W. A. (2021). Cost and affordability of nutritious diets at retail prices: Evidence from 177 countries. *Food Policy, 99*. doi:10.1016/j.foodpol.2020.101983

Barrett, D. (2009a). Dairy. *Australian Commodities*, *16*(3), 473–479. PMID:19933542

Barrett, D. (2009b). Dairy. [*Australian Commodities*, *16*(4), 635–641. PMID:19933542

Barrett, D., & Murray, C. (2013). Dairy. *Agricultural Commodities*, *3*(3), 102–108. PMID:23832907

Bjornlund, H., & Rossini, P. (2007). Fundamentals determining prices in the market for water entitlements: An Australian case study. *International Journal of Water Resources Development, 23*(3), 537–553. doi:10.1080/07900620701486020

Cadilhon, J.-J., Pham, N. D., & Maass, B. L. (2016). The tanga dairy platform: Fostering innovations for more efficient dairy chain coordination in Tanzania. *International Journal on Food System Dynamics, 7*(2), 81–91. doi:10.18461/ijfsd.v7i2.723

Castellari, E., Moro, D., Platoni, S., & Sckokai, P. (2018). Retailers' strategies and food price dynamics: Evidence from dairy scanner data. *Food Policy, 74*, 212–224. doi:10.1016/j.foodpol.2017.12.005

Charlebois, S., Langenbacher, W., & Tamilia, R. D. (2007). The Canadian Dairy Commission: An empirical survey on its relevance in today's civil society. *The International Food and Agribusiness Management Review*, *10*(1), 81–99.

Chavas, J.-P., & Kim, K. (2005). Cointegration relationships and hedonic pricing of differentiated commodities: An application to price dynamics in the US dairy sector. *Applied Economics, 37*(16), 1813–1827. doi:10.1080/00036840500215311

Dong, F. (2006). The outlook for Asian dairy markets: The role of demographics, income, and prices. *Food Policy, 31*(3), 260–271. doi:10.1016/j.foodpol.2006.02.007

Foltz, J. D. (2004). Entry, exit, and farm size: Assessing an experiment in dairy price policy. *American Journal of Agricultural Economics, 86*(3), 594–604. doi:10.1111/j.0002-9092.2004.00603.x

Gibson, M., Pereira, J. P., Slade, R., & Rogelj, J. (2022). *Agent-Based Modelling of Future Dairy and Plant-Based Milk Consumption for UK Climate Targets. JASSS, 25(2).* doi:10.18564/jasss.4801

Gilbert, M. S., Pantophlet, A. J., van den Borne, J. J. G. C., Hendriks, W. H., Schols, H. A., & Gerrits, W. J. J. (2016). Effects of replacing lactose from milk replacer by glucose, fructose, or glycerol on energy partitioning in veal calves. *Journal of Dairy Science, 99*(2), 1121–1132. doi:10.3168/jds.2015-10062

Howden, M., McCarthy, O., & Hyde, M. (2016). The EU dairy industry. *Agricultural Commodities, 6*(2), 126–139.

Karelakis, C. (2019). Redefining integration of the European dairy markets: Do they converge? *Agricultural Economics Review*, *20*(1), 27–41.

Kharin, S. (2015). Vertical price transmission along the diary supply chain in Russia. *Studies in Agricultural Economics, 117*(2), 80–85. doi:10.7896/j.1517

Liu, Y., & Rabinowitz, A. N. (2021). The impact of the COVID-19 pandemic on retail dairy prices. *Agribusiness, 37*(1), 108–121. doi:10.1002/agr.21687

McCarthy, O. (2015). Dairy. *Agricultural Commodities*, *5*(4), 118–125.

Morell, M., & Söderhäll, M. (2019). Smallholders' and large estates' reaction to changed market conditions 1860–1910. *Scandinavian Economic History Review, 67*(3), 312–331. doi:10.1080/03585522.2019.1620850

Muhammad, A., & Kilmer, R. L. (2008). The impact of EU export subsidy reductions on U.S. dairy exports. *Agribusiness, 24*(4), 557–574. doi:10.1002/agr.20179

Murray, C. (2014). Dairy. *Agricultural Commodities*, *4*(1), 105–113. PMID:25262190

Olipra, J. (2020). Leading properties of GDT auctions for dairy prices. *British Food Journal, 122*(7), 2303–2328. doi:10.1108/BFJ-06-2018-0404

Ordofa, G., Zemedu, L., & Tegegne, B. (2021). Structure conduct and performance of dairy market in Ada'a Berga district, Ethiopia. *Cogent Food and Agriculture, 7*(1). doi:10.1080/23311932.2021.1918878

Philippidis, G., Ferrer Pérez, H., Gracia de Rentería, P., M'barek, R., & Sanjuán López, A. I. (2021). Eating your greens: A global sustainability assessment. *Resources, Conservation and Recycling, 168.* doi:10.1016/j.resconrec.2021.105460

Rezitis, A. N., & Tremma, O. A. (2022). The linkage between international dairy commodity prices and volatility: A panel-GARCH analysis. *Journal of Agribusiness in Developing and Emerging Economies*. doi:10.1108/JADEE-02-2022-0027

Rude, J., & An, H. (2013). Trans-pacific partnership: Implications for the canadian industrial dairy sector. *Canadian Public Policy, 39*(3), 393–410. doi:10.3138/CPP.39.3.393

Scopus. (2023). *Scopus Database*. https://www.scopus.com/search/form.uri?display=basic#basic

SIMA. (2023). *Several Statistics and Information*. https://www.gpp.pt/index.php/sima/sima-2018

Stata. (2023). *Statistical software for data science*. https://www.stata.com/

StataCorp. (2017a). *Stata 15 Base Reference Manual*. Stata Press.

StataCorp. (2017b). *Stata Statistical Software: Release 15*. StataCorp LLC.

Stewart, H., & Blayney, D. P. (2011). Retail dairy prices fluctuate with the farm value of milk. *Agricultural and Resource Economics Review, 40*(2), 201–217. doi:10.1017/S1068280500008017

Sturm, R., & Datar, A. (2011). Regional price differences and food consumption frequency among elementary school children. *Public Health, 125*(3), 136–141. doi:10.1016/j.puhe.2010.11.016

Swinkels, J. M., Rooijendijk, J. G. A., Zadoks, R. N., & Hogeveen, H. (2005). Use of partial budgeting to determine the economic benefits of antibiotic treatment of chronic subclinical mastitis caused by Streptococcus uberis or Streptococcus dysgalactiae. *Journal of Dairy Research, 72*(1), 75–85. doi:10.1017/S0022029904000603

Zmami, M., & Ben-Salha, O. (2019). *Does oil price drive world food prices? Evidence from linear and nonlinear ARDL modeling. Economies, 7(1)*. doi:10.3390/economies7010012

Chapter 13
COVID-19 and Conflicts' Impact on Portuguese Livestock Production Markets

ABSTRACT

Livestock production is important for the sustainability of certain regions of Portugal and the food security of the Portuguese population. Nonetheless, the associated markets have specific particularities worldwide, which justify particular policies and supports for these sectors designed, for example, by the European Union institutions. This chapter proposes to assess the impacts of the COVID-19 pandemic and the Russia-Ukraine conflict on Portuguese livestock producer prices. Data were considered from SIMA (Sistema de Informação de Mercados Agrícolas) for the weeks of 2017-2022, namely for cow and country chicken carcasses. There are some signs of impacts on the producer prices from the most recent events, namely the Russia-Ukraine crisis.

INTRODUCTION

The livestock markets are vulnerable to animal diseases (Barigye et al., 2022), the respective particularities and conditions of transmission (Fielding et al., 2021). Sometimes these diseases are transmitted through trade (Ezanno et al., 2022), contributing to their spread (Knight et al., 2022), with consequences in the farms' productivity (ElAshmawy et al., 2023) and prices (Jean-Pierre et al., 2022). The respective treatments with antimicrobials (Adebowale et al., 2022) also affect in some circumstances these markets.

The dimensions related to animal disease and sanitary risks were also the focus of studies for the following countries: Japan (Fujimoto et al., 2020); Nigeria (Limon et al., 2020; Funsho-Sanni et al., 2022); United Arab Emirates (Hassan et al., 2020); Kenya (Murungi et al., 2021); United States (Nichols et al., 2022); Uganda (Okello et al., 2021); Rwanda (Shyaka et al., 2022).

The livestock markets are important for farmers, particularly in developing countries (Onduso et al., 2020). In these markets, the availability of information is fundamental (Afolabi & Tiamiyu, 2021), namely for animal health investigation (George et al., 2021). The same importance must be given to

DOI: 10.4018/978-1-6684-8923-9.ch013

the institutional services (Mohamed Sala et al., 2020) and the conditions to sell and cull the animals (Soliman, 2021).

Animal welfare is another concern (Creutzinger et al., 2021) worldwide (Windsor, 2021), specifically during transport (Bravo et al., 2020) and in livestock markets (Sánchez-Hidalgo et al., 2020). The risks in the production processes (Çevrimli et al., 2022) have deserved also special attention. In some cases, animals are not in condition to be transported, because of health problems (Stojkov et al., 2020a). These and others factors compromise the conditions of the animals that arrive at livestock markets (Stojkov et al., 2020b). Some of these dimensions influence the buyers' decisions and the respective prices practised (Rosen et al., 2020).

Taking into account the perspectives presented before, this study aims to analyse the Portuguese livestock production markets considering data from the national agricultural market information (SIMA, 2023) from 2017 to 2022 and using the Stata (StataCorp, 2017a, 2017b; Stata, 2023) software to obtain some findings. For the literature review, the Scopus (2023) database was analysed.

In the following section, the options made to choose the Portuguese livestock markets were based on a previous analysis of the markets and the more relevant changes in the respective prices practised. For the normalised values, the following formula was considered: $(x_i - x_{minimum})/(x_{maximum} - x_{minimum})$.

COW FOR SLAUGHTER (CRUZ-CHAROLÊS, CARCASS) PRICES

The producer prices of cow carcass "Cow*For slaughter*Cruz. Charolês*EUR/kg P. Carcass" (Beira Litoral) were 2.5-2.8 euros/kg (the averages prices were 2.6 euros, with an average logarithm of 0.9, average growth rates of 0.2% and an average normalised value of 0.1) in 2017 (Table 1), 2.8-3.1 euros/kg (the averages were: 3.0 euros for prices; 1.1 for logarithm; 0.1% for growth rates; and 0.3 for normalised value) in 2018 (Table 2), 2.9 euros/kg (the averages were: prices of 2.9 euros; logarithm of 1.1; growth rates of 0.0%; and a normalised value of 0.3) in 2019 (Table 3), 2.3-2.9 euros/kg (the averages were: prices of 2.5 euros; logarithm of 0.9; growth rates of -0.4%; and a normalised value of 0.1) in 2020 (Table 4), 2.3-3.0 euros/kg (the averages were: prices of 2.6 euros; logarithm of 0.9; growth rates of 0.7%; and a normalised value of 0.1) in 2021 (Table 5) and 3.0-4.5 euros/kg (the averages were: prices of 4.1 euros; logarithm of 1.4; growth rates of 0.9%; and a normalised value of 0.8) in 2022 (Table 6). The Russia-Ukraine crisis had an impact on these prices.

COW FOR SLAUGHTER (TURINA, CARCASS) PRICES

The Portuguese cow "Cow*For slaughter*Turina*EUR/kg P. Carcass" producer prices over the period 2017-2022 for the Beira Litoral region were, in euros/kg, 2.0-2.4 (average prices of 2.1 euros and average growth rates of 0.3%) in 2017 (Table 7), 2.3 (2018 and 2019, Tables 8-9), 1.9-2.3 (average prices of 2.0 euros and average growth rates of -0.4%) in 2020 (Table 10), 1.9-2.5 (average prices of 2.1 euros and average growth rates of 0.6%) in 2021 (Table 11) and 2.5-3.5 (average prices of 3.3 euros and average growth rates of 0.7%) in 2022 (Table 12).

Table 1. Cow (Cruz-Charolês, Beira Litoral (region)) for slaughter prices, euro/kg of the carcass (2017)

Period	Month	Value	Logarithm	Growth Rates (%)	Normalised Value for All Periods
02/01/2017	1	2.500	0.916		0.100
09/01/2017	1	2.500	0.916	0.000	0.100
16/01/2017	1	2.500	0.916	0.000	0.100
23/01/2017	1	2.500	0.916	0.000	0.100
30/01/2017	1	2.500	0.916	0.000	0.100
06/02/2017	2	2.500	0.916	0.000	0.100
13/02/2017	2	2.500	0.916	0.000	0.100
20/02/2017	2	2.500	0.916	0.000	0.100
27/02/2017	2	2.500	0.916	0.000	0.100
06/03/2017	3	2.500	0.916	0.000	0.100
13/03/2017	3	2.500	0.916	0.000	0.100
20/03/2017	3	2.500	0.916	0.000	0.100
27/03/2017	3	2.500	0.916	0.000	0.100
03/04/2017	4	2.500	0.916	0.000	0.100
10/04/2017	4	2.500	0.916	0.000	0.100
17/04/2017	4	2.500	0.916	0.000	0.100
24/04/2017	4	2.500	0.916	0.000	0.100
01/05/2017	5	2.500	0.916	0.000	0.100
08/05/2017	5	2.500	0.916	0.000	0.100
15/05/2017	5	2.500	0.916	0.000	0.100
22/05/2017	5	2.500	0.916	0.000	0.100
29/05/2017	5	2.500	0.916	0.000	0.100
05/06/2017	6	2.500	0.916	0.000	0.100
12/06/2017	6	2.500	0.916	0.000	0.100
19/06/2017	6	2.500	0.916	0.000	0.100
26/06/2017	6	2.500	0.916	0.000	0.100
03/07/2017	7	2.500	0.916	0.000	0.100
10/07/2017	7	2.500	0.916	0.000	0.100
17/07/2017	7	2.500	0.916	0.000	0.100
24/07/2017	7	2.500	0.916	0.000	0.100
31/07/2017	7	2.500	0.916	0.000	0.100
07/08/2017	8	2.500	0.916	0.000	0.100
14/08/2017	8	2.500	0.916	0.000	0.100
21/08/2017	8	2.500	0.916	0.000	0.100
28/08/2017	8	2.500	0.916	0.000	0.100
04/09/2017	9	2.500	0.916	0.000	0.100
11/09/2017	9	2.500	0.916	0.000	0.100

continued on following page

Table 1. Continued

Period	Month	Value	Logarithm	Growth Rates (%)	Normalised Value for All Periods
18/09/2017	9	2.500	0.916	0.000	0.100
25/09/2017	9	2.500	0.916	0.000	0.100
02/10/2017	10	2.500	0.916	0.000	0.100
09/10/2017	10	2.500	0.916	0.000	0.100
16/10/2017	10	2.500	0.916	0.000	0.100
23/10/2017	10	2.500	0.916	0.000	0.100
30/10/2017	10	2.800	1.030	12.000	0.200
06/11/2017	11	2.800	1.030	0.000	0.200
13/11/2017	11	2.800	1.030	0.000	0.200
20/11/2017	11	2.800	1.030	0.000	0.200
27/11/2017	11	2.800	1.030	0.000	0.200
04/12/2017	12	2.800	1.030	0.000	0.200
11/12/2017	12	2.800	1.030	0.000	0.200
18/12/2017	12	2.800	1.030	0.000	0.200
25/12/2017	12	2.800	1.030	0.000	0.200

Table 2. Cow (Cruz-Charolês, Beira Litoral (region)) for slaughter prices, euro/kg of the carcass (2018)

Period	Month	Value	Logarithm	Growth Rates (%)	Normalised Value for All Periods
01/01/2018	1	2.800	1.030		0.200
08/01/2018	1	2.800	1.030	0.000	0.200
15/01/2018	1	2.800	1.030	0.000	0.200
22/01/2018	1	3.100	1.131	10.714	0.400
29/01/2018	1	3.100	1.131	0.000	0.400
05/02/2018	2	3.100	1.131	0.000	0.400
12/02/2018	2	3.100	1.131	0.000	0.400
19/02/2018	2	3.100	1.131	0.000	0.400
26/02/2018	2	3.100	1.131	0.000	0.400
05/03/2018	3	3.100	1.131	0.000	0.400
12/03/2018	3	3.100	1.131	0.000	0.400
19/03/2018	3	3.100	1.131	0.000	0.400
26/03/2018	3	3.100	1.131	0.000	0.400
02/04/2018	4	3.100	1.131	0.000	0.400
09/04/2018	4	3.100	1.131	0.000	0.400
16/04/2018	4	3.100	1.131	0.000	0.400
23/04/2018	4	3.100	1.131	0.000	0.400

continued on following page

Table 2. Continued

Period	Month	Value	Logarithm	Growth Rates (%)	Normalised Value for All Periods
30/04/2018	4	3.100	1.131	0.000	0.400
07/05/2018	5	3.000	1.099	-3.226	0.300
14/05/2018	5	3.000	1.099	0.000	0.300
21/05/2018	5	3.000	1.099	0.000	0.300
28/05/2018	5	3.000	1.099	0.000	0.300
04/06/2018	6	3.000	1.099	0.000	0.300
11/06/2018	6	3.000	1.099	0.000	0.300
18/06/2018	6	3.000	1.099	0.000	0.300
25/06/2018	6	3.000	1.099	0.000	0.300
02/07/2018	7	3.000	1.099	0.000	0.300
09/07/2018	7	3.000	1.099	0.000	0.300
16/07/2018	7	3.000	1.099	0.000	0.300
23/07/2018	7	3.000	1.099	0.000	0.300
30/07/2018	7	3.000	1.099	0.000	0.300
06/08/2018	8	3.000	1.099	0.000	0.300
13/08/2018	8	3.000	1.099	0.000	0.300
20/08/2018	8	3.000	1.099	0.000	0.300
27/08/2018	8	3.000	1.099	0.000	0.300
03/09/2018	9	3.000	1.099	0.000	0.300
10/09/2018	9	3.000	1.099	0.000	0.300
17/09/2018	9	3.000	1.099	0.000	0.300
24/09/2018	9	3.000	1.099	0.000	0.300
01/10/2018	10	3.000	1.099	0.000	0.300
08/10/2018	10	2.900	1.065	-3.333	0.300
15/10/2018	10	2.900	1.065	0.000	0.300
22/10/2018	10	2.900	1.065	0.000	0.300
29/10/2018	10	2.900	1.065	0.000	0.300
05/11/2018	11	2.900	1.065	0.000	0.300
12/11/2018	11	2.900	1.065	0.000	0.300
19/11/2018	11	2.900	1.065	0.000	0.300
26/11/2018	11	2.900	1.065	0.000	0.300
03/12/2018	12	2.900	1.065	0.000	0.300
10/12/2018	12	2.900	1.065	0.000	0.300
17/12/2018	12	2.900	1.065	0.000	0.300
24/12/2018	12	2.900	1.065	0.000	0.300
31/12/2018	12	2.900	1.065	0.000	0.300

Table 3. Cow (Cruz-Charolês, Beira Litoral (region)) for slaughter prices, euro/kg of the carcass (2019)

Period	Month	Value	Logarithm	Growth Rates (%)	Normalised Value for All Periods
07/01/2019	1	2.900	1.065		0.300
14/01/2019	1	2.900	1.065	0.000	0.300
21/01/2019	1	2.900	1.065	0.000	0.300
28/01/2019	1	2.900	1.065	0.000	0.300
04/02/2019	2	2.900	1.065	0.000	0.300
11/02/2019	2	2.900	1.065	0.000	0.300
18/02/2019	2	2.900	1.065	0.000	0.300
25/02/2019	2	2.900	1.065	0.000	0.300
04/03/2019	3	2.900	1.065	0.000	0.300
11/03/2019	3	2.900	1.065	0.000	0.300
18/03/2019	3	2.900	1.065	0.000	0.300
25/03/2019	3	2.900	1.065	0.000	0.300
01/04/2019	4	2.900	1.065	0.000	0.300
08/04/2019	4	2.900	1.065	0.000	0.300
15/04/2019	4	2.900	1.065	0.000	0.300
22/04/2019	4	2.900	1.065	0.000	0.300
29/04/2019	4	2.900	1.065	0.000	0.300
06/05/2019	5	2.900	1.065	0.000	0.300
13/05/2019	5	2.900	1.065	0.000	0.300
20/05/2019	5	2.900	1.065	0.000	0.300
27/05/2019	5	2.900	1.065	0.000	0.300
03/06/2019	6	2.900	1.065	0.000	0.300
10/06/2019	6	2.900	1.065	0.000	0.300
17/06/2019	6	2.900	1.065	0.000	0.300
24/06/2019	6	2.900	1.065	0.000	0.300
01/07/2019	7	2.900	1.065	0.000	0.300
08/07/2019	7	2.900	1.065	0.000	0.300
15/07/2019	7	2.900	1.065	0.000	0.300
22/07/2019	7	2.900	1.065	0.000	0.300
29/07/2019	7	2.900	1.065	0.000	0.300
05/08/2019	8	2.900	1.065	0.000	0.300
12/08/2019	8	2.900	1.065	0.000	0.300
19/08/2019	8	2.900	1.065	0.000	0.300
26/08/2019	8	2.900	1.065	0.000	0.300
02/09/2019	9	2.900	1.065	0.000	0.300
09/09/2019	9	2.900	1.065	0.000	0.300
16/09/2019	9	2.900	1.065	0.000	0.300

continued on following page

Table 3. Continued

Period	Month	Value	Logarithm	Growth Rates (%)	Normalised Value for All Periods
23/09/2019	9	2.900	1.065	0.000	0.300
30/09/2019	9	2.900	1.065	0.000	0.300
07/10/2019	10	2.900	1.065	0.000	0.300
14/10/2019	10	2.900	1.065	0.000	0.300
21/10/2019	10	2.900	1.065	0.000	0.300
28/10/2019	10	2.900	1.065	0.000	0.300
04/11/2019	11	2.900	1.065	0.000	0.300
11/11/2019	11	2.900	1.065	0.000	0.300
18/11/2019	11	2.900	1.065	0.000	0.300
25/11/2019	11	2.900	1.065	0.000	0.300
02/12/2019	12	2.900	1.065	0.000	0.300
09/12/2019	12	2.900	1.065	0.000	0.300
16/12/2019	12	2.900	1.065	0.000	0.300
23/12/2019	12	2.900	1.065	0.000	0.300
30/12/2019	12	2.900	1.065	0.000	0.300

Table 4. Cow (Cruz-Charolês, Beira Litoral (region)) for slaughter prices, euro/kg of the carcass (2020)

Period	Month	Value	Logarithm	Growth Rates (%)	Normalised Value for All Periods
06/01/2020	1	2.900	1.065		0.300
13/01/2020	1	2.900	1.065	0.000	0.300
20/01/2020	1	2.900	1.065	0.000	0.300
27/01/2020	1	2.900	1.065	0.000	0.300
03/02/2020	2	2.900	1.065	0.000	0.300
10/02/2020	2	2.900	1.065	0.000	0.300
17/02/2020	2	2.900	1.065	0.000	0.300
24/02/2020	2	2.900	1.065	0.000	0.300
02/03/2020	3	2.900	1.065	0.000	0.300
09/03/2020	3	2.900	1.065	0.000	0.300
16/03/2020	3	2.900	1.065	0.000	0.300
23/03/2020	3	2.800	1.030	-3.448	0.200
30/03/2020	3	2.800	1.030	0.000	0.200
06/04/2020	4	2.800	1.030	0.000	0.200
13/04/2020	4	2.500	0.916	-10.714	0.100
20/04/2020	4	2.500	0.916	0.000	0.100
27/04/2020	4	2.500	0.916	0.000	0.100

continued on following page

Table 4. Continued

Period	Month	Value	Logarithm	Growth Rates (%)	Normalised Value for All Periods
04/05/2020	5	2.300	0.833	-8.000	0.000
11/05/2020	5	2.300	0.833	0.000	0.000
18/05/2020	5	2.300	0.833	0.000	0.000
25/05/2020	5	2.300	0.833	0.000	0.000
01/06/2020	6	2.300	0.833	0.000	0.000
08/06/2020	6	2.300	0.833	0.000	0.000
15/06/2020	6	2.300	0.833	0.000	0.000
22/06/2020	6	2.300	0.833	0.000	0.000
29/06/2020	6	2.300	0.833	0.000	0.000
06/07/2020	7	2.300	0.833	0.000	0.000
13/07/2020	7	2.300	0.833	0.000	0.000
20/07/2020	7	2.300	0.833	0.000	0.000
27/07/2020	7	2.300	0.833	0.000	0.000
03/08/2020	8	2.300	0.833	0.000	0.000
10/08/2020	8	2.300	0.833	0.000	0.000
17/08/2020	8	2.300	0.833	0.000	0.000
24/08/2020	8	2.300	0.833	0.000	0.000
31/08/2020	8	2.300	0.833	0.000	0.000
07/09/2020	9	2.300	0.833	0.000	0.000
14/09/2020	9	2.300	0.833	0.000	0.000
21/09/2020	9	2.300	0.833	0.000	0.000
28/09/2020	9	2.300	0.833	0.000	0.000
05/10/2020	10	2.300	0.833	0.000	0.000
12/10/2020	10	2.300	0.833	0.000	0.000
19/10/2020	10	2.300	0.833	0.000	0.000
26/10/2020	10	2.300	0.833	0.000	0.000
02/11/2020	11	2.300	0.833	0.000	0.000
09/11/2020	11	2.300	0.833	0.000	0.000
16/11/2020	11	2.300	0.833	0.000	0.000
23/11/2020	11	2.300	0.833	0.000	0.000
30/11/2020	11	2.300	0.833	0.000	0.000
07/12/2020	12	2.300	0.833	0.000	0.000
14/12/2020	12	2.300	0.833	0.000	0.000
21/12/2020	12	2.300	0.833	0.000	0.000
28/12/2020	12	2.300	0.833	0.000	0.000

Table 5. Cow (Cruz-Charolês, Beira Litoral (region)) for slaughter prices, euro/kg of the carcass (2021)

Period	Month	Value	Logarithm	Growth Rates (%)	Normalised Value for All Periods
04/01/2021	1	2.300	0.833		0.000
11/01/2021	1	2.300	0.833	0.000	0.000
18/01/2021	1	2.300	0.833	0.000	0.000
25/01/2021	1	2.300	0.833	0.000	0.000
01/02/2021	2	2.300	0.833	0.000	0.000
08/02/2021	2	2.300	0.833	0.000	0.000
15/02/2021	2	2.300	0.833	0.000	0.000
22/02/2021	2	2.300	0.833	0.000	0.000
01/03/2021	3	2.300	0.833	0.000	0.000
08/03/2021	3	2.300	0.833	0.000	0.000
15/03/2021	3	2.300	0.833	0.000	0.000
22/03/2021	3	2.300	0.833	0.000	0.000
29/03/2021	3	2.300	0.833	0.000	0.000
05/04/2021	4	2.300	0.833	0.000	0.000
12/04/2021	4	2.300	0.833	0.000	0.000
19/04/2021	4	2.300	0.833	0.000	0.000
26/04/2021	4	2.300	0.833	0.000	0.000
03/05/2021	5	2.300	0.833	0.000	0.000
10/05/2021	5	2.300	0.833	0.000	0.000
17/05/2021	5	2.300	0.833	0.000	0.000
24/05/2021	5	2.300	0.833	0.000	0.000
31/05/2021	5	2.300	0.833	0.000	0.000
07/06/2021	6	2.300	0.833	0.000	0.000
14/06/2021	6	2.300	0.833	0.000	0.000
21/06/2021	6	2.300	0.833	0.000	0.000
28/06/2021	6	2.300	0.833	0.000	0.000
05/07/2021	7	2.300	0.833	0.000	0.000
12/07/2021	7	2.300	0.833	0.000	0.000
19/07/2021	7	2.300	0.833	0.000	0.000
26/07/2021	7	3.000	1.099	30.435	0.300
02/08/2021	8	3.000	1.099	0.000	0.300
09/08/2021	8	3.000	1.099	0.000	0.300
16/08/2021	8	3.000	1.099	0.000	0.300
23/08/2021	8	3.000	1.099	0.000	0.300
30/08/2021	8	3.000	1.099	0.000	0.300
06/09/2021	9	3.000	1.099	0.000	0.300
13/09/2021	9	3.000	1.099	0.000	0.300

continued on following page

Table 5. Continued

Period	Month	Value	Logarithm	Growth Rates (%)	Normalised Value for All Periods
20/09/2021	9	3.000	1.099	0.000	0.300
27/09/2021	9	3.000	1.099	0.000	0.300
04/10/2021	10	3.000	1.099	0.000	0.300
11/10/2021	10	3.000	1.099	0.000	0.300
18/10/2021	10	3.000	1.099	0.000	0.300
25/10/2021	10	3.000	1.099	0.000	0.300
01/11/2021	11	3.000	1.099	0.000	0.300
08/11/2021	11	3.000	1.099	0.000	0.300
15/11/2021	11	3.000	1.099	0.000	0.300
22/11/2021	11	2.500	0.916	-16.667	0.100
29/11/2021	11	3.000	1.099	20.000	0.300
06/12/2021	12	3.000	1.099	0.000	0.300
13/12/2021	12	3.000	1.099	0.000	0.300
20/12/2021	12	3.000	1.099	0.000	0.300
27/12/2021	12	3.000	1.099	0.000	0.300

Table 6. Cow (Cruz-Charolês, Beira Litoral (region)) for slaughter prices, euro/kg of the carcass (2022)

Period	Month	Value	Logarithm	Growth Rates (%)	Normalised Value for All Periods
03/01/2022	1	3.000	1.099		0.300
10/01/2022	1	3.000	1.099	0.000	0.300
17/01/2022	1	3.000	1.099	0.000	0.300
24/01/2022	1	3.000	1.099	0.000	0.300
31/01/2022	1	3.000	1.099	0.000	0.300
07/02/2022	2	3.000	1.099	0.000	0.300
14/02/2022	2	3.000	1.099	0.000	0.300
21/02/2022	2	3.000	1.099	0.000	0.300
28/02/2022	2	3.000	1.099	0.000	0.300
07/03/2022	3	3.800	1.335	26.667	0.700
14/03/2022	3	3.800	1.335	0.000	0.700
21/03/2022	3	3.900	1.361	2.632	0.700
28/03/2022	3	3.900	1.361	0.000	0.700
04/04/2022	4	3.900	1.361	0.000	0.700
11/04/2022	4	3.900	1.361	0.000	0.700
18/04/2022	4	3.900	1.361	0.000	0.700
25/04/2022	4	3.900	1.361	0.000	0.700

continued on following page

Table 6. Continued

Period	Month	Value	Logarithm	Growth Rates (%)	Normalised Value for All Periods
02/05/2022	5	3.900	1.361	0.000	0.700
09/05/2022	5	4.200	1.435	7.692	0.900
16/05/2022	5	4.200	1.435	0.000	0.900
23/05/2022	5	4.200	1.435	0.000	0.900
30/05/2022	5	4.200	1.435	0.000	0.900
06/06/2022	6	4.200	1.435	0.000	0.900
13/06/2022	6	4.200	1.435	0.000	0.900
20/06/2022	6	4.200	1.435	0.000	0.900
27/06/2022	6	4.200	1.435	0.000	0.900
04/07/2022	7	4.200	1.435	0.000	0.900
11/07/2022	7	4.500	1.504	7.143	1.000
18/07/2022	7	4.500	1.504	0.000	1.000
25/07/2022	7	4.500	1.504	0.000	1.000
01/08/2022	8	4.500	1.504	0.000	1.000
08/08/2022	8	4.500	1.504	0.000	1.000
15/08/2022	8	4.500	1.504	0.000	1.000
22/08/2022	8	4.500	1.504	0.000	1.000
29/08/2022	8	4.500	1.504	0.000	1.000
05/09/2022	9	4.500	1.504	0.000	1.000
12/09/2022	9	4.500	1.504	0.000	1.000
19/09/2022	9	4.500	1.504	0.000	1.000
26/09/2022	9	4.500	1.504	0.000	1.000
03/10/2022	10	4.500	1.504	0.000	1.000
10/10/2022	10	4.500	1.504	0.000	1.000
17/10/2022	10	4.500	1.504	0.000	1.000
24/10/2022	10	4.400	1.482	-2.222	1.000
31/10/2022	10	4.500	1.504	2.273	1.000
07/11/2022	11	4.500	1.504	0.000	1.000
14/11/2022	11	4.500	1.504	0.000	1.000
21/11/2022	11	4.500	1.504	0.000	1.000
28/11/2022	11	4.500	1.504	0.000	1.000
05/12/2022	12	4.500	1.504	0.000	1.000
12/12/2022	12	4.500	1.504	0.000	1.000
19/12/2022	12	4.500	1.504	0.000	1.000
26/12/2022	12	4.500	1.504	0.000	1.000

Table 7. Cow (Turina, Beira Litoral (region)) for slaughter prices, euro/kg of the carcass (2017)

Period	Month	Value	Logarithm	Growth Rates (%)	Normalised Value for All Periods
02/01/2017	1	2.000	0.693		0.100
09/01/2017	1	2.000	0.693	0.000	0.100
16/01/2017	1	2.000	0.693	0.000	0.100
23/01/2017	1	2.000	0.693	0.000	0.100
30/01/2017	1	2.000	0.693	0.000	0.100
06/02/2017	2	2.000	0.693	0.000	0.100
13/02/2017	2	2.000	0.693	0.000	0.100
20/02/2017	2	2.000	0.693	0.000	0.100
27/02/2017	2	2.000	0.693	0.000	0.100
06/03/2017	3	2.000	0.693	0.000	0.100
13/03/2017	3	2.000	0.693	0.000	0.100
20/03/2017	3	2.000	0.693	0.000	0.100
27/03/2017	3	2.000	0.693	0.000	0.100
03/04/2017	4	2.000	0.693	0.000	0.100
10/04/2017	4	2.000	0.693	0.000	0.100
17/04/2017	4	2.000	0.693	0.000	0.100
24/04/2017	4	2.000	0.693	0.000	0.100
01/05/2017	5	2.000	0.693	0.000	0.100
08/05/2017	5	2.000	0.693	0.000	0.100
15/05/2017	5	2.000	0.693	0.000	0.100
22/05/2017	5	2.000	0.693	0.000	0.100
29/05/2017	5	2.000	0.693	0.000	0.100
05/06/2017	6	2.000	0.693	0.000	0.100
12/06/2017	6	2.000	0.693	0.000	0.100
19/06/2017	6	2.000	0.693	0.000	0.100
26/06/2017	6	2.000	0.693	0.000	0.100
03/07/2017	7	2.000	0.693	0.000	0.100
10/07/2017	7	2.000	0.693	0.000	0.100
17/07/2017	7	2.000	0.693	0.000	0.100
24/07/2017	7	2.000	0.693	0.000	0.100
31/07/2017	7	2.000	0.693	0.000	0.100
07/08/2017	8	2.000	0.693	0.000	0.100
14/08/2017	8	2.000	0.693	0.000	0.100
21/08/2017	8	2.000	0.693	0.000	0.100
28/08/2017	8	2.000	0.693	0.000	0.100
04/09/2017	9	2.000	0.693	0.000	0.100
11/09/2017	9	2.000	0.693	0.000	0.100

continued on following page

Table 7. Continued

Period	Month	Value	Logarithm	Growth Rates (%)	Normalised Value for All Periods
18/09/2017	9	2.000	0.693	0.000	0.100
25/09/2017	9	2.000	0.693	0.000	0.100
02/10/2017	10	2.000	0.693	0.000	0.100
09/10/2017	10	2.000	0.693	0.000	0.100
16/10/2017	10	2.000	0.693	0.000	0.100
23/10/2017	10	2.000	0.693	0.000	0.100
30/10/2017	10	2.400	0.875	20.000	0.300
06/11/2017	11	2.300	0.833	-4.167	0.300
13/11/2017	11	2.300	0.833	0.000	0.300
20/11/2017	11	2.300	0.833	0.000	0.300
27/11/2017	11	2.300	0.833	0.000	0.300
04/12/2017	12	2.300	0.833	0.000	0.300
11/12/2017	12	2.300	0.833	0.000	0.300
18/12/2017	12	2.300	0.833	0.000	0.300
25/12/2017	12	2.300	0.833	0.000	0.300

Table 8. Cow (Turina, Beira Litoral (region)) for slaughter prices, euro/kg of the carcass (2018)

Period	Month	Value	Logarithm	Growth Rates (%)	Normalised Value for All Periods
01/01/2018	1	2.300	0.833		0.300
08/01/2018	1	2.300	0.833	0.000	0.300
15/01/2018	1	2.300	0.833	0.000	0.300
22/01/2018	1	2.300	0.833	0.000	0.300
29/01/2018	1	2.300	0.833	0.000	0.300
05/02/2018	2	2.300	0.833	0.000	0.300
12/02/2018	2	2.300	0.833	0.000	0.300
19/02/2018	2	2.300	0.833	0.000	0.300
26/02/2018	2	2.300	0.833	0.000	0.300
05/03/2018	3	2.300	0.833	0.000	0.300
12/03/2018	3	2.300	0.833	0.000	0.300
19/03/2018	3	2.300	0.833	0.000	0.300
26/03/2018	3	2.300	0.833	0.000	0.300
02/04/2018	4	2.300	0.833	0.000	0.300
09/04/2018	4	2.300	0.833	0.000	0.300
16/04/2018	4	2.300	0.833	0.000	0.300
23/04/2018	4	2.300	0.833	0.000	0.300

continued on following page

Table 8. Continued

Period	Month	Value	Logarithm	Growth Rates (%)	Normalised Value for All Periods
30/04/2018	4	2.300	0.833	0.000	0.300
07/05/2018	5	2.300	0.833	0.000	0.300
14/05/2018	5	2.300	0.833	0.000	0.300
21/05/2018	5	2.300	0.833	0.000	0.300
28/05/2018	5	2.300	0.833	0.000	0.300
04/06/2018	6	2.300	0.833	0.000	0.300
11/06/2018	6	2.300	0.833	0.000	0.300
18/06/2018	6	2.300	0.833	0.000	0.300
25/06/2018	6	2.300	0.833	0.000	0.300
02/07/2018	7	2.300	0.833	0.000	0.300
09/07/2018	7	2.300	0.833	0.000	0.300
16/07/2018	7	2.300	0.833	0.000	0.300
23/07/2018	7	2.300	0.833	0.000	0.300
30/07/2018	7	2.300	0.833	0.000	0.300
06/08/2018	8	2.300	0.833	0.000	0.300
13/08/2018	8	2.300	0.833	0.000	0.300
20/08/2018	8	2.300	0.833	0.000	0.300
27/08/2018	8	2.300	0.833	0.000	0.300
03/09/2018	9	2.300	0.833	0.000	0.300
10/09/2018	9	2.300	0.833	0.000	0.300
17/09/2018	9	2.300	0.833	0.000	0.300
24/09/2018	9	2.300	0.833	0.000	0.300
01/10/2018	10	2.300	0.833	0.000	0.300
08/10/2018	10	2.300	0.833	0.000	0.300
15/10/2018	10	2.300	0.833	0.000	0.300
22/10/2018	10	2.300	0.833	0.000	0.300
29/10/2018	10	2.300	0.833	0.000	0.300
05/11/2018	11	2.300	0.833	0.000	0.300
12/11/2018	11	2.300	0.833	0.000	0.300
19/11/2018	11	2.300	0.833	0.000	0.300
26/11/2018	11	2.300	0.833	0.000	0.300
03/12/2018	12	2.300	0.833	0.000	0.300
10/12/2018	12	2.300	0.833	0.000	0.300
17/12/2018	12	2.300	0.833	0.000	0.300
24/12/2018	12	2.300	0.833	0.000	0.300
31/12/2018	12	2.300	0.833	0.000	0.300

Table 9. Cow (Turina, Beira Litoral (region)) for slaughter prices, euro/kg of the carcass (2019)

Period	Month	Value	Logarithm	Growth Rates (%)	Normalised Value for All Periods
07/01/2019	1	2.300	0.833		0.300
14/01/2019	1	2.300	0.833	0.000	0.300
21/01/2019	1	2.300	0.833	0.000	0.300
28/01/2019	1	2.300	0.833	0.000	0.300
04/02/2019	2	2.300	0.833	0.000	0.300
11/02/2019	2	2.300	0.833	0.000	0.300
18/02/2019	2	2.300	0.833	0.000	0.300
25/02/2019	2	2.300	0.833	0.000	0.300
04/03/2019	3	2.300	0.833	0.000	0.300
11/03/2019	3	2.300	0.833	0.000	0.300
18/03/2019	3	2.300	0.833	0.000	0.300
25/03/2019	3	2.300	0.833	0.000	0.300
01/04/2019	4	2.300	0.833	0.000	0.300
08/04/2019	4	2.300	0.833	0.000	0.300
15/04/2019	4	2.300	0.833	0.000	0.300
22/04/2019	4	2.300	0.833	0.000	0.300
29/04/2019	4	2.300	0.833	0.000	0.300
06/05/2019	5	2.300	0.833	0.000	0.300
13/05/2019	5	2.300	0.833	0.000	0.300
20/05/2019	5	2.300	0.833	0.000	0.300
27/05/2019	5	2.300	0.833	0.000	0.300
03/06/2019	6	2.300	0.833	0.000	0.300
10/06/2019	6	2.300	0.833	0.000	0.300
17/06/2019	6	2.300	0.833	0.000	0.300
24/06/2019	6	2.300	0.833	0.000	0.300
01/07/2019	7	2.300	0.833	0.000	0.300
08/07/2019	7	2.300	0.833	0.000	0.300
15/07/2019	7	2.300	0.833	0.000	0.300
22/07/2019	7	2.300	0.833	0.000	0.300
29/07/2019	7	2.300	0.833	0.000	0.300
05/08/2019	8	2.300	0.833	0.000	0.300
12/08/2019	8	2.300	0.833	0.000	0.300
19/08/2019	8	2.300	0.833	0.000	0.300
26/08/2019	8	2.300	0.833	0.000	0.300
02/09/2019	9	2.300	0.833	0.000	0.300
09/09/2019	9	2.300	0.833	0.000	0.300
16/09/2019	9	2.300	0.833	0.000	0.300

continued on following page

Table 9. Continued

Period	Month	Value	Logarithm	Growth Rates (%)	Normalised Value for All Periods
23/09/2019	9	2.300	0.833	0.000	0.300
30/09/2019	9	2.300	0.833	0.000	0.300
07/10/2019	10	2.300	0.833	0.000	0.300
14/10/2019	10	2.300	0.833	0.000	0.300
21/10/2019	10	2.300	0.833	0.000	0.300
28/10/2019	10	2.300	0.833	0.000	0.300
04/11/2019	11	2.300	0.833	0.000	0.300
11/11/2019	11	2.300	0.833	0.000	0.300
18/11/2019	11	2.300	0.833	0.000	0.300
25/11/2019	11	2.300	0.833	0.000	0.300
02/12/2019	12	2.300	0.833	0.000	0.300
09/12/2019	12	2.300	0.833	0.000	0.300
16/12/2019	12	2.300	0.833	0.000	0.300
23/12/2019	12	2.300	0.833	0.000	0.300
30/12/2019	12	2.300	0.833	0.000	0.300

Table 10. Cow (Turina, Beira Litoral (region)) for slaughter prices, euro/kg of the carcass (2020)

Period	Month	Value	Logarithm	Growth Rates (%)	Normalised Value for All Periods
06/01/2020	1	2.300	0.833		0.300
13/01/2020	1	2.300	0.833	0.000	0.300
20/01/2020	1	2.300	0.833	0.000	0.300
27/01/2020	1	2.300	0.833	0.000	0.300
03/02/2020	2	2.300	0.833	0.000	0.300
10/02/2020	2	2.300	0.833	0.000	0.300
17/02/2020	2	2.300	0.833	0.000	0.300
24/02/2020	2	2.300	0.833	0.000	0.300
02/03/2020	3	2.300	0.833	0.000	0.300
09/03/2020	3	2.300	0.833	0.000	0.300
16/03/2020	3	2.300	0.833	0.000	0.300
23/03/2020	3	2.300	0.833	0.000	0.300
30/03/2020	3	2.300	0.833	0.000	0.300
06/04/2020	4	2.300	0.833	0.000	0.300
13/04/2020	4	2.000	0.693	-13.043	0.100
20/04/2020	4	2.000	0.693	0.000	0.100
27/04/2020	4	2.000	0.693	0.000	0.100
04/05/2020	5	1.900	0.642	-5.000	0.000

continued on following page

Table 10. Continued

Period	Month	Value	Logarithm	Growth Rates (%)	Normalised Value for All Periods
11/05/2020	5	1.900	0.642	0.000	0.000
18/05/2020	5	1.900	0.642	0.000	0.000
25/05/2020	5	1.900	0.642	0.000	0.000
01/06/2020	6	1.900	0.642	0.000	0.000
08/06/2020	6	1.900	0.642	0.000	0.000
15/06/2020	6	1.900	0.642	0.000	0.000
22/06/2020	6	1.900	0.642	0.000	0.000
29/06/2020	6	1.900	0.642	0.000	0.000
06/07/2020	7	1.900	0.642	0.000	0.000
13/07/2020	7	1.900	0.642	0.000	0.000
20/07/2020	7	1.900	0.642	0.000	0.000
27/07/2020	7	1.900	0.642	0.000	0.000
03/08/2020	8	1.900	0.642	0.000	0.000
10/08/2020	8	1.900	0.642	0.000	0.000
17/08/2020	8	1.900	0.642	0.000	0.000
24/08/2020	8	1.900	0.642	0.000	0.000
31/08/2020	8	1.900	0.642	0.000	0.000
07/09/2020	9	1.900	0.642	0.000	0.000
14/09/2020	9	1.900	0.642	0.000	0.000
21/09/2020	9	1.900	0.642	0.000	0.000
28/09/2020	9	1.900	0.642	0.000	0.000
05/10/2020	10	1.900	0.642	0.000	0.000
12/10/2020	10	1.900	0.642	0.000	0.000
19/10/2020	10	1.900	0.642	0.000	0.000
26/10/2020	10	1.900	0.642	0.000	0.000
02/11/2020	11	1.900	0.642	0.000	0.000
09/11/2020	11	1.900	0.642	0.000	0.000
16/11/2020	11	1.900	0.642	0.000	0.000
23/11/2020	11	1.900	0.642	0.000	0.000
30/11/2020	11	1.900	0.642	0.000	0.000
07/12/2020	12	1.900	0.642	0.000	0.000
14/12/2020	12	1.900	0.642	0.000	0.000
21/12/2020	12	1.900	0.642	0.000	0.000
28/12/2020	12	1.900	0.642	0.000	0.000

Table 11. Cow (Turina, Beira Litoral (region)) for slaughter prices, euro/kg of the carcass (2021)

Period	Month	Value	Logarithm	Growth Rates (%)	Normalised Value for All Periods
04/01/2021	1	1.900	0.642		0.000
11/01/2021	1	1.900	0.642	0.000	0.000
18/01/2021	1	1.900	0.642	0.000	0.000
25/01/2021	1	1.900	0.642	0.000	0.000
01/02/2021	2	1.900	0.642	0.000	0.000
08/02/2021	2	1.900	0.642	0.000	0.000
15/02/2021	2	1.900	0.642	0.000	0.000
22/02/2021	2	1.900	0.642	0.000	0.000
01/03/2021	3	1.900	0.642	0.000	0.000
08/03/2021	3	1.900	0.642	0.000	0.000
15/03/2021	3	1.900	0.642	0.000	0.000
22/03/2021	3	1.900	0.642	0.000	0.000
29/03/2021	3	1.900	0.642	0.000	0.000
05/04/2021	4	1.900	0.642	0.000	0.000
12/04/2021	4	1.900	0.642	0.000	0.000
19/04/2021	4	1.900	0.642	0.000	0.000
26/04/2021	4	1.900	0.642	0.000	0.000
03/05/2021	5	1.900	0.642	0.000	0.000
10/05/2021	5	1.900	0.642	0.000	0.000
17/05/2021	5	1.900	0.642	0.000	0.000
24/05/2021	5	1.900	0.642	0.000	0.000
31/05/2021	5	1.900	0.642	0.000	0.000
07/06/2021	6	1.900	0.642	0.000	0.000
14/06/2021	6	1.900	0.642	0.000	0.000
21/06/2021	6	1.900	0.642	0.000	0.000
28/06/2021	6	1.900	0.642	0.000	0.000
05/07/2021	7	1.900	0.642	0.000	0.000
12/07/2021	7	1.900	0.642	0.000	0.000
19/07/2021	7	1.900	0.642	0.000	0.000
26/07/2021	7	2.100	0.742	10.526	0.100
02/08/2021	8	2.100	0.742	0.000	0.100
09/08/2021	8	2.100	0.742	0.000	0.100
16/08/2021	8	2.100	0.742	0.000	0.100
23/08/2021	8	2.100	0.742	0.000	0.100
30/08/2021	8	2.100	0.742	0.000	0.100
06/09/2021	9	2.100	0.742	0.000	0.100
13/09/2021	9	2.100	0.742	0.000	0.100

continued on following page

Table 11. Continued

Period	Month	Value	Logarithm	Growth Rates (%)	Normalised Value for All Periods
20/09/2021	9	2.100	0.742	0.000	0.100
27/09/2021	9	2.100	0.742	0.000	0.100
04/10/2021	10	2.100	0.742	0.000	0.100
11/10/2021	10	2.100	0.742	0.000	0.100
18/10/2021	10	2.100	0.742	0.000	0.100
25/10/2021	10	2.500	0.916	19.048	0.400
01/11/2021	11	2.500	0.916	0.000	0.400
08/11/2021	11	2.500	0.916	0.000	0.400
15/11/2021	11	2.500	0.916	0.000	0.400
22/11/2021	11	2.500	0.916	0.000	0.400
29/11/2021	11	2.500	0.916	0.000	0.400
06/12/2021	12	2.500	0.916	0.000	0.400
13/12/2021	12	2.500	0.916	0.000	0.400
20/12/2021	12	2.500	0.916	0.000	0.400
27/12/2021	12	2.500	0.916	0.000	0.400

Table 12. Cow (Turina, Beira Litoral (region)) for slaughter prices, euro/kg of the carcass (2022)

Period	Month	Value	Logarithm	Growth Rates (%)	Normalised Value for All Periods
03/01/2022	1	2.500	0.916		0.400
10/01/2022	1	2.500	0.916	0.000	0.400
17/01/2022	1	2.500	0.916	0.000	0.400
24/01/2022	1	2.500	0.916	0.000	0.400
31/01/2022	1	2.500	0.916	0.000	0.400
07/02/2022	2	2.500	0.916	0.000	0.400
14/02/2022	2	2.500	0.916	0.000	0.400
21/02/2022	2	3.300	1.194	32.000	0.900
28/02/2022	2	3.300	1.194	0.000	0.900
07/03/2022	3	3.300	1.194	0.000	0.900
14/03/2022	3	3.300	1.194	0.000	0.900
21/03/2022	3	3.300	1.194	0.000	0.900
28/03/2022	3	3.300	1.194	0.000	0.900
04/04/2022	4	3.300	1.194	0.000	0.900
11/04/2022	4	3.300	1.194	0.000	0.900
18/04/2022	4	3.300	1.194	0.000	0.900
25/04/2022	4	3.300	1.194	0.000	0.900

continued on following page

Table 12. Continued

Period	Month	Value	Logarithm	Growth Rates (%)	Normalised Value for All Periods
02/05/2022	5	3.300	1.194	0.000	0.900
09/05/2022	5	3.300	1.194	0.000	0.900
16/05/2022	5	3.300	1.194	0.000	0.900
23/05/2022	5	3.300	1.194	0.000	0.900
30/05/2022	5	3.300	1.194	0.000	0.900
06/06/2022	6	3.300	1.194	0.000	0.900
13/06/2022	6	3.300	1.194	0.000	0.900
20/06/2022	6	3.300	1.194	0.000	0.900
27/06/2022	6	3.300	1.194	0.000	0.900
04/07/2022	7	3.300	1.194	0.000	0.900
11/07/2022	7	3.300	1.194	0.000	0.900
18/07/2022	7	3.300	1.194	0.000	0.900
25/07/2022	7	3.300	1.194	0.000	0.900
01/08/2022	8	3.300	1.194	0.000	0.900
08/08/2022	8	3.300	1.194	0.000	0.900
15/08/2022	8	3.300	1.194	0.000	0.900
22/08/2022	8	3.300	1.194	0.000	0.900
29/08/2022	8	3.300	1.194	0.000	0.900
05/09/2022	9	3.300	1.194	0.000	0.900
12/09/2022	9	3.300	1.194	0.000	0.900
19/09/2022	9	3.500	1.253	6.061	1.000
26/09/2022	9	3.500	1.253	0.000	1.000
03/10/2022	10	3.500	1.253	0.000	1.000
10/10/2022	10	3.500	1.253	0.000	1.000
17/10/2022	10	3.500	1.253	0.000	1.000
24/10/2022	10	3.500	1.253	0.000	1.000
31/10/2022	10	3.500	1.253	0.000	1.000
07/11/2022	11	3.500	1.253	0.000	1.000
14/11/2022	11	3.500	1.253	0.000	1.000
21/11/2022	11	3.500	1.253	0.000	1.000
28/11/2022	11	3.500	1.253	0.000	1.000
05/12/2022	12	3.500	1.253	0.000	1.000
12/12/2022	12	3.500	1.253	0.000	1.000
19/12/2022	12	3.500	1.253	0.000	1.000
26/12/2022	12	3.500	1.253	0.000	1.000

COUNTRY CHICKEN (2.3 A 2.5 KG, CARCASS) PRICES

Table 13. Country chicken (2.3 a 2.5 kg) carcass prices, euro/kg (2019-2020, Beira Litoral (region and market))

Period	Month	Value	Logarithm	Growth Rates (%)	Normalised Value for all Periods and markets
30/12/2019	12	2.300	0.833		0.300
06/01/2020	1	2.100	0.742	-8.696	0.200
13/01/2020	1	2.100	0.742	0.000	0.200
20/01/2020	1	2.200	0.788	4.762	0.200
27/01/2020	1	2.200	0.788	0.000	0.200
03/02/2020	2	2.200	0.788	0.000	0.200
10/02/2020	2	2.100	0.742	-4.545	0.200
17/02/2020	2	2.100	0.742	0.000	0.200
24/02/2020	2	2.100	0.742	0.000	0.200
02/03/2020	3	2.200	0.788	4.762	0.200
09/03/2020	3	2.200	0.788	0.000	0.200
16/03/2020	3	2.600	0.956	18.182	0.400
23/03/2020	3	2.600	0.956	0.000	0.400
30/03/2020	3	2.600	0.956	0.000	0.400
06/04/2020	4	2.600	0.956	0.000	0.400
13/04/2020	4	2.600	0.956	0.000	0.400
20/04/2020	4	2.600	0.956	0.000	0.400
27/04/2020	4	2.600	0.956	0.000	0.400
04/05/2020	5	2.600	0.956	0.000	0.400
11/05/2020	5	2.600	0.956	0.000	0.400
18/05/2020	5	2.600	0.956	0.000	0.400
25/05/2020	5	2.600	0.956	0.000	0.400
01/06/2020	6	2.600	0.956	0.000	0.400
08/06/2020	6	2.600	0.956	0.000	0.400
15/06/2020	6	2.600	0.956	0.000	0.400
22/06/2020	6	2.600	0.956	0.000	0.400
29/06/2020	6	2.500	0.916	-3.846	0.400
06/07/2020	7	2.500	0.916	0.000	0.400
13/07/2020	7	2.500	0.916	0.000	0.400
20/07/2020	7	2.500	0.916	0.000	0.400
27/07/2020	7	2.500	0.916	0.000	0.400
03/08/2020	8	2.400	0.875	-4.000	0.300
10/08/2020	8	2.400	0.875	0.000	0.300
17/08/2020	8	2.400	0.875	0.000	0.300
24/08/2020	8	2.400	0.875	0.000	0.300
31/08/2020	8	2.400	0.875	0.000	0.300

continued on following page

Table 13. Continued

Period	Month	Value	Logarithm	Growth Rates (%)	Normalised Value for all Periods and markets
07/09/2020	9	2.300	0.833	-4.167	0.300
14/09/2020	9	2.300	0.833	0.000	0.300
21/09/2020	9	2.300	0.833	0.000	0.300
28/09/2020	9	2.200	0.788	-4.348	0.200
05/10/2020	10	2.200	0.788	0.000	0.200
12/10/2020	10	2.200	0.788	0.000	0.200
19/10/2020	10	2.200	0.788	0.000	0.200
26/10/2020	10	2.200	0.788	0.000	0.200
02/11/2020	11	2.200	0.788	0.000	0.200
09/11/2020	11	2.200	0.788	0.000	0.200
16/11/2020	11	2.200	0.788	0.000	0.200
23/11/2020	11	2.200	0.788	0.000	0.200
30/11/2020	11	2.200	0.788	0.000	0.200
07/12/2020	12	2.300	0.833	4.545	0.300
14/12/2020	12	2.300	0.833	0.000	0.300
21/12/2020	12	2.300	0.833	0.000	0.300
28/12/2020	12	2.300	0.833	0.000	0.300

Table 14. Country chicken (2.3 a 2.5 kg) carcass prices, euro/kg (2021, Beira Litoral (region and market))

Period	Month	Value	Logarithm	Growth Rates (%)	Normalised Value for All Periods and Markets
04/01/2021	1	2.300	0.833		0.300
11/01/2021	1	2.300	0.833	0.000	0.300
18/01/2021	1	2.400	0.875	4.348	0.300
25/01/2021	1	2.400	0.875	0.000	0.300
01/02/2021	2	2.300	0.833	-4.167	0.300
08/02/2021	2	2.300	0.833	0.000	0.300
15/02/2021	2	2.300	0.833	0.000	0.300
22/02/2021	2	2.300	0.833	0.000	0.300
01/03/2021	3	2.300	0.833	0.000	0.300
08/03/2021	3	2.300	0.833	0.000	0.300
15/03/2021	3	2.300	0.833	0.000	0.300
22/03/2021	3	2.300	0.833	0.000	0.300
29/03/2021	3	2.300	0.833	0.000	0.300
05/04/2021	4	2.300	0.833	0.000	0.300

continued on following page

Table 14. Continued

Period	Month	Value	Logarithm	Growth Rates (%)	Normalised Value for All Periods and Markets
12/04/2021	4	2.200	0.788	-4.348	0.200
19/04/2021	4	2.200	0.788	0.000	0.200
26/04/2021	4	2.200	0.788	0.000	0.200
03/05/2021	5	2.200	0.788	0.000	0.200
10/05/2021	5	2.200	0.788	0.000	0.200
17/05/2021	5	2.200	0.788	0.000	0.200
24/05/2021	5	2.200	0.788	0.000	0.200
31/05/2021	5	2.200	0.788	0.000	0.200
07/06/2021	6	2.200	0.788	0.000	0.200
14/06/2021	6	2.200	0.788	0.000	0.200
21/06/2021	6	2.200	0.788	0.000	0.200
28/06/2021	6	2.100	0.742	-4.545	0.200
05/07/2021	7	2.100	0.742	0.000	0.200
12/07/2021	7	2.100	0.742	0.000	0.200
19/07/2021	7	2.100	0.742	0.000	0.200
26/07/2021	7	2.100	0.742	0.000	0.200
02/08/2021	8	2.100	0.742	0.000	0.200
09/08/2021	8	2.000	0.693	-4.762	0.100
16/08/2021	8	2.000	0.693	0.000	0.100
23/08/2021	8	2.000	0.693	0.000	0.100
30/08/2021	8	2.000	0.693	0.000	0.100
06/09/2021	9	2.000	0.693	0.000	0.100
13/09/2021	9	2.000	0.693	0.000	0.100
20/09/2021	9	2.000	0.693	0.000	0.100
27/09/2021	9	2.000	0.693	0.000	0.100
04/10/2021	10	2.000	0.693	0.000	0.100
11/10/2021	10	2.000	0.693	0.000	0.100
18/10/2021	10	2.000	0.693	0.000	0.100
25/10/2021	10	2.000	0.693	0.000	0.100
01/11/2021	11	2.000	0.693	0.000	0.100
08/11/2021	11	2.000	0.693	0.000	0.100
15/11/2021	11	2.100	0.742	5.000	0.200
22/11/2021	11	2.300	0.833	9.524	0.300
29/11/2021	11	2.500	0.916	8.696	0.400
06/12/2021	12	2.500	0.916	0.000	0.400
13/12/2021	12	2.600	0.956	4.000	0.400
20/12/2021	12	2.600	0.956	0.000	0.400
27/12/2021	12	2.600	0.956	0.000	0.400

Table 15. Country chicken (2.3 a 2.5 kg) carcass prices, euro/kg (2022, Beira Litoral (region and market))

Period	Month	Value	Logarithm	Growth Rates (%)	Normalised Value for All Periods and markets
03/01/2022	1	2.600	0.956		0.400
10/01/2022	1	2.700	0.993	3.846	0.500
17/01/2022	1	2.700	0.993	0.000	0.500
24/01/2022	1	2.700	0.993	0.000	0.500
31/01/2022	1	2.700	0.993	0.000	0.500
07/02/2022	2	2.700	0.993	0.000	0.500
14/02/2022	2	2.700	0.993	0.000	0.500
21/02/2022	2	2.700	0.993	0.000	0.500
28/02/2022	2	2.800	1.030	3.704	0.500
07/03/2022	3	2.900	1.065	3.571	0.600
14/03/2022	3	3.100	1.131	6.897	0.700
21/03/2022	3	3.200	1.163	3.226	0.700
28/03/2022	3	3.200	1.163	0.000	0.700
04/04/2022	4	3.200	1.163	0.000	0.700
11/04/2022	4	3.200	1.163	0.000	0.700
18/04/2022	4	3.300	1.194	3.125	0.800
25/04/2022	4	3.300	1.194	0.000	0.800
02/05/2022	5	3.300	1.194	0.000	0.800
09/05/2022	5	3.300	1.194	0.000	0.800
16/05/2022	5	3.300	1.194	0.000	0.800
23/05/2022	5	3.300	1.194	0.000	0.800
30/05/2022	5	3.300	1.194	0.000	0.800
06/06/2022	6	3.300	1.194	0.000	0.800
13/06/2022	6	3.300	1.194	0.000	0.800
20/06/2022	6	3.300	1.194	0.000	0.800
27/06/2022	6	3.300	1.194	0.000	0.800
04/07/2022	7	3.200	1.163	-3.030	0.700
11/07/2022	7	3.200	1.163	0.000	0.700
18/07/2022	7	3.200	1.163	0.000	0.700
25/07/2022	7	3.200	1.163	0.000	0.700
01/08/2022	8	3.200	1.163	0.000	0.700
08/08/2022	8	3.200	1.163	0.000	0.700
15/08/2022	8	3.200	1.163	0.000	0.700
22/08/2022	8	3.200	1.163	0.000	0.700
29/08/2022	8	3.200	1.163	0.000	0.700
05/09/2022	9	3.200	1.163	0.000	0.700
12/09/2022	9	3.200	1.163	0.000	0.700

continued on following page

Table 15. Continued

Period	Month	Value	Logarithm	Growth Rates (%)	Normalised Value for All Periods and markets
19/09/2022	9	3.200	1.163	0.000	0.700
26/09/2022	9	3.200	1.163	0.000	0.700
03/10/2022	10	3.200	1.163	0.000	0.700
10/10/2022	10	3.200	1.163	0.000	0.700
17/10/2022	10	3.200	1.163	0.000	0.700
24/10/2022	10	3.200	1.163	0.000	0.700
31/10/2022	10	3.300	1.194	3.125	0.800
07/11/2022	11	3.300	1.194	0.000	0.800
14/11/2022	11	3.300	1.194	0.000	0.800
21/11/2022	11	3.500	1.253	6.061	0.900
28/11/2022	11	3.600	1.281	2.857	0.900
05/12/2022	12	3.600	1.281	0.000	0.900
12/12/2022	12	3.600	1.281	0.000	0.900
19/12/2022	12	3.700	1.308	2.778	1.000
26/12/2022	12	3.700	1.308	0.000	1.000

Table 16. Country chicken (2.3 a 2.5 kg) carcass prices, euro/kg (2017, Beira Litoral (region) and Dão-Lafões (market))

Period	Month	Value	Logarithm	Growth Rates (%)	Normalised Value for All Periods and Markets
02/01/2017	1	2.200	0.788		0.200
09/01/2017	1	2.100	0.742	-4.545	0.200
16/01/2017	1	2.000	0.693	-4.762	0.100
23/01/2017	1	2.000	0.693	0.000	0.100
30/01/2017	1	2.000	0.693	0.000	0.100
06/02/2017	2	2.000	0.693	0.000	0.100
13/02/2017	2	2.000	0.693	0.000	0.100
20/02/2017	2	2.000	0.693	0.000	0.100
27/02/2017	2	2.000	0.693	0.000	0.100
06/03/2017	3	1.800	0.588	-10.000	0.000
13/03/2017	3	1.800	0.588	0.000	0.000
20/03/2017	3	1.800	0.588	0.000	0.000
27/03/2017	3	1.800	0.588	0.000	0.000
03/04/2017	4	1.800	0.588	0.000	0.000
10/04/2017	4	1.800	0.588	0.000	0.000

continued on following page

Table 16. Continued

Period	Month	Value	Logarithm	Growth Rates (%)	Normalised Value for All Periods and Markets
17/04/2017	4	1.800	0.588	0.000	0.000
24/04/2017	4	1.800	0.588	0.000	0.000
01/05/2017	5	1.800	0.588	0.000	0.000
08/05/2017	5	2.100	0.742	16.667	0.200
15/05/2017	5	2.200	0.788	4.762	0.200
22/05/2017	5	2.200	0.788	0.000	0.200
29/05/2017	5	2.200	0.788	0.000	0.200
05/06/2017	6	2.200	0.788	0.000	0.200
12/06/2017	6	2.200	0.788	0.000	0.200
19/06/2017	6	2.200	0.788	0.000	0.200
26/06/2017	6	2.100	0.742	-4.545	0.200
03/07/2017	7	2.100	0.742	0.000	0.200
10/07/2017	7	2.100	0.742	0.000	0.200
17/07/2017	7	2.100	0.742	0.000	0.200
24/07/2017	7	2.100	0.742	0.000	0.200
31/07/2017	7	2.100	0.742	0.000	0.200
07/08/2017	8	2.100	0.742	0.000	0.200
14/08/2017	8	2.100	0.742	0.000	0.200
21/08/2017	8	2.100	0.742	0.000	0.200
28/08/2017	8	2.100	0.742	0.000	0.200
04/09/2017	9	2.100	0.742	0.000	0.200
11/09/2017	9	2.100	0.742	0.000	0.200
18/09/2017	9	2.100	0.742	0.000	0.200
25/09/2017	9	2.100	0.742	0.000	0.200
02/10/2017	10	2.100	0.742	0.000	0.200
09/10/2017	10	2.200	0.788	4.762	0.200
16/10/2017	10	2.200	0.788	0.000	0.200
23/10/2017	10	2.200	0.788	0.000	0.200
30/10/2017	10	2.300	0.833	4.545	0.300
06/11/2017	11	2.400	0.875	4.348	0.300
13/11/2017	11	2.600	0.956	8.333	0.400
20/11/2017	11	2.600	0.956	0.000	0.400
27/11/2017	11	2.600	0.956	0.000	0.400
04/12/2017	12	2.600	0.956	0.000	0.400
11/12/2017	12	2.600	0.956	0.000	0.400
18/12/2017	12	2.600	0.956	0.000	0.400
25/12/2017	12	2.600	0.956	0.000	0.400

Table 17. Country chicken (2.3 a 2.5 kg) carcass prices, euro/kg (2018, Beira Litoral (region) and Dão-Lafões (market))

Period	Month	Value	Logarithm	Growth Rates (%)	Normalised Value for All Periods and Markets
01/01/2018	1	2.600	0.956		0.400
08/01/2018	1	2.600	0.956	0.000	0.400
15/01/2018	1	2.600	0.956	0.000	0.400
22/01/2018	1	2.600	0.956	0.000	0.400
29/01/2018	1	2.600	0.956	0.000	0.400
05/02/2018	2	2.600	0.956	0.000	0.400
12/02/2018	2	2.600	0.956	0.000	0.400
19/02/2018	2	2.600	0.956	0.000	0.400
26/02/2018	2	2.600	0.956	0.000	0.400
05/03/2018	3	2.600	0.956	0.000	0.400
12/03/2018	3	2.600	0.956	0.000	0.400
19/03/2018	3	2.600	0.956	0.000	0.400
26/03/2018	3	2.600	0.956	0.000	0.400
02/04/2018	4	2.600	0.956	0.000	0.400
09/04/2018	4	2.600	0.956	0.000	0.400
16/04/2018	4	2.600	0.956	0.000	0.400
23/04/2018	4	2.500	0.916	-3.846	0.400
30/04/2018	4	2.400	0.875	-4.000	0.300
07/05/2018	5	2.400	0.875	0.000	0.300
14/05/2018	5	2.400	0.875	0.000	0.300
21/05/2018	5	2.400	0.875	0.000	0.300
28/05/2018	5	2.200	0.788	-8.333	0.200
04/06/2018	6	2.200	0.788	0.000	0.200
11/06/2018	6	2.200	0.788	0.000	0.200
18/06/2018	6	2.200	0.788	0.000	0.200
25/06/2018	6	2.100	0.742	-4.545	0.200
02/07/2018	7	2.100	0.742	0.000	0.200
09/07/2018	7	2.100	0.742	0.000	0.200
16/07/2018	7	2.100	0.742	0.000	0.200
23/07/2018	7	2.100	0.742	0.000	0.200
30/07/2018	7	2.200	0.788	4.762	0.200
06/08/2018	8	2.200	0.788	0.000	0.200
13/08/2018	8	2.300	0.833	4.545	0.300
20/08/2018	8	2.300	0.833	0.000	0.300
27/08/2018	8	2.400	0.875	4.348	0.300
03/09/2018	9	2.400	0.875	0.000	0.300

continued on following page

Table 17. Continued

Period	Month	Value	Logarithm	Growth Rates (%)	Normalised Value for All Periods and Markets
10/09/2018	9	2.500	0.916	4.167	0.400
17/09/2018	9	2.500	0.916	0.000	0.400
24/09/2018	9	2.600	0.956	4.000	0.400
01/10/2018	10	2.600	0.956	0.000	0.400
08/10/2018	10	2.600	0.956	0.000	0.400
15/10/2018	10	2.700	0.993	3.846	0.500
22/10/2018	10	2.800	1.030	3.704	0.500
29/10/2018	10	2.800	1.030	0.000	0.500
05/11/2018	11	2.800	1.030	0.000	0.500
12/11/2018	11	2.800	1.030	0.000	0.500
19/11/2018	11	2.800	1.030	0.000	0.500
26/11/2018	11	2.800	1.030	0.000	0.500
03/12/2018	12	2.800	1.030	0.000	0.500
10/12/2018	12	2.800	1.030	0.000	0.500
17/12/2018	12	2.800	1.030	0.000	0.500
24/12/2018	12	2.800	1.030	0.000	0.500
31/12/2018	12	2.800	1.030	0.000	0.500

Table 18. Country chicken (2.3 a 2.5 kg) carcass prices, euro/kg (2019, Beira Litoral (region) and Dão-Lafões (market))

Period	Month	Value	Logarithm	Growth Rates (%)	Normalised Value for All Periods and Markets
07/01/2019	1	2.700	0.993		0.500
14/01/2019	1	2.700	0.993	0.000	0.500
21/01/2019	1	2.600	0.956	-3.704	0.400
28/01/2019	1	2.500	0.916	-3.846	0.400
04/02/2019	2	2.500	0.916	0.000	0.400
11/02/2019	2	2.300	0.833	-8.000	0.300
18/02/2019	2	2.300	0.833	0.000	0.300
25/02/2019	2	2.300	0.833	0.000	0.300
04/03/2019	3	2.300	0.833	0.000	0.300
11/03/2019	3	2.300	0.833	0.000	0.300
18/03/2019	3	2.300	0.833	0.000	0.300
25/03/2019	3	2.300	0.833	0.000	0.300
01/04/2019	4	2.300	0.833	0.000	0.300
08/04/2019	4	2.200	0.788	-4.348	0.200

continued on following page

Table 18. Continued

Period	Month	Value	Logarithm	Growth Rates (%)	Normalised Value for All Periods and Markets
15/04/2019	4	2.200	0.788	0.000	0.200
22/04/2019	4	2.200	0.788	0.000	0.200
29/04/2019	4	2.200	0.788	0.000	0.200
06/05/2019	5	2.200	0.788	0.000	0.200
13/05/2019	5	2.200	0.788	0.000	0.200
20/05/2019	5	2.200	0.788	0.000	0.200
27/05/2019	5	2.200	0.788	0.000	0.200
03/06/2019	6	2.200	0.788	0.000	0.200
10/06/2019	6	2.200	0.788	0.000	0.200
17/06/2019	6	2.200	0.788	0.000	0.200
24/06/2019	6	2.200	0.788	0.000	0.200
01/07/2019	7	2.200	0.788	0.000	0.200
08/07/2019	7	2.200	0.788	0.000	0.200
15/07/2019	7	2.100	0.742	-4.545	0.200
22/07/2019	7	2.100	0.742	0.000	0.200
29/07/2019	7	2.000	0.693	-4.762	0.100
05/08/2019	8	2.000	0.693	0.000	0.100
12/08/2019	8	2.000	0.693	0.000	0.100
19/08/2019	8	2.000	0.693	0.000	0.100
26/08/2019	8	2.000	0.693	0.000	0.100
02/09/2019	9	2.000	0.693	0.000	0.100
09/09/2019	9	2.100	0.742	5.000	0.200
16/09/2019	9	2.100	0.742	0.000	0.200
23/09/2019	9	2.100	0.742	0.000	0.200
30/09/2019	9	2.100	0.742	0.000	0.200
07/10/2019	10	2.200	0.788	4.762	0.200
14/10/2019	10	2.200	0.788	0.000	0.200
21/10/2019	10	2.100	0.742	-4.545	0.200
28/10/2019	10	2.100	0.742	0.000	0.200
04/11/2019	11	2.100	0.742	0.000	0.200
11/11/2019	11	2.100	0.742	0.000	0.200
18/11/2019	11	2.100	0.742	0.000	0.200
25/11/2019	11	2.100	0.742	0.000	0.200
02/12/2019	12	2.100	0.742	0.000	0.200
09/12/2019	12	2.200	0.788	4.762	0.200
16/12/2019	12	2.300	0.833	4.545	0.300
23/12/2019	12	2.300	0.833	0.000	0.300

The changes in the Portuguese producer prices for country chicken "Country Chicken*2.3 a 2.5 kg*EUR/kg P. Carcass" in the Beira Litoral region were modest in the years 2019-2021 (Tables 13-14) and relevant in 2022 (Table 15). In fact, the prices were 2.3 euros at the end of 2019. The average prices were 2.4 euros in 2020 (average growth rate of 0.1%), 2.2 euros in 2021 (average growth rate of 0.3%) and 3.2 in 2022 (average growth rate of 0.7%). A similar pattern was found for the "Country Chicken*2.3 a 2.5 kg*EUR/kg P. Carcass" in the Dão-Lafões market from the Beira Litoral region (Tables 16-18), over the years 2017-2019.

CONCLUSION

Considering the externalities of the Covid-19 pandemic and the Russia-Ukraine conflict on several socioeconomic dynamics, this study aimed to analyse the impacts of these international shocks on livestock production prices over the period 2017-2022, with weekly data. To achieve these objectives, statistical information from the Portuguese SIMA (Sistema de Informação de Mercados Agrícolas) was considered for cow and country chicken carcasses prices.

The livestock markets have several vulnerabilities, some of them associated with animal diseases that may spread worldwide through international trade. The treatments for the diseases are not free of consequences for the stakeholders in the respective supply chains, particularly for the health of the consumers. In the frameworks, the availability of information is crucial, namely for scientific research, as well as the availability of extension services and the conditions of selling and managing the animals over the chain. Animal welfare over the chain is another concern, specifically during transport. The animal health and conditions during the transport impact the animal welfare, the buying decisions and the respective selling prices.

In the Portuguese scenario, the average prices for Cow*For slaughter*Cruz. Charolês*EUR/kg P. Carcass" (Beira Litoral) increased significantly only after the Russia-Ukraine conflict. The impacts of the pandemic seem to be irrelevant. The same happened for "Cow*For slaughter*Turina*EUR/kg P. Carcass" (Beira Litoral) and "Country Chicken*2.3 a 2.5 kg*EUR/kg P. Carcass" (Beira Litoral). These are relevant insights, namely for policymakers and public institutions.

This research highlighted that the pandemic had an irrelevant impact on the prices of the livestock sectors here analysed. On the other hand, these implications were visible after the Russia-Ukraine crisis. In future research, it is suggested to spread this approach to other Portuguese livestock markets. In terms of policy recommendations, it would be important to identify the real causes for the price increase in the year 2022 and create policy measures to mitigate these trends.

ACKNOWLEDGMENT

This work is funded by National Funds through the FCT - Foundation for Science and Technology, I.P., within the scope of the project Refª UIDB/00681/2020. Furthermore we would like to thank the CERNAS Research Centre and the Polytechnic Institute of Viseu for their support.

REFERENCES

Adebowale, O., Makanjuola, M., Bankole, N., Olasoju, M., Alamu, A., Kperegbeyi, E., Oladejo, O., Fasanmi, O., Adeyemo, O., & Fasina, F. O. (2022). Multi-Drug Resistant Escherichia coli, Biosecurity and Anti-Microbial Use in Live Bird Markets, Abeokuta, Nigeria. *Antibiotics, 11*(2). doi:10.3390/antibiotics11020253

Afolabi, S. O., & Tiamiyu, M. A. (2021). Information Use Behaviour of Sedentary Livestock Farmers in Egbeda Local Government Area of Oyo State, Nigeria. *Library Philosophy and Practice, 2021*, 1–15.

Barigye, R., Hassan, N. A. D., Abdalla-Alfaki, I. M., Al Alawi, H. K. K., Aldhaheri, A. M. B., Ghanim, F. M., Alkhateri, M. A. S., Alalawi, N. S. A., Mohamed, M. E. H., & Mohteshamuddin, K. (2022). Pilot serosurvey of Coxiella burnetii in domesticated small ruminants in the United Arab Emirates. *Tropical Animal Health and Production, 54*(3). doi:10.1007/s11250-022-03150-6

Bravo, V. M., Knowles, T. G., & Gallo, C. (2020). Transport, associated handling procedures and behaviour of calves marketed through chilean auction markets. *Animals, 10*(11), 1–13. doi:10.3390/ani10112170

ÇevrimliM. B.MatB.TekindalM. A.GünlüA. (2022). Determination of risk factors in cattle and small ruminant livestock enterprises from the perspective of animal life insurance: A case study in Turkey. *Arquivo Brasileiro de Medicina Veterinaria e Zootecnia, 74*(1), 21–32. doi:10.1590/1678-4162-12524

Creutzinger, K., Pempek, J., Habing, G., Proudfoot, K., Locke, S., Wilson, D., & Renaud, D. (2021). Perspectives on the Management of Surplus Dairy Calves in the United States and Canada. *Frontiers in Veterinary Science, 8*. doi:10.3389/fvets.2021.661453

ElAshmawy, W. R., Aly, S. S., & Farouk, M. M. (2023). Decision tree risk analysis for FMD outbreak prevention in Egyptian feedlots. *Preventive Veterinary Medicine, 211*. doi:10.1016/j.prevetmed.2022.105820

Ezanno, P., Arnoux, S., Joly, A., & Vermesse, R. (2022). Rewiring cattle trade movements helps to control bovine paratuberculosis at a regional scale. *Preventive Veterinary Medicine, 198*. doi:10.1016/j.prevetmed.2021.105529

Fielding, H. R., McKinley, T. J., Delahay, R. J., Silk, M. J., & McDonald, R. A. (2021). Characterization of potential superspreader farms for bovine tuberculosis: A review. *Veterinary Medicine and Science, 7*(2), 310–321. doi:10.1002/vms3.358

Fujimoto, Y., Ito, H., Higuchi, H., Ohno, H., & Makita, K. (2020). A case-control study of herd- and cow-level risk factors associated with an outbreak of Mycoplasma mastitis in Nemuro, Japan. *Preventive Veterinary Medicine, 177*. doi:10.1016/j.prevetmed.2020.104946

Funsho-Sanni, O. O., Ella, E., Samuel Sanni, O., Inabo, H., Abdulkarim luka, S., & Eleyi Ameh, R. (2022). Serological evaluation of Newcastle disease protection among broilers at a live bird market in Kano, Northwest Nigeria, and its epidemiological significance. *Journal of Immunoassay and Immunochemistry, 43*(5), 534–545. doi:10.1080/15321819.2022.2029744

George, J., Häsler, B., Komba, E., Sindato, C., Rweyemamu, M., & Mlangwa, J. (2021). Towards an integrated animal health surveillance system in Tanzania: Making better use of existing and potential data sources for early warning surveillance. *BMC Veterinary Research, 17*(1). doi:10.1186/s12917-021-02789-x

Hassan, N. A. D., Mohteshamuddin, K., Anthony, A., Al Aiyan, A., Mohamed, M. E. H., Abdalla Alfaki, I. M., & Barigye, R. (2020). Serological evidence of enzootic bovine leukosis in the periurban dairy cattle production system of Al Ain, United Arab Emirates. *Tropical Animal Health and Production, 52*(5), 2327–2332. doi:10.1007/s11250-020-02262-1

Jean-Pierre, R. P., Hagerman, A. D., & Rich, K. M. (2022). An analysis of African Swine Fever consequences on rural economies and smallholder swine producers in Haiti. *Frontiers in Veterinary Science, 9*. doi:10.3389/fvets.2022.960344

Knight, M. A., Hutchings, M. R., White, P. C. L., Davidson, R. S., & Marion, G. (2022). A mechanistic model captures livestock trading, disease dynamics, and compensatory behaviour in response to control measures. *Journal of Theoretical Biology, 539*. doi:10.1016/j.jtbi.2022.111059

Limon, G., Gamawa, A. A., Ahmed, A. I., Lyons, N. A., & Beard, P. M. (2020). Epidemiological Characteristics and Economic Impact of Lumpy Skin Disease, Sheeppox and Goatpox Among Subsistence Farmers in Northeast Nigeria. *Frontiers in Veterinary Science, 7*. doi:10.3389/fvets.2020.00008

Mohamed Sala, S., Otieno, D. J., Nzuma, J., & Mureithi, S. M. (2020). Determinants of pastoralists' participation in commercial fodder markets for livelihood resilience in drylands of northern Kenya: Case of Isiolo. *Pastoralism, 10*(1). doi:10.1186/s13570-020-00166-1

Murungi, M. K., Muloi, D. M., Muinde, P., Githigia, S. M., Akoko, J., Fèvre, E. M., Rushton, J., & Alarcon, P. (2021). The Nairobi Pork Value Chain: Mapping and Assessment of Governance, Challenges, and Food Safety Issues. *Frontiers in Veterinary Science, 8*. doi:10.3389/fvets.2021.581376

Nichols, M., Gollarza, L., Sockett, D., Aulik, N., Patton, E., Francois Watkins, L. K., Gambino-Shirley, K. J., Folster, J. P., Chen, J. C., Tagg, K. A., Stapleton, G. S., Trees, E., Ellison, Z., Lombard, J., Morningstar-Shaw, B., Schlater, L., Elbadawi, L., & Klos, R. (2022). Outbreak of Multidrug-Resistant Salmonella Heidelberg Infections Linked to Dairy Calf Exposure, United States, 2015-2018. *Foodborne Pathogens and Disease, 19*(3), 199–208. doi:10.1089/fpd.2021.0077

Okello, W. O., MacLeod, E. T., Muhanguzi, D., Waiswa, C., Shaw, A. P., & Welburn, S. C. (2021). Critical Linkages Between Livestock Production, Livestock Trade and Potential Spread of Human African Trypanosomiasis in Uganda: Bioeconomic Herd Modeling and Livestock Trade Analysis. *Frontiers in Veterinary Science, 8*. doi:10.3389/fvets.2021.611141

Onduso, R., Onono, J. O., & Ombui, J. N. (2020). Assessment of structure and performance of cattle markets in western Kenya. *Tropical Animal Health and Production, 52*(2), 725–732. doi:10.1007/s11250-019-02062-2

Rosen, K., Llonch, P., & Langford, F. M. (2020). The welfare of culled dairy cattle sent to market and its impact on economic return. *Animal Welfare, 29*(4), 359–370. doi:10.7120/09627286.29.4.359

Sánchez-Hidalgo, M., Bravo, V., & Gallo, C. (2020). Behavior and Health Indicators to Assess Cull Cow's Welfare in Livestock Markets. *Frontiers in Veterinary Science, 7*. doi:10.3389/fvets.2020.00471

Scopus. (2023). *Scopus Database*. https://www.scopus.com/search/form.uri?display=basic#basic

Shyaka, A., Quinnell, R. J., Rujeni, N., & Fèvre, E. M. (2022). Using a Value Chain Approach to Map the Pig Production System in Rwanda, Its Governance, and Sanitary Risks. *Frontiers in Veterinary Science, 8*. doi:10.3389/fvets.2021.720553

SIMA. (2023). *Several Statistics and Information*. https://www.gpp.pt/index.php/sima/sima-2018

Soliman, I. (2021). A model for prediction of the buffalo and cattle male calves' live weight. *Journal of Buffalo Science, 10*, 67–74. doi:10.6000/1927-520X.2021.10.10

Stata. (2023). *Statistical software for data science*. https://www.stata.com/

StataCorp. (2017a). *Stata 15 Base Reference Manual*. Stata Press.

StataCorp. (2017b). *Stata Statistical Software: Release 15*. StataCorp LLC.

Stojkov, J., von Keyserlingk, M. A. G., Duffield, T., & Fraser, D. (2020a). Fitness for transport of cull dairy cows at livestock markets. *Journal of Dairy Science, 103*(3), 2650–2661. doi:10.3168/jds.2019-17454

Stojkov, J., von Keyserlingk, M. A. G., Duffield, T., & Fraser, D. (2020b). Management of cull dairy cows: Culling decisions, duration of transport, and effect on cow condition. *Journal of Dairy Science, 103*(3), 2636–2649. doi:10.3168/jds.2019-17435

Windsor, P. A. (2021). Progress With Livestock Welfare in Extensive Production Systems: Lessons From Australia. *Frontiers in Veterinary Science, 8*. doi:10.3389/fvets.2021.674482

Chapter 14
Implications of Recent Events on Portuguese Fruit Production Markets

ABSTRACT

The fresh fruit markets have several particularities because of the perishability of these agricultural products, which brings additional pressure on the farmers, namely the smaller ones, to sell quickly after harvesting. These contexts have implications on the dynamics of the markets and in the definition of the prices. The question here is if the COVID-19 pandemic and the recent conflicts had consequences on these dynamics. The evolution of the fresh fruit producer prices in the Portuguese regions and markets was assessed. A special focus was given to the apple markets, considering their importance for the diet of the Portuguese population. For that, data from the SIMA (Sistema de Informação de Mercados Agrícolas) were considered for the weeks from 2017 to 2022. The implications from the most recent occurrences—if, in fact, had relevance—seem to have been slight.

INTRODUCTION

The agricultural markets are increasingly open to international trade. Consequently, they are influenced by the behaviour of the respective demand and supply (Jovicich et al., 2005). The fruit market dimensions, namely the prices, give an idea about the dynamics, returns and alternatives for the different parts involved (Adelaja et al., 1996).

The fruit prices fluctuate seasonally over the year, due to changes in the volume supplied during and outside the harvest period (Capocasa et al., 2021). These prices present lower volatility compared to vegetable ones (Uçak et al., 2022), for example, in some contexts. In general, the prices are higher during the off-season (Medina et al., 2018), such as happens with the 'Tahiti' limes (Spósito & Filho, 2003) and strawberries (Verdial et al., 2007).

These prices are often interrelated with the quality (Cittadini et al., 2007), size (Inglese et al., 2009) of the fruits, cost-benefit relation (Manolova & Kolev, 2012) and the decisions to collect in specific

DOI: 10.4018/978-1-6684-8923-9.ch014

circumstances (Ghanbari et al., 2020). Nonetheless, the quality is perceived differently by the diverse actors in the fruit chains (Djekic et al., 2019). Some fruits have traditionally higher prices (Kodrat et al., 2018). The level of fruit prices also influences the diet of the population (Lei & Shimokawa, 2020).

The agricultural practices are important for the farms' profitability (Alan Walters, 2020), specifically those related to irrigation management and harvest dates (Ballester et al., 2018). The harvest activities represent a relevant part of the total production costs of some fruits, such as citrus (Sanders, 2005).

The disease incidence in the fruits and respective plants affects the revenues of the farmers. The control of these diseases and pest is not usually an easy task (Márquez-Berber et al., 2012). This is true for strawberries (Baggio et al., 2021) and watermelon (Pereira et al., 2017), for example. The research and development programmes may bring relevant contributions to the fruit markets (Tsvakirai et al., 2018).

This research intends to bring more highlights about the Portuguese fruit prices in the producer markets with data from 2017 until 2022 from the national agricultural markets information system (SIMA, 2023). The Scopus database was surveyed (2023). Some data were assessed through the Stata (StataCorp, 2017a, 2017b; Stata, 2023) software. Normalised values equal to $(x_i\text{-}x_{minimum})/(x_{maximum}\text{-}x_{minimum})$.

Considering the great volume of information available for the fruit markets, particularly about prices, this chapter focused mainly on Portuguese apple prices.

ROYAL GALA APPLE PRICES

The Portuguese producer prices (euros/kg) for apples "Royal Gala*SE*I*65-70 mm" (Tables 1-2), "Royal Gala*SE*I*70-75 mm" (Tables 3-4), "Royal Gala*SE*I*75-80 mm" (Tables 5-6), "Royal Gala*SE*II*65-70 mm" (Tables 7-8), "Royal Gala*SE*II*70-75 mm" (Tables 9-11) and "Royal Gala*SE*II*75-80 mm" (Tables 12-13) did not change significantly, including in 2022. The average prices for apple "Royal Gala*SE*I*65-70 mm" were 0.8 euros in 2017 (logarithm of -0.3 and normalised value of 0.5), 0.8 euros in 2018 (logarithm of -0.2 and normalised value of 0.5), 0.7 euros in 2019 (logarithm of -0.3 and normalised value of 0.4), 0.9 euros in 2020 (logarithm of -0.1 and normalised value of 0.7), 0.9 in 2021 (logarithm of -0.1 and normalised value of 0.7) and 0.9 euros in 2022 (logarithm of -0.1 and normalised value of 0.7). For apple "Royal Gala*SE*I*70-75 mm" the average prices in 2017-2022 were, respectively the following: 0.8 euros; 1.0 euros; 0.8 euros; 1.1 euros; 1.2 euros; and 1.2 euros. For apple "Royal Gala*SE*I*75-80 mm", the average prices (euros/kg) were (2017-2022): 0.9; 1.0; 1.0; 1.2; 1.1; and 1.1. The average prices for apple "Royal Gala*SE*II*65-70 mm", for the years considered were the following: 0.6; 0.8; 0.8; 0.9; 0.8; and 0.6. A similar pattern was found for the apple "Royal Gala*SE*II*70-75 mm" and "Royal Gala*SE*II*75-80 mm".

*Table 1. Apple prices (Royal Gala*SE*I*65-70 mm, Ribatejo e Oeste (region) and Oeste (market)), for 2017-2019*

Period	Month	Year	Value	Logarithm	Normalised Value for All Periods
27/03/2017	3	2017	0.900	-0.105	0.600
03/04/2017	4	2017	0.900	-0.105	0.600
10/04/2017	4	2017	0.900	-0.105	0.600
17/04/2017	4	2017	0.900	-0.105	0.600
24/04/2017	4	2017	0.900	-0.105	0.600
01/05/2017	5	2017	0.700	-0.357	0.400
08/05/2017	5	2017	0.700	-0.357	0.400
08/05/2017	5	2017	0.700	-0.357	0.400
15/05/2017	5	2017	0.700	-0.357	0.400
22/05/2017	5	2017	0.700	-0.357	0.400
12/06/2017	6	2017	0.600	-0.511	0.300
27/11/2017	11	2017	0.700	-0.357	0.400
02/04/2018	4	2018	0.900	-0.105	0.600
16/04/2018	4	2018	0.400	-0.916	0.000
18/06/2018	6	2018	0.800	-0.223	0.500
25/06/2018	6	2018	0.800	-0.223	0.500
20/08/2018	8	2018	1.000	0.000	0.800
17/09/2018	9	2018	1.000	0.000	0.800
24/09/2018	9	2018	1.000	0.000	0.800
01/10/2018	10	2018	1.000	0.000	0.800
08/10/2018	10	2018	0.900	-0.105	0.600
15/10/2018	10	2018	0.900	-0.105	0.600
22/10/2018	10	2018	0.900	-0.105	0.600
29/10/2018	10	2018	0.900	-0.105	0.600
05/11/2018	11	2018	0.600	-0.511	0.300
12/11/2018	11	2018	0.700	-0.357	0.400
19/11/2018	11	2018	0.900	-0.105	0.600
26/11/2018	11	2018	0.900	-0.105	0.600
03/12/2018	12	2018	0.900	-0.105	0.600
10/12/2018	12	2018	0.800	-0.223	0.500
17/12/2018	12	2018	0.800	-0.223	0.500
24/12/2018	12	2018	0.700	-0.357	0.400
31/12/2018	12	2018	0.700	-0.357	0.400
07/01/2019	1	2019	0.700	-0.357	0.400
14/01/2019	1	2019	0.700	-0.357	0.400
21/01/2019	1	2019	0.700	-0.357	0.400
28/01/2019	1	2019	0.700	-0.357	0.400

Continued on following page

Table 1. Continued

Period	Month	Year	Value	Logarithm	Normalised Value for All Periods
04/02/2019	2	2019	0.700	-0.357	0.400
11/02/2019	2	2019	0.700	-0.357	0.400
18/02/2019	2	2019	0.700	-0.357	0.400
25/02/2019	2	2019	0.700	-0.357	0.400
04/03/2019	3	2019	0.700	-0.357	0.400
11/03/2019	3	2019	0.700	-0.357	0.400
18/03/2019	3	2019	0.700	-0.357	0.400
25/03/2019	3	2019	0.700	-0.357	0.400
01/04/2019	4	2019	0.700	-0.357	0.400
08/04/2019	4	2019	0.700	-0.357	0.400
29/04/2019	4	2019	0.700	-0.357	0.400
21/10/2019	10	2019	0.800	-0.223	0.500
28/10/2019	10	2019	0.800	-0.223	0.500
11/11/2019	11	2019	0.800	-0.223	0.500
18/11/2019	11	2019	0.900	-0.105	0.600
25/11/2019	11	2019	0.700	-0.357	0.400
02/12/2019	12	2019	0.700	-0.357	0.400

*Table 2. Apple prices (Royal Gala*SE*I*65-70 mm, Ribatejo e Oeste (region) and Oeste (market)), for 2020-2022*

Period	Month	Year	Value	Logarithm	Normalised Value for All Periods
13/01/2020	1	2020	0.800	-0.223	0.500
20/01/2020	1	2020	0.700	-0.357	0.400
27/01/2020	1	2020	0.700	-0.357	0.400
17/02/2020	2	2020	0.700	-0.357	0.400
24/02/2020	2	2020	0.900	-0.105	0.600
02/03/2020	3	2020	0.800	-0.223	0.500
09/03/2020	3	2020	0.800	-0.223	0.500
23/03/2020	3	2020	0.900	-0.105	0.600
30/03/2020	3	2020	1.100	0.095	0.900
13/04/2020	4	2020	0.800	-0.223	0.500
27/04/2020	4	2020	0.900	-0.105	0.600
11/05/2020	5	2020	0.700	-0.357	0.400
14/09/2020	9	2020	1.100	0.095	0.900
21/09/2020	9	2020	1.100	0.095	0.900
28/09/2020	9	2020	1.100	0.095	0.900
05/10/2020	10	2020	1.100	0.095	0.900
12/10/2020	10	2020	1.100	0.095	0.900
19/10/2020	10	2020	1.100	0.095	0.900
26/10/2020	10	2020	1.000	0.000	0.800
02/11/2020	11	2020	1.000	0.000	0.800
09/11/2020	11	2020	1.000	0.000	0.800
16/11/2020	11	2020	1.000	0.000	0.800
23/11/2020	11	2020	1.000	0.000	0.800
30/11/2020	11	2020	1.000	0.000	0.800
07/12/2020	12	2020	1.000	0.000	0.800
14/12/2020	12	2020	1.000	0.000	0.800
18/01/2021	1	2021	0.600	-0.511	0.300
08/03/2021	3	2021	0.600	-0.511	0.300
29/03/2021	3	2021	1.100	0.095	0.900
29/03/2021	3	2021	1.100	0.095	0.900
05/04/2021	4	2021	1.000	0.000	0.800
12/04/2021	4	2021	1.000	0.000	0.800
06/09/2021	9	2021	1.000	0.000	0.800
13/09/2021	9	2021	1.000	0.000	0.800
20/09/2021	9	2021	1.000	0.000	0.800
04/10/2021	10	2021	1.000	0.000	0.800
18/10/2021	10	2021	1.000	0.000	0.800

Continued on following page

Table 2. Continued

Period	Month	Year	Value	Logarithm	Normalised Value for All Periods
25/10/2021	10	2021	1.000	0.000	0.800
01/11/2021	11	2021	1.000	0.000	0.800
08/11/2021	11	2021	0.800	-0.223	0.500
24/01/2022	1	2022	0.800	-0.223	0.500
31/01/2022	1	2022	0.800	-0.223	0.500
07/02/2022	2	2022	1.000	0.000	0.800
14/02/2022	2	2022	0.800	-0.223	0.500
21/02/2022	2	2022	1.000	0.000	0.800
28/02/2022	2	2022	1.000	0.000	0.800
15/08/2022	8	2022	1.200	0.182	1.000
22/08/2022	8	2022	1.100	0.095	0.900
29/08/2022	8	2022	1.100	0.095	0.900
05/09/2022	9	2022	1.000	0.000	0.800
12/09/2022	9	2022	0.900	-0.105	0.600
19/09/2022	9	2022	0.900	-0.105	0.600
26/09/2022	9	2022	0.900	-0.105	0.600
03/10/2022	10	2022	0.900	-0.105	0.600
10/10/2022	10	2022	0.900	-0.105	0.600
17/10/2022	10	2022	0.900	-0.105	0.600
24/10/2022	10	2022	0.900	-0.105	0.600
31/10/2022	10	2022	0.900	-0.105	0.600
07/11/2022	11	2022	0.900	-0.105	0.600
14/11/2022	11	2022	0.900	-0.105	0.600
21/11/2022	11	2022	0.900	-0.105	0.600
05/12/2022	12	2022	0.900	-0.105	0.600
12/12/2022	12	2022	0.900	-0.105	0.600
19/12/2022	12	2022	0.900	-0.105	0.600
26/12/2022	12	2022	0.900	-0.105	0.600

*Table 3. Apple prices (Royal Gala*SE*I*70-75 mm, Ribatejo e Oeste (region) and Oeste (market)), for 2017-2019*

Period	Month	Year	Value	Logarithm	Normalised Value for All Periods
27/03/2017	3	2017	0.800	-0.223	0.200
03/04/2017	4	2017	0.800	-0.223	0.200
10/04/2017	4	2017	0.800	-0.223	0.200
17/04/2017	4	2017	0.800	-0.223	0.200
24/04/2017	4	2017	0.800	-0.223	0.200
01/05/2017	5	2017	0.900	-0.105	0.300
08/05/2017	5	2017	0.900	-0.105	0.300
08/05/2017	5	2017	0.900	-0.105	0.300
15/05/2017	5	2017	0.900	-0.105	0.300
22/05/2017	5	2017	0.900	-0.105	0.300
27/11/2017	11	2017	0.700	-0.357	0.100
02/04/2018	4	2018	1.000	0.000	0.400
16/04/2018	4	2018	0.700	-0.357	0.100
18/06/2018	6	2018	0.600	-0.511	0.000
25/06/2018	6	2018	1.000	0.000	0.400
20/08/2018	8	2018	1.300	0.262	0.800
27/08/2018	8	2018	1.100	0.095	0.600
17/09/2018	9	2018	1.200	0.182	0.700
24/09/2018	9	2018	1.200	0.182	0.700
01/10/2018	10	2018	1.000	0.000	0.400
08/10/2018	10	2018	1.100	0.095	0.600
15/10/2018	10	2018	1.000	0.000	0.400
22/10/2018	10	2018	1.000	0.000	0.400
29/10/2018	10	2018	1.000	0.000	0.400
05/11/2018	11	2018	1.000	0.000	0.400
12/11/2018	11	2018	1.000	0.000	0.400
19/11/2018	11	2018	1.000	0.000	0.400
26/11/2018	11	2018	1.000	0.000	0.400
03/12/2018	12	2018	1.000	0.000	0.400
10/12/2018	12	2018	1.000	0.000	0.400
17/12/2018	12	2018	1.000	0.000	0.400
24/12/2018	12	2018	0.700	-0.357	0.100
31/12/2018	12	2018	0.900	-0.105	0.300
07/01/2019	1	2019	0.900	-0.105	0.300
14/01/2019	1	2019	0.700	-0.357	0.100
21/01/2019	1	2019	0.900	-0.105	0.300
28/01/2019	1	2019	0.900	-0.105	0.300

Continued on following page

Table 3. Continued

Period	Month	Year	Value	Logarithm	Normalised Value for All Periods
04/03/2019	3	2019	0.800	-0.223	0.200
01/04/2019	4	2019	0.800	-0.223	0.200
08/04/2019	4	2019	0.800	-0.223	0.200
21/10/2019	10	2019	1.100	0.095	0.600
28/10/2019	10	2019	1.100	0.095	0.600
11/11/2019	11	2019	0.800	-0.223	0.200
16/12/2019	12	2019	0.700	-0.357	0.100
23/12/2019	12	2019	0.700	-0.357	0.100
30/12/2019	12	2019	0.700	-0.357	0.100

*Table 4. Apple prices (Royal Gala*SE*I*70-75 mm, Ribatejo e Oeste (region) and Oeste (market)), for 2020-2022*

Period	Month	Year	Value	Logarithm	Normalised Value for All Periods
06/01/2020	1	2020	0.700	-0.357	0.100
13/01/2020	1	2020	1.100	0.095	0.600
20/01/2020	1	2020	0.700	-0.357	0.100
24/02/2020	2	2020	1.100	0.095	0.600
02/03/2020	3	2020	0.900	-0.105	0.300
09/03/2020	3	2020	1.100	0.095	0.600
23/03/2020	3	2020	1.100	0.095	0.600
30/03/2020	3	2020	1.100	0.095	0.600
13/04/2020	4	2020	0.900	-0.105	0.300
27/04/2020	4	2020	0.800	-0.223	0.200
11/05/2020	5	2020	1.100	0.095	0.600
21/09/2020	9	2020	1.200	0.182	0.700
28/09/2020	9	2020	1.200	0.182	0.700
05/10/2020	10	2020	1.200	0.182	0.700
12/10/2020	10	2020	1.200	0.182	0.700
19/10/2020	10	2020	1.200	0.182	0.700
26/10/2020	10	2020	1.200	0.182	0.700
02/11/2020	11	2020	1.200	0.182	0.700
09/11/2020	11	2020	1.200	0.182	0.700
16/11/2020	11	2020	1.200	0.182	0.700
23/11/2020	11	2020	1.200	0.182	0.700
30/11/2020	11	2020	1.200	0.182	0.700
07/12/2020	12	2020	1.200	0.182	0.700
14/12/2020	12	2020	1.200	0.182	0.700
01/03/2021	3	2021	0.600	-0.511	0.000
08/03/2021	3	2021	1.200	0.182	0.700
29/03/2021	3	2021	1.300	0.262	0.800
29/03/2021	3	2021	1.200	0.182	0.700
05/04/2021	4	2021	1.200	0.182	0.700
12/04/2021	4	2021	1.200	0.182	0.700
06/09/2021	9	2021	1.200	0.182	0.700
13/09/2021	9	2021	1.200	0.182	0.700
20/09/2021	9	2021	1.200	0.182	0.700
04/10/2021	10	2021	1.200	0.182	0.700
18/10/2021	10	2021	1.200	0.182	0.700
25/10/2021	10	2021	1.200	0.182	0.700
01/11/2021	11	2021	1.200	0.182	0.700

Continued on following page

Table 4. Continued

Period	Month	Year	Value	Logarithm	Normalised Value for All Periods
08/11/2021	11	2021	1.000	0.000	0.400
24/01/2022	1	2022	0.800	-0.223	0.200
31/01/2022	1	2022	0.800	-0.223	0.200
07/02/2022	2	2022	1.200	0.182	0.700
14/02/2022	2	2022	0.800	-0.223	0.200
21/02/2022	2	2022	1.200	0.182	0.700
28/02/2022	2	2022	1.200	0.182	0.700
07/03/2022	3	2022	0.700	-0.357	0.100
15/08/2022	8	2022	1.500	0.405	1.000
22/08/2022	8	2022	1.400	0.336	0.900
29/08/2022	8	2022	1.400	0.336	0.900
05/09/2022	9	2022	1.400	0.336	0.900
12/09/2022	9	2022	1.300	0.262	0.800
19/09/2022	9	2022	1.300	0.262	0.800
26/09/2022	9	2022	1.300	0.262	0.800
03/10/2022	10	2022	1.300	0.262	0.800
10/10/2022	10	2022	1.300	0.262	0.800
17/10/2022	10	2022	1.300	0.262	0.800
24/10/2022	10	2022	1.300	0.262	0.800
31/10/2022	10	2022	1.200	0.182	0.700
07/11/2022	11	2022	1.200	0.182	0.700
14/11/2022	11	2022	1.200	0.182	0.700
21/11/2022	11	2022	1.200	0.182	0.700
05/12/2022	12	2022	1.200	0.182	0.700
12/12/2022	12	2022	1.200	0.182	0.700
19/12/2022	12	2022	1.200	0.182	0.700
26/12/2022	12	2022	1.200	0.182	0.700

*Table 5. Apple prices (Royal Gala*SE*I*75-80 mm, Ribatejo e Oeste (region) and Oeste (market)), for 2017-2019*

Period	Month	Year	Value	Logarithm	Normalised Value for All Periods
27/03/2017	3	2017	0.900	-0.105	0.300
03/04/2017	4	2017	0.900	-0.105	0.300
10/04/2017	4	2017	0.900	-0.105	0.300
17/04/2017	4	2017	0.900	-0.105	0.300
24/04/2017	4	2017	0.900	-0.105	0.300
01/05/2017	5	2017	0.900	-0.105	0.300
08/05/2017	5	2017	0.900	-0.105	0.300
08/05/2017	5	2017	0.900	-0.105	0.300
15/05/2017	5	2017	0.900	-0.105	0.300
22/05/2017	5	2017	0.900	-0.105	0.300
02/04/2018	4	2018	1.000	0.000	0.400
18/06/2018	6	2018	1.100	0.095	0.600
25/06/2018	6	2018	0.900	-0.105	0.300
20/08/2018	8	2018	1.400	0.336	1.000
27/08/2018	8	2018	1.100	0.095	0.600
17/09/2018	9	2018	1.200	0.182	0.700
24/09/2018	9	2018	1.100	0.095	0.600
01/10/2018	10	2018	1.100	0.095	0.600
08/10/2018	10	2018	1.100	0.095	0.600
15/10/2018	10	2018	1.100	0.095	0.600
22/10/2018	10	2018	1.000	0.000	0.400
29/10/2018	10	2018	1.000	0.000	0.400
05/11/2018	11	2018	1.100	0.095	0.600
12/11/2018	11	2018	1.000	0.000	0.400
19/11/2018	11	2018	1.000	0.000	0.400
26/11/2018	11	2018	1.000	0.000	0.400
03/12/2018	12	2018	1.000	0.000	0.400
10/12/2018	12	2018	1.000	0.000	0.400
17/12/2018	12	2018	1.000	0.000	0.400
24/12/2018	12	2018	0.800	-0.223	0.100
31/12/2018	12	2018	0.900	-0.105	0.300
07/01/2019	1	2019	0.900	-0.105	0.300
14/01/2019	1	2019	0.900	-0.105	0.300
21/01/2019	1	2019	0.900	-0.105	0.300
28/01/2019	1	2019	0.900	-0.105	0.300
04/03/2019	3	2019	0.900	-0.105	0.300
01/04/2019	4	2019	0.900	-0.105	0.300
21/10/2019	10	2019	1.100	0.095	0.600
28/10/2019	10	2019	1.100	0.095	0.600
11/11/2019	11	2019	1.100	0.095	0.600
25/11/2019	11	2019	0.800	-0.223	0.100

*Table 6. Apple prices (Royal Gala*SE*I*75-80 mm, Ribatejo e Oeste (region) and Oeste (market)), for 2020-2022*

Period	Month	Year	Value	Logarithm	Normalised Value for All Periods
13/01/2020	1	2020	1.100	0.095	0.600
24/02/2020	2	2020	1.100	0.095	0.600
02/03/2020	3	2020	1.000	0.000	0.400
09/03/2020	3	2020	1.100	0.095	0.600
23/03/2020	3	2020	1.100	0.095	0.600
30/03/2020	3	2020	1.000	0.000	0.400
14/09/2020	9	2020	1.200	0.182	0.700
21/09/2020	9	2020	1.200	0.182	0.700
28/09/2020	9	2020	1.200	0.182	0.700
05/10/2020	10	2020	1.200	0.182	0.700
12/10/2020	10	2020	1.200	0.182	0.700
19/10/2020	10	2020	1.200	0.182	0.700
26/10/2020	10	2020	1.200	0.182	0.700
02/11/2020	11	2020	1.200	0.182	0.700
09/11/2020	11	2020	1.200	0.182	0.700
16/11/2020	11	2020	1.200	0.182	0.700
23/11/2020	11	2020	1.200	0.182	0.700
30/11/2020	11	2020	1.200	0.182	0.700
07/12/2020	12	2020	1.200	0.182	0.700
14/12/2020	12	2020	1.200	0.182	0.700
01/02/2021	2	2021	1.200	0.182	0.700
08/02/2021	2	2021	0.800	-0.223	0.100
15/02/2021	2	2021	0.800	-0.223	0.100
22/02/2021	2	2021	0.800	-0.223	0.100
01/03/2021	3	2021	1.200	0.182	0.700
29/03/2021	3	2021	1.200	0.182	0.700
05/04/2021	4	2021	1.200	0.182	0.700
12/04/2021	4	2021	1.200	0.182	0.700
06/09/2021	9	2021	1.200	0.182	0.700
13/09/2021	9	2021	1.200	0.182	0.700
20/09/2021	9	2021	1.200	0.182	0.700
04/10/2021	10	2021	1.200	0.182	0.700
18/10/2021	10	2021	1.200	0.182	0.700
25/10/2021	10	2021	1.200	0.182	0.700
01/11/2021	11	2021	1.200	0.182	0.700
22/11/2021	11	2021	0.700	-0.357	0.000
31/01/2022	1	2022	0.900	-0.105	0.300
07/02/2022	2	2022	1.200	0.182	0.700
28/02/2022	2	2022	1.200	0.182	0.700

*Table 7. Apple prices (Royal Gala*SE*II*65-70 mm, Ribatejo e Oeste (region) and Oeste (market)), for 2017-2019*

Period	Month	Year	Value	Logarithm	Normalised Value for All Periods
02/01/2017	1	2017	0.600	-0.511	0.300
09/01/2017	1	2017	0.600	-0.511	0.300
16/01/2017	1	2017	0.600	-0.511	0.300
23/01/2017	1	2017	0.600	-0.511	0.300
30/01/2017	1	2017	0.600	-0.511	0.300
06/02/2017	2	2017	0.600	-0.511	0.300
13/02/2017	2	2017	0.600	-0.511	0.300
20/02/2017	2	2017	0.600	-0.511	0.300
27/02/2017	2	2017	0.600	-0.511	0.300
06/03/2017	3	2017	0.600	-0.511	0.300
13/03/2017	3	2017	0.600	-0.511	0.300
20/03/2017	3	2017	0.600	-0.511	0.300
09/04/2018	4	2018	0.800	-0.223	0.500
16/04/2018	4	2018	0.800	-0.223	0.500
30/04/2018	4	2018	0.800	-0.223	0.500
07/05/2018	5	2018	0.700	-0.357	0.400
14/05/2018	5	2018	0.800	-0.223	0.500
21/05/2018	5	2018	0.600	-0.511	0.300
04/06/2018	6	2018	0.600	-0.511	0.300
11/06/2018	6	2018	1.100	0.095	0.800
18/06/2018	6	2018	0.900	-0.105	0.600
02/07/2018	7	2018	1.000	0.000	0.700
06/08/2018	8	2018	1.300	0.262	1.000
24/09/2018	9	2018	0.800	-0.223	0.500
01/10/2018	10	2018	0.700	-0.357	0.400
08/10/2018	10	2018	0.600	-0.511	0.300
15/10/2018	10	2018	0.700	-0.357	0.400
22/10/2018	10	2018	0.900	-0.105	0.600
05/11/2018	11	2018	0.700	-0.357	0.400
12/11/2018	11	2018	0.700	-0.357	0.400
19/11/2018	11	2018	0.900	-0.105	0.600
26/11/2018	11	2018	0.900	-0.105	0.600
03/12/2018	12	2018	0.900	-0.105	0.600
10/12/2018	12	2018	0.700	-0.357	0.400
17/12/2018	12	2018	0.700	-0.357	0.400
21/01/2019	1	2019	0.900	-0.105	0.600
04/02/2019	2	2019	0.900	-0.105	0.600

Continued on following page

Table 7. Continued

Period	Month	Year	Value	Logarithm	Normalised Value for All Periods
11/02/2019	2	2019	0.700	-0.357	0.400
18/02/2019	2	2019	0.800	-0.223	0.500
25/02/2019	2	2019	0.800	-0.223	0.500
04/03/2019	3	2019	0.900	-0.105	0.600
11/03/2019	3	2019	0.800	-0.223	0.500
18/03/2019	3	2019	0.700	-0.357	0.400
25/03/2019	3	2019	0.800	-0.223	0.500
01/04/2019	4	2019	0.900	-0.105	0.600
08/04/2019	4	2019	0.800	-0.223	0.500
15/04/2019	4	2019	0.800	-0.223	0.500
22/04/2019	4	2019	0.900	-0.105	0.600
21/10/2019	10	2019	0.800	-0.223	0.500
04/11/2019	11	2019	0.900	-0.105	0.600
18/11/2019	11	2019	0.800	-0.223	0.500
25/11/2019	11	2019	0.800	-0.223	0.500

*Table 8. Apple prices (Royal Gala*SE*II*65-70 mm, Ribatejo e Oeste (region) and Oeste (market)), for 2020-2022*

Period	Month	Year	Value	Logarithm	Normalised Value for All Periods
03/02/2020	2	2020	0.700	-0.357	0.400
10/02/2020	2	2020	0.800	-0.223	0.500
17/02/2020	2	2020	0.800	-0.223	0.500
02/03/2020	3	2020	0.700	-0.357	0.400
09/03/2020	3	2020	0.800	-0.223	0.500
16/03/2020	3	2020	0.800	-0.223	0.500
30/03/2020	3	2020	0.900	-0.105	0.600
06/04/2020	4	2020	0.600	-0.511	0.300
13/04/2020	4	2020	0.300	-1.204	0.000
20/04/2020	4	2020	0.900	-0.105	0.600
27/04/2020	4	2020	0.900	-0.105	0.600
11/05/2020	5	2020	0.900	-0.105	0.600
01/06/2020	6	2020	1.100	0.095	0.800
14/09/2020	9	2020	1.100	0.095	0.800
21/09/2020	9	2020	1.100	0.095	0.800
28/09/2020	9	2020	1.100	0.095	0.800
05/10/2020	10	2020	1.100	0.095	0.800
12/10/2020	10	2020	1.100	0.095	0.800
19/10/2020	10	2020	1.100	0.095	0.800
26/10/2020	10	2020	1.000	0.000	0.700
02/11/2020	11	2020	1.000	0.000	0.700
09/11/2020	11	2020	0.900	-0.105	0.600
16/11/2020	11	2020	1.000	0.000	0.700
23/11/2020	11	2020	1.000	0.000	0.700
30/11/2020	11	2020	1.000	0.000	0.700
07/12/2020	12	2020	1.000	0.000	0.700
14/12/2020	12	2020	1.000	0.000	0.700
04/01/2021	1	2021	0.700	-0.357	0.400
11/01/2021	1	2021	0.600	-0.511	0.300
01/02/2021	2	2021	0.700	-0.357	0.400
08/02/2021	2	2021	0.600	-0.511	0.300
15/02/2021	2	2021	0.600	-0.511	0.300
01/03/2021	3	2021	1.000	0.000	0.700
08/03/2021	3	2021	0.700	-0.357	0.400
15/03/2021	3	2021	0.800	-0.223	0.500
22/03/2021	3	2021	0.900	-0.105	0.600
29/03/2021	3	2021	0.900	-0.105	0.600

Continued on following page

Table 8. Continued

Period	Month	Year	Value	Logarithm	Normalised Value for All Periods
29/03/2021	3	2021	0.900	-0.105	0.600
05/04/2021	4	2021	1.000	0.000	0.700
12/04/2021	4	2021	0.900	-0.105	0.600
19/04/2021	4	2021	0.800	-0.223	0.500
26/04/2021	4	2021	1.000	0.000	0.700
03/05/2021	5	2021	0.900	-0.105	0.600
23/08/2021	8	2021	1.000	0.000	0.700
30/08/2021	8	2021	0.900	-0.105	0.600
06/09/2021	9	2021	1.000	0.000	0.700
13/09/2021	9	2021	1.000	0.000	0.700
20/09/2021	9	2021	1.000	0.000	0.700
04/10/2021	10	2021	1.000	0.000	0.700
11/10/2021	10	2021	0.700	-0.357	0.400
18/10/2021	10	2021	1.000	0.000	0.700
25/10/2021	10	2021	0.900	-0.105	0.600
01/11/2021	11	2021	0.900	-0.105	0.600
15/11/2021	11	2021	0.900	-0.105	0.600
22/11/2021	11	2021	0.800	-0.223	0.500
29/11/2021	11	2021	0.800	-0.223	0.500
06/12/2021	12	2021	0.700	-0.357	0.400
13/12/2021	12	2021	0.700	-0.357	0.400
20/12/2021	12	2021	0.600	-0.511	0.300
27/12/2021	12	2021	0.600	-0.511	0.300
03/01/2022	1	2022	1.000	0.000	0.700
10/01/2022	1	2022	0.600	-0.511	0.300
17/01/2022	1	2022	0.600	-0.511	0.300
24/01/2022	1	2022	0.600	-0.511	0.300
31/01/2022	1	2022	0.600	-0.511	0.300
07/02/2022	2	2022	0.900	-0.105	0.600
14/02/2022	2	2022	0.900	-0.105	0.600
21/02/2022	2	2022	1.000	0.000	0.700
28/02/2022	2	2022	1.000	0.000	0.700
07/03/2022	3	2022	0.600	-0.511	0.300
14/03/2022	3	2022	0.500	-0.693	0.200
21/03/2022	3	2022	0.600	-0.511	0.300
28/03/2022	3	2022	0.600	-0.511	0.300
04/04/2022	4	2022	0.500	-0.693	0.200
11/04/2022	4	2022	0.600	-0.511	0.300

Continued on following page

Table 8. Continued

Period	Month	Year	Value	Logarithm	Normalised Value for All Periods
18/04/2022	4	2022	0.600	-0.511	0.300
25/04/2022	4	2022	0.600	-0.511	0.300
02/05/2022	5	2022	0.700	-0.357	0.400
09/05/2022	5	2022	0.500	-0.693	0.200
16/05/2022	5	2022	0.500	-0.693	0.200
23/05/2022	5	2022	0.500	-0.693	0.200
06/06/2022	6	2022	0.400	-0.916	0.100
13/06/2022	6	2022	0.400	-0.916	0.100

*Table 9. Apple prices (Royal Gala*SE*II*70-75 mm, Ribatejo e Oeste (region) and Oeste (market)), for 2017-2019*

Period	Month	Year	Value	Logarithm	Normalised Value for All Periods
02/01/2017	1	2017	0.700	-0.357	0.100
09/01/2017	1	2017	0.700	-0.357	0.100
16/01/2017	1	2017	0.700	-0.357	0.100
23/01/2017	1	2017	0.700	-0.357	0.100
30/01/2017	1	2017	0.700	-0.357	0.100
06/02/2017	2	2017	0.700	-0.357	0.100
13/02/2017	2	2017	0.700	-0.357	0.100
20/02/2017	2	2017	0.700	-0.357	0.100
27/02/2017	2	2017	0.700	-0.357	0.100
06/03/2017	3	2017	0.700	-0.357	0.100
13/03/2017	3	2017	0.700	-0.357	0.100
20/03/2017	3	2017	0.700	-0.357	0.100
02/04/2018	4	2018	0.800	-0.223	0.300
09/04/2018	4	2018	1.000	0.000	0.500
16/04/2018	4	2018	1.000	0.000	0.500
30/04/2018	4	2018	1.000	0.000	0.500
07/05/2018	5	2018	0.800	-0.223	0.300
14/05/2018	5	2018	1.000	0.000	0.500
11/06/2018	6	2018	1.300	0.262	0.900
18/06/2018	6	2018	1.000	0.000	0.500
09/07/2018	7	2018	1.400	0.336	1.000
16/07/2018	7	2018	0.600	-0.511	0.000
23/07/2018	7	2018	0.700	-0.357	0.100
06/08/2018	8	2018	1.400	0.336	1.000
17/09/2018	9	2018	1.100	0.095	0.600
24/09/2018	9	2018	1.000	0.000	0.500
01/10/2018	10	2018	0.800	-0.223	0.300
08/10/2018	10	2018	0.900	-0.105	0.400
15/10/2018	10	2018	0.900	-0.105	0.400
22/10/2018	10	2018	1.100	0.095	0.600
29/10/2018	10	2018	0.800	-0.223	0.300
05/11/2018	11	2018	0.800	-0.223	0.300
12/11/2018	11	2018	0.700	-0.357	0.100
19/11/2018	11	2018	1.100	0.095	0.600
26/11/2018	11	2018	1.000	0.000	0.500
03/12/2018	12	2018	1.000	0.000	0.500
10/12/2018	12	2018	0.800	-0.223	0.300

Continued on following page

Table 9. Continued

Period	Month	Year	Value	Logarithm	Normalised Value for All Periods
17/12/2018	12	2018	0.800	-0.223	0.300
21/01/2019	1	2019	1.000	0.000	0.500
04/02/2019	2	2019	1.100	0.095	0.600
25/02/2019	2	2019	1.100	0.095	0.600
04/03/2019	3	2019	1.100	0.095	0.600
11/03/2019	3	2019	1.000	0.000	0.500
18/03/2019	3	2019	0.700	-0.357	0.100
25/03/2019	3	2019	1.100	0.095	0.600
01/04/2019	4	2019	1.200	0.182	0.800
08/04/2019	4	2019	1.200	0.182	0.800
15/04/2019	4	2019	1.100	0.095	0.600
22/04/2019	4	2019	1.100	0.095	0.600
21/10/2019	10	2019	1.100	0.095	0.600
04/11/2019	11	2019	1.100	0.095	0.600
18/11/2019	11	2019	1.100	0.095	0.600
25/11/2019	11	2019	1.100	0.095	0.600

*Table 10. Apple prices (Royal Gala*SE*II*70-75 mm, Ribatejo e Oeste (region) and Oeste (market)), for 2020-2021*

Period	Month	Year	Value	Logarithm	Normalised Value for All Periods
03/02/2020	2	2020	0.900	-0.105	0.400
10/02/2020	2	2020	1.000	0.000	0.500
02/03/2020	3	2020	0.700	-0.357	0.100
09/03/2020	3	2020	0.900	-0.105	0.400
16/03/2020	3	2020	0.800	-0.223	0.300
30/03/2020	3	2020	1.000	0.000	0.500
06/04/2020	4	2020	1.000	0.000	0.500
18/05/2020	5	2020	0.700	-0.357	0.100
14/09/2020	9	2020	1.000	0.000	0.500
21/09/2020	9	2020	1.100	0.095	0.600
28/09/2020	9	2020	1.000	0.000	0.500
05/10/2020	10	2020	0.600	-0.511	0.000
12/10/2020	10	2020	0.700	-0.357	0.100
19/10/2020	10	2020	1.000	0.000	0.500
26/10/2020	10	2020	1.000	0.000	0.500
02/11/2020	11	2020	1.000	0.000	0.500
09/11/2020	11	2020	0.900	-0.105	0.400
23/11/2020	11	2020	1.200	0.182	0.800
07/12/2020	12	2020	1.100	0.095	0.600
14/12/2020	12	2020	1.100	0.095	0.600
21/12/2020	12	2020	1.100	0.095	0.600
28/12/2020	12	2020	1.100	0.095	0.600
04/01/2021	1	2021	1.100	0.095	0.600
11/01/2021	1	2021	1.000	0.000	0.500
18/01/2021	1	2021	1.000	0.000	0.500
25/01/2021	1	2021	0.700	-0.357	0.100
01/02/2021	2	2021	0.700	-0.357	0.100
08/02/2021	2	2021	1.000	0.000	0.500
15/02/2021	2	2021	1.000	0.000	0.500
01/03/2021	3	2021	0.600	-0.511	0.000
08/03/2021	3	2021	0.700	-0.357	0.100
15/03/2021	3	2021	0.700	-0.357	0.100
22/03/2021	3	2021	0.700	-0.357	0.100
29/03/2021	3	2021	0.700	-0.357	0.100
29/03/2021	3	2021	0.700	-0.357	0.100
12/04/2021	4	2021	0.700	-0.357	0.100
03/05/2021	5	2021	1.200	0.182	0.800

Continued on following page

Table 10. Continued

Period	Month	Year	Value	Logarithm	Normalised Value for All Periods
09/08/2021	8	2021	1.000	0.000	0.500
16/08/2021	8	2021	1.000	0.000	0.500
23/08/2021	8	2021	1.200	0.182	0.800
30/08/2021	8	2021	1.200	0.182	0.800
27/09/2021	9	2021	1.200	0.182	0.800
04/10/2021	10	2021	1.200	0.182	0.800
11/10/2021	10	2021	1.100	0.095	0.600
18/10/2021	10	2021	1.100	0.095	0.600
25/10/2021	10	2021	1.100	0.095	0.600
01/11/2021	11	2021	1.100	0.095	0.600
15/11/2021	11	2021	0.900	-0.105	0.400
22/11/2021	11	2021	0.900	-0.105	0.400
29/11/2021	11	2021	0.900	-0.105	0.400
06/12/2021	12	2021	0.800	-0.223	0.300
13/12/2021	12	2021	0.700	-0.357	0.100
20/12/2021	12	2021	0.700	-0.357	0.100
27/12/2021	12	2021	0.700	-0.357	0.100

*Table 11. Apple prices (Royal Gala*SE*II*70-75 mm, Ribatejo e Oeste (region) and Oeste (market)), for 2022*

Period	Month	Year	Value	Logarithm	Normalised Value for All Periods
03/01/2022	1	2022	0.700	-0.357	0.100
10/01/2022	1	2022	0.600	-0.511	0.000
17/01/2022	1	2022	0.600	-0.511	0.000
24/01/2022	1	2022	0.700	-0.357	0.100
31/01/2022	1	2022	0.700	-0.357	0.100
07/02/2022	2	2022	0.700	-0.357	0.100
14/02/2022	2	2022	0.700	-0.357	0.100
21/02/2022	2	2022	0.700	-0.357	0.100
28/02/2022	2	2022	0.700	-0.357	0.100
07/03/2022	3	2022	0.700	-0.357	0.100
14/03/2022	3	2022	0.700	-0.357	0.100
21/03/2022	3	2022	0.700	-0.357	0.100
28/03/2022	3	2022	0.700	-0.357	0.100
04/04/2022	4	2022	0.700	-0.357	0.100
11/04/2022	4	2022	0.700	-0.357	0.100
18/04/2022	4	2022	0.700	-0.357	0.100
25/04/2022	4	2022	0.700	-0.357	0.100
02/05/2022	5	2022	0.600	-0.511	0.000
09/05/2022	5	2022	0.900	-0.105	0.400
16/05/2022	5	2022	0.900	-0.105	0.400
23/05/2022	5	2022	0.900	-0.105	0.400
29/08/2022	8	2022	1.200	0.182	0.800
05/09/2022	9	2022	1.100	0.095	0.600
12/09/2022	9	2022	1.000	0.000	0.500
19/09/2022	9	2022	1.000	0.000	0.500
26/09/2022	9	2022	1.000	0.000	0.500
03/10/2022	10	2022	1.000	0.000	0.500
10/10/2022	10	2022	0.900	-0.105	0.400
17/10/2022	10	2022	1.000	0.000	0.500
24/10/2022	10	2022	1.000	0.000	0.500
31/10/2022	10	2022	1.000	0.000	0.500
07/11/2022	11	2022	1.000	0.000	0.500
14/11/2022	11	2022	1.000	0.000	0.500
21/11/2022	11	2022	1.000	0.000	0.500
05/12/2022	12	2022	1.000	0.000	0.500
12/12/2022	12	2022	1.000	0.000	0.500
19/12/2022	12	2022	1.000	0.000	0.500
26/12/2022	12	2022	1.000	0.000	0.500

*Table 12. Apple prices (Royal Gala*SE*II*75-80 mm, Ribatejo e Oeste (region) and Oeste (market)), for 2017-2019*

Period	Month	Year	Value	Logarithm	Normalised Value for All Periods
02/01/2017	1	2017	0.800	-0.223	0.200
09/01/2017	1	2017	0.800	-0.223	0.200
16/01/2017	1	2017	0.800	-0.223	0.200
23/01/2017	1	2017	0.800	-0.223	0.200
30/01/2017	1	2017	0.800	-0.223	0.200
06/02/2017	2	2017	0.800	-0.223	0.200
13/02/2017	2	2017	0.800	-0.223	0.200
20/02/2017	2	2017	0.800	-0.223	0.200
27/02/2017	2	2017	0.800	-0.223	0.200
06/03/2017	3	2017	0.800	-0.223	0.200
13/03/2017	3	2017	0.800	-0.223	0.200
20/03/2017	3	2017	0.800	-0.223	0.200
27/11/2017	11	2017	0.800	-0.223	0.200
09/04/2018	4	2018	1.000	0.000	0.400
16/04/2018	4	2018	1.000	0.000	0.400
30/04/2018	4	2018	1.000	0.000	0.400
07/05/2018	5	2018	0.800	-0.223	0.200
14/05/2018	5	2018	1.000	0.000	0.400
04/06/2018	6	2018	1.000	0.000	0.400
18/06/2018	6	2018	1.000	0.000	0.400
09/07/2018	7	2018	1.500	0.405	1.000
06/08/2018	8	2018	1.400	0.336	0.900
24/09/2018	9	2018	1.000	0.000	0.400
01/10/2018	10	2018	0.900	-0.105	0.300
08/10/2018	10	2018	0.800	-0.223	0.200
15/10/2018	10	2018	0.900	-0.105	0.300
22/10/2018	10	2018	1.100	0.095	0.600
05/11/2018	11	2018	0.900	-0.105	0.300
12/11/2018	11	2018	0.800	-0.223	0.200
19/11/2018	11	2018	1.300	0.262	0.800
26/11/2018	11	2018	1.200	0.182	0.700
10/12/2018	12	2018	0.900	-0.105	0.300
17/12/2018	12	2018	0.900	-0.105	0.300
21/01/2019	1	2019	1.100	0.095	0.600
25/02/2019	2	2019	1.100	0.095	0.600
04/03/2019	3	2019	1.300	0.262	0.800
11/03/2019	3	2019	1.000	0.000	0.400

Continued on following page

Table 12. Continued

Period	Month	Year	Value	Logarithm	Normalised Value for All Periods
18/03/2019	3	2019	0.800	-0.223	0.200
25/03/2019	3	2019	1.100	0.095	0.600
21/10/2019	10	2019	1.000	0.000	0.400
04/11/2019	11	2019	1.100	0.095	0.600
18/11/2019	11	2019	1.100	0.095	0.600
25/11/2019	11	2019	1.000	0.000	0.400

*Table 13. Apple prices (Royal Gala*SE*II*75-80 mm, Ribatejo e Oeste (region) and Oeste (market)), for 2020-2022*

Period	Month	Year	Value	Logarithm	Normalised Value for All Periods
03/02/2020	2	2020	0.900	-0.105	0.300
10/02/2020	2	2020	0.900	-0.105	0.300
09/03/2020	3	2020	0.900	-0.105	0.300
16/03/2020	3	2020	0.900	-0.105	0.300
30/03/2020	3	2020	0.900	-0.105	0.300
06/04/2020	4	2020	1.000	0.000	0.400
20/04/2020	4	2020	1.000	0.000	0.400
27/04/2020	4	2020	1.100	0.095	0.600
11/05/2020	5	2020	1.000	0.000	0.400
21/09/2020	9	2020	0.800	-0.223	0.200
05/10/2020	10	2020	0.800	-0.223	0.200
26/10/2020	10	2020	1.000	0.000	0.400
02/11/2020	11	2020	1.000	0.000	0.400
11/01/2021	1	2021	1.000	0.000	0.400
18/01/2021	1	2021	1.000	0.000	0.400
01/03/2021	3	2021	1.200	0.182	0.700
15/03/2021	3	2021	0.800	-0.223	0.200
22/03/2021	3	2021	0.800	-0.223	0.200
29/03/2021	3	2021	0.800	-0.223	0.200
29/03/2021	3	2021	0.800	-0.223	0.200
12/04/2021	4	2021	0.800	-0.223	0.200
19/04/2021	4	2021	0.700	-0.357	0.100
26/04/2021	4	2021	0.800	-0.223	0.200
03/05/2021	5	2021	1.200	0.182	0.700
09/08/2021	8	2021	1.200	0.182	0.700
27/09/2021	9	2021	0.800	-0.223	0.200
06/12/2021	12	2021	0.600	-0.511	0.000
13/12/2021	12	2021	0.600	-0.511	0.000
27/12/2021	12	2021	0.600	-0.511	0.000
03/01/2022	1	2022	1.100	0.095	0.600
10/01/2022	1	2022	0.600	-0.511	0.000
17/01/2022	1	2022	0.600	-0.511	0.000
07/03/2022	3	2022	0.700	-0.357	0.100
14/03/2022	3	2022	0.600	-0.511	0.000
21/03/2022	3	2022	0.700	-0.357	0.100
28/03/2022	3	2022	0.700	-0.357	0.100
04/04/2022	4	2022	0.700	-0.357	0.100

Continued on following page

Table 13. Continued

Period	Month	Year	Value	Logarithm	Normalised Value for All Periods
11/04/2022	4	2022	0.700	-0.357	0.100
18/04/2022	4	2022	0.700	-0.357	0.100
25/04/2022	4	2022	0.700	-0.357	0.100
02/05/2022	5	2022	0.700	-0.357	0.100
09/05/2022	5	2022	0.600	-0.511	0.000
16/05/2022	5	2022	0.600	-0.511	0.000
23/05/2022	5	2022	0.600	-0.511	0.000

Table 14. Average apple prices, for 2017-2019

Year	Month	Value	Normalised Value for All Periods
2017	1	0.7	0.1
2017	2	0.7	0.1
2017	3	0.7	0.1
2017	4	0.7	0.1
2017	5	0.7	0.1
2017	6	0.7	0.1
2017	8	0.6	0.0
2017	9	0.8	0.2
2017	10	0.7	0.1
2017	11	0.6	0.0
2017	12	0.6	0.0
2018	1	0.6	0.0
2018	2	0.6	0.0
2018	3	0.6	0.0
2018	4	0.6	0.0
2018	5	0.6	0.0
2018	6	0.7	0.1
2018	7	0.7	0.1
2018	8	0.9	0.3
2018	9	0.9	0.3
2018	10	0.8	0.2
2018	11	0.7	0.1
2018	12	0.7	0.1
2019	1	0.7	0.1
2019	2	0.7	0.1
2019	3	0.7	0.1
2019	4	0.7	0.1
2019	5	0.7	0.1
2019	6	0.6	0.0
2019	7	0.7	0.1
2019	9	0.9	0.3
2019	10	0.7	0.1
2019	11	0.7	0.1
2019	12	0.6	0.0

APPLE PRICES

Considering average Portuguese producer prices for the different varieties of apple, the results presented in Tables 14-15 confirm the findings obtained in the previous section for the apple "Royal Gala". In fact, apple producer prices do not change significantly over the period 2017-2022. The average prices, for the years 2017-2022, were, respectively, the following: 0.7 euros (normalised value of 0.1); 0.7 euros (normalised value of 0.1); 0.7 euros (normalised value of 0.1); 0.8 euros (normalised value of 0.2); 0.9 euros (normalised value of 0.3); and 0.9 euros (normalised value of 0.3).

Table 15. Average apple prices, for 2020-2022

Year	Month	Value	Normalised Value for All Periods
2020	1	0.6	0.0
2020	2	0.6	0.0
2020	3	0.6	0.0
2020	4	0.6	0.0
2020	5	0.6	0.0
2020	6	0.6	0.0
2020	7	0.6	0.0
2020	8	1.7	1.0
2020	9	0.9	0.3
2020	10	0.8	0.2
2020	11	0.9	0.3
2020	12	0.9	0.3
2021	1	0.8	0.2
2021	2	0.9	0.3
2021	3	0.9	0.3
2021	4	0.9	0.3
2021	5	0.9	0.3
2021	6	0.8	0.2
2021	7	0.9	0.3
2021	8	1.1	0.5
2021	9	0.8	0.2
2021	10	0.8	0.2
2021	11	0.8	0.2
2021	12	0.8	0.2
2022	1	0.8	0.2
2022	2	0.8	0.2
2022	3	0.8	0.2
2022	4	0.8	0.2
2022	5	0.8	0.2
2022	6	0.9	0.3
2022	7	0.8	0.2
2022	8	0.8	0.2
2022	9	1.0	0.4
2022	10	0.9	0.3
2022	11	0.9	0.3
2022	12	0.9	0.3

FRUIT PRICES

The average Portuguese producer fruit prices (chestnut, cherry, orange, apple, melon, blueberry, strawberry, peach and grape) do not seem also to be affected significantly by the shocks verified since the Covid-19 pandemic and the Russia-Ukraine conflict, such as highlighted in Tables 16-17. The average prices were 0.9 euros in 2017, 1.0 euros in 2018, 1.1 euros in 2019, 1.1 euros in 2020, 1.3 euros in 2021 and 1.4 euros in 2022. There is a slight increase in the prices in 2021 and 2022, but seems that they follow a similar trend as that verified for the previous years.

Table 16. Average fruit (chestnut, cherry, orange, apple, melon, blueberry, strawberry, peach, and grape) prices, for 2017-2019

Year	Month	Value	Normalised Value for All Periods
2017	1	0.8	0.1
2017	2	0.8	0.1
2017	3	0.8	0.1
2017	4	0.8	0.1
2017	5	0.9	0.1
2017	6	1.0	0.2
2017	7	1.6	0.6
2017	8	1.2	0.4
2017	9	1.0	0.2
2017	10	0.9	0.1
2017	11	0.7	0.0
2017	12	0.7	0.0
2018	1	0.7	0.0
2018	2	0.8	0.1
2018	3	0.8	0.1
2018	4	0.8	0.1
2018	5	0.8	0.1
2018	6	1.0	0.2
2018	7	1.4	0.5
2018	8	1.4	0.5
2018	9	1.4	0.5
2018	10	1.2	0.4
2018	11	0.9	0.1
2018	12	0.8	0.1
2019	1	0.8	0.1
2019	2	0.8	0.1
2019	3	0.8	0.1
2019	4	0.8	0.1
2019	5	1.0	0.2
2019	6	1.1	0.3
2019	7	1.8	0.8
2019	8	1.6	0.6
2019	9	1.5	0.6
2019	10	1.2	0.4
2019	11	0.9	0.1
2019	12	0.7	0.0

Table 17. Average fruit (chestnut, cherry, orange, apple, melon, blueberry, strawberry, peach, and grape) prices, for 2020-2022

Year	Month	Value	Normalised Value for All Periods
2020	1	0.7	0.0
2020	2	0.7	0.0
2020	3	0.7	0.0
2020	4	0.8	0.1
2020	5	1.0	0.2
2020	6	1.4	0.5
2020	7	1.9	0.9
2020	8	1.9	0.9
2020	9	1.4	0.5
2020	10	1.1	0.3
2020	11	1.0	0.2
2020	12	1.1	0.3
2021	1	1.0	0.2
2021	2	1.1	0.3
2021	3	1.1	0.3
2021	4	1.2	0.4
2021	5	1.4	0.5
2021	6	1.7	0.7
2021	7	2.1	1.0
2021	8	1.9	0.9
2021	9	1.2	0.4
2021	10	1.1	0.3
2021	11	1.1	0.3
2021	12	1.1	0.3
2022	1	1.1	0.3
2022	2	1.0	0.2
2022	3	1.1	0.3
2022	4	1.1	0.3
2022	5	1.3	0.4
2022	6	1.5	0.6
2022	7	1.7	0.7
2022	8	1.9	0.9
2022	9	2.0	0.9
2022	10	1.8	0.8
2022	11	1.4	0.5
2022	12	1.4	0.5

CONCLUSION

This chapter proposed to investigate the consequences of the most recent international events on the fresh fruit producer prices in the Portuguese context, considering data from the SIMA (Sistema de Informação de Mercados Agrícolas) for the years 2017-2022 disaggregated weekly. Considering the current trends in the fruit prices in the retailers, it seemed important to understand the behaviour of these prices in the producer markets. These assessments may be interesting for governments, public institutions and international organisations with a focus on these issues.

In globalised economies, international trade opens interesting opportunities for several economic sectors, but also become them more vulnerable to exogenous shocks, with consequences on the dynamics of the markets and adjustments in the respective prices. The agricultural prices fluctuate over the year, because of the supply volume in-season and off-season. The fruit sector is not an exception to these frameworks. In any case, the fruit prices are interrelated with the quality and present less volatility than that verified for the vegetables, for example. These prices are also a consequence of the costs that occurred in the production process, where those associated with the harvest activities represent a relevant portion of the total production costs.

The Portuguese producer prices for apples "Royal Gala*SE*I*65-70 mm", "Royal Gala*SE*I*70-75 mm", "Royal Gala*SE*I*75-80 mm", "Royal Gala*SE*II*65-70 mm", "Royal Gala*SE*II*70-75 mm" and "Royal Gala*SE*II*75-80 mm" did not change relevantly after the two most recent international shocks (Covid-19 pandemic and Russia-Ukraine crisis). These trends were confirmed for all Portuguese apple prices and fruit sector prices.

The insights here obtained reveal that the impacts of the international exogenous shocks did not promote relevant changes in Portuguese fruit prices. For future research, it is suggested to assess these trends with other data, namely replicating the disaggregated assessment carried out for the apple "Royal Gala" to other varieties and other fresh fruits. Nonetheless, the findings obtained for all apple prices and fruit sector prices have their importance and robustness and reflect the trends in Portuguese fresh fruit prices. Another important topic of research could be to compare these prices in the producer markets with those practised in the retailers. In terms of policy recommendations, it is suggested to better monitor these markets, namely in the retailers, considering the importance of these products for human health. Higher prices bring constraints for the consumption of the lower-income population.

ACKNOWLEDGMENT

This work is funded by National Funds through the FCT - Foundation for Science and Technology, I.P., within the scope of the project Refª UIDB/00681/2020. Furthermore we would like to thank the CERNAS Research Centre and the Polytechnic Institute of Viseu for their support.

REFERENCES

Adelaja, A. O., Brumfield, R. G., & Schilling, B. J. (1996). Share analysis of drivers of growth in the ornamental industry: A case study of New Jersey. In Acta Hortic. (Vol. 429, p. 307). International Society for Horticultural Science. doi:10.17660/ActaHortic.1996.429.38

Alan Walters, S. (2020). Impact of nitrogen fertilizer on yield and associated revenues in no-tillage pumpkin. *HortScience, 55*(4), 522–527. doi:10.21273/HORTSCI14512-19

Baggio, J. S., Cordova, L. G., Seijo, T. E., Noling, J. W., Whitaker, V. M., & Peres, N. A. (2021). Cultivar Selection Is an Effective and Economic Strategy for Managing Charcoal Rot of Strawberry in Florida. *Plant Disease, 105*(8). doi:10.1094/PDIS-10-20-2250-RE

Ballester, C., Buesa, I., Soler, E., Besada, C., Salvador, A., Bonet, L., & Intrigliolo, D. S. (2018). Post-harvest regulated deficit irrigation in early- and intermediate-maturing loquat trees. *Agricultural Water Management, 205*, 1–8. doi:10.1016/j.agwat.2018.04.030

Capocasa, F., Balducci, F., Mazzoni, L., Marcellini, M., Qaderi, R., & Mezzetti, B. (2021). Preliminary results of soilless cultivated strawberry cultivars in the autumn-spring cycle in the mid-Adriatic area. In Acta Hortic. (Vol. 1309, p. 596). International Society for Horticultural Science. doi:10.17660/ActaHortic.2021.1309.85

Cittadini, E. D., Van Keulen, H., Peri, P. L., & De Ridder, N. (2007). Designing a 'target-tree' for maximizing gross value of product in Patagonian sweet cherry orchards. *International Journal of Fruit Science, 6*(3), 3–22. doi:10.1300/J492v06n03_02

Djekic, I., Radivojevic, D., & Milivojevic, J. (2019). Quality perception throughout the apple fruit chain. *Journal of Food Measurement and Characterization, 13*(4), 3106–3118. doi:10.1007/s11694-019-00233-1

Ghanbari, S., Vaezin, S. M. H., Shamekhi, T., Eastin, I. L., Lovrić, N., & Aghai, M. M. (2020). The Economic and Biological Benefits of Non-wood Forest Products to Local Communities in Iran. *Economic Botany, 74*(1), 59–73. doi:10.1007/s12231-019-09478-9

Inglese, P., Gugliuzza, G., & Liguori, G. (2009). Fruit production of cultivated cacti: A short overview on plant ecophysiology and C budget. In Acta Hortic. (Vol. 811, p. 86). International Society for Horticultural Science. doi:10.17660/ActaHortic.2009.811.6

Jovicich, E., VanSickle, J. J., Cantliffe, D. J., & Stoffella, P. J. (2005). Greenhouse-grown colored peppers: A profitable alternative for vegetable production in Florida? *HortTechnology, 15*(2), 355–369. doi:10.21273/HORTTECH.15.2.0355

Kodrat, K. F., Sinulingga, S., Napitupulu, H. L., & Hadiguna, R. A. (2018). Passion fruit supply chain simulation using Monte Carlo method. *International Journal of Civil Engineering and Technology, 9*(13), 1164–1173.

Lei, L., & Shimokawa, S. (2020). Promoting dietary guidelines and environmental sustainability in China. *China Economic Review, 59*. doi:10.1016/j.chieco.2017.08.001

Manolova, V., & Kolev, K. (2012). Economic results from growing of cherry in different levels of intensification. *Acta Horticulturae, 981*(1), 719–724. https://doi.org/10.17660/actahortic.2013.981.115

Márquez-Berber, S. R., Torcuato-Calderón, C., Almaguer-Vargas, G., Colinas-León, M. T., & Gardezi, A. K. (2012). Cactus pear (Opuntia albicarpa and O. megacantha) production system in axapusco, Estado de Mexico. Problems and alternatives. *Revista Chapingo Serie Horticultura, 18*(1), 81–93.

Medina, R. B., Cantuarias-Avilés, T. E., Angolini, S. F., & Da Silva, S. R. (2018). Performance of 'emerald' and 'jewel' blueberry cultivars under no-chill incidence. *Pesquisa Agropecuaria Tropical, 48*(2), 147–152. Scopus. doi:10.1590/1983-40632018v4852093

Pereira, P. S., Sarmento, R. A., Galdino, T. V. S., Lima, C. H. O., dos Santos, F. A., Silva, J., dos Santos, G. R., & Picanço, M. C. (2017). Economic injury levels and sequential sampling plans for Frankliniella schultzei in watermelon crops. *Pest Management Science, 73*(7), 1438–1445. doi:10.1002/ps.4475

Sanders, K. F. (2005). Orange harvesting systems review. *Biosystems Engineering, 90*(2), 115–125. doi:10.1016/j.biosystemseng.2004.10.006

Scopus. (2023). *Scopus Database*. https://www.scopus.com/search/form.uri?display=basic#basic

SIMA. (2023). *Several Statistics and Information*. https://www.gpp.pt/index.php/sima/sima-2018

Spósito, M. B., & Filho, F. D. A. A. M. (2003). 'Tahiti' lime fruit set related to gibberellic acid application on out-of-season flowering and the accumulation of degree days. *Fruits, 58*(3), 151–156. doi:10.1051/fruits:2003003

Stata. (2023). *Statistical software for data science*. https://www.stata.com/

StataCorp. (2017a). *Stata 15 Base Reference Manual*. Stata Press.

StataCorp. (2017b). *Stata Statistical Software: Release 15*. StataCorp LLC.

Tsvakirai, C. Z., Liebenberg, F., & Kirsten, J. F. (2018). Does research and development (R&D) investment lead to economic growth? Evidence from the South African peach and nectarine industry. *African Journal of Science, Technology, Innovation and Development, 10*(4), 463–472. doi:10.1080/20421338.2018.1473826

Uçak, H., Yelgen, E., & Arı, Y. (2022). The Role of Energy on the Price Volatility of Fruits and Vegetables: Evidence from Turkey. *Bio-Based and Applied Economics, 11*(1), 37–54. doi:10.36253/bae-10896

Verdial, M. F., Neto, J. T., Minami, K., Scarpare Filho, J. A., Christoffoleti, P. J., Scarpare, F. V., Barela, J. F., Del Aguila, J. S., & Kluge, R. A. (2007). Vernalization on five cultivars of strawberry. *Ciência Rural, 37*(4), 976–981. doi:10.15900103-84782007000400009

Chapter 15
Current Realities for Portuguese Vegetable and Cereal Production Markets

ABSTRACT

Vegetable and cereal production dynamics impact socioeconomic conditions, agricultural performance, and the food industry structure. Specifically, cereals play a determinant role in the sustainability of societies, considering that they are used by several food industries and their importance for the human diet. Taking into account the importance of potatoes and wheat for the Portuguese contexts, a special focus was given to these two products. In this way, it analysed the potential consequences of the pandemic and recent crises on the potatoes and wheat producer prices, considering statistical information for the weeks of 2017-2022 from the SIMA (Sistema de Informação de Mercados Agrícolas). The impacts seem to be higher on the wheat prices than on the potatoes, and these implications came from namely in 2022.

INTRODUCTION

Cereal markets and respective chains have a crucial contribution to food security and the whole food industry, considering their use for the production of many other products. In this way, cereal prices influence other markets (De Roest et al., 2008), including the supply of livestock feed (Olaizola et al., 2015) and household consumption (Sy et al., 2022). They affect also agricultural decisions (Doole & Weetman, 2009), options (Xu & Mola-Yudego, 2021) and resource allocation (Dzanku, 2015). Effects on pulses production are an example of these implications from the cereal markets (Ghelfi & Palmieri, 2017).

The cereal markets contribute significantly to inflation in developing countries (Durevall et al., 2013). The food and agricultural supply must adjust to the input prices (Deleuran et al., 2013). Wheat and sorghum are among the most important cereals (Abay et al., 2023). These markets are impacted by several variables, but speculation, in some circumstances is among the main causal variables (Andreosso-O'Callaghan & Zolin, 2010). The competition for land between biofuel crops and food production is

DOI: 10.4018/978-1-6684-8923-9.ch015

another factor that affects the commodities prices (Müller et al., 2008), as well as the climate changes in certain cases (Parry et al., 2004) and the agricultural practices (Salo et al., 2007).

The increase in cereal prices after the pandemic promoted the interest in better management of food waste (Ben Rejeb et al., 2022). The cereals prices were also impacted by the 2008 international crisis (Shenggen Fan, 2010), showing their vulnerability to international shocks, and consequent implications welfare of the citizens (Rocchi et al., 2013). The volatility in the cereal markets is another topic of interest (Ott, 2014).

The cereal markets are affected by the European Union decisions, specifically in the framework of the Common Agricultural Policy (CAP) and respective reforms (Benoit & Laignel, 2014) over the years after its creation in the fifties (Popović et al., 2020). Generally, agricultural policies impact the agricultural markets, including cereal prices (García et al., 2008), and food security (Rosegrant et al., 2005).

Potato is another important crop for food security worldwide and a raw material for processed food. The expansion of this crop across the world was decisive to improve the nutritional condition of the population (Lutaladio & Prakash, 2010).

These motivations and the current conjuncture justify the present research about the impacts of the recent events (the Covid-19 pandemic and the Russia-Ukraine crisis, particularly) on Portuguese vegetable and cereal prices. For that, data from the national agricultural markets information system (SIMA, 2023) were considered for the last years. Scopus (2023) was surveyed and Stata (StataCorp, 2017a, 2017b; Stata, 2023) software was taken into account to obtain some results. Normalised values=$(x_i-x_{minimum})/(x_{maximum}-x_{minimum})$.

POTATO PRODUCER PRICES AT THE BEGINNING OF THE YEAR FOR THE PERIOD CONSIDERED

For January and February, over the period 2017-2022, in general, the prices were higher for "Potato*Primor/Nova Branca*SE*Medium*Bag" (from Algarve), after for "Potato*Conservação Branca*SP*Medium*Bag" (from Trás-os-Montes) and finally the producer prices were lower for "Potato*Conservação Vermelha*SP*Medium*Bag" (from Trás-os-Montes). In some cases, the two potatoes from Trás-os-Montes have the same prices (Tables 1-6). The prices had some fluctuations over the period taken into account, but they did not seem to be caused by the international shocks. Maybe, in some cases, a slight decrease in 2020 and some increase in 2022. Nonetheless, some of these values were already verified before 2020. In fact, the average prices for these three potatoes ("Potato*Conservação Branca*SP*Medium*Bag", "Potato*Conservação Vermelha*SP*Medium*Bag" and "Potato*Primor/Nova Branca*SE*Medium*Bag") were 0.4 euros/kg in 2017 (normalized value of 0.4), 0.3 euros/kg in 2018 (normalized value of 0.2), 0.5 euros/kg in 2019 (normalized value of 0.6), 0.4 euros/kg in 2020 (normalized value of 0.4), 0.5 euros/kg in 2021 (normalized value of 0.5) and 0.6 euros/kg in 2022 (normalized value of 0.6).

Table 1. Potato producer prices at the beginning of the year for 2017

Product	Region	Market	Period	Value	Normalised Value
					For All Periods in Each Product
Potato*Conservação Branca*SP*Medium*Bag*EUR/kg	Trás-os-Montes	Chaves	02/01/2017	0.4	0.5
Potato*Conservação Branca*SP*Medium*Bag*EUR/kg	Trás-os-Montes	Chaves	09/01/2017	0.4	0.5
Potato*Conservação Branca*SP*Medium*Bag*EUR/kg	Trás-os-Montes	Chaves	16/01/2017	0.4	0.5
Potato*Conservação Branca*SP*Medium*Bag*EUR/kg	Trás-os-Montes	Chaves	23/01/2017	0.4	0.5
Potato*Conservação Branca*SP*Medium*Bag*EUR/kg	Trás-os-Montes	Chaves	30/01/2017	0.4	0.5
Potato*Conservação Branca*SP*Medium*Bag*EUR/kg	Trás-os-Montes	Chaves	06/02/2017	0.4	0.5
Potato*Conservação Branca*SP*Medium*Bag*EUR/kg	Trás-os-Montes	Chaves	13/02/2017	0.4	0.5
Potato*Conservação Branca*SP*Medium*Bag*EUR/kg	Trás-os-Montes	Chaves	20/02/2017	0.4	0.5
Potato*Conservação Branca*SP*Medium*Bag*EUR/kg	Trás-os-Montes	Chaves	27/02/2017	0.4	0.5
Potato*Conservação Vermelha*SP*Medium*Bag*EUR/kg	Trás-os-Montes	Chaves	02/01/2017	0.3	0.4
Potato*Conservação Vermelha*SP*Medium*Bag*EUR/kg	Trás-os-Montes	Chaves	09/01/2017	0.3	0.4
Potato*Conservação Vermelha*SP*Medium*Bag*EUR/kg	Trás-os-Montes	Chaves	16/01/2017	0.3	0.4
Potato*Conservação Vermelha*SP*Medium*Bag*EUR/kg	Trás-os-Montes	Chaves	23/01/2017	0.3	0.4
Potato*Conservação Vermelha*SP*Medium*Bag*EUR/kg	Trás-os-Montes	Chaves	30/01/2017	0.3	0.4
Potato*Conservação Vermelha*SP*Medium*Bag*EUR/kg	Trás-os-Montes	Chaves	06/02/2017	0.3	0.4
Potato*Conservação Vermelha*SP*Medium*Bag*EUR/kg	Trás-os-Montes	Chaves	13/02/2017	0.3	0.4
Potato*Conservação Vermelha*SP*Medium*Bag*EUR/kg	Trás-os-Montes	Chaves	20/02/2017	0.3	0.4
Potato*Conservação Vermelha*SP*Medium*Bag*EUR/kg	Trás-os-Montes	Chaves	27/02/2017	0.3	0.4
Potato*Primor/Nova Branca*SE*Medium*Bag*EUR/kg	Algarve	Algarve	02/01/2017	0.6	0.3
Potato*Primor/Nova Branca*SE*Medium*Bag*EUR/kg	Algarve	Algarve	09/01/2017	0.6	0.3
Potato*Primor/Nova Branca*SE*Medium*Bag*EUR/kg	Algarve	Algarve	16/01/2017	0.6	0.3
Potato*Primor/Nova Branca*SE*Medium*Bag*EUR/kg	Algarve	Algarve	23/01/2017	0.6	0.3

Continued on following page

Table 1. Continued

Product	Region	Market	Period	Value	Normalised Value
					For All Periods in Each Product
Potato*Primor/Nova Branca*SE*Medium*Bag*EUR/kg	Algarve	Algarve	30/01/2017	0.6	0.3
Potato*Primor/Nova Branca*SE*Medium*Bag*EUR/kg	Algarve	Algarve	06/02/2017	0.6	0.3
Potato*Primor/Nova Branca*SE*Medium*Bag*EUR/kg	Algarve	Algarve	13/02/2017	0.6	0.3
Potato*Primor/Nova Branca*SE*Medium*Bag*EUR/kg	Algarve	Algarve	20/02/2017	0.6	0.3
Potato*Primor/Nova Branca*SE*Medium*Bag*EUR/kg	Algarve	Algarve	27/02/2017	0.6	0.3

Table 2. Potato producer prices at the beginning of the year for 2018

Product	Region	Market	Period	Value	Normalised Value
					For All Periods in Each Product
Potato*Conservação Branca*SP*Medium*Bag*EUR/kg	Trás-os-Montes	Chaves	01/01/2018	0.2	0.0
Potato*Conservação Branca*SP*Medium*Bag*EUR/kg	Trás-os-Montes	Chaves	08/01/2018	0.2	0.0
Potato*Conservação Branca*SP*Medium*Bag*EUR/kg	Trás-os-Montes	Chaves	15/01/2018	0.2	0.0
Potato*Conservação Branca*SP*Medium*Bag*EUR/kg	Trás-os-Montes	Chaves	22/01/2018	0.2	0.0
Potato*Conservação Branca*SP*Medium*Bag*EUR/kg	Trás-os-Montes	Chaves	29/01/2018	0.2	0.0
Potato*Conservação Branca*SP*Medium*Bag*EUR/kg	Trás-os-Montes	Chaves	05/02/2018	0.2	0.0
Potato*Conservação Branca*SP*Medium*Bag*EUR/kg	Trás-os-Montes	Chaves	12/02/2018	0.2	0.0
Potato*Conservação Branca*SP*Medium*Bag*EUR/kg	Trás-os-Montes	Chaves	19/02/2018	0.2	0.0
Potato*Conservação Branca*SP*Medium*Bag*EUR/kg	Trás-os-Montes	Chaves	26/02/2018	0.2	0.0
Potato*Conservação Vermelha*SP*Medium*Bag*EUR/kg	Trás-os-Montes	Chaves	01/01/2018	0.2	0.2
Potato*Conservação Vermelha*SP*Medium*Bag*EUR/kg	Trás-os-Montes	Chaves	08/01/2018	0.2	0.2
Potato*Conservação Vermelha*SP*Medium*Bag*EUR/kg	Trás-os-Montes	Chaves	15/01/2018	0.2	0.2
Potato*Conservação Vermelha*SP*Medium*Bag*EUR/kg	Trás-os-Montes	Chaves	22/01/2018	0.1	0.0
Potato*Conservação Vermelha*SP*Medium*Bag*EUR/kg	Trás-os-Montes	Chaves	29/01/2018	0.1	0.0
Potato*Conservação Vermelha*SP*Medium*Bag*EUR/kg	Trás-os-Montes	Chaves	05/02/2018	0.1	0.0
Potato*Conservação Vermelha*SP*Medium*Bag*EUR/kg	Trás-os-Montes	Chaves	12/02/2018	0.1	0.0
Potato*Conservação Vermelha*SP*Medium*Bag*EUR/kg	Trás-os-Montes	Chaves	19/02/2018	0.1	0.0
Potato*Conservação Vermelha*SP*Medium*Bag*EUR/kg	Trás-os-Montes	Chaves	26/02/2018	0.1	0.0
Potato*Primor/Nova Branca*SE*Medium*Bag*EUR/kg	Algarve	Algarve	01/01/2018	0.7	0.5
Potato*Primor/Nova Branca*SE*Medium*Bag*EUR/kg	Algarve	Algarve	08/01/2018	0.7	0.5
Potato*Primor/Nova Branca*SE*Medium*Bag*EUR/kg	Algarve	Algarve	15/01/2018	0.7	0.5
Potato*Primor/Nova Branca*SE*Medium*Bag*EUR/kg	Algarve	Algarve	22/01/2018	0.7	0.5
Potato*Primor/Nova Branca*SE*Medium*Bag*EUR/kg	Algarve	Algarve	29/01/2018	0.7	0.5
Potato*Primor/Nova Branca*SE*Medium*Bag*EUR/kg	Algarve	Algarve	05/02/2018	0.7	0.5
Potato*Primor/Nova Branca*SE*Medium*Bag*EUR/kg	Algarve	Algarve	12/02/2018	0.7	0.5
Potato*Primor/Nova Branca*SE*Medium*Bag*EUR/kg	Algarve	Algarve	19/02/2018	0.7	0.5
Potato*Primor/Nova Branca*SE*Medium*Bag*EUR/kg	Algarve	Algarve	26/02/2018	0.7	0.5

Table 3. Potato producer prices at the beginning of the year for 2019

Product	Region	Market	Period	Value	Normalised Value
					For All Periods in Each Product
Potato*Conservação Branca*SP*Medium*Bag*EUR/kg	Trás-os-Montes	Chaves	07/01/2019	0.4	0.5
Potato*Conservação Branca*SP*Medium*Bag*EUR/kg	Trás-os-Montes	Chaves	14/01/2019	0.4	0.5
Potato*Conservação Branca*SP*Medium*Bag*EUR/kg	Trás-os-Montes	Chaves	21/01/2019	0.4	0.5
Potato*Conservação Branca*SP*Medium*Bag*EUR/kg	Trás-os-Montes	Chaves	28/01/2019	0.5	0.8
Potato*Conservação Branca*SP*Medium*Bag*EUR/kg	Trás-os-Montes	Chaves	04/02/2019	0.5	0.8
Potato*Conservação Branca*SP*Medium*Bag*EUR/kg	Trás-os-Montes	Chaves	11/02/2019	0.5	0.8
Potato*Conservação Branca*SP*Medium*Bag*EUR/kg	Trás-os-Montes	Chaves	18/02/2019	0.5	0.8
Potato*Conservação Branca*SP*Medium*Bag*EUR/kg	Trás-os-Montes	Chaves	25/02/2019	0.5	0.8
Potato*Conservação Vermelha*SP*Medium*Bag*EUR/kg	Trás-os-Montes	Chaves	07/01/2019	0.3	0.4
Potato*Conservação Vermelha*SP*Medium*Bag*EUR/kg	Trás-os-Montes	Chaves	14/01/2019	0.3	0.4
Potato*Conservação Vermelha*SP*Medium*Bag*EUR/kg	Trás-os-Montes	Chaves	21/01/2019	0.4	0.6
Potato*Conservação Vermelha*SP*Medium*Bag*EUR/kg	Trás-os-Montes	Chaves	28/01/2019	0.4	0.6
Potato*Conservação Vermelha*SP*Medium*Bag*EUR/kg	Trás-os-Montes	Chaves	04/02/2019	0.4	0.6
Potato*Conservação Vermelha*SP*Medium*Bag*EUR/kg	Trás-os-Montes	Chaves	11/02/2019	0.4	0.6
Potato*Conservação Vermelha*SP*Medium*Bag*EUR/kg	Trás-os-Montes	Chaves	18/02/2019	0.4	0.6
Potato*Conservação Vermelha*SP*Medium*Bag*EUR/kg	Trás-os-Montes	Chaves	25/02/2019	0.4	0.6
Potato*Primor/Nova Branca*SE*Medium*Bag*EUR/kg	Algarve	Algarve	07/01/2019	0.8	0.7
Potato*Primor/Nova Branca*SE*Medium*Bag*EUR/kg	Algarve	Algarve	14/01/2019	0.8	0.7
Potato*Primor/Nova Branca*SE*Medium*Bag*EUR/kg	Algarve	Algarve	21/01/2019	0.8	0.7
Potato*Primor/Nova Branca*SE*Medium*Bag*EUR/kg	Algarve	Algarve	28/01/2019	0.8	0.7
Potato*Primor/Nova Branca*SE*Medium*Bag*EUR/kg	Algarve	Algarve	04/02/2019	0.8	0.7
Potato*Primor/Nova Branca*SE*Medium*Bag*EUR/kg	Algarve	Algarve	11/02/2019	0.8	0.7

Continued on following page

Table 3. Continued

Product	Region	Market	Period	Value	Normalised Value
					For All Periods in Each Product
Potato*Primor/Nova Branca*SE*Medium*Bag*EUR/kg	Algarve	Algarve	18/02/2019	0.8	0.7
Potato*Primor/Nova Branca*SE*Medium*Bag*EUR/kg	Algarve	Algarve	25/02/2019	0.8	0.7
Potato*Primor/Nova Branca*SE*Small*Bag*EUR/kg	Algarve	Algarve	07/01/2019	0.8	0.6
Potato*Primor/Nova Branca*SE*Small*Bag*EUR/kg	Algarve	Algarve	14/01/2019	0.8	0.6
Potato*Primor/Nova Branca*SE*Small*Bag*EUR/kg	Algarve	Algarve	21/01/2019	0.8	0.6
Potato*Primor/Nova Branca*SE*Small*Bag*EUR/kg	Algarve	Algarve	28/01/2019	0.8	0.6
Potato*Primor/Nova Branca*SE*Small*Bag*EUR/kg	Algarve	Algarve	04/02/2019	0.8	0.6
Potato*Primor/Nova Branca*SE*Small*Bag*EUR/kg	Algarve	Algarve	11/02/2019	0.8	0.6
Potato*Primor/Nova Branca*SE*Small*Bag*EUR/kg	Algarve	Algarve	18/02/2019	0.8	0.6
Potato*Primor/Nova Branca*SE*Small*Bag*EUR/kg	Algarve	Algarve	25/02/2019	0.8	0.6

Table 4. Potato producer prices at the beginning of the year for 2020

Product	Region	Market	Period	Value	Normalised Value for All Periods in Each Product
Potato*Conservação Branca*SP*Medium*Bag*EUR/kg	Trás-os-Montes	Chaves	06/01/2020	0.2	0.0
Potato*Conservação Branca*SP*Medium*Bag*EUR/kg	Trás-os-Montes	Chaves	13/01/2020	0.2	0.0
Potato*Conservação Branca*SP*Medium*Bag*EUR/kg	Trás-os-Montes	Chaves	20/01/2020	0.2	0.0
Potato*Conservação Branca*SP*Medium*Bag*EUR/kg	Trás-os-Montes	Chaves	27/01/2020	0.3	0.3
Potato*Conservação Branca*SP*Medium*Bag*EUR/kg	Trás-os-Montes	Chaves	03/02/2020	0.3	0.3
Potato*Conservação Branca*SP*Medium*Bag*EUR/kg	Trás-os-Montes	Chaves	10/02/2020	0.3	0.3
Potato*Conservação Branca*SP*Medium*Bag*EUR/kg	Trás-os-Montes	Chaves	17/02/2020	0.3	0.3
Potato*Conservação Branca*SP*Medium*Bag*EUR/kg	Trás-os-Montes	Chaves	24/02/2020	0.3	0.3
Potato*Conservação Vermelha*SP*Medium*Bag*EUR/kg	Trás-os-Montes	Chaves	06/01/2020	0.2	0.2
Potato*Conservação Vermelha*SP*Medium*Bag*EUR/kg	Trás-os-Montes	Chaves	13/01/2020	0.2	0.2
Potato*Conservação Vermelha*SP*Medium*Bag*EUR/kg	Trás-os-Montes	Chaves	20/01/2020	0.2	0.2
Potato*Conservação Vermelha*SP*Medium*Bag*EUR/kg	Trás-os-Montes	Chaves	27/01/2020	0.2	0.2
Potato*Conservação Vermelha*SP*Medium*Bag*EUR/kg	Trás-os-Montes	Chaves	03/02/2020	0.2	0.2
Potato*Conservação Vermelha*SP*Medium*Bag*EUR/kg	Trás-os-Montes	Chaves	10/02/2020	0.2	0.2
Potato*Conservação Vermelha*SP*Medium*Bag*EUR/kg	Trás-os-Montes	Chaves	17/02/2020	0.2	0.2
Potato*Conservação Vermelha*SP*Medium*Bag*EUR/kg	Trás-os-Montes	Chaves	24/02/2020	0.2	0.2
Potato*Primor/Nova Branca*SE*Medium*Bag*EUR/kg	Algarve	Algarve	06/01/2020	0.8	0.7
Potato*Primor/Nova Branca*SE*Medium*Bag*EUR/kg	Algarve	Algarve	13/01/2020	0.8	0.7
Potato*Primor/Nova Branca*SE*Medium*Bag*EUR/kg	Algarve	Algarve	20/01/2020	0.8	0.7
Potato*Primor/Nova Branca*SE*Medium*Bag*EUR/kg	Algarve	Algarve	27/01/2020	0.8	0.7
Potato*Primor/Nova Branca*SE*Medium*Bag*EUR/kg	Algarve	Algarve	03/02/2020	0.8	0.7
Potato*Primor/Nova Branca*SE*Medium*Bag*EUR/kg	Algarve	Algarve	10/02/2020	0.8	0.7
Potato*Primor/Nova Branca*SE*Medium*Bag*EUR/kg	Algarve	Algarve	17/02/2020	0.8	0.7
Potato*Primor/Nova Branca*SE*Medium*Bag*EUR/kg	Algarve	Algarve	24/02/2020	0.8	0.7

Table 5. Potato producer prices at the beginning of the year for 2021

Product	Region	Market	Period	Value	Normalised Value
					For All Periods in Each Product
Potato*Conservação Branca*SP*Medium*Bag*EUR/kg	Trás-os-Montes	Chaves	04/01/2021	0.5	0.8
Potato*Conservação Branca*SP*Medium*Bag*EUR/kg	Trás-os-Montes	Chaves	11/01/2021	0.5	0.8
Potato*Conservação Branca*SP*Medium*Bag*EUR/kg	Trás-os-Montes	Chaves	18/01/2021	0.5	0.8
Potato*Conservação Branca*SP*Medium*Bag*EUR/kg	Trás-os-Montes	Chaves	25/01/2021	0.4	0.5
Potato*Conservação Branca*SP*Medium*Bag*EUR/kg	Trás-os-Montes	Chaves	01/02/2021	0.4	0.5
Potato*Conservação Branca*SP*Medium*Bag*EUR/kg	Trás-os-Montes	Chaves	08/02/2021	0.4	0.5
Potato*Conservação Branca*SP*Medium*Bag*EUR/kg	Trás-os-Montes	Chaves	15/02/2021	0.4	0.5
Potato*Conservação Branca*SP*Medium*Bag*EUR/kg	Trás-os-Montes	Chaves	22/02/2021	0.4	0.5
Potato*Conservação Vermelha*SP*Medium*Bag*EUR/kg	Trás-os-Montes	Chaves	04/01/2021	0.4	0.6
Potato*Conservação Vermelha*SP*Medium*Bag*EUR/kg	Trás-os-Montes	Chaves	11/01/2021	0.4	0.6
Potato*Conservação Vermelha*SP*Medium*Bag*EUR/kg	Trás-os-Montes	Chaves	18/01/2021	0.4	0.6
Potato*Conservação Vermelha*SP*Medium*Bag*EUR/kg	Trás-os-Montes	Chaves	25/01/2021	0.4	0.6
Potato*Conservação Vermelha*SP*Medium*Bag*EUR/kg	Trás-os-Montes	Chaves	01/02/2021	0.4	0.6
Potato*Conservação Vermelha*SP*Medium*Bag*EUR/kg	Trás-os-Montes	Chaves	08/02/2021	0.4	0.6
Potato*Conservação Vermelha*SP*Medium*Bag*EUR/kg	Trás-os-Montes	Chaves	15/02/2021	0.4	0.6
Potato*Conservação Vermelha*SP*Medium*Bag*EUR/kg	Trás-os-Montes	Chaves	22/02/2021	0.4	0.6
Potato*Primor/Nova Branca*SE*Medium*Bag*EUR/kg	Algarve	Algarve	04/01/2021	0.6	0.3
Potato*Primor/Nova Branca*SE*Medium*Bag*EUR/kg	Algarve	Algarve	11/01/2021	0.6	0.3
Potato*Primor/Nova Branca*SE*Medium*Bag*EUR/kg	Algarve	Algarve	18/01/2021	0.6	0.3
Potato*Primor/Nova Branca*SE*Medium*Bag*EUR/kg	Algarve	Algarve	25/01/2021	0.6	0.3
Potato*Primor/Nova Branca*SE*Medium*Bag*EUR/kg	Algarve	Algarve	01/02/2021	0.6	0.3
Potato*Primor/Nova Branca*SE*Medium*Bag*EUR/kg	Algarve	Algarve	08/02/2021	0.6	0.3
Potato*Primor/Nova Branca*SE*Medium*Bag*EUR/kg	Algarve	Algarve	15/02/2021	0.6	0.3
Potato*Primor/Nova Branca*SE*Medium*Bag*EUR/kg	Algarve	Algarve	22/02/2021	0.6	0.3

Table 6. Potato producer prices at the beginning of the year for 2022

Product	Region	Market	Period	Value	Normalised Value
					For All Periods in Each Product
Potato*Conservação Branca*SP*Medium*Bag*EUR/kg	Trás-os-Montes	Chaves	03/01/2022	0.4	0.5
Potato*Conservação Branca*SP*Medium*Bag*EUR/kg	Trás-os-Montes	Chaves	10/01/2022	0.4	0.5
Potato*Conservação Branca*SP*Medium*Bag*EUR/kg	Trás-os-Montes	Chaves	17/01/2022	0.4	0.5
Potato*Conservação Branca*SP*Medium*Bag*EUR/kg	Trás-os-Montes	Chaves	24/01/2022	0.4	0.5
Potato*Conservação Branca*SP*Medium*Bag*EUR/kg	Trás-os-Montes	Chaves	31/01/2022	0.4	0.5
Potato*Conservação Branca*SP*Medium*Bag*EUR/kg	Trás-os-Montes	Chaves	07/02/2022	0.4	0.5
Potato*Conservação Branca*SP*Medium*Bag*EUR/kg	Trás-os-Montes	Chaves	14/02/2022	0.4	0.5
Potato*Conservação Branca*SP*Medium*Bag*EUR/kg	Trás-os-Montes	Chaves	21/02/2022	0.4	0.5
Potato*Conservação Branca*SP*Medium*Bag*EUR/kg	Trás-os-Montes	Chaves	28/02/2022	0.4	0.5
Potato*Conservação Vermelha*SP*Medium*Bag*EUR/kg	Trás-os-Montes	Chaves	03/01/2022	0.4	0.6
Potato*Conservação Vermelha*SP*Medium*Bag*EUR/kg	Trás-os-Montes	Chaves	10/01/2022	0.4	0.6
Potato*Conservação Vermelha*SP*Medium*Bag*EUR/kg	Trás-os-Montes	Chaves	17/01/2022	0.4	0.6
Potato*Conservação Vermelha*SP*Medium*Bag*EUR/kg	Trás-os-Montes	Chaves	24/01/2022	0.4	0.6
Potato*Conservação Vermelha*SP*Medium*Bag*EUR/kg	Trás-os-Montes	Chaves	31/01/2022	0.4	0.6
Potato*Conservação Vermelha*SP*Medium*Bag*EUR/kg	Trás-os-Montes	Chaves	07/02/2022	0.4	0.6
Potato*Conservação Vermelha*SP*Medium*Bag*EUR/kg	Trás-os-Montes	Chaves	14/02/2022	0.4	0.6
Potato*Conservação Vermelha*SP*Medium*Bag*EUR/kg	Trás-os-Montes	Chaves	21/02/2022	0.4	0.6
Potato*Conservação Vermelha*SP*Medium*Bag*EUR/kg	Trás-os-Montes	Chaves	28/02/2022	0.4	0.6
Potato*Primor/Nova Branca*SE*Medium*Bag*EUR/kg	Algarve	Algarve	03/01/2022	0.9	0.8
Potato*Primor/Nova Branca*SE*Medium*Bag*EUR/kg	Algarve	Algarve	10/01/2022	0.9	0.8
Potato*Primor/Nova Branca*SE*Medium*Bag*EUR/kg	Algarve	Algarve	17/01/2022	0.9	0.8
Potato*Primor/Nova Branca*SE*Medium*Bag*EUR/kg	Algarve	Algarve	24/01/2022	0.9	0.8

Continued on following page

Table 6. Continued

Product	Region	Market	Period	Value	Normalised Value
					For All Periods in Each Product
Potato*Primor/Nova Branca*SE*Medium*Bag*EUR/kg	Algarve	Algarve	31/01/2022	0.9	0.8
Potato*Primor/Nova Branca*SE*Medium*Bag*EUR/kg	Algarve	Algarve	07/02/2022	0.9	0.8
Potato*Primor/Nova Branca*SE*Medium*Bag*EUR/kg	Algarve	Algarve	14/02/2022	0.9	0.8
Potato*Primor/Nova Branca*SE*Medium*Bag*EUR/kg	Algarve	Algarve	21/02/2022	0.9	0.8
Potato*Primor/Nova Branca*SE*Medium*Bag*EUR/kg	Algarve	Algarve	28/02/2022	0.9	0.8
Potato*Primor/Nova Branca*SE*Small*Bag*EUR/kg	Algarve	Algarve	03/01/2022	0.6	0.2
Potato*Primor/Nova Branca*SE*Small*Bag*EUR/kg	Algarve	Algarve	10/01/2022	0.6	0.2
Potato*Primor/Nova Branca*SE*Small*Bag*EUR/kg	Algarve	Algarve	17/01/2022	0.6	0.2
Potato*Primor/Nova Branca*SE*Small*Bag*EUR/kg	Algarve	Algarve	24/01/2022	0.6	0.2
Potato*Primor/Nova Branca*SE*Small*Bag*EUR/kg	Algarve	Algarve	31/01/2022	0.6	0.2
Potato*Primor/Nova Branca*SE*Small*Bag*EUR/kg	Algarve	Algarve	07/02/2022	0.6	0.2
Potato*Primor/Nova Branca*SE*Small*Bag*EUR/kg	Algarve	Algarve	14/02/2022	0.6	0.2
Potato*Primor/Nova Branca*SE*Small*Bag*EUR/kg	Algarve	Algarve	21/02/2022	0.6	0.2
Potato*Primor/Nova Branca*SE*Small*Bag*EUR/kg	Algarve	Algarve	28/02/2022	0.6	0.2

Table 7. Potato producer prices in the middle of the year for 2017

Product	Region	Market	Period	Value	Normalised Value
					For All Periods in Each Product
Potato*Conservação Branca*SP*Medium*Bag*EUR/kg	Trás-os-Montes	Chaves	06/03/2017	0.4	0.5
Potato*Conservação Branca*SP*Medium*Bag*EUR/kg	Trás-os-Montes	Chaves	13/03/2017	0.4	0.5
Potato*Conservação Branca*SP*Medium*Bag*EUR/kg	Trás-os-Montes	Chaves	20/03/2017	0.4	0.5
Potato*Conservação Branca*SP*Medium*Bag*EUR/kg	Trás-os-Montes	Chaves	27/03/2017	0.4	0.5
Potato*Conservação Branca*SP*Medium*Bag*EUR/kg	Trás-os-Montes	Chaves	03/04/2017	0.4	0.5
Potato*Conservação Branca*SP*Medium*Bag*EUR/kg	Trás-os-Montes	Chaves	10/04/2017	0.4	0.5
Potato*Conservação Branca*SP*Medium*Bag*EUR/kg	Trás-os-Montes	Chaves	17/04/2017	0.4	0.5
Potato*Conservação Branca*SP*Medium*Bag*EUR/kg	Trás-os-Montes	Chaves	24/04/2017	0.4	0.5
Potato*Conservação Vermelha*SP*Medium*Bag*EUR/kg	Trás-os-Montes	Chaves	06/03/2017	0.3	0.4
Potato*Conservação Vermelha*SP*Medium*Bag*EUR/kg	Trás-os-Montes	Chaves	13/03/2017	0.3	0.4
Potato*Conservação Vermelha*SP*Medium*Bag*EUR/kg	Trás-os-Montes	Chaves	20/03/2017	0.3	0.4
Potato*Conservação Vermelha*SP*Medium*Bag*EUR/kg	Trás-os-Montes	Chaves	27/03/2017	0.3	0.4
Potato*Conservação Vermelha*SP*Medium*Bag*EUR/kg	Trás-os-Montes	Chaves	03/04/2017	0.3	0.4
Potato*Conservação Vermelha*SP*Medium*Bag*EUR/kg	Trás-os-Montes	Chaves	10/04/2017	0.3	0.4
Potato*Conservação Vermelha*SP*Medium*Bag*EUR/kg	Trás-os-Montes	Chaves	17/04/2017	0.3	0.4
Potato*Conservação Vermelha*SP*Medium*Bag*EUR/kg	Trás-os-Montes	Chaves	24/04/2017	0.3	0.4
Potato*Primor/Nova Branca*SE*Medium*Bag*EUR/kg	Algarve	Algarve	06/03/2017	0.6	0.3
Potato*Primor/Nova Branca*SE*Medium*Bag*EUR/kg	Algarve	Algarve	13/03/2017	0.6	0.3
Potato*Primor/Nova Branca*SE*Medium*Bag*EUR/kg	Algarve	Algarve	20/03/2017	0.6	0.3
Potato*Primor/Nova Branca*SE*Medium*Bag*EUR/kg	Algarve	Algarve	27/03/2017	0.6	0.3
Potato*Primor/Nova Branca*SE*Medium*Bag*EUR/kg	Algarve	Algarve	03/04/2017	0.6	0.3

Continued on following page

Table 7. Continued

Product	Region	Market	Period	Value	Normalised Value
					For All Periods in Each Product
Potato*Primor/Nova Branca*SE*Medium*Bag*EUR/kg	Algarve	Algarve	10/04/2017	0.6	0.3
Potato*Primor/Nova Branca*SE*Medium*Bag*EUR/kg	Algarve	Algarve	17/04/2017	0.7	0.5
Potato*Primor/Nova Branca*SE*Medium*Bag*EUR/kg	Algarve	Algarve	24/04/2017	0.7	0.5
Potato*Primor/Nova Branca*SE*Medium*Bag*EUR/kg	Algarve	Algarve	01/05/2017	0.7	0.5
Potato*Primor/Nova Branca*SE*Medium*Bag*EUR/kg	Algarve	Algarve	08/05/2017	0.7	0.5
Potato*Primor/Nova Branca*SE*Medium*Bag*EUR/kg	Algarve	Algarve	15/05/2017	0.7	0.5
Potato*Primor/Nova Branca*SE*Medium*Bag*EUR/kg	Algarve	Algarve	22/05/2017	0.6	0.3
Potato*Primor/Nova Branca*SE*Medium*Bag*EUR/kg	Algarve	Algarve	29/05/2017	0.6	0.3
Potato*Primor/Nova Branca*SE*Medium*Bag*EUR/kg	Algarve	Algarve	05/06/2017	0.6	0.3
Potato*Primor/Nova Branca*SE*Medium*Bag*EUR/kg	Algarve	Algarve	12/06/2017	0.6	0.3
Potato*Primor/Nova Branca*SE*Medium*Bag*EUR/kg	Algarve	Algarve	19/06/2017	0.6	0.3
Potato*Primor/Nova Branca*SE*Medium*Bag*EUR/kg	Algarve	Algarve	26/06/2017	0.6	0.3
Potato*Primor/Nova Branca*SE*Medium*Bag*EUR/kg	Algarve	Algarve	03/07/2017	0.5	0.2
Potato*Primor/Nova Branca*SE*Medium*Bag*EUR/kg	Algarve	Algarve	10/07/2017	0.5	0.2
Potato*Primor/Nova Branca*SE*Medium*Bag*EUR/kg	Algarve	Algarve	17/07/2017	0.5	0.2
Potato*Primor/Nova Branca*SE*Medium*Bag*EUR/kg	Algarve	Algarve	24/07/2017	0.5	0.2
Potato*Primor/Nova Branca*SE*Medium*Bag*EUR/kg	Algarve	Algarve	31/07/2017	0.5	0.2

Table 8. Potato producer prices in the middle of the year for 2018

Product	Region	Market	Period	Value	Normalised Value
					For All Periods in Each Product
Potato*Conservação Branca*SP*Medium*Bag*EUR/kg	Trás-os-Montes	Chaves	05/03/2018	0.2	0.0
Potato*Conservação Branca*SP*Medium*Bag*EUR/kg	Trás-os-Montes	Chaves	12/03/2018	0.2	0.0
Potato*Conservação Branca*SP*Medium*Bag*EUR/kg	Trás-os-Montes	Chaves	19/03/2018	0.2	0.0
Potato*Conservação Branca*SP*Medium*Bag*EUR/kg	Trás-os-Montes	Chaves	26/03/2018	0.2	0.0
Potato*Conservação Branca*SP*Medium*Bag*EUR/kg	Trás-os-Montes	Chaves	02/04/2018	0.2	0.0
Potato*Conservação Branca*SP*Medium*Bag*EUR/kg	Trás-os-Montes	Chaves	09/04/2018	0.2	0.0
Potato*Conservação Branca*SP*Medium*Bag*EUR/kg	Trás-os-Montes	Chaves	16/04/2018	0.2	0.0
Potato*Conservação Branca*SP*Medium*Bag*EUR/kg	Trás-os-Montes	Chaves	23/04/2018	0.2	0.0
Potato*Conservação Branca*SP*Medium*Bag*EUR/kg	Trás-os-Montes	Chaves	30/04/2018	0.2	0.0
Potato*Conservação Branca*SP*Medium*Bag*EUR/kg	Trás-os-Montes	Chaves	07/05/2018	0.2	0.0
Potato*Conservação Branca*SP*Medium*Bag*EUR/kg	Trás-os-Montes	Chaves	14/05/2018	0.2	0.0
Potato*Conservação Vermelha*SP*Medium*Bag*EUR/kg	Trás-os-Montes	Chaves	05/03/2018	0.1	0.0
Potato*Conservação Vermelha*SP*Medium*Bag*EUR/kg	Trás-os-Montes	Chaves	12/03/2018	0.1	0.0
Potato*Conservação Vermelha*SP*Medium*Bag*EUR/kg	Trás-os-Montes	Chaves	19/03/2018	0.1	0.0
Potato*Conservação Vermelha*SP*Medium*Bag*EUR/kg	Trás-os-Montes	Chaves	26/03/2018	0.1	0.0
Potato*Conservação Vermelha*SP*Medium*Bag*EUR/kg	Trás-os-Montes	Chaves	02/04/2018	0.1	0.0
Potato*Conservação Vermelha*SP*Medium*Bag*EUR/kg	Trás-os-Montes	Chaves	09/04/2018	0.1	0.0
Potato*Conservação Vermelha*SP*Medium*Bag*EUR/kg	Trás-os-Montes	Chaves	16/04/2018	0.1	0.0
Potato*Conservação Vermelha*SP*Medium*Bag*EUR/kg	Trás-os-Montes	Chaves	23/04/2018	0.1	0.0
Potato*Conservação Vermelha*SP*Medium*Bag*EUR/kg	Trás-os-Montes	Chaves	30/04/2018	0.1	0.0
Potato*Conservação Vermelha*SP*Medium*Bag*EUR/kg	Trás-os-Montes	Chaves	07/05/2018	0.1	0.0
Potato*Conservação Vermelha*SP*Medium*Bag*EUR/kg	Trás-os-Montes	Chaves	14/05/2018	0.1	0.0

Continued on following page

Table 8. Continued

Product	Region	Market	Period	Value	Normalised Value
					For All Periods in Each Product
Potato*Primor/Nova Branca*SE*Medium*Bag*EUR/kg	Algarve	Algarve	05/03/2018	0.7	0.5
Potato*Primor/Nova Branca*SE*Medium*Bag*EUR/kg	Algarve	Algarve	12/03/2018	0.7	0.5
Potato*Primor/Nova Branca*SE*Medium*Bag*EUR/kg	Algarve	Algarve	19/03/2018	0.7	0.5
Potato*Primor/Nova Branca*SE*Medium*Bag*EUR/kg	Algarve	Algarve	26/03/2018	0.6	0.3
Potato*Primor/Nova Branca*SE*Medium*Bag*EUR/kg	Algarve	Algarve	02/04/2018	0.6	0.3
Potato*Primor/Nova Branca*SE*Medium*Bag*EUR/kg	Algarve	Algarve	09/04/2018	0.6	0.3
Potato*Primor/Nova Branca*SE*Medium*Bag*EUR/kg	Algarve	Algarve	16/04/2018	0.6	0.3
Potato*Primor/Nova Branca*SE*Medium*Bag*EUR/kg	Algarve	Algarve	23/04/2018	0.6	0.3
Potato*Primor/Nova Branca*SE*Medium*Bag*EUR/kg	Algarve	Algarve	30/04/2018	0.6	0.3
Potato*Primor/Nova Branca*SE*Medium*Bag*EUR/kg	Algarve	Algarve	07/05/2018	0.6	0.3
Potato*Primor/Nova Branca*SE*Medium*Bag*EUR/kg	Algarve	Algarve	14/05/2018	0.6	0.3
Potato*Primor/Nova Branca*SE*Medium*Bag*EUR/kg	Algarve	Algarve	21/05/2018	0.6	0.3
Potato*Primor/Nova Branca*SE*Medium*Bag*EUR/kg	Algarve	Algarve	28/05/2018	0.6	0.3
Potato*Primor/Nova Branca*SE*Medium*Bag*EUR/kg	Algarve	Algarve	04/06/2018	0.6	0.3
Potato*Primor/Nova Branca*SE*Medium*Bag*EUR/kg	Algarve	Algarve	11/06/2018	0.6	0.3
Potato*Primor/Nova Branca*SE*Medium*Bag*EUR/kg	Algarve	Algarve	18/06/2018	0.6	0.3
Potato*Primor/Nova Branca*SE*Medium*Bag*EUR/kg	Algarve	Algarve	25/06/2018	0.6	0.3
Potato*Primor/Nova Branca*SE*Medium*Bag*EUR/kg	Algarve	Algarve	02/07/2018	0.6	0.3
Potato*Primor/Nova Branca*SE*Medium*Bag*EUR/kg	Algarve	Algarve	09/07/2018	0.6	0.3
Potato*Primor/Nova Branca*SE*Medium*Bag*EUR/kg	Algarve	Algarve	16/07/2018	0.6	0.3
Potato*Primor/Nova Branca*SE*Medium*Bag*EUR/kg	Algarve	Algarve	23/07/2018	0.6	0.3
Potato*Primor/Nova Branca*SE*Medium*Bag*EUR/kg	Algarve	Algarve	30/07/2018	0.6	0.3

Table 9. Potato producer prices in the middle of the year for 2019

Product	Region	Market	Period	Value	Normalised Value
					For All Periods in Each Product
Potato*Conservação Branca*SP*Medium*Bag*EUR/kg	Trás-os-Montes	Chaves	04/03/2019	0.5	0.8
Potato*Conservação Branca*SP*Medium*Bag*EUR/kg	Trás-os-Montes	Chaves	11/03/2019	0.5	0.8
Potato*Conservação Branca*SP*Medium*Bag*EUR/kg	Trás-os-Montes	Chaves	18/03/2019	0.5	0.8
Potato*Conservação Branca*SP*Medium*Bag*EUR/kg	Trás-os-Montes	Chaves	25/03/2019	0.5	0.8
Potato*Conservação Branca*SP*Medium*Bag*EUR/kg	Trás-os-Montes	Chaves	01/04/2019	0.5	0.8
Potato*Conservação Branca*SP*Medium*Bag*EUR/kg	Trás-os-Montes	Chaves	08/04/2019	0.5	0.8
Potato*Conservação Branca*SP*Medium*Bag*EUR/kg	Trás-os-Montes	Chaves	15/04/2019	0.5	0.8
Potato*Conservação Vermelha*SP*Medium*Bag*EUR/kg	Trás-os-Montes	Chaves	04/03/2019	0.4	0.6
Potato*Conservação Vermelha*SP*Medium*Bag*EUR/kg	Trás-os-Montes	Chaves	11/03/2019	0.4	0.6
Potato*Conservação Vermelha*SP*Medium*Bag*EUR/kg	Trás-os-Montes	Chaves	18/03/2019	0.4	0.6
Potato*Conservação Vermelha*SP*Medium*Bag*EUR/kg	Trás-os-Montes	Chaves	25/03/2019	0.4	0.6
Potato*Conservação Vermelha*SP*Medium*Bag*EUR/kg	Trás-os-Montes	Chaves	01/04/2019	0.4	0.6
Potato*Conservação Vermelha*SP*Medium*Bag*EUR/kg	Trás-os-Montes	Chaves	08/04/2019	0.4	0.6
Potato*Conservação Vermelha*SP*Medium*Bag*EUR/kg	Trás-os-Montes	Chaves	15/04/2019	0.4	0.6
Potato*Primor/Nova Branca*SE*Medium*Bag*EUR/kg	Algarve	Algarve	04/03/2019	0.8	0.7
Potato*Primor/Nova Branca*SE*Medium*Bag*EUR/kg	Algarve	Algarve	11/03/2019	0.8	0.7
Potato*Primor/Nova Branca*SE*Medium*Bag*EUR/kg	Algarve	Algarve	18/03/2019	0.8	0.7
Potato*Primor/Nova Branca*SE*Medium*Bag*EUR/kg	Algarve	Algarve	25/03/2019	0.8	0.7
Potato*Primor/Nova Branca*SE*Medium*Bag*EUR/kg	Algarve	Algarve	01/04/2019	0.8	0.7
Potato*Primor/Nova Branca*SE*Medium*Bag*EUR/kg	Algarve	Algarve	08/04/2019	0.8	0.7
Potato*Primor/Nova Branca*SE*Medium*Bag*EUR/kg	Algarve	Algarve	15/04/2019	0.8	0.7
Potato*Primor/Nova Branca*SE*Medium*Bag*EUR/kg	Algarve	Algarve	22/04/2019	0.8	0.7

Continued on following page

Table 9. Continued

Product	Region	Market	Period	Value	Normalised Value
					For All Periods in Each Product
Potato*Primor/Nova Branca*SE*Medium*Bag*EUR/kg	Algarve	Algarve	29/04/2019	0.8	0.7
Potato*Primor/Nova Branca*SE*Medium*Bag*EUR/kg	Algarve	Algarve	06/05/2019	0.8	0.7
Potato*Primor/Nova Branca*SE*Medium*Bag*EUR/kg	Algarve	Algarve	13/05/2019	0.8	0.7
Potato*Primor/Nova Branca*SE*Medium*Bag*EUR/kg	Algarve	Algarve	20/05/2019	0.8	0.7
Potato*Primor/Nova Branca*SE*Medium*Bag*EUR/kg	Algarve	Algarve	27/05/2019	0.7	0.5
Potato*Primor/Nova Branca*SE*Medium*Bag*EUR/kg	Algarve	Algarve	03/06/2019	0.7	0.5
Potato*Primor/Nova Branca*SE*Medium*Bag*EUR/kg	Algarve	Algarve	10/06/2019	0.7	0.5
Potato*Primor/Nova Branca*SE*Medium*Bag*EUR/kg	Algarve	Algarve	17/06/2019	0.6	0.3
Potato*Primor/Nova Branca*SE*Medium*Bag*EUR/kg	Algarve	Algarve	24/06/2019	0.6	0.3
Potato*Primor/Nova Branca*SE*Medium*Bag*EUR/kg	Algarve	Algarve	01/07/2019	0.6	0.3
Potato*Primor/Nova Branca*SE*Medium*Bag*EUR/kg	Algarve	Algarve	08/07/2019	0.6	0.3
Potato*Primor/Nova Branca*SE*Medium*Bag*EUR/kg	Algarve	Algarve	15/07/2019	0.6	0.3
Potato*Primor/Nova Branca*SE*Medium*Bag*EUR/kg	Algarve	Algarve	22/07/2019	0.6	0.3
Potato*Primor/Nova Branca*SE*Medium*Bag*EUR/kg	Algarve	Algarve	29/07/2019	0.5	0.2
Potato*Primor/Nova Branca*SE*Small*Bag*EUR/kg	Algarve	Algarve	04/03/2019	0.8	0.6
Potato*Primor/Nova Branca*SE*Small*Bag*EUR/kg	Algarve	Algarve	11/03/2019	0.8	0.6
Potato*Primor/Nova Branca*SE*Small*Bag*EUR/kg	Algarve	Algarve	18/03/2019	0.8	0.6
Potato*Primor/Nova Branca*SE*Small*Bag*EUR/kg	Algarve	Algarve	25/03/2019	0.8	0.6
Potato*Primor/Nova Branca*SE*Small*Bag*EUR/kg	Algarve	Algarve	01/04/2019	0.8	0.6
Potato*Primor/Nova Branca*SE*Small*Bag*EUR/kg	Algarve	Algarve	08/04/2019	0.8	0.6
Potato*Primor/Nova Branca*SE*Small*Bag*EUR/kg	Algarve	Algarve	15/04/2019	0.8	0.6
Potato*Primor/Nova Branca*SE*Small*Bag*EUR/kg	Algarve	Algarve	22/04/2019	0.8	0.6
Potato*Primor/Nova Branca*SE*Small*Bag*EUR/kg	Algarve	Algarve	29/04/2019	0.8	0.6
Potato*Primor/Nova Branca*SE*Small*Bag*EUR/kg	Algarve	Algarve	06/05/2019	0.8	0.6
Potato*Primor/Nova Branca*SE*Small*Bag*EUR/kg	Algarve	Algarve	13/05/2019	0.8	0.6
Potato*Primor/Nova Branca*SE*Small*Bag*EUR/kg	Algarve	Algarve	20/05/2019	0.8	0.6
Potato*Primor/Nova Branca*SE*Small*Bag*EUR/kg	Algarve	Algarve	27/05/2019	0.7	0.4

Continued on following page

Table 9. Continued

Product	Region	Market	Period	Value	Normalised Value
					For All Periods in Each Product
Potato*Primor/Nova Branca*SE*Small*Bag*EUR/kg	Algarve	Algarve	03/06/2019	0.7	0.4
Potato*Primor/Nova Branca*SE*Small*Bag*EUR/kg	Algarve	Algarve	10/06/2019	0.7	0.4
Potato*Primor/Nova Branca*SE*Small*Bag*EUR/kg	Algarve	Algarve	17/06/2019	0.6	0.2
Potato*Primor/Nova Branca*SE*Small*Bag*EUR/kg	Algarve	Algarve	24/06/2019	0.6	0.2
Potato*Primor/Nova Branca*SE*Small*Bag*EUR/kg	Algarve	Algarve	01/07/2019	0.6	0.2
Potato*Primor/Nova Branca*SE*Small*Bag*EUR/kg	Algarve	Algarve	08/07/2019	0.6	0.2
Potato*Primor/Nova Branca*SE*Small*Bag*EUR/kg	Algarve	Algarve	15/07/2019	0.6	0.2
Potato*Primor/Nova Branca*SE*Small*Bag*EUR/kg	Algarve	Algarve	22/07/2019	0.6	0.2
Potato*Primor/Nova Branca*SE*Small*Bag*EUR/kg	Algarve	Algarve	29/07/2019	0.5	0.0

Table 10. Potato producer prices in the middle of the year for 2020

Product	Region	Market	Period	Value	Normalised Value for All Periods in Each Product
Potato*Conservação Branca*SP*Medium*Bag*EUR/kg	Trás-os-Montes	Chaves	02/03/2020	0.3	0.3
Potato*Conservação Branca*SP*Medium*Bag*EUR/kg	Trás-os-Montes	Chaves	09/03/2020	0.3	0.3
Potato*Conservação Branca*SP*Medium*Bag*EUR/kg	Trás-os-Montes	Chaves	16/03/2020	0.3	0.3
Potato*Conservação Branca*SP*Medium*Bag*EUR/kg	Trás-os-Montes	Chaves	23/03/2020	0.3	0.3
Potato*Conservação Branca*SP*Medium*Bag*EUR/kg	Trás-os-Montes	Chaves	30/03/2020	0.3	0.3
Potato*Conservação Branca*SP*Medium*Bag*EUR/kg	Trás-os-Montes	Chaves	06/04/2020	0.3	0.3
Potato*Conservação Branca*SP*Medium*Bag*EUR/kg	Trás-os-Montes	Chaves	13/04/2020	0.3	0.3
Potato*Conservação Branca*SP*Medium*Bag*EUR/kg	Trás-os-Montes	Chaves	20/04/2020	0.3	0.3
Potato*Conservação Vermelha*SP*Medium*Bag*EUR/kg	Trás-os-Montes	Chaves	02/03/2020	0.2	0.2
Potato*Conservação Vermelha*SP*Medium*Bag*EUR/kg	Trás-os-Montes	Chaves	09/03/2020	0.2	0.2
Potato*Conservação Vermelha*SP*Medium*Bag*EUR/kg	Trás-os-Montes	Chaves	16/03/2020	0.2	0.2
Potato*Conservação Vermelha*SP*Medium*Bag*EUR/kg	Trás-os-Montes	Chaves	23/03/2020	0.2	0.2
Potato*Conservação Vermelha*SP*Medium*Bag*EUR/kg	Trás-os-Montes	Chaves	30/03/2020	0.2	0.2
Potato*Conservação Vermelha*SP*Medium*Bag*EUR/kg	Trás-os-Montes	Chaves	06/04/2020	0.2	0.2
Potato*Conservação Vermelha*SP*Medium*Bag*EUR/kg	Trás-os-Montes	Chaves	13/04/2020	0.2	0.2
Potato*Conservação Vermelha*SP*Medium*Bag*EUR/kg	Trás-os-Montes	Chaves	20/04/2020	0.2	0.2
Potato*Primor/Nova Branca*SE*Medium*Bag*EUR/kg	Algarve	Algarve	02/03/2020	0.8	0.7
Potato*Primor/Nova Branca*SE*Medium*Bag*EUR/kg	Algarve	Algarve	09/03/2020	0.8	0.7
Potato*Primor/Nova Branca*SE*Medium*Bag*EUR/kg	Algarve	Algarve	16/03/2020	0.8	0.7
Potato*Primor/Nova Branca*SE*Medium*Bag*EUR/kg	Algarve	Algarve	23/03/2020	0.8	0.7
Potato*Primor/Nova Branca*SE*Medium*Bag*EUR/kg	Algarve	Algarve	30/03/2020	1.0	1.0
Potato*Primor/Nova Branca*SE*Medium*Bag*EUR/kg	Algarve	Algarve	06/04/2020	1.0	1.0
Potato*Primor/Nova Branca*SE*Medium*Bag*EUR/kg	Algarve	Algarve	13/04/2020	1.0	1.0
Potato*Primor/Nova Branca*SE*Medium*Bag*EUR/kg	Algarve	Algarve	20/04/2020	1.0	1.0
Potato*Primor/Nova Branca*SE*Medium*Bag*EUR/kg	Algarve	Algarve	27/04/2020	1.0	1.0
Potato*Primor/Nova Branca*SE*Medium*Bag*EUR/kg	Algarve	Algarve	04/05/2020	1.0	1.0
Potato*Primor/Nova Branca*SE*Medium*Bag*EUR/kg	Algarve	Algarve	11/05/2020	1.0	1.0
Potato*Primor/Nova Branca*SE*Medium*Bag*EUR/kg	Algarve	Algarve	18/05/2020	1.0	1.0
Potato*Primor/Nova Branca*SE*Medium*Bag*EUR/kg	Algarve	Algarve	25/05/2020	1.0	1.0
Potato*Primor/Nova Branca*SE*Medium*Bag*EUR/kg	Algarve	Algarve	01/06/2020	0.9	0.8
Potato*Primor/Nova Branca*SE*Medium*Bag*EUR/kg	Algarve	Algarve	08/06/2020	0.8	0.7

Continued on following page

Table 10. Continued

Product	Region	Market	Period	Value	Normalised Value for All Periods in Each Product
Potato*Primor/Nova Branca*SE*Medium*Bag*EUR/kg	Algarve	Algarve	15/06/2020	0.8	0.7
Potato*Primor/Nova Branca*SE*Medium*Bag*EUR/kg	Algarve	Algarve	22/06/2020	0.7	0.5
Potato*Primor/Nova Branca*SE*Medium*Bag*EUR/kg	Algarve	Algarve	29/06/2020	0.7	0.5
Potato*Primor/Nova Branca*SE*Medium*Bag*EUR/kg	Algarve	Algarve	06/07/2020	0.7	0.5
Potato*Primor/Nova Branca*SE*Medium*Bag*EUR/kg	Algarve	Algarve	13/07/2020	0.7	0.5
Potato*Primor/Nova Branca*SE*Medium*Bag*EUR/kg	Algarve	Algarve	20/07/2020	0.7	0.5
Potato*Primor/Nova Branca*SE*Medium*Bag*EUR/kg	Algarve	Algarve	27/07/2020	0.7	0.5

Table 11. Potato producer prices in the middle of the year for 2021

Product	Region	Market	Period	Value	Normalised Value
					For All Periods in Each Product
Potato*Conservação Branca*SP*Medium*Bag*EUR/kg	Trás-os-Montes	Chaves	01/03/2021	0.4	0.5
Potato*Conservação Branca*SP*Medium*Bag*EUR/kg	Trás-os-Montes	Chaves	08/03/2021	0.4	0.5
Potato*Conservação Branca*SP*Medium*Bag*EUR/kg	Trás-os-Montes	Chaves	15/03/2021	0.4	0.5
Potato*Conservação Branca*SP*Medium*Bag*EUR/kg	Trás-os-Montes	Chaves	22/03/2021	0.4	0.5
Potato*Conservação Branca*SP*Medium*Bag*EUR/kg	Trás-os-Montes	Chaves	29/03/2021	0.4	0.5
Potato*Conservação Branca*SP*Medium*Bag*EUR/kg	Trás-os-Montes	Chaves	05/04/2021	0.4	0.5
Potato*Conservação Vermelha*SP*Medium*Bag*EUR/kg	Trás-os-Montes	Chaves	01/03/2021	0.4	0.6
Potato*Conservação Vermelha*SP*Medium*Bag*EUR/kg	Trás-os-Montes	Chaves	08/03/2021	0.4	0.6
Potato*Conservação Vermelha*SP*Medium*Bag*EUR/kg	Trás-os-Montes	Chaves	15/03/2021	0.4	0.6
Potato*Conservação Vermelha*SP*Medium*Bag*EUR/kg	Trás-os-Montes	Chaves	22/03/2021	0.4	0.6
Potato*Conservação Vermelha*SP*Medium*Bag*EUR/kg	Trás-os-Montes	Chaves	29/03/2021	0.4	0.6
Potato*Conservação Vermelha*SP*Medium*Bag*EUR/kg	Trás-os-Montes	Chaves	05/04/2021	0.4	0.6
Potato*Primor/Nova Branca*SE*Medium*Bag*EUR/kg	Algarve	Algarve	01/03/2021	0.6	0.3
Potato*Primor/Nova Branca*SE*Medium*Bag*EUR/kg	Algarve	Algarve	08/03/2021	0.6	0.3
Potato*Primor/Nova Branca*SE*Medium*Bag*EUR/kg	Algarve	Algarve	15/03/2021	0.6	0.3
Potato*Primor/Nova Branca*SE*Medium*Bag*EUR/kg	Algarve	Algarve	22/03/2021	0.6	0.3
Potato*Primor/Nova Branca*SE*Medium*Bag*EUR/kg	Algarve	Algarve	29/03/2021	0.6	0.3
Potato*Primor/Nova Branca*SE*Medium*Bag*EUR/kg	Algarve	Algarve	05/04/2021	0.8	0.7
Potato*Primor/Nova Branca*SE*Medium*Bag*EUR/kg	Algarve	Algarve	12/04/2021	0.9	0.8
Potato*Primor/Nova Branca*SE*Medium*Bag*EUR/kg	Algarve	Algarve	19/04/2021	0.9	0.8
Potato*Primor/Nova Branca*SE*Medium*Bag*EUR/kg	Algarve	Algarve	26/04/2021	0.9	0.8
Potato*Primor/Nova Branca*SE*Medium*Bag*EUR/kg	Algarve	Algarve	03/05/2021	0.9	0.8

Continued on following page

Table 11. Continued

Product	Region	Market	Period	Value	Normalised Value
					For All Periods in Each Product
Potato*Primor/Nova Branca*SE*Medium*Bag*EUR/kg	Algarve	Algarve	10/05/2021	0.8	0.7
Potato*Primor/Nova Branca*SE*Medium*Bag*EUR/kg	Algarve	Algarve	17/05/2021	0.8	0.7
Potato*Primor/Nova Branca*SE*Medium*Bag*EUR/kg	Algarve	Algarve	24/05/2021	0.8	0.7
Potato*Primor/Nova Branca*SE*Medium*Bag*EUR/kg	Algarve	Algarve	31/05/2021	0.8	0.7
Potato*Primor/Nova Branca*SE*Medium*Bag*EUR/kg	Algarve	Algarve	07/06/2021	0.8	0.7
Potato*Primor/Nova Branca*SE*Medium*Bag*EUR/kg	Algarve	Algarve	14/06/2021	0.8	0.7
Potato*Primor/Nova Branca*SE*Medium*Bag*EUR/kg	Algarve	Algarve	21/06/2021	0.8	0.7
Potato*Primor/Nova Branca*SE*Medium*Bag*EUR/kg	Algarve	Algarve	28/06/2021	0.7	0.5
Potato*Primor/Nova Branca*SE*Medium*Bag*EUR/kg	Algarve	Algarve	05/07/2021	0.7	0.5
Potato*Primor/Nova Branca*SE*Medium*Bag*EUR/kg	Algarve	Algarve	12/07/2021	0.7	0.5
Potato*Primor/Nova Branca*SE*Medium*Bag*EUR/kg	Algarve	Algarve	19/07/2021	0.7	0.5
Potato*Primor/Nova Branca*SE*Medium*Bag*EUR/kg	Algarve	Algarve	26/07/2021	0.7	0.5

POTATO PRODUCER PRICES IN THE MIDDLE OF THE YEAR FOR THE PERIOD CONSIDERED

A similar pattern was found for the months from March until July, in the years 2017-2022, to those identified in the previous section for the first two months of the years analysed. In fact, generally, the producer prices were higher for the potatoes from Algarve ("Potato*Primor/Nova Branca*SE*Medium*Bag"). In the potatoes from Trás-os-Montes, the prices are often higher for the "Potato*Conservação Branca*SP*Medium*Bag" and in the same cases equal (Tables 7-12). Again, the prices present some fluctuations, however, they do not seem a consequence of the Covid-19 pandemic and Russia-Ukraine conflict. The average prices for these potatoes, in the period considered, were 0.5 euros/kg in 2017, 0.4 euros/kg in 2018, 0.6 euros/kg in 2019, 0.6 euros/kg in 2020, 0.6 euros/kg in 2021 and 0.6 euros/kg in 2022.

Table 12. Potato producer prices in the middle of the year for 2022

Product	Region	Market	Period	Value	Normalised Value
					For All Periods in Each Product
Potato*Conservação Branca*SP*Medium*Bag*EUR/kg	Trás-os-Montes	Chaves	07/03/2022	0.4	0.5
Potato*Conservação Branca*SP*Medium*Bag*EUR/kg	Trás-os-Montes	Chaves	14/03/2022	0.4	0.5
Potato*Conservação Branca*SP*Medium*Bag*EUR/kg	Trás-os-Montes	Chaves	21/03/2022	0.4	0.5
Potato*Conservação Branca*SP*Medium*Bag*EUR/kg	Trás-os-Montes	Chaves	28/03/2022	0.4	0.5
Potato*Conservação Branca*SP*Medium*Bag*EUR/kg	Trás-os-Montes	Chaves	04/04/2022	0.4	0.5
Potato*Conservação Branca*SP*Medium*Bag*EUR/kg	Trás-os-Montes	Chaves	11/04/2022	0.4	0.5
Potato*Conservação Branca*SP*Medium*Bag*EUR/kg	Trás-os-Montes	Chaves	18/04/2022	0.4	0.5
Potato*Conservação Branca*SP*Medium*Bag*EUR/kg	Trás-os-Montes	Chaves	25/04/2022	0.4	0.5
Potato*Conservação Vermelha*SP*Medium*Bag*EUR/kg	Trás-os-Montes	Chaves	07/03/2022	0.4	0.6
Potato*Conservação Vermelha*SP*Medium*Bag*EUR/kg	Trás-os-Montes	Chaves	14/03/2022	0.4	0.6
Potato*Conservação Vermelha*SP*Medium*Bag*EUR/kg	Trás-os-Montes	Chaves	21/03/2022	0.4	0.6
Potato*Conservação Vermelha*SP*Medium*Bag*EUR/kg	Trás-os-Montes	Chaves	28/03/2022	0.4	0.6
Potato*Conservação Vermelha*SP*Medium*Bag*EUR/kg	Trás-os-Montes	Chaves	04/04/2022	0.4	0.6
Potato*Conservação Vermelha*SP*Medium*Bag*EUR/kg	Trás-os-Montes	Chaves	11/04/2022	0.4	0.6
Potato*Conservação Vermelha*SP*Medium*Bag*EUR/kg	Trás-os-Montes	Chaves	18/04/2022	0.4	0.6
Potato*Conservação Vermelha*SP*Medium*Bag*EUR/kg	Trás-os-Montes	Chaves	25/04/2022	0.4	0.6
Potato*Primor/Nova Branca*SE*Medium*Bag*EUR/kg	Algarve	Algarve	07/03/2022	0.9	0.8
Potato*Primor/Nova Branca*SE*Medium*Bag*EUR/kg	Algarve	Algarve	14/03/2022	0.9	0.8
Potato*Primor/Nova Branca*SE*Medium*Bag*EUR/kg	Algarve	Algarve	21/03/2022	0.9	0.8
Potato*Primor/Nova Branca*SE*Medium*Bag*EUR/kg	Algarve	Algarve	28/03/2022	0.9	0.8
Potato*Primor/Nova Branca*SE*Medium*Bag*EUR/kg	Algarve	Algarve	04/04/2022	0.9	0.8
Potato*Primor/Nova Branca*SE*Medium*Bag*EUR/kg	Algarve	Algarve	11/04/2022	0.9	0.8
Potato*Primor/Nova Branca*SE*Medium*Bag*EUR/kg	Algarve	Algarve	18/04/2022	0.9	0.8
Potato*Primor/Nova Branca*SE*Medium*Bag*EUR/kg	Algarve	Algarve	25/04/2022	0.9	0.8
Potato*Primor/Nova Branca*SE*Medium*Bag*EUR/kg	Algarve	Algarve	02/05/2022	0.8	0.7
Potato*Primor/Nova Branca*SE*Medium*Bag*EUR/kg	Algarve	Algarve	09/05/2022	0.8	0.7
Potato*Primor/Nova Branca*SE*Medium*Bag*EUR/kg	Algarve	Algarve	16/05/2022	0.8	0.7
Potato*Primor/Nova Branca*SE*Medium*Bag*EUR/kg	Algarve	Algarve	23/05/2022	0.8	0.7
Potato*Primor/Nova Branca*SE*Medium*Bag*EUR/kg	Algarve	Algarve	30/05/2022	0.8	0.7
Potato*Primor/Nova Branca*SE*Medium*Bag*EUR/kg	Algarve	Algarve	06/06/2022	0.8	0.7
Potato*Primor/Nova Branca*SE*Medium*Bag*EUR/kg	Algarve	Algarve	13/06/2022	0.8	0.7
Potato*Primor/Nova Branca*SE*Medium*Bag*EUR/kg	Algarve	Algarve	20/06/2022	0.8	0.7
Potato*Primor/Nova Branca*SE*Medium*Bag*EUR/kg	Algarve	Algarve	27/06/2022	0.8	0.7
Potato*Primor/Nova Branca*SE*Medium*Bag*EUR/kg	Algarve	Algarve	04/07/2022	0.8	0.7
Potato*Primor/Nova Branca*SE*Medium*Bag*EUR/kg	Algarve	Algarve	11/07/2022	0.8	0.7

Continued on following page

Table 12. Continued

Product	Region	Market	Period	Value	Normalised Value
					For All Periods in Each Product
Potato*Primor/Nova Branca*SE*Medium*Bag*EUR/kg	Algarve	Algarve	18/07/2022	0.8	0.7
Potato*Primor/Nova Branca*SE*Medium*Bag*EUR/kg	Algarve	Algarve	25/07/2022	0.8	0.7
Potato*Primor/Nova Branca*SE*Small*Bag*EUR/kg	Algarve	Algarve	07/03/2022	0.6	0.2
Potato*Primor/Nova Branca*SE*Small*Bag*EUR/kg	Algarve	Algarve	14/03/2022	0.6	0.2
Potato*Primor/Nova Branca*SE*Small*Bag*EUR/kg	Algarve	Algarve	21/03/2022	0.6	0.2
Potato*Primor/Nova Branca*SE*Small*Bag*EUR/kg	Algarve	Algarve	28/03/2022	0.6	0.2
Potato*Primor/Nova Branca*SE*Small*Bag*EUR/kg	Algarve	Algarve	04/04/2022	0.6	0.2
Potato*Primor/Nova Branca*SE*Small*Bag*EUR/kg	Algarve	Algarve	11/04/2022	0.6	0.2
Potato*Primor/Nova Branca*SE*Small*Bag*EUR/kg	Algarve	Algarve	18/04/2022	0.6	0.2
Potato*Primor/Nova Branca*SE*Small*Bag*EUR/kg	Algarve	Algarve	25/04/2022	0.6	0.2
Potato*Primor/Nova Branca*SE*Small*Bag*EUR/kg	Algarve	Algarve	02/05/2022	0.6	0.2
Potato*Primor/Nova Branca*SE*Small*Bag*EUR/kg	Algarve	Algarve	09/05/2022	0.6	0.2
Potato*Primor/Nova Branca*SE*Small*Bag*EUR/kg	Algarve	Algarve	16/05/2022	0.6	0.2
Potato*Primor/Nova Branca*SE*Small*Bag*EUR/kg	Algarve	Algarve	23/05/2022	0.6	0.2
Potato*Primor/Nova Branca*SE*Small*Bag*EUR/kg	Algarve	Algarve	30/05/2022	0.6	0.2
Potato*Primor/Nova Branca*SE*Small*Bag*EUR/kg	Algarve	Algarve	06/06/2022	0.6	0.2
Potato*Primor/Nova Branca*SE*Small*Bag*EUR/kg	Algarve	Algarve	13/06/2022	0.6	0.2
Potato*Primor/Nova Branca*SE*Small*Bag*EUR/kg	Algarve	Algarve	20/06/2022	0.6	0.2
Potato*Primor/Nova Branca*SE*Small*Bag*EUR/kg	Algarve	Algarve	27/06/2022	0.6	0.2
Potato*Primor/Nova Branca*SE*Small*Bag*EUR/kg	Algarve	Algarve	04/07/2022	0.6	0.2
Potato*Primor/Nova Branca*SE*Small*Bag*EUR/kg	Algarve	Algarve	11/07/2022	0.6	0.2
Potato*Primor/Nova Branca*SE*Small*Bag*EUR/kg	Algarve	Algarve	18/07/2022	0.6	0.2
Potato*Primor/Nova Branca*SE*Small*Bag*EUR/kg	Algarve	Algarve	25/07/2022	0.6	0.2

POTATO PRODUCER PRICES AT THE END OF THE YEAR FOR THE PERIOD CONSIDERED

The producer prices for the last months (August to December) of the years taken into account are presented in Tables 13-18. The average prices for potatoes "Potato*Conservação Branca*SP*Medium*Bag", "Potato*Conservação Vermelha*SP*Medium*Bag and "Potato*Primor/Nova Branca*SE*Medium*Bag" were, in euros/kg, in 2017-2022, respectively the following: 0.3; 0.4; 0.4; 0.5; 0.5; and 0.8. The prices verified at the end of 2022 seem, in fact, different from the other identified in the previous years, meaning that the Russia-Ukraine crisis appears to have impacted these prices.

Table 13. Potato producer prices at the end of the year for 2017

Product	Region	Market	Period	Value	Normalised Value For All Periods in Each Product
Potato*Conservação Branca*SP*Medium*Bag*EUR/kg	Trás-os-Montes	Chaves	02/10/2017	0.2	0.0
Potato*Conservação Branca*SP*Medium*Bag*EUR/kg	Trás-os-Montes	Chaves	09/10/2017	0.2	0.0
Potato*Conservação Branca*SP*Medium*Bag*EUR/kg	Trás-os-Montes	Chaves	16/10/2017	0.2	0.0
Potato*Conservação Branca*SP*Medium*Bag*EUR/kg	Trás-os-Montes	Chaves	23/10/2017	0.2	0.0
Potato*Conservação Branca*SP*Medium*Bag*EUR/kg	Trás-os-Montes	Chaves	30/10/2017	0.2	0.0
Potato*Conservação Branca*SP*Medium*Bag*EUR/kg	Trás-os-Montes	Chaves	06/11/2017	0.2	0.0
Potato*Conservação Branca*SP*Medium*Bag*EUR/kg	Trás-os-Montes	Chaves	13/11/2017	0.2	0.0
Potato*Conservação Branca*SP*Medium*Bag*EUR/kg	Trás-os-Montes	Chaves	20/11/2017	0.2	0.0
Potato*Conservação Branca*SP*Medium*Bag*EUR/kg	Trás-os-Montes	Chaves	27/11/2017	0.2	0.0
Potato*Conservação Branca*SP*Medium*Bag*EUR/kg	Trás-os-Montes	Chaves	04/12/2017	0.2	0.0
Potato*Conservação Branca*SP*Medium*Bag*EUR/kg	Trás-os-Montes	Chaves	11/12/2017	0.2	0.0
Potato*Conservação Branca*SP*Medium*Bag*EUR/kg	Trás-os-Montes	Chaves	18/12/2017	0.2	0.0
Potato*Conservação Branca*SP*Medium*Bag*EUR/kg	Trás-os-Montes	Chaves	25/12/2017	0.2	0.0
Potato*Conservação Vermelha*SP*Medium*Bag*EUR/kg	Trás-os-Montes	Chaves	02/10/2017	0.2	0.2
Potato*Conservação Vermelha*SP*Medium*Bag*EUR/kg	Trás-os-Montes	Chaves	09/10/2017	0.2	0.2
Potato*Conservação Vermelha*SP*Medium*Bag*EUR/kg	Trás-os-Montes	Chaves	16/10/2017	0.2	0.2
Potato*Conservação Vermelha*SP*Medium*Bag*EUR/kg	Trás-os-Montes	Chaves	23/10/2017	0.2	0.2
Potato*Conservação Vermelha*SP*Medium*Bag*EUR/kg	Trás-os-Montes	Chaves	30/10/2017	0.2	0.2
Potato*Conservação Vermelha*SP*Medium*Bag*EUR/kg	Trás-os-Montes	Chaves	06/11/2017	0.2	0.2
Potato*Conservação Vermelha*SP*Medium*Bag*EUR/kg	Trás-os-Montes	Chaves	13/11/2017	0.2	0.2
Potato*Conservação Vermelha*SP*Medium*Bag*EUR/kg	Trás-os-Montes	Chaves	20/11/2017	0.2	0.2
Potato*Conservação Vermelha*SP*Medium*Bag*EUR/kg	Trás-os-Montes	Chaves	27/11/2017	0.2	0.2
Potato*Conservação Vermelha*SP*Medium*Bag*EUR/kg	Trás-os-Montes	Chaves	04/12/2017	0.2	0.2
Potato*Conservação Vermelha*SP*Medium*Bag*EUR/kg	Trás-os-Montes	Chaves	11/12/2017	0.2	0.2
Potato*Conservação Vermelha*SP*Medium*Bag*EUR/kg	Trás-os-Montes	Chaves	18/12/2017	0.2	0.2
Potato*Conservação Vermelha*SP*Medium*Bag*EUR/kg	Trás-os-Montes	Chaves	25/12/2017	0.2	0.2
Potato*Primor/Nova Branca*SE*Medium*Bag*EUR/kg	Algarve	Algarve	07/08/2017	0.5	0.2

Continued on following page

Table 13. Continued

Product	Region	Market	Period	Value	Normalised Value
					For All Periods in Each Product
Potato*Primor/Nova Branca*SE*Medium*Bag*EUR/kg	Algarve	Algarve	14/08/2017	0.5	0.2
Potato*Primor/Nova Branca*SE*Medium*Bag*EUR/kg	Algarve	Algarve	21/08/2017	0.5	0.2
Potato*Primor/Nova Branca*SE*Medium*Bag*EUR/kg	Algarve	Algarve	06/11/2017	0.7	0.5
Potato*Primor/Nova Branca*SE*Medium*Bag*EUR/kg	Algarve	Algarve	13/11/2017	0.7	0.5
Potato*Primor/Nova Branca*SE*Medium*Bag*EUR/kg	Algarve	Algarve	20/11/2017	0.7	0.5
Potato*Primor/Nova Branca*SE*Medium*Bag*EUR/kg	Algarve	Algarve	27/11/2017	0.7	0.5
Potato*Primor/Nova Branca*SE*Medium*Bag*EUR/kg	Algarve	Algarve	04/12/2017	0.7	0.5
Potato*Primor/Nova Branca*SE*Medium*Bag*EUR/kg	Algarve	Algarve	11/12/2017	0.7	0.5
Potato*Primor/Nova Branca*SE*Medium*Bag*EUR/kg	Algarve	Algarve	18/12/2017	0.7	0.5
Potato*Primor/Nova Branca*SE*Medium*Bag*EUR/kg	Algarve	Algarve	25/12/2017	0.7	0.5

Table 14. Potato producer prices at the end of the year for 2018

Product	Region	Market	Period	Value	Normalised Value
					For All Periods in Each Product
Potato*Conservação Branca*SP*Medium*Bag*EUR/kg	Trás-os-Montes	Chaves	15/10/2018	0.3	0.3
Potato*Conservação Branca*SP*Medium*Bag*EUR/kg	Trás-os-Montes	Chaves	22/10/2018	0.3	0.3
Potato*Conservação Branca*SP*Medium*Bag*EUR/kg	Trás-os-Montes	Chaves	29/10/2018	0.3	0.3
Potato*Conservação Branca*SP*Medium*Bag*EUR/kg	Trás-os-Montes	Chaves	05/11/2018	0.3	0.3
Potato*Conservação Branca*SP*Medium*Bag*EUR/kg	Trás-os-Montes	Chaves	12/11/2018	0.3	0.3
Potato*Conservação Branca*SP*Medium*Bag*EUR/kg	Trás-os-Montes	Chaves	19/11/2018	0.3	0.3
Potato*Conservação Branca*SP*Medium*Bag*EUR/kg	Trás-os-Montes	Chaves	26/11/2018	0.3	0.3
Potato*Conservação Branca*SP*Medium*Bag*EUR/kg	Trás-os-Montes	Chaves	03/12/2018	0.3	0.3
Potato*Conservação Branca*SP*Medium*Bag*EUR/kg	Trás-os-Montes	Chaves	10/12/2018	0.3	0.3
Potato*Conservação Branca*SP*Medium*Bag*EUR/kg	Trás-os-Montes	Chaves	17/12/2018	0.3	0.3
Potato*Conservação Branca*SP*Medium*Bag*EUR/kg	Trás-os-Montes	Chaves	24/12/2018	0.3	0.3
Potato*Conservação Branca*SP*Medium*Bag*EUR/kg	Trás-os-Montes	Chaves	31/12/2018	0.3	0.3
Potato*Conservação Vermelha*SP*Medium*Bag*EUR/kg	Trás-os-Montes	Chaves	15/10/2018	0.2	0.2
Potato*Conservação Vermelha*SP*Medium*Bag*EUR/kg	Trás-os-Montes	Chaves	22/10/2018	0.2	0.2
Potato*Conservação Vermelha*SP*Medium*Bag*EUR/kg	Trás-os-Montes	Chaves	29/10/2018	0.2	0.2
Potato*Conservação Vermelha*SP*Medium*Bag*EUR/kg	Trás-os-Montes	Chaves	05/11/2018	0.2	0.2
Potato*Conservação Vermelha*SP*Medium*Bag*EUR/kg	Trás-os-Montes	Chaves	12/11/2018	0.3	0.4
Potato*Conservação Vermelha*SP*Medium*Bag*EUR/kg	Trás-os-Montes	Chaves	19/11/2018	0.3	0.4
Potato*Conservação Vermelha*SP*Medium*Bag*EUR/kg	Trás-os-Montes	Chaves	26/11/2018	0.3	0.4
Potato*Conservação Vermelha*SP*Medium*Bag*EUR/kg	Trás-os-Montes	Chaves	03/12/2018	0.3	0.4
Potato*Conservação Vermelha*SP*Medium*Bag*EUR/kg	Trás-os-Montes	Chaves	10/12/2018	0.3	0.4
Potato*Conservação Vermelha*SP*Medium*Bag*EUR/kg	Trás-os-Montes	Chaves	17/12/2018	0.3	0.4

Continued on following page

Table 14. Continued

Product	Region	Market	Period	Value	Normalised Value For All Periods in Each Product
Potato*Conservação Vermelha*SP*Medium*Bag*EUR/kg	Trás-os-Montes	Chaves	24/12/2018	0.3	0.4
Potato*Conservação Vermelha*SP*Medium*Bag*EUR/kg	Trás-os-Montes	Chaves	31/12/2018	0.3	0.4
Potato*Primor/Nova Branca*SE*Medium*Bag*EUR/kg	Algarve	Algarve	06/08/2018	0.6	0.3
Potato*Primor/Nova Branca*SE*Medium*Bag*EUR/kg	Algarve	Algarve	13/08/2018	0.6	0.3
Potato*Primor/Nova Branca*SE*Medium*Bag*EUR/kg	Algarve	Algarve	26/11/2018	1.0	1.0
Potato*Primor/Nova Branca*SE*Medium*Bag*EUR/kg	Algarve	Algarve	03/12/2018	1.0	1.0
Potato*Primor/Nova Branca*SE*Medium*Bag*EUR/kg	Algarve	Algarve	10/12/2018	1.0	1.0
Potato*Primor/Nova Branca*SE*Medium*Bag*EUR/kg	Algarve	Algarve	17/12/2018	1.0	1.0
Potato*Primor/Nova Branca*SE*Medium*Bag*EUR/kg	Algarve	Algarve	24/12/2018	0.8	0.7
Potato*Primor/Nova Branca*SE*Medium*Bag*EUR/kg	Algarve	Algarve	31/12/2018	0.8	0.7
Potato*Primor/Nova Branca*SE*Small*Bag*EUR/kg	Algarve	Algarve	26/11/2018	1.0	1.0
Potato*Primor/Nova Branca*SE*Small*Bag*EUR/kg	Algarve	Algarve	03/12/2018	1.0	1.0
Potato*Primor/Nova Branca*SE*Small*Bag*EUR/kg	Algarve	Algarve	10/12/2018	1.0	1.0
Potato*Primor/Nova Branca*SE*Small*Bag*EUR/kg	Algarve	Algarve	17/12/2018	1.0	1.0
Potato*Primor/Nova Branca*SE*Small*Bag*EUR/kg	Algarve	Algarve	24/12/2018	0.8	0.6
Potato*Primor/Nova Branca*SE*Small*Bag*EUR/kg	Algarve	Algarve	31/12/2018	0.8	0.6

Table 15. Potato producer prices at the end of the year for 2019

Product	Region	Market	Period	Value	Normalised Value
					For All Periods in Each Product
Potato*Conservação Branca*SP*Medium*Bag*EUR/kg	Trás-os-Montes	Chaves	21/10/2019	0.2	0.0
Potato*Conservação Branca*SP*Medium*Bag*EUR/kg	Trás-os-Montes	Chaves	28/10/2019	0.2	0.0
Potato*Conservação Branca*SP*Medium*Bag*EUR/kg	Trás-os-Montes	Chaves	04/11/2019	0.2	0.0
Potato*Conservação Branca*SP*Medium*Bag*EUR/kg	Trás-os-Montes	Chaves	11/11/2019	0.2	0.0
Potato*Conservação Branca*SP*Medium*Bag*EUR/kg	Trás-os-Montes	Chaves	18/11/2019	0.2	0.0
Potato*Conservação Branca*SP*Medium*Bag*EUR/kg	Trás-os-Montes	Chaves	25/11/2019	0.2	0.0
Potato*Conservação Branca*SP*Medium*Bag*EUR/kg	Trás-os-Montes	Chaves	02/12/2019	0.2	0.0
Potato*Conservação Branca*SP*Medium*Bag*EUR/kg	Trás-os-Montes	Chaves	09/12/2019	0.2	0.0
Potato*Conservação Branca*SP*Medium*Bag*EUR/kg	Trás-os-Montes	Chaves	16/12/2019	0.2	0.0
Potato*Conservação Branca*SP*Medium*Bag*EUR/kg	Trás-os-Montes	Chaves	23/12/2019	0.2	0.0
Potato*Conservação Branca*SP*Medium*Bag*EUR/kg	Trás-os-Montes	Chaves	30/12/2019	0.2	0.0
Potato*Conservação Vermelha*SP*Medium*Bag*EUR/kg	Trás-os-Montes	Chaves	21/10/2019	0.2	0.2
Potato*Conservação Vermelha*SP*Medium*Bag*EUR/kg	Trás-os-Montes	Chaves	28/10/2019	0.2	0.2
Potato*Conservação Vermelha*SP*Medium*Bag*EUR/kg	Trás-os-Montes	Chaves	04/11/2019	0.2	0.2
Potato*Conservação Vermelha*SP*Medium*Bag*EUR/kg	Trás-os-Montes	Chaves	11/11/2019	0.2	0.2
Potato*Conservação Vermelha*SP*Medium*Bag*EUR/kg	Trás-os-Montes	Chaves	18/11/2019	0.2	0.2
Potato*Conservação Vermelha*SP*Medium*Bag*EUR/kg	Trás-os-Montes	Chaves	25/11/2019	0.2	0.2
Potato*Conservação Vermelha*SP*Medium*Bag*EUR/kg	Trás-os-Montes	Chaves	02/12/2019	0.2	0.2
Potato*Conservação Vermelha*SP*Medium*Bag*EUR/kg	Trás-os-Montes	Chaves	09/12/2019	0.2	0.2
Potato*Conservação Vermelha*SP*Medium*Bag*EUR/kg	Trás-os-Montes	Chaves	16/12/2019	0.2	0.2
Potato*Conservação Vermelha*SP*Medium*Bag*EUR/kg	Trás-os-Montes	Chaves	23/12/2019	0.2	0.2
Potato*Conservação Vermelha*SP*Medium*Bag*EUR/kg	Trás-os-Montes	Chaves	30/12/2019	0.2	0.2
Potato*Primor/Nova Branca*SE*Medium*Bag*EUR/kg	Algarve	Algarve	05/08/2019	0.5	0.2
Potato*Primor/Nova Branca*SE*Medium*Bag*EUR/kg	Algarve	Algarve	12/08/2019	0.5	0.2
Potato*Primor/Nova Branca*SE*Medium*Bag*EUR/kg	Algarve	Algarve	19/08/2019	0.5	0.2
Potato*Primor/Nova Branca*SE*Medium*Bag*EUR/kg	Algarve	Algarve	26/08/2019	0.5	0.2
Potato*Primor/Nova Branca*SE*Medium*Bag*EUR/kg	Algarve	Algarve	02/09/2019	0.5	0.2
Potato*Primor/Nova Branca*SE*Medium*Bag*EUR/kg	Algarve	Algarve	09/09/2019	0.5	0.2
Potato*Primor/Nova Branca*SE*Medium*Bag*EUR/kg	Algarve	Algarve	16/09/2019	0.5	0.2

Continued on following page

Table 15. Continued

Product	Region	Market	Period	Value	Normalised Value
					For All Periods in Each Product
Potato*Primor/Nova Branca*SE*Medium*Bag*EUR/kg	Algarve	Algarve	23/09/2019	0.5	0.2
Potato*Primor/Nova Branca*SE*Medium*Bag*EUR/kg	Algarve	Algarve	30/09/2019	0.5	0.2
Potato*Primor/Nova Branca*SE*Medium*Bag*EUR/kg	Algarve	Algarve	07/10/2019	0.5	0.2
Potato*Primor/Nova Branca*SE*Medium*Bag*EUR/kg	Algarve	Algarve	14/10/2019	0.5	0.2
Potato*Primor/Nova Branca*SE*Medium*Bag*EUR/kg	Algarve	Algarve	25/11/2019	1.0	1.0
Potato*Primor/Nova Branca*SE*Medium*Bag*EUR/kg	Algarve	Algarve	02/12/2019	1.0	1.0
Potato*Primor/Nova Branca*SE*Medium*Bag*EUR/kg	Algarve	Algarve	09/12/2019	1.0	1.0
Potato*Primor/Nova Branca*SE*Medium*Bag*EUR/kg	Algarve	Algarve	16/12/2019	1.0	1.0
Potato*Primor/Nova Branca*SE*Medium*Bag*EUR/kg	Algarve	Algarve	23/12/2019	0.8	0.7
Potato*Primor/Nova Branca*SE*Medium*Bag*EUR/kg	Algarve	Algarve	30/12/2019	0.8	0.7
Potato*Primor/Nova Branca*SE*Small*Bag*EUR/kg	Algarve	Algarve	05/08/2019	0.5	0.0
Potato*Primor/Nova Branca*SE*Small*Bag*EUR/kg	Algarve	Algarve	12/08/2019	0.5	0.0
Potato*Primor/Nova Branca*SE*Small*Bag*EUR/kg	Algarve	Algarve	19/08/2019	0.5	0.0
Potato*Primor/Nova Branca*SE*Small*Bag*EUR/kg	Algarve	Algarve	26/08/2019	0.5	0.0
Potato*Primor/Nova Branca*SE*Small*Bag*EUR/kg	Algarve	Algarve	02/09/2019	0.5	0.0
Potato*Primor/Nova Branca*SE*Small*Bag*EUR/kg	Algarve	Algarve	09/09/2019	0.5	0.0
Potato*Primor/Nova Branca*SE*Small*Bag*EUR/kg	Algarve	Algarve	16/09/2019	0.5	0.0
Potato*Primor/Nova Branca*SE*Small*Bag*EUR/kg	Algarve	Algarve	23/09/2019	0.5	0.0
Potato*Primor/Nova Branca*SE*Small*Bag*EUR/kg	Algarve	Algarve	30/09/2019	0.5	0.0
Potato*Primor/Nova Branca*SE*Small*Bag*EUR/kg	Algarve	Algarve	07/10/2019	0.5	0.0
Potato*Primor/Nova Branca*SE*Small*Bag*EUR/kg	Algarve	Algarve	14/10/2019	0.5	0.0

Table 16. Potato producer prices at the end of the year for 2020

Product	Region	Market	Period	Value	Normalised Value For All Periods in Each Product
Potato*Conservação Branca*SP*Medium*Bag*EUR/kg	Trás-os-Montes	Chaves	30/11/2020	0.4	0.5
Potato*Conservação Branca*SP*Medium*Bag*EUR/kg	Trás-os-Montes	Chaves	07/12/2020	0.4	0.5
Potato*Conservação Branca*SP*Medium*Bag*EUR/kg	Trás-os-Montes	Chaves	14/12/2020	0.4	0.5
Potato*Conservação Branca*SP*Medium*Bag*EUR/kg	Trás-os-Montes	Chaves	21/12/2020	0.4	0.5
Potato*Conservação Branca*SP*Medium*Bag*EUR/kg	Trás-os-Montes	Chaves	28/12/2020	0.4	0.5
Potato*Conservação Vermelha*SP*Medium*Bag*EUR/kg	Trás-os-Montes	Chaves	30/11/2020	0.3	0.4
Potato*Conservação Vermelha*SP*Medium*Bag*EUR/kg	Trás-os-Montes	Chaves	07/12/2020	0.3	0.4
Potato*Conservação Vermelha*SP*Medium*Bag*EUR/kg	Trás-os-Montes	Chaves	14/12/2020	0.3	0.4
Potato*Conservação Vermelha*SP*Medium*Bag*EUR/kg	Trás-os-Montes	Chaves	21/12/2020	0.3	0.4
Potato*Conservação Vermelha*SP*Medium*Bag*EUR/kg	Trás-os-Montes	Chaves	28/12/2020	0.3	0.4
Potato*Primor/Nova Branca*SE*Medium*Bag*EUR/kg	Algarve	Algarve	03/08/2020	0.7	0.5
Potato*Primor/Nova Branca*SE*Medium*Bag*EUR/kg	Algarve	Algarve	10/08/2020	0.7	0.5
Potato*Primor/Nova Branca*SE*Medium*Bag*EUR/kg	Algarve	Algarve	17/08/2020	0.7	0.5
Potato*Primor/Nova Branca*SE*Medium*Bag*EUR/kg	Algarve	Algarve	24/08/2020	0.7	0.5
Potato*Primor/Nova Branca*SE*Medium*Bag*EUR/kg	Algarve	Algarve	26/10/2020	0.5	0.2
Potato*Primor/Nova Branca*SE*Medium*Bag*EUR/kg	Algarve	Algarve	02/11/2020	0.5	0.2
Potato*Primor/Nova Branca*SE*Medium*Bag*EUR/kg	Algarve	Algarve	09/11/2020	0.5	0.2
Potato*Primor/Nova Branca*SE*Medium*Bag*EUR/kg	Algarve	Algarve	16/11/2020	0.4	0.0
Potato*Primor/Nova Branca*SE*Medium*Bag*EUR/kg	Algarve	Algarve	23/11/2020	0.4	0.0
Potato*Primor/Nova Branca*SE*Medium*Bag*EUR/kg	Algarve	Algarve	30/11/2020	0.4	0.0
Potato*Primor/Nova Branca*SE*Medium*Bag*EUR/kg	Algarve	Algarve	07/12/2020	0.4	0.0
Potato*Primor/Nova Branca*SE*Medium*Bag*EUR/kg	Algarve	Algarve	14/12/2020	0.4	0.0
Potato*Primor/Nova Branca*SE*Medium*Bag*EUR/kg	Algarve	Algarve	21/12/2020	0.6	0.3
Potato*Primor/Nova Branca*SE*Medium*Bag*EUR/kg	Algarve	Algarve	28/12/2020	0.6	0.3

Table 17. Potato producer prices at the end of the year for 2021

Product	Region	Market	Period	Value	Normalised Value
					For All Periods in Each Product
Potato*Conservação Branca*SP*Medium*Bag*EUR/kg	Trás-os-Montes	Chaves	08/11/2021	0.4	0.5
Potato*Conservação Branca*SP*Medium*Bag*EUR/kg	Trás-os-Montes	Chaves	15/11/2021	0.4	0.5
Potato*Conservação Branca*SP*Medium*Bag*EUR/kg	Trás-os-Montes	Chaves	22/11/2021	0.4	0.5
Potato*Conservação Branca*SP*Medium*Bag*EUR/kg	Trás-os-Montes	Chaves	29/11/2021	0.4	0.5
Potato*Conservação Branca*SP*Medium*Bag*EUR/kg	Trás-os-Montes	Chaves	06/12/2021	0.4	0.5
Potato*Conservação Branca*SP*Medium*Bag*EUR/kg	Trás-os-Montes	Chaves	13/12/2021	0.4	0.5
Potato*Conservação Branca*SP*Medium*Bag*EUR/kg	Trás-os-Montes	Chaves	20/12/2021	0.4	0.5
Potato*Conservação Branca*SP*Medium*Bag*EUR/kg	Trás-os-Montes	Chaves	27/12/2021	0.4	0.5
Potato*Conservação Vermelha*SP*Medium*Bag*EUR/kg	Trás-os-Montes	Chaves	08/11/2021	0.4	0.6
Potato*Conservação Vermelha*SP*Medium*Bag*EUR/kg	Trás-os-Montes	Chaves	15/11/2021	0.4	0.6
Potato*Conservação Vermelha*SP*Medium*Bag*EUR/kg	Trás-os-Montes	Chaves	22/11/2021	0.4	0.6
Potato*Conservação Vermelha*SP*Medium*Bag*EUR/kg	Trás-os-Montes	Chaves	29/11/2021	0.4	0.6
Potato*Conservação Vermelha*SP*Medium*Bag*EUR/kg	Trás-os-Montes	Chaves	06/12/2021	0.4	0.6
Potato*Conservação Vermelha*SP*Medium*Bag*EUR/kg	Trás-os-Montes	Chaves	13/12/2021	0.4	0.6
Potato*Conservação Vermelha*SP*Medium*Bag*EUR/kg	Trás-os-Montes	Chaves	20/12/2021	0.4	0.6
Potato*Conservação Vermelha*SP*Medium*Bag*EUR/kg	Trás-os-Montes	Chaves	27/12/2021	0.4	0.6
Potato*Primor/Nova Branca*SE*Medium*Bag*EUR/kg	Algarve	Algarve	02/08/2021	0.7	0.5
Potato*Primor/Nova Branca*SE*Medium*Bag*EUR/kg	Algarve	Algarve	22/11/2021	0.9	0.8
Potato*Primor/Nova Branca*SE*Medium*Bag*EUR/kg	Algarve	Algarve	29/11/2021	0.9	0.8
Potato*Primor/Nova Branca*SE*Medium*Bag*EUR/kg	Algarve	Algarve	06/12/2021	0.9	0.8
Potato*Primor/Nova Branca*SE*Medium*Bag*EUR/kg	Algarve	Algarve	13/12/2021	0.9	0.8
Potato*Primor/Nova Branca*SE*Medium*Bag*EUR/kg	Algarve	Algarve	20/12/2021	0.9	0.8
Potato*Primor/Nova Branca*SE*Medium*Bag*EUR/kg	Algarve	Algarve	27/12/2021	0.9	0.8
Potato*Primor/Nova Branca*SE*Small*Bag*EUR/kg	Algarve	Algarve	22/11/2021	0.6	0.2
Potato*Primor/Nova Branca*SE*Small*Bag*EUR/kg	Algarve	Algarve	29/11/2021	0.6	0.2
Potato*Primor/Nova Branca*SE*Small*Bag*EUR/kg	Algarve	Algarve	06/12/2021	0.6	0.2
Potato*Primor/Nova Branca*SE*Small*Bag*EUR/kg	Algarve	Algarve	13/12/2021	0.6	0.2
Potato*Primor/Nova Branca*SE*Small*Bag*EUR/kg	Algarve	Algarve	20/12/2021	0.6	0.2
Potato*Primor/Nova Branca*SE*Small*Bag*EUR/kg	Algarve	Algarve	27/12/2021	0.6	0.2

Table 18. Potato producer prices at the end of the year for 2022

Product	Region	Market	Period	Value	Normalised Value
					For All Periods in Each Product
Potato*Conservação Branca*SP*Medium*Bag*EUR/kg	Trás-os-Montes	Chaves	21/11/2022	0.6	1.0
Potato*Conservação Branca*SP*Medium*Bag*EUR/kg	Trás-os-Montes	Chaves	28/11/2022	0.6	1.0
Potato*Conservação Branca*SP*Medium*Bag*EUR/kg	Trás-os-Montes	Chaves	05/12/2022	0.6	1.0
Potato*Conservação Branca*SP*Medium*Bag*EUR/kg	Trás-os-Montes	Chaves	12/12/2022	0.6	1.0
Potato*Conservação Branca*SP*Medium*Bag*EUR/kg	Trás-os-Montes	Chaves	19/12/2022	0.6	1.0
Potato*Conservação Branca*SP*Medium*Bag*EUR/kg	Trás-os-Montes	Chaves	26/12/2022	0.6	1.0
Potato*Conservação Vermelha*SP*Medium*Bag*EUR/kg	Trás-os-Montes	Chaves	21/11/2022	0.6	1.0
Potato*Conservação Vermelha*SP*Medium*Bag*EUR/kg	Trás-os-Montes	Chaves	28/11/2022	0.6	1.0
Potato*Conservação Vermelha*SP*Medium*Bag*EUR/kg	Trás-os-Montes	Chaves	05/12/2022	0.6	1.0
Potato*Conservação Vermelha*SP*Medium*Bag*EUR/kg	Trás-os-Montes	Chaves	12/12/2022	0.6	1.0
Potato*Conservação Vermelha*SP*Medium*Bag*EUR/kg	Trás-os-Montes	Chaves	19/12/2022	0.6	1.0
Potato*Conservação Vermelha*SP*Medium*Bag*EUR/kg	Trás-os-Montes	Chaves	26/12/2022	0.6	1.0
Potato*Primor/Nova Branca*SE*Medium*Bag*EUR/kg	Algarve	Algarve	01/08/2022	0.8	0.7
Potato*Primor/Nova Branca*SE*Medium*Bag*EUR/kg	Algarve	Algarve	14/11/2022	1.0	1.0
Potato*Primor/Nova Branca*SE*Medium*Bag*EUR/kg	Algarve	Algarve	21/11/2022	1.0	1.0
Potato*Primor/Nova Branca*SE*Medium*Bag*EUR/kg	Algarve	Algarve	28/11/2022	1.0	1.0
Potato*Primor/Nova Branca*SE*Medium*Bag*EUR/kg	Algarve	Algarve	05/12/2022	1.0	1.0
Potato*Primor/Nova Branca*SE*Medium*Bag*EUR/kg	Algarve	Algarve	12/12/2022	1.0	1.0
Potato*Primor/Nova Branca*SE*Medium*Bag*EUR/kg	Algarve	Algarve	19/12/2022	1.0	1.0
Potato*Primor/Nova Branca*SE*Medium*Bag*EUR/kg	Algarve	Algarve	26/12/2022	1.0	1.0
Potato*Primor/Nova Branca*SE*Small*Bag*EUR/kg	Algarve	Algarve	01/08/2022	0.6	0.2

CEREAL PRICES

The changes verified in the Portuguese cereal (durum and soft wheat) producer prices appear more relevant (Tables 19-20) and here seems that the Russia-Ukraine conflict had a clear impact on the tonne value, namely in the durum wheat. In general, the prices for durum wheat are, on average, higher than those for soft wheat. On the other hand, the average prices (euros/tonne) for durum wheat were 230 in 2017, 217 in 2018, 229 in 2019, 230 in 2020, 273 in 2021 and 494 in 2022. For soft wheat, the average prices were 175 euros/tonne in 2017, 220 euros/tonne in 2019, 205 euros/tonne in 2020, 240 euros/tonne in 2021 and 382 euros/tonne in 2022.

Table 19. Cereal producer prices for 2017-2020

Product	Region	Period	Value	Normalised Value for All Periods in Each Product
Durum Wheat*EUR/Tonne	Alentejo	03/07/2017	230	0.1
Durum Wheat*EUR/Tonne	Alentejo	10/07/2017	230	0.1
Durum Wheat*EUR/Tonne	Alentejo	17/07/2017	230	0.1
Durum Wheat*EUR/Tonne	Alentejo	24/07/2017	230	0.1
Durum Wheat*EUR/Tonne	Alentejo	31/07/2017	230	0.1
Durum Wheat*EUR/Tonne	Alentejo	07/08/2017	230	0.1
Durum Wheat*EUR/Tonne	Alentejo	09/07/2018	210	0.0
Durum Wheat*EUR/Tonne	Alentejo	16/07/2018	210	0.0
Durum Wheat*EUR/Tonne	Alentejo	23/07/2018	210	0.0
Durum Wheat*EUR/Tonne	Alentejo	30/07/2018	210	0.0
Durum Wheat*EUR/Tonne	Alentejo	06/08/2018	220	0.0
Durum Wheat*EUR/Tonne	Alentejo	13/08/2018	220	0.0
Durum Wheat*EUR/Tonne	Alentejo	20/08/2018	220	0.0
Durum Wheat*EUR/Tonne	Alentejo	27/08/2018	220	0.0
Durum Wheat*EUR/Tonne	Alentejo	03/09/2018	220	0.0
Durum Wheat*EUR/Tonne	Alentejo	10/09/2018	220	0.0
Durum Wheat*EUR/Tonne	Alentejo	17/09/2018	220	0.0
Durum Wheat*EUR/Tonne	Alentejo	24/09/2018	220	0.0
Durum Wheat*EUR/Tonne	Alentejo	01/10/2018	220	0.0
Durum Wheat*EUR/Tonne	Alentejo	08/07/2019	220	0.0
Durum Wheat*EUR/Tonne	Alentejo	15/07/2019	230	0.1
Durum Wheat*EUR/Tonne	Alentejo	22/07/2019	230	0.1
Durum Wheat*EUR/Tonne	Alentejo	29/07/2019	230	0.1
Durum Wheat*EUR/Tonne	Alentejo	05/08/2019	230	0.1
Durum Wheat*EUR/Tonne	Alentejo	12/08/2019	230	0.1
Durum Wheat*EUR/Tonne	Alentejo	19/08/2019	230	0.1
Durum Wheat*EUR/Tonne	Alentejo	20/07/2020	230	0.1
Durum Wheat*EUR/Tonne	Alentejo	27/07/2020	230	0.1
Durum Wheat*EUR/Tonne	Alentejo	03/08/2020	230	0.1
Durum Wheat*EUR/Tonne	Alentejo	10/08/2020	230	0.1
Durum Wheat*EUR/Tonne	Alentejo	17/08/2020	230	0.1
Durum Wheat*EUR/Tonne	Alentejo	24/08/2020	230	0.1
Durum Wheat*EUR/Tonne	Alentejo	31/08/2020	230	0.1
Durum Wheat*EUR/Tonne	Alentejo	07/09/2020	230	0.1
Durum Wheat*EUR/Tonne	Alentejo	14/09/2020	230	0.1
Durum Wheat*EUR/Tonne	Alentejo	21/09/2020	230	0.1
Durum Wheat*EUR/Tonne	Alentejo	28/09/2020	230	0.1
Soft wheat*EUR/Tonne	Alentejo	19/06/2017	160	0.0

Continued on following page

Table 19. Continued

Product	Region	Period	Value	Normalised Value for All Periods in Each Product
Soft wheat*EUR/Tonne	Alentejo	26/06/2017	160	0.0
Soft wheat*EUR/Tonne	Alentejo	03/07/2017	180	0.1
Soft wheat*EUR/Tonne	Alentejo	10/07/2017	180	0.1
Soft wheat*EUR/Tonne	Alentejo	17/07/2017	180	0.1
Soft wheat*EUR/Tonne	Alentejo	24/07/2017	180	0.1
Soft wheat*EUR/Tonne	Alentejo	31/07/2017	180	0.1
Soft wheat*EUR/Tonne	Alentejo	07/08/2017	180	0.1
Soft wheat*EUR/Tonne	Alentejo	08/07/2019	220	0.3
Soft wheat*EUR/Tonne	Alentejo	15/07/2019	220	0.3
Soft wheat*EUR/Tonne	Alentejo	22/07/2019	220	0.3
Soft wheat*EUR/Tonne	Alentejo	29/07/2019	220	0.3
Soft wheat*EUR/Tonne	Alentejo	05/08/2019	220	0.3
Soft wheat*EUR/Tonne	Alentejo	12/08/2019	220	0.3
Soft wheat*EUR/Tonne	Alentejo	19/08/2019	220	0.3
Soft wheat*EUR/Tonne	Alentejo	06/07/2020	205	0.2
Soft wheat*EUR/Tonne	Alentejo	13/07/2020	205	0.2
Soft wheat*EUR/Tonne	Alentejo	20/07/2020	205	0.2
Soft wheat*EUR/Tonne	Alentejo	27/07/2020	205	0.2
Soft wheat*EUR/Tonne	Alentejo	03/08/2020	205	0.2
Soft wheat*EUR/Tonne	Alentejo	10/08/2020	205	0.2
Soft wheat*EUR/Tonne	Alentejo	17/08/2020	205	0.2
Soft wheat*EUR/Tonne	Alentejo	24/08/2020	205	0.2
Soft wheat*EUR/Tonne	Alentejo	31/08/2020	205	0.2
Soft wheat*EUR/Tonne	Alentejo	07/09/2020	205	0.2
Soft wheat*EUR/Tonne	Alentejo	14/09/2020	205	0.2
Soft wheat*EUR/Tonne	Alentejo	21/09/2020	205	0.2
Soft wheat*EUR/Tonne	Alentejo	28/09/2020	205	0.2

Table 20. Cereal producer prices for 2021-2022

Product	Region	Period	Value	Normalised Value for All Periods in Each Product
Durum Wheat*EUR/Tonne	Alentejo	05/07/2021	270	0.2
Durum Wheat*EUR/Tonne	Alentejo	12/07/2021	273	0.2
Durum Wheat*EUR/Tonne	Alentejo	19/07/2021	273	0.2
Durum Wheat*EUR/Tonne	Alentejo	26/07/2021	273	0.2
Durum Wheat*EUR/Tonne	Alentejo	02/08/2021	273	0.2
Durum Wheat*EUR/Tonne	Alentejo	23/08/2021	274	0.2
Durum Wheat*EUR/Tonne	Alentejo	30/08/2021	274	0.2
Durum Wheat*EUR/Tonne	Alentejo	06/09/2021	274	0.2
Durum Wheat*EUR/Tonne	Alentejo	13/09/2021	274	0.2
Durum Wheat*EUR/Tonne	Alentejo	20/09/2021	274	0.2
Durum Wheat*EUR/Tonne	Alentejo	27/09/2021	274	0.2
Durum Wheat*EUR/Tonne	Alentejo	04/10/2021	274	0.2
Durum Wheat*EUR/Tonne	Alentejo	11/10/2021	274	0.2
Durum Wheat*EUR/Tonne	Alentejo	13/06/2022	420	0.7
Durum Wheat*EUR/Tonne	Alentejo	20/06/2022	500	1.0
Durum Wheat*EUR/Tonne	Alentejo	27/06/2022	500	1.0
Durum Wheat*EUR/Tonne	Alentejo	04/07/2022	500	1.0
Durum Wheat*EUR/Tonne	Alentejo	11/07/2022	500	1.0
Durum Wheat*EUR/Tonne	Alentejo	18/07/2022	500	1.0
Durum Wheat*EUR/Tonne	Alentejo	25/07/2022	500	1.0
Durum Wheat*EUR/Tonne	Alentejo	01/08/2022	500	1.0
Durum Wheat*EUR/Tonne	Alentejo	15/08/2022	500	1.0
Durum Wheat*EUR/Tonne	Alentejo	22/08/2022	500	1.0
Durum Wheat*EUR/Tonne	Alentejo	29/08/2022	500	1.0
Durum Wheat*EUR/Tonne	Alentejo	05/09/2022	500	1.0
Durum Wheat*EUR/Tonne	Alentejo	12/09/2022	500	1.0
Soft wheat*EUR/Tonne	Alentejo	05/07/2021	235	0.3
Soft wheat*EUR/Tonne	Alentejo	12/07/2021	235	0.3
Soft wheat*EUR/Tonne	Alentejo	19/07/2021	235	0.3
Soft wheat*EUR/Tonne	Alentejo	26/07/2021	235	0.3
Soft wheat*EUR/Tonne	Alentejo	02/08/2021	235	0.3
Soft wheat*EUR/Tonne	Alentejo	09/08/2021	235	0.3
Soft wheat*EUR/Tonne	Alentejo	16/08/2021	235	0.3
Soft wheat*EUR/Tonne	Alentejo	23/08/2021	240	0.3
Soft wheat*EUR/Tonne	Alentejo	30/08/2021	235	0.3
Soft wheat*EUR/Tonne	Alentejo	06/09/2021	235	0.3
Soft wheat*EUR/Tonne	Alentejo	13/09/2021	235	0.3
Soft wheat*EUR/Tonne	Alentejo	20/09/2021	235	0.3

Continued on following page

Table 20. Continued

Product	Region	Period	Value	Normalised Value for All Periods in Each Product
Soft wheat*EUR/Tonne	Alentejo	27/09/2021	260	0.4
Soft wheat*EUR/Tonne	Alentejo	04/10/2021	260	0.4
Soft wheat*EUR/Tonne	Alentejo	11/10/2021	260	0.4
Soft wheat*EUR/Tonne	Alentejo	13/06/2022	350	0.8
Soft wheat*EUR/Tonne	Alentejo	20/06/2022	380	1.0
Soft wheat*EUR/Tonne	Alentejo	27/06/2022	380	1.0
Soft wheat*EUR/Tonne	Alentejo	04/07/2022	380	1.0
Soft wheat*EUR/Tonne	Alentejo	11/07/2022	380	1.0
Soft wheat*EUR/Tonne	Alentejo	18/07/2022	380	1.0
Soft wheat*EUR/Tonne	Alentejo	25/07/2022	380	1.0
Soft wheat*EUR/Tonne	Alentejo	01/08/2022	380	1.0
Soft wheat*EUR/Tonne	Alentejo	15/08/2022	390	1.0
Soft wheat*EUR/Tonne	Alentejo	22/08/2022	390	1.0
Soft wheat*EUR/Tonne	Alentejo	29/08/2022	390	1.0
Soft wheat*EUR/Tonne	Alentejo	05/09/2022	390	1.0
Soft wheat*EUR/Tonne	Alentejo	12/09/2022	390	1.0

CONCLUSION

This contribution proposed to bring additional insights into the Portuguese vegetable and cereal producer prices evolution after the pandemic and Russia-Ukraine crisis. For that, statistical information was considered from the Portuguese agricultural markets information system for the period 2017-2022, with data disaggregated weekly. Considering the volume of information available in this database, this research focused on potatoes, durum wheat and soft wheat.

Vegetables and cereals are fundaments for the food sovereignty and food security of each country. In this way, it is crucial to assess the trends in these markets and how they influence the consumption by the citizens. These agricultural products are often used to produce other products in the food industry. In this perspective, any change in the vegetable and cereal markets has implications in other markets over the chain, contributing, in certain circumstances, to food inflation (particularly the cereal prices). Wheat is one of the most important cereals. The cereal markets are often the targets of speculation in the prices and are affected by the competition for other land uses, including by biofuel crops. These markets are vulnerable to international shocks and some volatility. The agricultural policies, specifically in the European Union CAP framework, also have effects on the dynamics of the vegetable and cereal markets.

The data for the Portuguese framework reveal that the potato ("Potato*Conservação Branca*SP*Medium*Bag", "Potato*Conservação Vermelha*SP*Medium*Bag" and "Potato*Primor/Nova Branca*SE*Medium*Bag") and wheat (durum and soft) prices changed, namely in the last months of 2022, showing that the Russia-Ukraine conflict had impacts on these prices (more than the pandemic shocks). For example, the average potatoes prices increased from 0.5 euros/kg in 2021 to 0.8 euros/kg in

2022, the durum wheat average price increased from 273 euros/tonne in 2021 to 494 euros/tonne in 2022 and the soft wheat prices increased, on average, from 240 euros/tonne in 2021 to 382 euros/tonne in 2022.

This study highlighted the vulnerability of the vegetable and cereal producer markets, in the Portuguese context, to the recent international shocks, particularly to the Russia-Ukraine conflict. It could be important, however, to understand in future research which part of the price increase in these markets is due to direct implications and which part is a consequence of eventual speculation. Governments and public institutions have here a fundamental role to mitigate eventual consequences on food security.

ACKNOWLEDGMENT

This work is funded by National Funds through the FCT - Foundation for Science and Technology, I.P., within the scope of the project Refª UIDB/00681/2020. Furthermore we would like to thank the CERNAS Research Centre and the Polytechnic Institute of Viseu for their support.

REFERENCES

Abay, K. A., Abdelfattah, L., Breisinger, C., & Siddig, K. (2023). Evaluating cereal market (dis)integration in less developed and fragile markets: The case of Sudan. *Food Policy, 114*. doi:10.1016/j.foodpol.2022.102399

Andreosso-O'Callaghan, B., & Zolin, M. B. (2010). Long-term Cereal Price Changes: How Important is the Speculative Element? *Transition Studies Review, 17*(4), 624–637. doi:10.1007/s11300-010-0178-7

Ben Rejeb, I., Charfi, I., Baraketi, S., Hached, H., & Gargouri, M. (2022). Bread Surplus: A Cumulative Waste or a Staple Material for High-Value Products? *Molecules, 27*(23). doi:10.3390/molecules27238410

Benoit, M., & Laignel, G. (2014). Sheep-for-meat farming systems in French semi-upland area. Adapting to new context: Increased concentrates and energy prices, and new agricultural policy. *International Journal of Sustainable Development, 17*(1), 35–48. doi:10.1504/IJSD.2014.058435

De Roest, K., Menghi, A., & Trindade, H. (2008). Adaptation of Mediterranean dairy systems to the new context of feed prices and European regulation. *Fourrages (Versailles), 2008*(196), 447–459.

Deleuran, L. C., Kristensen, K., Gislum, R., & Boelt, B. (2013). Optimizing the number of consecutive seed harvests in red fescue (Festuca rubra L.) and perennial ryegrass (Lolium perenne L.) for yield, yield components and economic return. *Acta Agriculturae Scandinavica Section B: Soil and Plant Science, 63*(1), 1–10. doi:10.1080/09064710.2012.703229

Doole, G. J., & Weetman, E. (2009). Tactical management of pasture fallows in Western Australian cropping systems. *Agricultural Systems, 102*(1–3), 24–32. doi:10.1016/j.agsy.2009.06.002

Durevall, D., Loening, J. L., & Ayalew Birru, Y. (2013). Inflation dynamics and food prices in Ethiopia. *Journal of Development Economics, 104*, 89–106. doi:10.1016/j.jdeveco.2013.05.002

Dzanku, F. M. (2015). Household-specific food price differentials and high-value crop production in rural Ghana. *Food Policy, 57*, 73–82. doi:10.1016/j.foodpol.2015.09.004

García, E. J., Damián, M. A. M., & Cruz, A. K. (2008). Support price policy versus direct payments to producers: Analisys of producers' welfare. *Revista Fitotecnia Mexicana, 31*(4), 383–389.

Ghelfi, R., & Palmieri, A. (2017). Pulses production in Italy: Trade, marketing and. doi:10.4081/ija.2017.891

Lutaladio, N., & Prakash, A. (2010). The potato: Its history and economic development. *Cahiers de Nutrition et de Dietetique, 45*(6), S5–S16. doi:10.1016/S0007-9960(10)70003-1

Müller, A., Schmidhuber, J., Hoogeveen, J., & Steduto, P. (2008). Some insights in the effect of growing bio-energy demand on global food security and natural resources. *Water Policy, 10*(Suppl. 1), 83–94. doi:10.2166/wp.2008.053

Olaizola, A. M., Ameen, F., & Manrique, E. (2015). Potential strategies of adaptation of mixed sheep-crop systems to changes in the economic environment in a Mediterranean mountain area. *Livestock Science, 176*, 166–180. doi:10.1016/j.livsci.2015.03.012

Ott, H. (2014). Volatility in cereal prices: Intra- versus inter-annual volatility. *Journal of Agricultural Economics, 65*(3), 557–578. doi:10.1111/1477-9552.12073

Parry, M. L., Rosenzweig, C., Iglesias, A., Livermore, M., & Fischer, G. (2004). Effects of climate change on global food production under SRES emissions and socio-economic scenarios. *Global Environmental Change, 14*(1), 53–67. doi:10.1016/j.gloenvcha.2003.10.008

Popović, G., Erić, O., & Bjelić, J. (2020). Factor analysis of prices and agricultural production in the European union. *Economics, 8*(1), 73–81. doi:10.2478/eoik-2020-0001

Rocchi, B., Romano, D., & Hamza, R. (2013). Agriculture reform and food crisis in Syria: Impacts on poverty and inequality. *Food Policy, 43*, 190–203. doi:10.1016/j.foodpol.2013.09.009

Rosegrant, M. W., Cline, S. A., Li, W., Sulser, T. B., & Valmonte-Santos, R. A. (2005). Looking ahead long-term prospects for Africa's agricultural development and food security. Food, Agriculture, and the Environment Discussion Paper, 41, 1–71.

Salo, T., Eskelinen, J., Jauhiainen, L., & Kartio, M. (2007). Reduced fertiliser use and changes in cereal grain weight, test weight and protein content in Finland in 1990-2005. *Agricultural and Food Science, 16*(4), 407–420. doi:10.2137/145960607784125375

Scopus. (2023). *Scopus Database*. https://www.scopus.com/search/form.uri?display=basic#basic

Shenggen Fan, D. H. (2010). Reflections on the global food crisis: How did it happen? How has it hurt? And how can we prevent the next one? *Research Report of the International Food Policy Research Institute, 165*, 1–140.

SIMA. (2023). *Several Statistics and Information*. https://www.gpp.pt/index.php/sima/sima-2018

Stata. (2023). *Statistical software for data science*. https://www.stata.com/

StataCorp. (2017a). *Stata 15 Base Reference Manual*. Stata Press.

StataCorp. (2017b). *Stata Statistical Software: Release 15*. StataCorp LLC.

Sy, M. A. T., Massaoud, W., Oumarou, D. H., Issoufou, B., Simeoné, C., & Abdouramane, B. (2022). Impact of cereal production and cereal prices on food consumption and changes in household livelihoods in Burkina Faso. *African Journal of Food, Agriculture, Nutrition and Development, 22*(1), 19252–19270. doi:10.18697/ajfand.106.20715

Xu, X., & Mola-Yudego, B. (2021). Where and when are plantations established? Land-use replacement patterns of fast-growing plantations on agricultural land. *Biomass and Bioenergy, 144*. doi:10.1016/j.biombioe.2020.105921

Compilation of References

Abay, K. A., Abdelfattah, L., Breisinger, C., & Siddig, K. (2023). Evaluating cereal market (dis)integration in less developed and fragile markets: The case of Sudan. *Food Policy, 114*. doi:10.1016/j.foodpol.2022.102399

Abbassi, W., Kumari, V., & Pandey, D. K. (2022). What makes firms vulnerable to the Russia–Ukraine crisis? *The Journal of Risk Finance*. doi:10.1108/JRF-05-2022-0108

Abdel-Wahab, T. I., Abdel-Wahab, S. I., Abdelwahab, E. I., & Adel, M. M. (2020). A step forward to control of COVID-19. *International Journal of Pharmaceutical Research, 12*, 1167–1190. doi:10.31838/ijpr/2020.SP1.191

Abhijit, D. A. S., Sivaram, M., & Thejesh, S. (2021). Economic impact of COVID-19 pandemic on dairy sector: A meta-analysis. [Scopus.]. *The Indian Journal of Animal Sciences, 91*(7), 582–594.

Abu Hatab, A. (2022). Africa's Food Security under the Shadow of the Russia-Ukraine Conflict. *Strategic Review for Southern Africa, 44*(1), 37–46. doi:10.35293rsa.v44i1.4083

Aday, S., & Aday, M. S. (2020). Impact of COVID-19 on the food supply chain. *Food Quality and Safety, 4*(4), 167–180. doi:10.1093/fqsafe/fyaa024

Adebowale, O., Makanjuola, M., Bankole, N., Olasoju, M., Alamu, A., Kperegbeyi, E., Oladejo, O., Fasanmi, O., Adeyemo, O., & Fasina, F. O. (2022). Multi-Drug Resistant Escherichia coli, Biosecurity and Anti-Microbial Use in Live Bird Markets, Abeokuta, Nigeria. *Antibiotics, 11*(2). doi:10.3390/antibiotics11020253

Adelaja, A. O., Brumfield, R. G., & Schilling, B. J. (1996). Share analysis of drivers of growth in the ornamental industry: A case study of New Jersey. In Acta Hortic. (Vol. 429, p. 307). International Society for Horticultural Science. doi:10.17660/ActaHortic.1996.429.38

Adelodun, B., Kareem, K. Y., Kumar, P., Kumar, V., Choi, K. S., Yadav, K. K., Yadav, A., El-Denglawey, A., Cabral-Pinto, M., Son, C. T., Krishnan, S., & Khan, N. A. (2021). Understanding the impacts of the COVID-19 pandemic on sustainable agri-food system and agroecosystem decarbonization nexus: A review. *Journal of Cleaner Production, 318*, 128451. doi:10.1016/j.jclepro.2021.128451 PMID:36570877

Adhikari, J., Timsina, J., Khadka, S. R., Ghale, Y., & Ojha, H. (2021). COVID-19 impacts on agriculture and food systems in Nepal: Implications for SDGs. *Agricultural Systems, 186*. Scopus. doi:10.1016/j.agsy.2020.102990

Adnan, N., & Nordin, S. M. (2021). How COVID 19 effect Malaysian paddy industry? Adoption of green fertilizer a potential resolution. *Environment, Development and Sustainability, 23*(6), 8089–8129. Scopus. doi:10.1007/s10668-020-00978-6

Afolabi, S. O., & Tiamiyu, M. A. (2021). Information Use Behaviour of Sedentary Livestock Farmers in Egbeda Local Government Area of Oyo State, Nigeria. *Library Philosophy and Practice, 2021*, 1–15.

Aharon, D. Y., & Qadan, M. (2022). Infection, invasion, and inflation: Recent lessons. *Finance Research Letters, 50*, 103307. doi:10.1016/j.frl.2022.103307

Ahmad Rizal, A. R., & Md Nordin, S. (2022). Getting ahead of the pandemic curve: A systematic review of critical determining factors for innovation adoption in ensuring food security. *Frontiers in Nutrition, 9*. Scopus. doi:10.3389/fnut.2022.986324

Ahmad, A., Ordoñez, J., Cartujo, P., & Martos, V. (2020). Remotely piloted aircraft (RPA) in agriculture: A pursuit of sustainability. *Agronomy, 11*(1). Scopus. doi:10.3390/agronomy11010007

Ahmed, S., Hasan, M. M., & Kamal, M. R. (2022). Russia–Ukraine crisis: The effects on the European stock market. *European Financial Management,* 1–41. doi:10.1111/eufm.12386

Ahoya, B., Kavle, J. A., Kiige, L., Gathi, C., Samburu, B., Maina, L., Ramirez, L., Wambu, R., & Codjia, P. (2022). *How COVID-19 affected food systems, health service delivery and maternal and infant nutrition practices: Implications for moving forward in Kenya. Maternal and Child Nutrition.* Scopus., doi:10.1111/mcn.13466

Aiswarya, A., & Bhagya, D. (2021). Effect of Covid 19 lockdown on the lifestyle and dietary diversity of women handloom workers. *Clinical Epidemiology and Global Health, 12*. Scopus. doi:10.1016/j.cegh.2021.100856

Akber, N., & Paltasingh, K. R. (2022). COVID-19 lockdown and prices of essential food items in India: Examining law of one price. *Journal of Agribusiness in Developing and Emerging Economies, 12*(3), 510–530. Scopus. doi:10.1108/JADEE-12-2021-0316

Akter, M. S., Talsma, E. F., Feskens, E. J. M., Thilsted, S. H., & Rasheed, S. (2022). How did the COVID-19 pandemic affect food environment, food purchase, and fish consumption among low-income urban households in Bangladesh—A path analysis. *Frontiers in Public Health, 10*. Scopus. doi:10.3389/fpubh.2022.994236

Akter, S. (2020). The impact of COVID-19 related 'stay-at-home' restrictions on food prices in Europe: Findings from a preliminary analysis. *Food Security, 12*(4), 719–725. doi:10.1007/s12571-020-01082-3

Alan Walters, S. (2020). Impact of nitrogen fertilizer on yield and associated revenues in no-tillage pumpkin. *HortScience, 55*(4), 522–527. doi:10.21273/HORTSCI14512-19

Alfaro-Navarro, J.-L., & Andrés-Martínez, M.-E. (2021). A longitudinal and cross-sectional analysis of the distribution of Common Agricultural Policy aids in European countries. *Agricultural Economics (Czech Republic), 67*(9), 351–362. Scopus. doi:10.17221/87/2021-AGRICECON

Ali, A. M., Kunugi, H., Abdelmageed, H. A., Mandour, A. S., Ahmed, M. E., Ahmad, S., & Hendawy, A. O. (2021). Vitamin k in covid-19—Potential anti-covid-19 properties of fermented milk fortified with bee honey as a natural source of vitamin k and probiotics. *Fermentation, 7*(4). Scopus. doi:10.3390/fermentation7040202

Ali, A., & Imran, M. (2021). National spatial data infrastructure vs. Cadastre system for economic development: Evidence from Pakistan. *Land, 10*(2), 1–17. Scopus. doi:10.3390/land10020188

Ali, M. S., Siddik, M. S., Nur-E-Alam, S. M., Khanam, T., & Khatun, M. M. (2020). Analysis of the problems and minimization initiatives addressed by bangladeshi farmers during the covid-19 pandemic. *Asian Journal of Agriculture and Rural Development, 10*(4), 773–781. Scopus. doi:10.18488/journal.ajard.2020.104.773.781

Allison, J., Jr., Burdine, K. H., Dillon, C., Smith, S. R., Butler, D. M., Bates, G. E., & Pighetti, G. M. (2021). Optimal forage and supplement balance for organic dairy farms in the Southeastern United States. *Agricultural Systems, 189*. doi:10.1016/j.agsy.2021.103048

Almeida, E. M. T. C., & Almeida, M. T. C. (2021). Attitude and behavioral changes of dairy consumers during the new coronavirus pandemic in Brazil. *International Journal of Dairy Science, 16*(2), 67–74. Scopus. doi:10.3923/ijds.2021.67.74

Al-Saidi, M., Saad, S. A. G., & Elagib, N. A. (2022). *From scenario to mounting risks: COVID-19's perils for development and supply security in the Sahel. Environment, Development and Sustainability*. Scopus., doi:10.100710668-022-02303-9

Altieri, M. A., & Nicholls, C. I. (2020). Agroecology: Challenges and opportunities for farming in the Anthropocene. *International Journal of Agriculture and Natural Resources, 47*(3), 204–215. Scopus. doi:10.7764/ijanr.v47i3.2281

Alyukov, M. (2022). Making Sense of the News in an Authoritarian Regime: Russian Television Viewers' Reception of the Russia-Ukraine Conflict. *Europe-Asia Studies, 74*(3), 337–359. doi:10.1080/09668136.2021.2016633

Amahmid, O., El Guamri, Y., Rakibi, Y., Ouizat, S., Yazidi, M., Razoki, B., Kaid Rassou, K., Asmama, S., Bouhoum, K., & Belghyti, D. (2022). Occurrence of SARS-CoV-2 in excreta, sewage, and environment: Epidemiological significance and potential risks. *International Journal of Environmental Health Research, 32*(8), 1686–1706. doi:10.1080/09603123.2021.1901865 PMID:33752527

Amin, F., Poespito Hadi, W., Zauhar, S., & Santoso Haryono, B. (2022). Determinants of post COVID-19 food security policy success. *International Journal of Disaster Resilience in the Built Environment, 13*(4), 440–450. Scopus. doi:10.1108/IJDRBE-11-2020-0118

Anderson, K. (2022). Agriculture in a more uncertain global trade environment. *Agricultural Economics (United Kingdom), 53*(4), 563–579. Scopus. doi:10.1111/agec.12726

Andreosso-O'Callaghan, B., & Zolin, M. B. (2010). Long-term Cereal Price Changes: How Important is the Speculative Element? *Transition Studies Review, 17*(4), 624–637. doi:10.1007/s11300-010-0178-7

Andrieu, N., Hossard, L., Graveline, N., Dugue, P., Guerra, P., & Chirinda, N. (2021). Covid-19 management by farmers and policymakers in Burkina Faso, Colombia and France: Lessons for climate action. *Agricultural Systems, 190*. Scopus. doi:10.1016/j.agsy.2021.103092

Andrusenko, N., Martynova, L., Sharko, V., Garbazhii, K., Hyrych, S., & Vasylyshyna, O. (2022). Changes in the Organic Products Market as a Result of the 2022 Events in Eastern Europe. *WSEAS Transactions on Environment and Development, 18*, 918–929. doi:10.37394/232015.2022.18.87

Anil, S. M., & Kumar, K. A. (2022). India and the UN. *International Journal of Early Childhood Special Education, 14*(5), 1083–1092. doi:10.9756/INTJECSE/V14I5.107

Antiushko, D., Bozhko, T., Osyka, V., Shapovalova, N., Lozova, T., Fil, M., Sienohonova, L., Dudla, I., Volodavchyk, V., & Sienohonova, H. (2022). Ensuring environmental protection and economic renewal through organic production certification. *Eastern-European Journal of Enterprise Technologies, 4*(13–118), 53–60. doi:10.15587/1729-4061.2022.262328

Apaliya, M. T., Kwaw, E., Osae, R., Alolga, R. N., Aikins, A. S. S., Otoo, G. S., Kaburi, S. A., Lamptey, F. P., & Amo-Broni, C. (2022). THE IMPACT OF COVID-19 ON FOOD SECURITY: GHANA IN REVIEW. *Journal of Food Technology Research, 9*(3), 160–175. Scopus. doi:10.18488/jftr.v9i3.3228

Araujo-Enciso, S. R., & Fellmann, T. (2020). Yield Variability and Harvest Failures in Russia, Ukraine and Kazakhstan and Their Possible Impact on Food Security in the Middle East and North Africa. *Journal of Agricultural Economics, 71*(2), 493–516. Scopus. doi:10.1111/1477-9552.12367

Arias, C., & Cox, T. L. (2001). Estimation of a US dairy sector model by maximum simulated likehood. *Applied Economics, 33*(9), 1201–1211. doi:10.1080/00036840010005797

Arroyo-Lambaer, D., Zambrano, L., Rivas, M. I., Vázquez-Mendoza, D. L., Figueroa, F., Puente-Uribe, M. B., Espinosa-García, A. C., Tapia-Palacios, M. A., Mazari-Hiriart, M., Revollo-Fernández, D., Jiménez-Serna, A., Covarrubias, M., & Sumano, C. (2022). Identifying Urban Agriculture Needs and Challenges for the Implementation of Green Labeling in Xochimilco, Mexico. *Frontiers in Sustainable Cities, 4.* Scopus. doi:10.3389/frsc.2022.892341

Arsene, M. B., & Nkulu Mwine Fyama, J. (2021). Potential threats to agricultural food production and farmers' coping strategies in the marshlands of Kabare in the Democratic Republic of Congo. *Cogent Food and Agriculture, 7*(1). Scopus. doi:10.1080/23311932.2021.1933747

Arshad, M. U., Zhao, Y., Hanif, O., & Fatima, F. (2022). Evolution of Overall Cotton Production and Its Determinants: Implications for Developing Countries Using Pakistan Case. *Sustainability (Switzerland), 14*(2). Scopus. doi:10.3390/su14020840

Asegie, A. M., Adisalem, S. T., & Eshetu, A. A. (2021). The effects of COVID-19 on livelihoods of rural households: South Wollo and Oromia Zones, Ethiopia. *Heliyon, 7*(12). Scopus. doi:10.1016/j.heliyon.2021.e08550

Attia, A. I., Reda, F. M., Patra, A. K., Elnesr, S. S., Attia, Y. A., & Alagawany, M. (2021). Date (Phoenix dactylifera l.) by-products: Chemical composition, nutritive value and applications in poultry nutrition, an updating review. *Animals, 11*(4). doi:10.3390/ani11041133

Auliah, A., Prayitno, G., Ari, I. R. D., Rahmawati, Wardani, L. E., & Meidiana, C. (2022). The Role of Social Capital Facing Pandemic COVID-19 in Tourism Village to Support Sustainable Agriculture (Empirical Evidence from Two Tourism Villages in Indonesia). *Economies, 10*(12). Scopus. doi:10.3390/economies10120320

Ayan, B., Güner, E., & Son-Turan, S. (2022). Blockchain Technology and Sustainability in Supply Chains and a Closer Look at Different Industries: A Mixed Method Approach. *Logistics, 6*(4). Scopus. doi:10.3390/logistics6040085

Aydin, E., & Demir, P. A. (2022). The Effect Of The Pandemic On The Consumption Of Animal Products: The Case Of Kafkas University Of Turkey. *Online Journal of Animal and Feed Research, 12*(1), 37–45. Scopus. doi:10.51227/ojafr.2022.6

Ayo-Lawal, R. A., Ilevbare, O. E., Omotoso, K. O., Omimakinde, E. A., & Ukwuoma, O. (2022). Social assistance and food security during covid-19 pandemic lock-down: Insights from Nigeria. *Future of Food: Journal on Food, Agriculture and Society, 10*(1), 1–14. Scopus. doi:10.17170/kobra-202110144902

Azenzem, R. (2022). Oilseeds development in Morocco in the current international context. *OCL - Oilseeds and Fats, Crops and Lipids, 29.* Scopus. doi:10.1051/ocl/2022032

Badruddoza, S., Carlson, A. C., & McCluskey, J. J. (2022). Long-term dynamics of US organic milk, eggs, and yogurt premiums. *Agribusiness, 38*(1), 45–72. doi:10.1002/agr.21723

Baggio, J. S., Cordova, L. G., Seijo, T. E., Noling, J. W., Whitaker, V. M., & Peres, N. A. (2021). Cultivar Selection Is an Effective and Economic Strategy for Managing Charcoal Rot of Strawberry in Florida. *Plant Disease, 105*(8). doi:10.1094/PDIS-10-20-2250-RE

Bahrami, H., McNairn, H., Mahdianpari, M., & Homayouni, S. (2022). A Meta-Analysis of Remote Sensing Technologies and Methodologies for Crop Characterization. *Remote Sensing, 14*(22). Scopus. doi:10.3390/rs14225633

Bai, Y., Alemu, R., Block, S. A., Headey, D., & Masters, W. A. (2021). Cost and affordability of nutritious diets at retail prices: Evidence from 177 countries. *Food Policy, 99.* doi:10.1016/j.foodpol.2020.101983

Bai, Y., Costlow, L., Ebel, A., Laves, S., Ueda, Y., Volin, N., Zamek, M., & Masters, W. A. (2022). Retail prices of nutritious food rose more in countries with higher COVID-19 case counts. *Nature Food, 3*(5), 325–330. Scopus. doi:10.1038/s43016-022-00502-1

Bailey, E. S., Curcic, M., & Sobsey, M. D. (2022). Persistence of Coronavirus Surrogates on Meat and Fish Products during Long-Term Storage. *Applied and Environmental Microbiology, 88*(12). doi:10.1128/aem.00504-22

Bairagi, S., Mishra, A. K., & Mottaleb, K. A. (2022). Impacts of the COVID-19 pandemic on food prices: Evidence from storable and perishable commodities in India. *PLoS ONE, 17*(3 March). Scopus. doi:10.1371/journal.pone.0264355

Balcilar, M., Gabauer, D., & Umar, Z. (2021). Crude Oil futures contracts and commodity markets: New evidence from a TVP-VAR extended joint connectedness approach. *Resources Policy, 73*, 102219. doi:10.1016/j.resourpol.2021.102219

Baležentis, T., Morkūnas, M., Žičkienė, A., Volkov, A., Ribašauskienė, E., & Štreimikienė, D. (2021). Policies for Rapid Mitigation of the Crisis' Effects on Agricultural Supply Chains: A Multi-Criteria Decision Support System with Monte Carlo Simulation. *Sustainability, 13*(21), 21. doi:10.3390u132111899

Ballester, C., Buesa, I., Soler, E., Besada, C., Salvador, A., Bonet, L., & Intrigliolo, D. S. (2018). Postharvest regulated deficit irrigation in early- and intermediate-maturing loquat trees. *Agricultural Water Management, 205*, 1–8. doi:10.1016/j.agwat.2018.04.030

Bamidele, O., & Amole, T. A. (2021). Impact of covid-19 on smallholder poultry farmers in nigeria. *Sustainability (Switzerland), 13*(20). doi:10.3390/su132011475

Ban, C., Bohle, D., & Naczyk, M. (2022). A perfect storm: COVID-19 and the reorganisation of the German meat industry. *Transfer, 28*(1), 101–118. doi:10.1177/10242589221081943

Bansal, E., Chaturvedi, K. R., Saroha, K., Kaur, J., Mehra, P., Ahmad, F., & Vynnychenko, N. (2020). Covid-19: Social media impact on agricultural prices and consumer's buying behavior. *International Journal of Agricultural and Statistics Sciences, 16*, 953–957.

Baptista, F., Lourenço, P., Fitas da Cruz, V., Silva, L. L., Silva, J. R., Correia, M., Picuno, P., Dimitriou, E., & Papadakis, G. (2021). Which are the best practices for MSc programmes in sustainable agriculture? *Journal of Cleaner Production, 303*. Scopus. doi:10.1016/j.jclepro.2021.126914

Barbosa, F. A. R. (2022). Limnology And The Sustainable Use Of Water In Brazil: Visions And Challenges. *Oecologia Australis, 26*(2), 112–127. Scopus. doi:10.4257/oeco.2022.2602.03

Barigye, R., Hassan, N. A. D., Abdalla-Alfaki, I. M., Al Alawi, H. K. K., Aldhaheri, A. M. B., Ghanim, F. M., Alkhateri, M. A. S., Alalawi, N. S. A., Mohamed, M. E. H., & Mohteshamuddin, K. (2022). Pilot serosurvey of Coxiella burnetii in domesticated small ruminants in the United Arab Emirates. *Tropical Animal Health and Production, 54*(3). doi:10.1007/s11250-022-03150-6

Barrett, D. (2009a). Dairy. *Australian Commodities, 16*(3), 473–479. PMID:19933542

Barrett, D., & Murray, C. (2013). Dairy. *Agricultural Commodities, 3*(3), 102–108. PMID:23832907

Bartol, T. (2023). Smallholders and small-scale agriculture: Mapping and visualization of knowledge domains and research trends. *Cogent Social Sciences, 9*(1). Scopus. doi:10.1080/23311886.2022.2161778

Bastin, B., Thompson, W., Benzinger, M. J., Crowley, E. S., Leonte, A.-M., Vandoros, E. J., Thomas, D., Hughes, A., Crabtree, D., Evans, K., & Sohier, D. (2022). Evaluation of the Thermo Scientific SureTect Salmonella Species PCR Assay in a Broad Range of Foods and Select Environmental Surfaces: Pre-Collaborative and Collaborative Study: First Action 2021.02. *Journal of AOAC International, 105*(1), 167–190. doi:10.1093/jaoacint/qsab122

Belasco, E. J., & Smith, V. (2022). The impact of policy design on payment concentration in Ad-hoc disaster Relief: Lessons from the Market Facilitation and Coronavirus food Assistance programs. *Food Policy, 106*. Scopus. doi:10.1016/j.foodpol.2021.102189

Belsey-Priebe, M., Lyons, D., & Buonocore, J. J. (2021). COVID-19′ s impact on american women's food insecurity foreshadows vulnerabilities to climate change. *International Journal of Environmental Research and Public Health, 18*(13). Scopus. doi:10.3390/ijerph18136867

Ben Hassen, T., El Bilali, H., Allahyari, M. S., Berjan, S., & Fotina, O. (2021). Food purchase and eating behavior during the COVID-19 pandemic: A cross-sectional survey of Russian adults. *Appetite, 165*. Scopus. doi:10.1016/j.appet.2021.105309

Ben Hassen, T., & El Bilali, H. (2022). Impacts of the Russia-Ukraine War on Global Food Security: Towards More Sustainable and Resilient Food Systems? *Foods, 11*(15), 2301. doi:10.3390/foods11152301 PMID:35954068

Ben Rejeb, I., Charfi, I., Baraketi, S., Hached, H., & Gargouri, M. (2022). Bread Surplus: A Cumulative Waste or a Staple Material for High-Value Products? *Molecules, 27*(23). doi:10.3390/molecules27238410

Benedek, Z., Fertö, I., Marreiros, C. G., De Aguiar, P. M., Pocol, C. B., Čechura, L., Poder, A., Pääso, P., & Bakucs, Z. (2021). Farm diversification as a potential success factor for small-scale farmers constrained by COVID-related lockdown. Contributions from a survey conducted in four European countries during the first wave of COVID-19. *PLoS ONE, 16*(5 May). Scopus. doi:10.1371/journal.pone.0251715

Benoit, M., & Laignel, G. (2014). Sheep-for-meat farming systems in French semi-upland area. Adapting to new context: Increased concentrates and energy prices, and new agricultural policy. *International Journal of Sustainable Development, 17*(1), 35–48. doi:10.1504/IJSD.2014.058435

Benzing, A., Piepho, H.-P., Malik, W. A., Finckh, M. R., Mittelhammer, M., Strempel, D., Jaschik, J., Neuendorff, J., Guamán, L., Mancheno, J., Melo, L., Pavón, O., Cangahuamín, R., & Ullauri, J.-C. (2021). Appropriate sampling methods and statistics can tell apart fraud from pesticide drift in organic farming. *Scientific Reports, 11*(1). doi:10.1038/s41598-021-93624-8

Beraich, M., Amzile, K., Laamire, J., Zirari, O., & Fadali, M. A. (2022). Volatility Spillover Effects of the US, European and Chinese Financial Markets in the Context of the Russia-Ukraine Conflict. *International Journal of Financial Studies, 10*(4), 95. doi:10.3390/ijfs10040095

Berhane, Y., Suderman, M., Babiuk, S., & Pickering, B. (2021). Susceptibility of turkeys, chickens and chicken embryos to SARS-CoV-2. *Transboundary and Emerging Diseases, 68*(6), 3038–3042. doi:10.1111/tbed.13970

Bezáková, M., & Bezák, P. (2022). Which sustainability objectives are difficult to achieve? The mid-term evaluation of predicted scenarios in remote mountain agricultural landscapes in Slovakia. *Land Use Policy, 115*. Scopus. doi:10.1016/j.landusepol.2022.106020

Bhandari, G., Lal, P., Chaudhary, U., Haritha, K., Malhotra, R., & Chandel, B. S. (2021). Assessing snowball effect of COVID-19 pandemic on Indian dairy sector. [Scopus.]. *The Indian Journal of Animal Sciences, 91*(12), 1011–1017.

Biadgilign, S., & Yigzaw, M. (2020). Covid-19 in ethiopia: Current situation, missed opportunities, and the risk of health system disruptions. *Pan African Medical Journal, 35*, 1–2. Scopus. doi:10.11604/pamj.supp.2020.35.2.23906

Bilous, S., Kyryliuk, Y., Kyryliuk, I., Proshchalykina, A., & Stachowiak, Z. (2021). Scenario forecasting of consumption of main organic livestock products in Ukraine. *Agricultural and Resource Economics, 7*(3), 22–43. doi:10.51599/are.2021.07.03.02

Bisht, I. S., Rana, J. C., Yadav, R., & Ahlawat, S. P. (2020). Mainstreaming agricultural biodiversity in traditional production landscapes for sustainable development: The Indian scenario. *Sustainability (Switzerland), 12*(24), 1–25. Scopus. doi:10.3390/su122410690

Biswal, J., Vijayalakshmy, K., & Rahman, H. (2020). Impact of COVID-19 and associated lockdown on livestock and poultry sectors in India. *Veterinary World, 13*(9), 1928–1933. Scopus. doi:10.14202/vetworld.2020.1928-1933

Bitsoumanou Nkounkou, J., & Temple, L. (2021). Resilience of the food sector in the face of the Covid-19 crisis and prospects for agricultural policies in the Republic of Congo. *Cahiers Agricultures, 30*. Scopus. doi:10.1051/cagri/2021025

Bjornlund, H., & Rossini, P. (2007). Fundamentals determining prices in the market for water entitlements: An Australian case study. *International Journal of Water Resources Development, 23*(3), 537–553. doi:10.1080/07900620701486020

Bliznashka, L., Ahun, M. N., Velthausz, D., Donco, R., Karuskina-Drivdale, S., Pinto, J., Yousafzai, A. K., & Jeong, J. (2022). Effects of COVID-19 on child health services utilization and delivery in rural Mozambique: A qualitative study. *Health Policy and Planning, 37*(6), 737–746. Scopus. doi:10.1093/heapol/czac009

Blume, L., Hoischen-Taubner, S., & Sundrum, A. (2021). Effects of alfalfa leaf mass as a part of organic feeding strategies on growth and slaughtering performance of dual-purpose roosters: A preliminary study. *European Poultry Science, 85*. doi:10.1399/eps.2021.341

Bogevska, Z., Berjan, S., El Bilali, H., Sadegh Allahyari, M., Radosavac, A., & Davitkovska, M. (2022). Exploring food shopping, consumption and waste habits in North Macedonia during the COVID-19 pandemic. *Socio-Economic Planning Sciences, 82*. Scopus. doi:10.1016/j.seps.2021.101150

Bonnel, G. (2022). Étudier le travail au ras du sol. Une enquête sur le travail en agriculture biologique. *BMS Bulletin of Sociological Methodology/ Bulletin de Methodologie Sociologique, 156*(1), 92–123. doi:10.1177/07591063221128597

Bórawski, P., Bełdycka-Bórawska, A., Parzonko, A., Rokicki, T., & Holden, L. (2022). Changes in the comparative advantage of Polish dairy products. *Agricultural Economics (Czech Republic), 68*(12), 464–475. Scopus. doi:10.17221/322/2022-AGRICECON

Botene, P. H. R., de Azevedo, A. T., & de Arruda Ignácio, P. S. (2021). Blockchain as an enabling technology in the COVID-19 pandemic: A systematic review. *Health and Technology, 11*(6), 1369–1382. Scopus. doi:10.1007/s12553-021-00593-z

Boughton, D., Goeb, J., Lambrecht, I., Headey, D., Takeshima, H., Mahrt, K., Masias, I., Goudet, S., Ragasa, C., Maredia, M. K., Minten, B., & Diao, X. (2021). Impacts of COVID-19 on agricultural production and food systems in late transforming Southeast Asia: The case of Myanmar. *Agricultural Systems, 188*. Scopus. doi:10.1016/j.agsy.2020.103026

Bravo, V. M., Knowles, T. G., & Gallo, C. (2020). Transport, associated handling procedures and behaviour of calves marketed through chilean auction markets. *Animals, 10*(11), 1–13. doi:10.3390/ani10112170

Brosch, C., & Cartanza, G. (2021). Covid-19 acutely impacted the delmarva poultry industry in early 2020. *Delaware Journal of Public Health, 7*(1). doi:10.32481/djph.2021.01.008

Bulut, E., Stout, A., Wemette, M., Llanos-Soto, S., Schell, R. C., Greiner Safi, A., Shapiro, M. A., Moroni, P., & Ivanek, R. (2021). How does public perception of antibiotic use on dairy farms contribute to self-reported purchasing of organic? *Journal of Food Science, 86*(5), 2045–2060. doi:10.1111/1750-3841.15720

Buquera, R. B., & Marques, P. E. M. (2021). Trust relationships involving organic food consumers: A case study in Sorocaba (in the state of São Paulo). *Revista de Economia e Sociologia Rural, 60*(Special Issue). Advance online publication. doi:10.1590/1806-9479.2021.251225

Buraczyńska, B. U., Żelazna, A. E., & Bojanowska, A. B. (2022). The Impact of the COVID-19 Pandemic on the Behaviours of Polish Consumers in the Organic Products Market. *Sustainability (Switzerland), 14*(19). doi:10.3390/su141911821

Burrone, S., Dingacci, G., Dia, M., Bamba, B., Tarchiani, V., Grieco, E., Zini, C., Di Vecchia, A., & Vignaroli, P. (2022). *The role of staple crop production during the Covid-19 outbreak. Evidence for women small producers in Senegal. Applied Economics*. Scopus., doi:10.1080/00036846.2022.2108749

Buzigi, E., & Onakuse, S. (2023). Food price volatility and socio-economic inequalities in poor food consumption status during coronavirus disease-2019 lockdown among slum and non-slum households in urban Nansana municipality, Uganda. *Nutrition Journal, 22*(1). Scopus. doi:10.1186/s12937-023-00836-x

Cabelkova, I., Smutka, L., Rotterova, S., Zhytna, O., Kluger, V., & Mares, D. (2022). The Sustainability of International Trade: The Impact of Ongoing Military Conflicts, Infrastructure, Common Language, and Economic Wellbeing in Post-Soviet Region. *Sustainability, 14*(17), 10840. doi:10.3390u141710840

Cadilhon, J.-J., Pham, N. D., & Maass, B. L. (2016). The tanga dairy platform: Fostering innovations for more efficient dairy chain coordination in Tanzania. *International Journal on Food System Dynamics, 7*(2), 81–91. doi:10.18461/ijfsd.v7i2.723

Cao, L., Li, T., Wang, R., & Zhu, J. (2021). Impact of COVID-19 on China's agricultural trade. *China Agricultural Economic Review, 13*(1), 1–21. Scopus. doi:10.1108/CAER-05-2020-0079

Cao, Y., & Cheng, S. (2021). Impact of COVID-19 outbreak on multi-scale asymmetric spillovers between food and oil prices. *Resources Policy, 74*, 102364. doi:10.1016/j.resourpol.2021.102364 PMID:34584328

Capocasa, F., Balducci, F., Mazzoni, L., Marcellini, M., Qaderi, R., & Mezzetti, B. (2021). Preliminary results of soilless cultivated strawberry cultivars in the autumn-spring cycle in the mid-Adriatic area. In Acta Hortic. (Vol. 1309, p. 596). International Society for Horticultural Science. doi:10.17660/ActaHortic.2021.1309.85

Cardoso, B., Cunha, L., Leiras, A., Gonçalves, P., Yoshizaki, H., de Brito, I. Junior, & Pedroso, F. (2021). Causal Impacts of Epidemics and Pandemics on Food Supply Chains: A Systematic Review. *Sustainability, 13*(17), 17. doi:10.3390u13179799

Cariappa, A. A., Acharya, K. K., Adhav, C. A., R, S., Ramasundaram, P., Kumar, A., Singh, S., & Singh, G. P. (2022). COVID-19 induced lockdown effect on wheat supply chain and prices in India – Insights from state interventions led resilience. *Socio-Economic Planning Sciences, 84*. doi:10.1016/j.seps.2022.101366

Carmona, I., Griffith, D. M., & Aguirre, I. (2021). Understanding the factors limiting organic consumption: The effect of marketing channel on produce price, availability, and price fairness. *Organic Agriculture, 11*(1), 89–103. doi:10.1007/s13165-020-00331-1

Carrière-Swallow, Y., Deb, P., Furceri, D., Jiménez, D., & Ostry, J. D. (2023). Shipping costs and inflation. *Journal of International Money and Finance, 130*. Scopus. doi:10.1016/j.jimonfin.2022.102771

Casado-Aranda, L.-A., Sánchez-Fernández, J., & Viedma-del-Jesús, M. I. (2021). Analysis of the scientific production of the effect of COVID-19 on the environment: A bibliometric study. *Environmental Research, 193*. Scopus. doi:10.1016/j.envres.2020.110416

Castellari, E., Moro, D., Platoni, S., & Sckokai, P. (2018). Retailers' strategies and food price dynamics: Evidence from dairy scanner data. *Food Policy, 74*, 212–224. doi:10.1016/j.foodpol.2017.12.005

Ceballos, F., Kannan, S., & Kramer, B. (2021). Crop prices, farm incomes, and food security during the COVID-19 pandemic in India: Phone-based producer survey evidence from Haryana State. *Agricultural Economics (United Kingdom), 52*(3), 525–542. doi:10.1111/agec.12633

Cechin, A., da Silva Araújo, V., & Amand, L. (2021). Exploring the synergy between Community Supported Agriculture and agroforestry: Institutional innovation from smallholders in a brazilian rural settlement. *Journal of Rural Studies, 81*, 246–258. doi:10.1016/j.jrurstud.2020.10.031

ÇevrimliM. B.MatB.TekindalM. A.GünlüA. (2022). Determination of risk factors in cattle and small ruminant livestock enterprises from the perspective of animal life insurance: A case study in Turkey. *Arquivo Brasileiro de Medicina Veterinaria e Zootecnia, 74*(1), 21–32. doi:10.1590/1678-4162-12524

Chaaya, C., Thambi, V. D., Sabuncu, O., Abedi, R., Osman, A. O. A., Uwishema, O., & Onyeaka, H. (2022). Ukraine—Russia crisis and its impacts on the mental health of Ukrainian young people during the COVID-19 pandemic. *Annals of Medicine and Surgery (London), 79*, 104033. doi:10.1016/j.amsu.2022.104033 PMID:35765517

Chaban, N., Elgström, O., & Knodt, M. (2019). Perceptions of EU mediation and mediation effectiveness: Comparing perspectives from Ukraine and the EU. *Cooperation and Conflict, 54*(4), 506–523. doi:10.1177/0010836718823813

Chandorkar, N., Tambe, S., Amin, P., & Madankar, C. (2021). *A systematic and comprehensive review on current understanding of the pharmacological actions, molecular mechanisms, and clinical implications of the genus Eucalyptus. Phytomedicine Plus, 1(4).* Scopus., doi:10.1016/j.phyplu.2021.100089

Chang, B. H., Gohar, R., Derindag, O. F., & Uche, E. (2022). COVID-19, lockdown measures and their impact on food and healthcare prices: Empirical evidence using a dynamic ARDL model. *Journal of Economic Studies*. Scopus. doi:10.1108/JES-12-2021-0627

Chari, F., Muzinda, O., Novukela, C., & Ngcamu, B. S. (2022). *Pandemic outbreaks and food supply chains in developing countries: A case of COVID-19 in Zimbabwe. Cogent Business and Management, 9(1).* Scopus., doi:10.1080/23311975.2022.2026188

Charlebois, S., Langenbacher, W., & Tamilia, R. D. (2007). The Canadian Dairy Commission: An empirical survey on its relevance in today's civil society. *The International Food and Agribusiness Management Review, 10*(1), 81–99.

Chatellier, V., Chaumet, J.-M., & Pouch, T. (2022). Covid-19, internationalagriculturaleconomy andanimalsectors: The case ofChina, the UnitedStates andthe EU. *Inra Productions Animales, 35*(1), 1–20. Scopus. doi:10.20870/productions-animales.2022.35.1.5376

Chavas, J.-P., & Kim, K. (2005). Cointegration relationships and hedonic pricing of differentiated commodities: An application to price dynamics in the US dairy sector. *Applied Economics, 37*(16), 1813–1827. doi:10.1080/00036840500215311

Chen, Q. (2021). Prospects for chinese-russian cooperation in the dairy sector. *R-Economy, 7*(1), 52–60. Scopus. doi:10.15826/recon.2021.7.1.005

Cheng, S., Li, X., & Cao, Y. (2023). Global evidence of the exposure-lag-response associations between temperature anomalies and food markets. *Journal of Environmental Management, 325*. Scopus. doi:10.1016/j.jenvman.2022.116592

Cheng, Y., Nadeem, M., Haq, S. U., Prisca, K., Aziz, B., Imran, M., & Shahbaz, P. (2022). Maintaining Quality of Life during the Pandemic: Managing Economic, Social, and Health Well-Being Amid the COVID-19 Crisis of Agricultural Entrepreneurs. *Sustainability (Switzerland), 14*(23). Scopus. doi:10.3390/su142315597

Chen, Z., Yan, B., & Kang, H. (2022). Dynamic correlation between crude oil and agricultural futures markets. *Review of Development Economics, 26*(3), 1798–1849. doi:10.1111/rode.12885

Cittadini, E. D., Van Keulen, H., Peri, P. L., & De Ridder, N. (2007). Designing a 'target-tree' for maximizing gross value of product in Patagonian sweet cherry orchards. *International Journal of Fruit Science, 6*(3), 3–22. doi:10.1300/J492v06n03_02

Clem, R. S. (2017). Clearing the Fog of War: Public versus official sources and geopolitical storylines in the Russia-Ukraine conflict. *Eurasian Geography and Economics*, *58*(6), 592–612. doi:10.1080/15387216.2018.1424006

Codjia, C. O., & Saghaian, S. H. (2022). Determinants of Food Expenditure Patterns: Evidence from U.S. Consumers in the Context of the COVID-19 Price Shocks. *Sustainability (Switzerland), 14*(13). Scopus. doi:10.3390/su14138156

Consavage Stanley, K., Harrigan, P. B., Serrano, E. L., & Kraak, V. I. (2022). A systematic scoping review of the literacy literature to develop a digital food and nutrition literacy model for low-income adults to make healthy choices in the online food retail ecosystem to reduce obesity risk. *Obesity Reviews, 23*(4). Scopus. doi:10.1111/obr.13414

Costantini, V., Morando, V., Olk, C., & Tausch, L. (2022). Fuelling the Fire: Rethinking European Policy in Times of Energy and Climate Crises. *Energies*, *15*(20), 7781. doi:10.3390/en15207781

Creutzinger, K., Pempek, J., Habing, G., Proudfoot, K., Locke, S., Wilson, D., & Renaud, D. (2021). Perspectives on the Management of Surplus Dairy Calves in the United States and Canada. *Frontiers in Veterinary Science, 8*. doi:10.3389/fvets.2021.661453

Cross, M. K. D., & Karolewski, I. P. (2017). What Type of Power has the EU Exercised in the Ukraine–Russia Crisis? A Framework of Analysis. *Journal of Common Market Studies*, *55*(1), 3–19. doi:10.1111/jcms.12442

Cruz Nieto, D. D., Rodriguez Espinoza, R. F., Legua Cárdenas, J. A., Natividad Huasupoma, D. E., Ipanaqué Roña, J. M., Abarca Rodríguez, J. J., & Ramos Pacheco, R. L. (2022). Analysis of the Yield of Radish (Raphanus Sativus L.) Fertilized with Compost Based on Organic Waste from Markets and its Relationship with its Stomatal Density. *Chemical Engineering Transactions, 94*, 1387–1392. doi:10.3303/CET2294231

Dagevos, H., & Verbeke, W. (2022). Meat consumption and flexitarianism in the Low Countries. *Meat Science, 192*. doi:10.1016/j.meatsci.2022.108894

Darand, M., Hassanizadeh, S., Marzban, A., Mirzaei, M., & Hosseinzadeh, M. (2022). The association between dairy products and the risk of COVID-19. *European Journal of Clinical Nutrition, 76*(11), 1583–1589. Scopus. doi:10.1038/s41430-022-01149-8

Darnhofer, I. (2021). Resilience or how do we enable agricultural systems to ride the waves of unexpected change? *Agricultural Systems, 187*. Scopus. doi:10.1016/j.agsy.2020.102997

Das, N. K., & Roy, A. (2022). COVID-19 and agri-food value chain: A systematic review and bibliometric mapping. *Journal of Agribusiness in Developing and Emerging Economies, 12*(3), 442–462. Scopus. doi:10.1108/JADEE-07-2021-0188

Das, P. K., & Samanta, I. (2021). Role of backyard poultry in South-East Asian countries: Post COVID-19 perspective. *World's Poultry Science Journal, 77*(2), 415–426. doi:10.1080/00439339.2021.1893620

Dasgupta, S., & Robinson, E. J. Z. (2022). Impact of COVID-19 on food insecurity using multiple waves of high frequency household surveys. *Scientific Reports, 12*(1). Scopus. doi:10.1038/s41598-022-05664-3

Dasgupta, K., & Finch, S. E. (2020). A Case of Pediatric Multisystem Inflammatory Syndrome Temporally Associated with COVID-19 in South Dakota. *South Dakota Medicine : The Journal of the South Dakota State Medical Association, 73*(6), 246–251. PMID:32580256

Das, S., Mandal, S., & Mohanty, S. (2021). Agribusiness Opportunities of Organic Agriculture in West Bengal – An Empirical Analysis. *Indian Journal of Agricultural Economics, 76*(3), 514–532.

de Melo Souza, T. C., Liboredo, J. C., Ferreira, L. G., Daniel, M. M., Di Renzo, L., Pivari, F., & Anastácio, L. R. (2022). Daily Habits of Brazilians at Different Moments of the COVID-19 Pandemic. *Nutrients, 14*(23). Scopus. doi:10.3390/nu14235136

De Roest, K., Menghi, A., & Trindade, H. (2008). Adaptation of Mediterranean dairy systems to the new context of feed prices and European regulation. *Fourrages (Versailles), 2008*(196), 447–459.

Debnath, F., Chakraborty, D., Deb, A., Saha, M., & Dutta, S. (2021). Increased human-animal interface & emerging zoonotic diseases: An enigma requiring multi-sectoral efforts to address. *Indian Journal of Medical Research, 153*(5), 577–584. Scopus. doi:10.4103/ijmr.IJMR_2971_20

Deleuran, L. C., Kristensen, K., Gislum, R., & Boelt, B. (2013). Optimizing the number of consecutive seed harvests in red fescue (Festuca rubra L.) and perennial ryegrass (Lolium perenne L.) for yield, yield components and economic return. *Acta Agriculturae Scandinavica Section B: Soil and Plant Science, 63*(1), 1–10. doi:10.1080/09064710.2012.703229

Deniz, A., & Ozgur, E. M. (2022). The establishment of the Ukrainian diaspora in Turkey through migrant associations: 'We feel as part of the diaspora from now on.'. *Southeast European and Black Sea Studies*, *22*(2), 243–261. doi:10.10 80/14683857.2022.2037820

Dentener, F., Emberson, L., Galmarini, S., Cappelli, G., Irimescu, A., Mihailescu, D., Van Dingenen, R., & Van Den Berg, M. (2020). Lower air pollution during COVID-19 lock-down: Improving models and methods estimating ozone impacts on crops: O3-crop impact during COVID-19. *Philosophical Transactions of the Royal Society A: Mathematical, Physical and Engineering Sciences, 378*(2183). Scopus. doi:10.1098/rsta.2020.0188

Détang-Dessendre, C., Guyomard, H., Réquillart, V., & Soler, L.-G. (2020). Changing agricultural systems and food diets to prevent and mitigate global health shocks. *Sustainability (Switzerland), 12*(16). Scopus. doi:10.3390/su12166462

Dhakal, J., Jia, M., Joyce, J. D., Moore, G. A., Ovissipour, R., & Bertke, A. S. (2021). Survival of severe acute respiratory syndrome coronavirus 2 (Sars-cov-2) and herpes simplex virus 1 (hsv-1) on foods stored at refrigerated temperature. *Foods, 10*(5). doi:10.3390/foods10051005

Dhama, K., Patel, S. K., Yatoo, M. I., Tiwari, R., Sharun, K., Dhama, J., Natesan, S., Malik, Y. S., Singh, K. P., & Harapan, H. (2021). SARS-CoV-2 existence in sewage and wastewater: A global public health concern? *Journal of Environmental Management, 280*. Scopus. doi:10.1016/j.jenvman.2020.111825

Diagourtas, G., Kounetas, K. E., & Simaki, V. (2022). *Consumer attitudes and sociodemographic profiles in purchasing organic food products: Evidence from a Greek and Swedish survey. British Food Journal*. doi:10.1108/BFJ-03-2022-0196

Dietrich, S., Giuffrida, V., Martorano, B., & Schmerzeck, G. (2022). COVID-19 policy responses, mobility, and food prices. *American Journal of Agricultural Economics, 104*(2), 569–588. Scopus. doi:10.1111/ajae.12278

Dimitri, C., & Nehring, R. (2022). Thirty years of organic dairy in the United States: The influences of farms, the market and the organic regulation. *Renewable Agriculture and Food Systems, 37*(6), 588–602. doi:10.1017/S1742170522000254

Djekic, I., Radivojevic, D., & Milivojevic, J. (2019). Quality perception throughout the apple fruit chain. *Journal of Food Measurement and Characterization, 13*(4), 3106–3118. doi:10.1007/s11694-019-00233-1

Djurle, A., Young, B., Berlin, A., Vågsholm, I., Blomström, A.-L., Nygren, J., & Kvarnheden, A. (2022). Addressing biohazards to food security in primary production. *Food Security, 14*(6), 1475–1497. Scopus. doi:10.1007/s12571-022-01296-7

Dokić, D., Gavran, M., Gregić, M., & Gantner, V. (2020). The Impact of Trade Balance of Agri-Food Products on the State's Ability to Withstand the Crisis. *HighTech and Innovation Journal, 1*(3), 107–111. Scopus. doi:10.28991/HIJ-2020-01-03-02

Dong, F. (2006). The outlook for Asian dairy markets: The role of demographics, income, and prices. *Food Policy, 31*(3), 260–271. doi:10.1016/j.foodpol.2006.02.007

Doole, G. J., & Weetman, E. (2009). Tactical management of pasture fallows in Western Australian cropping systems. *Agricultural Systems, 102*(1–3), 24–32. doi:10.1016/j.agsy.2009.06.002

Duflot, B. (2021). Which consequences of african swine fever on the global swine industry and other animal industries? *Bulletin de l'Academie Veterinaire de France, 174*. doi:10.3406/bavf.2021.70947

Dugué, P., Sraïri, M. T., Jamin, J.-Y., Vall, E., Mburu, J., Ndambi, A., Sall, C., Camara, A. D., Sow, A., Ba, K., Corniaux, C., Diaw, A., Seck, D., Vigne, M., Audouin, S., Rakotomalala, L. J. E., Rakotonoely, L. N., Ferreira, F. D., Véromalalanirina, E., … Duteurtre, G. (2021). Early effects of the COVID-19 outbreak on the African dairy industry: Cases of Burkina Faso, Kenya, Madagascar, and Senegal. *Cahiers Agricultures, 30*. Scopus. doi:10.1051/cagri/2020047

Durevall, D., Loening, J. L., & Ayalew Birru, Y. (2013). Inflation dynamics and food prices in Ethiopia. *Journal of Development Economics, 104*, 89–106. doi:10.1016/j.jdeveco.2013.05.002

Dylan, H., & Maguire, T. J. (2022). Secret Intelligence and Public Diplomacy in the Ukraine War. *Survival, 64*(4), 33–74. doi:10.1080/00396338.2022.2103257

Dzanku, F. M. (2015). Household-specific food price differentials and high-value crop production in rural Ghana. *Food Policy, 57*, 73–82. doi:10.1016/j.foodpol.2015.09.004

Edwards, R., Halimatussadiah, A., Moeis, F. R., & Maulia, R. F. (2022). Agriculture, Development and Sustainability in the Covid-19 Era. *Bulletin of Indonesian Economic Studies, 58*(1), 1–30. doi:10.1080/00074918.2022.2056935

Ejeromedoghene, O., Tesi, J. N., Uyanga, V. A., Adebayo, A. O., Nwosisi, M. C., Tesi, G. O., & Akinyeye, R. O. (2020). Food security and safety concerns in animal production and public health issues in Africa: A perspective of COVID-19 pandemic era. *Ethics, Medicine and Public Health, 15*. doi:10.1016/j.jemep.2020.100600

Ekstrand, K., Flanagan, A. J., Lin, I. E., Vejseli, B., Cole, A., Lally, A. P., Morris, R. L., & Morgan, K. N. (2021). Animal Transmission of SARS-CoV-2 and the Welfare of Animals during the COVID-19 Pandemic. *Animals (Basel), 11*(7), 7. Advance online publication. doi:10.3390/ani11072044 PMID:34359172

ElAshmawy, W. R., Aly, S. S., & Farouk, M. M. (2023). Decision tree risk analysis for FMD outbreak prevention in Egyptian feedlots. *Preventive Veterinary Medicine, 211*. doi:10.1016/j.prevetmed.2022.105820

Elbasuney, S., El-Sayyad, G. S., Attia, M. S., & Abdelaziz, A. M. (2022). Ferric Oxide Colloid: Towards Green Nano-Fertilizer for Tomato Plant with Enhanced Vegetative Growth and Immune Response Against Fusarium Wilt Disease. *Journal of Inorganic and Organometallic Polymers and Materials, 32*(11), 4270–4283. doi:10.100710904-022-02442-6 PMID:35910584

Elgström, O., Chaban, N., Knodt, M., Müller, P., & Pardo, S. (2018). Perceptions of the EU's Role in the Ukraine-Russia and the Israel-Palestine Conflicts: A Biased Mediator? *International Negotiation, 23*(2), 299–318. doi:10.1163/15718069-23021154

Éliás, B. A., & Jámbor, A. (2021). Food security and covid-19: A systematic review of the first-year experience. *Sustainability (Switzerland), 13*(9). Scopus. doi:10.3390/su13095294

Espinoza Lastra, Ó. R., Lluglla Luna, L. A., & Padilla Buñay, A. (2021). Post-covid-19 rural entrepreneurship: Considerations for a sustainable economic recovery. [Scopus.]. *Universidad y Sociedad, 13*(S1), 443–451.

Esteves, R. C., do Amaral Vendramini, A. L., & Accioly, F. (2021). A qualitative meta-synthesis study of the convergence between organic crop regulations in the United States, Brazil, and Europe. *Trends in Food Science and Technology, 107*, 343–357. doi:10.1016/j.tifs.2020.10.044

Eurostat. (2023). *Several Statistics and Information*. https://ec.europa.eu/eurostat

Ewert, B., & Loer, K. (2022). *COVID-19 as a Catalyst for Policy Change: The Role of Trigger Points and Spillover Effects. German Politics*. Scopus., doi:10.1080/09644008.2022.2088733

Ezanno, P., Arnoux, S., Joly, A., & Vermesse, R. (2022). Rewiring cattle trade movements helps to control bovine paratuberculosis at a regional scale. *Preventive Veterinary Medicine, 198*. doi:10.1016/j.prevetmed.2021.105529

Fang, P., Belton, B., Zhang, X., & Ei Win, H. (2021). Impacts of COVID-19 on Myanmar's chicken and egg sector, with implications for the sustainable development goals. *Agricultural Systems, 190*. doi:10.1016/j.agsy.2021.103094

Fang, Y., & Shao, Z. (2022). The Russia-Ukraine conflict and volatility risk of commodity markets. *Finance Research Letters, 50*, 103264. doi:10.1016/j.frl.2022.103264

FAOSTAT. (2023). *Several Statistics and Information*. FAO. https://www.fao.org/faostat/en/#home

Farias da Cruz, M., Baraúna Magno, M., Alves Jural, L., Pimentel, T. C., Masterson Tavares Pereira Ferreira, D., Almeida Esmerino, E., Luis Paiva Anciens Ramos, G., Vicente Gomila, J., Cristina Silva, M., Cruz, A. G. D., da Silva Fidalgo, T. K., & Cople Maia, L. (2022). Probiotics and dairy products in dentistry: A bibliometric and critical review of randomized clinical trials. *Food Research International, 157*. Scopus. doi:10.1016/j.foodres.2022.111228

Farias, L. F., Guimarães Soares, J. P., & de Sousa, R. L. (2022). The organic market and the prices practiced in the main commercialization channels in the city of Goiânia, GO. *Iheringia - Serie Botanica, 77*. doi:10.21826/2446-82312022v77e2022009

Feng, F., Jia, N., & Lin, F. (2023). *Quantifying the impact of Russia–Ukraine crisis on food security and trade pattern: Evidence from a structural general equilibrium trade model. China Agricultural Economic Review*. doi:10.1108/CAER-07-2022-0156

Fielding, H. R., McKinley, T. J., Delahay, R. J., Silk, M. J., & McDonald, R. A. (2021). Characterization of potential superspreader farms for bovine tuberculosis: A review. *Veterinary Medicine and Science, 7*(2), 310–321. doi:10.1002/vms3.358

Fink, M. (2022). Naval Blockade and the Russia-Ukraine Conflict. *Netherlands International Law Review, 69*(3), 411–437. doi:10.100740802-022-00228-8

Firmansyah, Afriani, H., & Hoesni, F. (2022). The Volatility Of Broiler Meat Prices In Traditional And Modern Markets Before, At The Beginning Of, And During The Covid-19 Pandemic In Jambi City. *Xinan Jiaotong Daxue Xuebao/Journal of Southwest Jiaotong University, 57*(3), 138–151. Scopus. doi:10.35741/issn.0258-2724.57.3.12

Fleming, E., Griffith, G., Mounter, S., Hartmann, M., & Simons, J. (2021). Food Value Chain Coordination in Practice: European and Australian Case Studies of the Creation of Chain Good Innovations*. *International Journal on Food System Dynamics, 12*(3), 196–205. doi:10.18461/ijfsd.v12i2.85

Flis, Z., Szczecina, J., & Molik, E. (2022). The role of sheep's milk bioactive substances in the prevention of metabolic and viral diseases. *Journal of Animal and Feed Sciences, 31*(3), 211–216. Scopus. doi:10.22358/jafs/151020/2022

Foltz, J. D. (2004). Entry, exit, and farm size: Assessing an experiment in dairy price policy. *American Journal of Agricultural Economics, 86*(3), 594–604. doi:10.1111/j.0002-9092.2004.00603.x

Fouotsa Manfouo, N. C., & Von Fintel, D. (2022). Investigating the effects of drought and lockdowns on smallholder and commercial agricultural production in KwaZulu-Natal using remotely sensed data. *Heliyon, 8*(11). Scopus. doi:10.1016/j.heliyon.2022.e11637

Franch, B., Vermote, E., Skakun, S., Santamaria-Artigas, A., Kalecinski, N., Roger, J.-C., Becker-Reshef, I., Barker, B., Justice, C., & Sobrino, J. A. (2021). The ARYA crop yield forecasting algorithm: Application to the main wheat exporting countries. *International Journal of Applied Earth Observation and Geoinformation, 104*. Scopus. doi:10.1016/j.jag.2021.102552

Fujimoto, Y., Ito, H., Higuchi, H., Ohno, H., & Makita, K. (2020). A case-control study of herd- and cow-level risk factors associated with an outbreak of Mycoplasma mastitis in Nemuro, Japan. *Preventive Veterinary Medicine, 177*. doi:10.1016/j.prevetmed.2020.104946

Fulber, J. P. C., & Kamen, A. A. (2022). Development and Scalable Production of Newcastle Disease Virus-Vectored Vaccines for Human and Veterinary Use. *Viruses, 14*(5). doi:10.3390/v14050975

Funsho-Sanni, O. O., Ella, E., Samuel Sanni, O., Inabo, H., Abdulkarim luka, S., & Eleyi Ameh, R. (2022). Serological evaluation of Newcastle disease protection among broilers at a live bird market in Kano, Northwest Nigeria, and its epidemiological significance. *Journal of Immunoassay and Immunochemistry, 43*(5), 534–545. doi:10.1080/15321819.2022.2029744

Gaffney, J., Girma, D., Kane, N. A., Llaca, V., Mace, E., Taylor, N., & Tibebu, R. (2022). Maximizing value of genetic sequence data requires an enabling environment and urgency. *Global Food Security, 33*. Scopus. doi:10.1016/j.gfs.2022.100619

Gaio, L. E., Stefanelli, N. O., Pimenta, T. Junior, Bonacim, C. A. G., & Gatsios, R. C. (2022). The impact of the Russia-Ukraine conflict on market efficiency: Evidence for the developed stock market. *Finance Research Letters, 50*, 103302. doi:10.1016/j.frl.2022.103302

García, E. J., Damián, M. A. M., & Cruz, A. K. (2008). Support price policy versus direct payments to producers: Analisys of producers' welfare. *Revista Fitotecnia Mexicana, 31*(4), 383–389.

Gascón, J. (2020). COVID-19, state of emergency and family farming in Spain: Rural markets in Barcelona and self-consumption orchards in Alcaine (Teruel). *Ager, 2020*(30), 177–206. Scopus. doi:10.4422/ager.2020.13

Gatto, M., & Islam, A. H. M. S. (2021). Impacts of COVID-19 on rural livelihoods in Bangladesh: Evidence using panel data. *PLoS ONE, 16*(11 November). Scopus. doi:10.1371/journal.pone.0259264

George, J., Häsler, B., Komba, E., Sindato, C., Rweyemamu, M., & Mlangwa, J. (2021). Towards an integrated animal health surveillance system in Tanzania: Making better use of existing and potential data sources for early warning surveillance. *BMC Veterinary Research, 17*(1). doi:10.1186/s12917-021-02789-x

Ghalibaf, M. B., Gholami, M., & Mohammadian, N. (2022). Stability of Food Security in Iran; Challenges and Ways Forward: A Narrative Review. [Scopus.]. *Iranian Journal of Public Health, 51*(12), 2654–2663. doi:10.18502/ijph.v51i12.11456 PMID:36742249

Ghanbari, S., Vaezin, S. M. H., Shamekhi, T., Eastin, I. L., Lovrić, N., & Aghai, M. M. (2020). The Economic and Biological Benefits of Non-wood Forest Products to Local Communities in Iran. *Economic Botany, 74*(1), 59–73. doi:10.1007/s12231-019-09478-9

Ghelfi, R., & Palmieri, A. (2017). Pulses production in Italy: Trade, marketing and. doi:10.4081/ija.2017.891

Ghosh-Jerath, S., Kapoor, R., Dhasmana, A., Singh, A., Downs, S., & Ahmed, S. (2022). Effect of COVID-19 Pandemic on Food Systems and Determinants of Resilience in Indigenous Communities of Jharkhand State, India: A Serial Cross-Sectional Study. *Frontiers in Sustainable Food Systems, 6*. doi:10.3389/fsufs.2022.724321

Gibson, M., Pereira, J. P., Slade, R., & Rogelj, J. (2022). *Agent-Based Modelling of Future Dairy and Plant-Based Milk Consumption for UK Climate Targets. JASSS, 25(2).* doi:10.18564/jasss.4801

Gilbert, M. S., Pantophlet, A. J., van den Borne, J. J. G. C., Hendriks, W. H., Schols, H. A., & Gerrits, W. J. J. (2016). Effects of replacing lactose from milk replacer by glucose, fructose, or glycerol on energy partitioning in veal calves. *Journal of Dairy Science, 99*(2), 1121–1132. doi:10.3168/jds.2015-10062

Goeb, J., Maredia, M., Win, K. Z., Masias, I., Lambrecht, I., Boughton, D., & Minten, B. (2022). Urban food prices under lockdown: Evidence from Myanmar's traditional food retail sector during COVID-19. *Applied Economics, 54*(47), 5412–5441. Scopus. doi:10.1080/00036846.2022.2044999

Goeb, J., Zone, P. P., Kham Synt, N. L., Zu, A. M., Tang, Y., & Minten, B. (2022). Food prices, processing, and shocks: Evidence from rice and COVID-19. *Journal of Agricultural Economics, 73*(2), 338–355. Scopus. doi:10.1111/1477-9552.12461

Gómez Serna, L. C., & Bernal Rivas, J. (2020). Food production, access and dietary diversity in agroecological farming families in times of COVID-19. [Scopus.]. *Agroalimentaria, 26*(51), 39–52.

Górska, P., Górna, I., Miechowicz, I., & Przysławski, J. (2021a). Changes in eating behaviour during sars-cov-2 pandemic among the inhabitants of five European countries. *Foods, 10*(7). doi:10.3390/foods10071624

Górska, P., Górna, I., Miechowicz, I., & Przysławski, J. (2021b). Changes in life situations during the sars-cov-2 virus pandemic and their impact on eating behaviors for residents of europe, australia as well as North and South America. *Nutrients, 13*(10). doi:10.3390/nu13103570

Grinberga-Zalite, G., Pilvere, I., Muska, A., & Kruzmetra, Z. (2021). Resilience of meat supply chains during and after covid-19 crisis. *Emerging Science Journal, 5*(1), 57–66. doi:10.28991/esj-2021-01257

Gruère, G., & Brooks, J. (2021). Viewpoint: Characterising early agricultural and food policy responses to the outbreak of COVID-19. *Food Policy, 100*. Scopus. doi:10.1016/j.foodpol.2020.102017

Gualtieri, P., Marchetti, M., Frank, G., Cianci, R., Bigioni, G., Colica, C., Soldati, L., Moia, A., De Lorenzo, A., & Di Renzo, L. (2023). Exploring the Sustainable Benefits of Adherence to the Mediterranean Diet during the COVID-19 Pandemic in Italy. *Nutrients, 15*(1). doi:10.3390/nu15010110

Guiné, R. P. F., Florença, S. G., Costa, D. T. V. A., Çelik, S., Ferreira, M., Cardoso, A. P., Çetin, S., & Costa, C. A. (2022). Comparative Study about the Consumption of Organic Food Products on Samples of Portuguese and Turkish Consumers under the COVID-19 Pandemic Context. *Agronomy, 12*(6). Scopus. doi:10.3390/agronomy12061385

Guo, J., & Tanaka, T. (2022). Do biofuel production and financial speculation in agricultural commodities influence African food prices? New evidence from a TVP-VAR extended joint connectedness approach. *Energy Economics, 116*. Scopus. doi:10.1016/j.eneco.2022.106422

Guo, J., Song, W., Ni, X., Zhou, K., Wu, J., Liu, W., Xia, W., He, F., Tong, X., Fan, G., Deng, Z., Liao, Z., Chen, H., & Chen, S. (2022). The Impact of the Closure of the Live Poultry Market due to COVID-19 on the Avian Influenza Virus in Nanchang, Jiangxi Province, China. *American Journal of Tropical Medicine and Hygiene, 106*(1), 127–131. doi:10.4269/ajtmh.21-0732

Gurbuz, I. B., & Ozkan, G. (2021). Will agriculture beat the odds against covid-19? The covid-19 outbreak and its effect on agricultural supply in Turkey. *New Medit, 20*(2), 15–25. Scopus. doi:10.30682/nm2102b

Gutiérrez-Moya, E., Adenso-Díaz, B., & Lozano, S. (2021). Analysis and vulnerability of the international wheat trade network. *Food Security, 13*(1), 113–128. Scopus. doi:10.1007/s12571-020-01117-9

Gutiérrez-Romero, R. (2022). Conflicts increased in Africa shortly after COVID-19 lockdowns, but welfare assistance reduced fatalities. *Economic Modelling, 116*. Scopus. doi:10.1016/j.econmod.2022.105991

Guzel-Seydim, Z. B., Gökırmaklı, Ç., & Greene, A. K. (2021). A comparison of milk kefir and water kefir: Physical, chemical, microbiological and functional properties. *Trends in Food Science and Technology, 113*, 42–53. Scopus. doi:10.1016/j.tifs.2021.04.041

Hafez, H. M., & Attia, Y. A. (2020). Challenges to the Poultry Industry: Current Perspectives and Strategic Future After the COVID-19 Outbreak. *Frontiers in Veterinary Science, 7*. Scopus. doi:10.3389/fvets.2020.00516

Hafez, H. M., Attia, Y. A., Bovera, F., Abd El-Hack, M. E., Khafaga, A. F., & de Oliveira, M. C. (2021). Influence of COVID-19 on the poultry production and environment. *Environmental Science and Pollution Research, 28*(33), 44833–44844. doi:10.1007/s11356-021-15052-5

Haji Ali, B., Shahin, M. S., Masoumi Sangani, M. M., Faghihinezhad, M., & Baghdadi, M. (2021). Wastewater aerosols produced during flushing toilets, WWTPs, and irrigation with reclaimed municipal wastewater as indirect exposure to SARS-CoV-2. *Journal of Environmental Chemical Engineering, 9*(5), 106201. doi:10.1016/j.jece.2021.106201 PMID:34405082

Hambardzumyan, G., & Gevorgyan, S. (2022). *The impact of COVID-19 on the small and medium dairy farms and comparative analysis of customers' behavior in Armenia. Future Foods, 5*. Scopus., doi:10.1016/j.fufo.2021.100110

Hamilton, H., Henry, R., Rounsevell, M., Moran, D., Cossar, F., Allen, K., Boden, L., & Alexander, P. (2020). Exploring global food system shocks, scenarios and outcomes. *Futures, 123*. Scopus. doi:10.1016/j.futures.2020.102601

Hammond, J., Siegal, K., Milner, D., Elimu, E., Vail, T., Cathala, P., Gatera, A., Karim, A., Lee, J.-E., Douxchamps, S., Tu, M. T., Ouma, E., Lukuyu, B., Lutakome, P., Leitner, S., Wanyama, I., Thi, T. P., Phuc, P. T. H., Herrero, M., & van Wijk, M. (2022). Perceived effects of COVID-19 restrictions on smallholder farmers: Evidence from seven lower- and middle-income countries. *Agricultural Systems, 198*. doi:10.1016/j.agsy.2022.103367

Han, G., & Grudens-Schuck, N. (2022). Motivations and Challenges for Adoption of Organic Grain Production: A Qualitative Study of Iowa Organic Farmers. *Foods, 11*(21). doi:10.3390/foods11213512

Haqiqi, I., & Bahalou Horeh, M. (2021). Assessment of COVID-19 impacts on U.S. counties using the immediate impact model of local agricultural production (IMLAP). *Agricultural Systems, 190*. Scopus. doi:10.1016/j.agsy.2021.103132

Haque, M. H., Islam, M. A., Karim, M. R., Kayesh, M. E. H., Sarker, S., Nazir, K. H. M. N. H., & Anwer, M. S. (2021). Coronavirus disease 2019 and future pandemics: Impacts on livestock health and production and possible mitigation measures. *Veterinary World, 14*(9), 2434–2443. Scopus. doi:10.14202/vetworld.2021.2434-2443

Harris, J., Depenbusch, L., Pal, A. A., Nair, R. M., & Ramasamy, S. (2020). Food system disruption: Initial livelihood and dietary effects of COVID-19 on vegetable producers in India. *Food Security, 12*(4), 841–851. Scopus. doi:10.1007/s12571-020-01064-5

Hart, T. G., Davids, Y. D., Rule, S., Tirivanhu, P., & Mtyingizane, S. (2022). The COVID-19 pandemic reveals an unprecedented rise in hunger: The South African Government was ill-prepared to meet the challenge. *Scientific African, 16*. Scopus. 10.1016/j.sciaf.2022.e01169

Hassan, M. M., El Zowalaty, M. E., Lundkvist, Å., Järhult, J. D., Khan Nayem, M. R., Tanzin, A. Z., Badsha, M. R., Khan, S. A., & Ashour, H. M. (2021). Residual antimicrobial agents in food originating from animals. *Trends in Food Science and Technology, 111*, 141–150. doi:10.1016/j.tifs.2021.01.075

Hassan, N. A. D., Mohteshamuddin, K., Anthony, A., Al Aiyan, A., Mohamed, M. E. H., Abdalla Alfaki, I. M., & Barigye, R. (2020). Serological evidence of enzootic bovine leukosis in the periurban dairy cattle production system of Al Ain, United Arab Emirates. *Tropical Animal Health and Production, 52*(5), 2327–2332. doi:10.1007/s11250-020-02262-1

Hauck, K., Szolnoki, G., & Pabst, E. (2021). Motivation factors for organic wines. An analysis from the perspective of German producers and retailers. *Wine Economics and Policy, 10*(2), 61–74. doi:10.36253/wep-9893

Head, J. R., Chanthavilay, P., Catton, H., Vongsitthi, A., Khamphouxay, K., & Simphaly, N. (2022). Changes in household food security, access to health services and income in northern Lao PDR during the COVID-19 pandemic: A cross-sectional survey. *BMJ Open, 12*(6). Scopus. doi:10.1136/bmjopen-2021-055935

Heaney, A. K., Head, J. R., Broen, K., Click, K., Taylor, J., Balmes, J. R., Zelner, J., & Remais, J. V. (2021). Coccidioidomycosis and COVID-19 co-infection, United States, 2020. *Emerging Infectious Diseases, 27*(5), 1266–1273. Scopus. doi:10.3201/eid2705.204661

Helfenstein, J., Bürgi, M., Debonne, N., Dimopoulos, T., Diogo, V., Dramstad, W., Edlinger, A., Garcia-Martin, M., Hernik, J., Kizos, T., Lausch, A., Levers, C., Mohr, F., Moreno, G., Pazur, R., Siegrist, M., Swart, R., Thenail, C., Verburg, P. H., … Herzog, F. (2022). Farmer surveys in Europe suggest that specialized, intensive farms were more likely to perceive negative impacts from COVID-19. *Agronomy for Sustainable Development, 42*(5). doi:10.1007/s13593-022-00820-5

Henrikson, A. K. (2022). The Trauma of Territorial Break-up: The Russia-Ukraine Conflict and Its International Management-Geopolitical Strategy and Diplomatic Therapy. *Geopoliticas-Revista De Estudios Sobre Espacio Y Poder, 13*(1), 15–40. doi:10.5209/geop.81880

Henry, F. J., Lawrence, B., & Nelson, M. (2022). *Comparative cost of diets for low-income families in the Caribbean. Revista Panamericana de Salud Publica/Pan American Journal of Public Health, 46*. Scopus., doi:10.26633/RPSP.2022.120

Hernández Castaño, C. (2021). Advances in the production of knowledge on biofertilizers and new economic scenarios for Peru. *Revista Venezolana de Gerencia, 26*(Special Issue 5), 708–721. Scopus. doi:10.52080/rvgluz.26.e5.45

Hernández-Vásquez, A., Visconti-Lopez, F. J., & Vargas-Fernández, R. (2022). Factors Associated with Food Insecurity in Latin America and the Caribbean Countries: A Cross-Sectional Analysis of 13 Countries. *Nutrients, 14*(15). Scopus. doi:10.3390/nu14153190

Hidalgo, G., Monticelli, J. M., Pedroso, J., Verschoore, J. R., & De Matos, C. A. (2022). The Influence of Formal Institution Agents on Coopetition in the Organic Food Industry. *Journal of Agricultural and Food Industrial Organization, 20*(2), 61–74. doi:10.1515/jafio-2019-0009

Higashi, R. T., Sood, A., Conrado, A. B., Shahan, K. L., Leonard, T., & Pruitt, S. L. (2022). Experiences of increased food insecurity, economic and psychological distress during the COVID-19 pandemic among Supplemental Nutrition Assistance Program-enrolled food pantry clients. *Public Health Nutrition, 25*(4), 1027–1037. Scopus. doi:10.1017/S1368980021004717

Hoanga, H. T. K., Thompson, W., & Kwon, S. (2021). Low-income household food consumption consequences of rice policy and pandemic impacts on income and price in Thailand. [Scopus.]. *Journal of Food Distribution Research, 52*(2), 76–100.

Hobbs, E. C., & Reid, T. J. (2021). Animals and SARS-CoV-2: Species susceptibility and viral transmission in experimental and natural conditions, and the potential implications for community transmission. *Transboundary and Emerging Diseases, 68*(4), 1850–1867. doi:10.1111/tbed.13885

HoteitM.MortadaH.Al-JawaldehA.IbrahimC.MansourR. (2022). COVID-19 home isolation and food consumption patterns: Investigating the correlates of poor dietary diversity in Lebanon: A cross-sectional study. *F1000Research, 11*. Scopus. doi:10.12688/f1000research.75761.1

Howden, M., McCarthy, O., & Hyde, M. (2016). The EU dairy industry. *Agricultural Commodities*, *6*(2), 126–139.

Huang, J., Cao, Y., & Zhong, P. (2022). Searching for a safe haven to crude oil: Green bond or precious metals? *Finance Research Letters*, *50*, 103303. doi:10.1016/j.frl.2022.103303

Huang, J., Li, Y., Suleman, M. T., & Zhang, H. (2021). Effects of Geopolitical Risks on Gold Market Return Dynamics: Evidence from a Nonparametric Causality-in-quantiles Approach. *Defence and Peace Economics*, 1–15. doi:10.1080/10242694.2021.2007333

Huang, M., Shao, W., & Wang, J. (2023). Correlations between the crude oil market and capital markets under the Russia-Ukraine conflict: A perspective of crude oil importing and exporting countries. *Resources Policy*, *80*, 103233. doi:10.1016/j.resourpol.2022.103233

Iannucci, G., & Sacchi, G. (2021). The evolution of organic market between third-party certification and participatory guarantee systems. *Bio-Based and Applied Economics, 10*(3), 239–251. doi:10.36253/bae-10470

Ibar-Alonso, R., Quiroga-Garcia, R., & Arenas-Parra, M. (2022). Opinion Mining of Green Energy Sentiment: A Russia-Ukraine Conflict Analysis. *Mathematics*, *10*(14), 2532. doi:10.3390/math10142532

Ibiyemi, T. E., & Oldewage-Theron, W. (2022). A review of food and nutrition security in Nigeria. *African Journal of Food, Agriculture, Nutrition and Development, 22*(1).[REMOVED HYPERLINK FIELD]

Ibiyemi, T. E., & Oldewage-Theron, W. (2022). A review of food and nutrition security in nigeria. *African Journal of Food, Agriculture, Nutrition and Development*, *22*(1).

Iheme, G. O., Adile, A. D., Egechizuorom, I. M., Kupoluyi, O. E., Ogbonna, O. C., Olah, L. E., Enuka, H. C., Idris, H., Asouzu, N. C., & Oyebamiji, E. A. (2022). Impact of COVID-19 pandemic on food price index in Nigeria. *Future of Food: Journal on Food, Agriculture and Society, 10*(3), 1–9. Scopus. doi:10.17170/kobra-202204136008

Ikehi, M. E., Ifeanyieze, F. O., Onu, F. M., Ejiofor, T. E., & Nwankwo, C. U. (2022). *Assessing climate change mitigation and adaptation strategies and agricultural innovation systems in the Niger Delta. GeoJournal*. Scopus., doi:10.100710708-022-10596-6

Inglese, P., Gugliuzza, G., & Liguori, G. (2009). Fruit production of cultivated cacti: A short overview on plant ecophysiology and C budget. In Acta Hortic. (Vol. 811, p. 86). International Society for Horticultural Science. doi:10.17660/ActaHortic.2009.811.6

Ingram, C., Downey, V., Roe, M., Chen, Y., Archibald, M., Kallas, K.-A., Kumar, J., Naughton, P., Uteh, C. O., Rojas-Chaves, A., Shrestha, S., Syed, S., Cléirigh Büttner, F., Buggy, C., & Perrotta, C. (2021). COVID-19 Prevention and Control Measures in Workplace Settings: A Rapid Review and Meta-Analysis. *International Journal of Environmental Research and Public Health*, *18*(15), 7847. doi:10.3390/ijerph18157847 PMID:34360142

Ivascu, L., Frank Ahimaz, D., Arulanandam, B. V., & Tirian, G.-O. (2021). The perception and degree of adoption by urbanites towards urban farming. *Sustainability (Switzerland), 13*(21). doi:10.3390/su132112151

Jacobs, M., McDade, T. R., Chaparro, M. V., & Corea, M. (2022). Sick, Hungry, and Vulnerable: Federal Stimulus and Food Security on Marginalized Populations During the COVID-19 Pandemic. *Journal of Racial and Ethnic Health Disparities*. Scopus. doi:10.1007/s40615-022-01447-8

Jagtap, S., Trollman, H., Trollman, F., Garcia-Garcia, G., Parra-Lopez, C., Duong, L., Martindale, W., Munekata, P. E. S., Lorenzo, J. M., Hdaifeh, A., Hassoun, A., Salonitis, K., & Afy-Shararah, M. (2022). The Russia-Ukraine Conflict: Its Implications for the Global Food Supply Chains. *Foods, 11*(14), 2098. doi:10.3390/foods11142098 PMID:35885340

Jean-Pierre, R. P., Hagerman, A. D., & Rich, K. M. (2022). An analysis of African Swine Fever consequences on rural economies and smallholder swine producers in Haiti. *Frontiers in Veterinary Science, 9*. doi:10.3389/fvets.2022.960344

Jha, P. K., Araya, A., Stewart, Z. P., Faye, A., Traore, H., Middendorf, B. J., & Prasad, P. V. V. (2021). Projecting potential impact of COVID-19 on major cereal crops in Senegal and Burkina Faso using crop simulation models. *Agricultural Systems, 190*. Scopus. doi:10.1016/j.agsy.2021.103107

Jha, A., Bhatt, N., Dixit, A. K., Singh, U. K., & Choudhary, L. R. (2022). Y Role of Religion and Politics in War Between Russia and Ukraine: A Study. *International Journal of Early Childhood Special Education, 14*(4), 1181–1185. doi:10.9756/INT-JECSE/V14I4.156

Jiménez, D. E., & Saldarriaga-Isaza, A. (2022). Economic impacts of agricultural policy responses to the outbreak of COVID-19. *Estudios Gerenciales, 38*(163), 200–210. Scopus. doi:10.18046/j.estger.2022.163.4924

Johansson, E., Martin, R., & Mapunda, K. M. (2023). *Participatory future visions of collaborative agroecological farmer-pastoralist systems in Tanzania. Agroecology and Sustainable Food Systems*. doi:10.1080/21683565.2023.2165592

Johnson, R. A. B., Hann, K., Leno, A., Timire, C., Bangura, A. J. A., George, M. I. Z., Davtyan, H., Satyanarayana, S., Nair, D., Mansaray, A. H. D., Bangura, F. I., Kanu, J. S., & Edwards, J. K. (2022). Pesticide Importation in Sierra Leone, 2010–2021: Implications for Food Production and Antimicrobial Resistance. *International Journal of Environmental Research and Public Health, 19*(8). Scopus. doi:10.3390/ijerph19084792

Joshi, T., Poudel, R. P., Kafle, K., Bhattarai, B., Prasai, B. P., & Adhikari, S. (2022). Assessing the Impacts of the COVID-19 Pandemics on Sustainable Development Goals in Nepal. *Frontiers in Sustainable Food Systems, 6*. Scopus. doi:10.3389/fsufs.2022.852759

Jouili, M., & Elloumi, M. (2022). Extraversion versus développement agricole autocentré: Le cas des pays du Maghreb. *Canadian Journal of Development Studies*. Scopus. doi:10.1080/02255189.2022.2092459

Jovicich, E., VanSickle, J. J., Cantliffe, D. J., & Stoffella, P. J. (2005). Greenhouse-grown colored peppers: A profitable alternative for vegetable production in Florida? *HortTechnology, 15*(2), 355–369. doi:10.21273/HORTTECH.15.2.0355

Just, M., & Echaust, K. (2022). Dynamic spillover transmission in agricultural commodity markets: What has changed after the COVID-19 threat? *Economics Letters, 217*, 110671. doi:10.1016/j.econlet.2022.110671

Kaicker, N., Gupta, A., & Gaiha, R. (2022). Covid-19 pandemic and food security in India: Can authorities alleviate the disproportionate burden on the disadvantaged? *Journal of Policy Modeling, 44*(5), 963–980. Scopus. doi:10.1016/j.jpolmod.2022.08.001

Karagöl, E. T., Akyüz, M., & Karul, C. (2021). *Is health management of the COVID-19 pandemic a cause of agricultural commodity prices? New evidences from bootstrap fourier causality test. Duzce Medical Journal, 23(Special Issue 1), 86–93*. Scopus., doi:10.18678/dtfd.879572

Karcher, E. L., Koltes, D., Wenner, B., & Wells, J. (2022). Sparking curiosity and engagement through online curriculum. *Poultry Science, 101*(2). doi:10.1016/j.psj.2021.101577

Karelakis, C. (2019). Redefining integration of the European dairy markets: Do they converge? *Agricultural Economics Review, 20*(1), 27–41.

Karolewski, I. P., & Cross, M. K. D. (2017). The EU's Power in the Russia–Ukraine Crisis: Enabled or Constrained? *Journal of Common Market Studies, 55*(1), 137–152. doi:10.1111/jcms.12446

Karoliina, R., Jyrki, A., Kalle, A., & Pasi, R. (2023). *The elements of resilience in the food system and means to enhance the stability of the food supply. Environment Systems and Decisions.* Scopus., doi:10.100710669-022-09889-5

Keep, S., Stevenson-Leggett, P., Dowgier, G., Foldes, K., Webb, I., Fones, A., Littolff, K., Everest, H., Britton, P., & Bickerton, E. (2022). A Temperature-Sensitive Recombinant of Avian Coronavirus Infectious Bronchitis Virus Provides Complete Protection against Homologous Challenge. *Journal of Virology, 96*(17). doi:10.1128/jvi.01100-22

Ker, A. P. (2020). Risk management in Canada's agricultural sector in light of COVID-19. *Canadian Journal of Agricultural Economics, 68*(2), 251–258. Scopus. doi:10.1111/cjag.12232

Keshavamurthy, R., Dixon, S., Pazdernik, K. T., & Charles, L. E. (2022). Predicting infectious disease for biopreparedness and response: A systematic review of machine learning and deep learning approaches. *One Health, 15*. Scopus. doi:10.1016/j.onehlt.2022.100439

Kh'ng, K., Chang, C.-C., & Hsu, S.-H. (2022). Implications of food and nutrition security on household food expenditure: The case of Malaysia. *Agriculture and Food Security, 11*(1). Scopus. doi:10.1186/s40066-022-00367-4

Khan, N., Ma, J., Kassem, H. S., Kazim, R., Ray, R. L., Ihtisham, M., & Zhang, S. (2022). Rural Farmers' Cognition and Climate Change Adaptation Impact on Cash Crop Productivity: Evidence from a Recent Study. *International Journal of Environmental Research and Public Health, 19*(19). Scopus. doi:10.3390/ijerph191912556

Kharin, S. (2015). Vertical price transmission along the diary supply chain in Russia. *Studies in Agricultural Economics, 117*(2), 80–85. doi:10.7896/j.1517

Khoso, K. K., Buarod, T., & Kaewpijit, J. (2021). Sustainable impact of organic farming: A social constructivist perspective. *Humanities, Arts and Social Sciences Studies, 21*(1), 150–161. doi:10.14456/hasss.2021.15

Kim, J.-O., & Mueller, C. W. (1978a). Factor Analysis: Statistical Methods and Practical Issues. *Sage (Atlanta, Ga.).*

Kim, J.-O., & Mueller, C. W. (1978b). *Introduction to Factor Analysis: What It Is and How To Do It.* SAGE Publications. doi:10.4135/9781412984652

Knight, M. A., Hutchings, M. R., White, P. C. L., Davidson, R. S., & Marion, G. (2022). A mechanistic model captures livestock trading, disease dynamics, and compensatory behaviour in response to control measures. *Journal of Theoretical Biology, 539*. doi:10.1016/j.jtbi.2022.111059

Koç, E., & Karayiğit, B. (2022). Assessment of Biofortification Approaches Used to Improve Micronutrient-Dense Plants That Are a Sustainable Solution to Combat Hidden Hunger. *Journal of Soil Science and Plant Nutrition, 22*(1), 475–500. Scopus. doi:10.1007/s42729-021-00663-1

Kodrat, K. F., Sinulingga, S., Napitupulu, H. L., & Hadiguna, R. A. (2018). Passion fruit supply chain simulation using Monte Carlo method. *International Journal of Civil Engineering and Technology, 9*(13), 1164–1173.

Kolluri, G., Tyagi, J. S., & Sasidhar, P. V. K. (2021). Research Note: Indian poultry industry vis-à-vis coronavirus disease 2019: A situation analysis report. *Poultry Science, 100*(3). doi:10.1016/j.psj.2020.11.011

Kondratieva, N. (2021). *EU agricultural digitalization decalogue. Istoriya, 12, 16–23.* Scopus., doi:10.18254/S207987840015224-5

Koo, J. H., Lee, S. W., Lee, G. E., & Pyeon, M. W. (2021). A Study on Smart Farmer Service Using Community Mapping. *Journal of the Korean Society of Surveying, Geodesy, Photogrammetry and Cartography, 39*(6), 419–427. Scopus. doi:10.7848/ksgpc.2021.39.6.419

Korkmaz Aslan, G., Kılınç, E., & Kartal, A. (2022). The effect of COVID-19 pandemic on lifestyle-related behaviours in Turkey: A web-based cross-sectional study. *International Journal of Nursing Practice, 28*(5). Scopus. doi:10.1111/ijn.13053

Kormych, B., & Malyarenko, T. (2022). From gray zone to conventional warfare: The Russia-Ukraine conflict in the Black Sea. *Small Wars & Insurgencies, 0*(0), 1–36. doi:10.1080/09592318.2022.2122278

Kumar, A., Anju, T., Kumar, S., Chhapekar, S. S., Sreedharan, S., Singh, S., Choi, S. R., Ramchiary, N., & Lim, Y. P. (2021). Integrating omics and gene editing tools for rapid improvement of traditional food plants for diversified and sustainable food security. *International Journal of Molecular Sciences, 22*(15). Scopus. doi:10.3390/ijms22158093

Kumar, A., Panwar, A., Batra, K., De, S., & Maan, S. (2021). In-silico analysis of angiotensin-converting enzyme 2 (ACE2) of livestock, pet and poultry animals to determine its susceptibility to SARS–CoV-2 infection. *Combinatorial Chemistry and High Throughput Screening, 24*(10), 1769–1783. doi:10.2174/1386207323666201110144542

Kumar, M., Kumar Choubey, V., Deepak, A., Gedam, V. V., & Raut, R. D. (2021). Life cycle assessment (LCA) of dairy processing industry: A case study of North India. *Journal of Cleaner Production, 326*. Scopus. doi:10.1016/j.jclepro.2021.129331

Kumar, P., Singh, S. S., Pandey, A. K., Singh, R. K., Srivastava, P. K., Kumar, M., Dubey, S. K., Sah, U., Nandan, R., Singh, S. K., Agrawal, P., Kushwaha, A., Rani, M., Biswas, J. K., & Drews, M. (2021). Multi-level impacts of the COVID-19 lockdown on agricultural systems in India: The case of Uttar Pradesh. *Agricultural Systems, 187*. Scopus. doi:10.1016/j.agsy.2020.103027

Kumar, N. P. (2022). *Modeling the impact of covid-19 on the farm produce availability and pricing in India. Interdisciplinary Journal of Information, Knowledge, and Management, 17, 35–63*. doi:10.28945/4897

Kurczewska, U. (2021). European solidarity and its limitations during the covid-19 pandemic. [Scopus.]. *European Journal of Science and Theology*, *17*(5), 1–13.

Kyfyak, V., Verbivska, L., Alioshkina, L., Galunets, N., Kucher, L., & Skrypnyk, S. (2022). The Influence of the Social and Economic Situation on Agribusiness. *WSEAS Transactions on Environment and Development, 18*, 1021–1035. Scopus. doi:10.37394/232015.2022.18.98

Lang, L. D., Behl, A., Dong, N. T., Temouri, Y., & Thu, N. H. (2022). Effect of social capital on agribusiness diversification intention in the emerging market. *Journal of Intellectual Capital*, *23*(1), 56–84. doi:10.1108/JIC-01-2021-0039

Lang, L. D., Behl, A., Dong, N. T., Thu, N. H., & Dewani, P. P. (2022). Social capital in agribusiness: An exploratory investigation from a supply chain perspective during the COVID-19 crisis. *International Journal of Logistics Management, 33*(4), 1437–1473. doi:10.1108/IJLM-01-2021-0039

Larsen, K. M., DeCicco, M., Hood, K., & Etter, A. J. (2022). Salmonella enterica frequency in backyard chickens in Vermont and biosecurity knowledge and practices of owners. *Frontiers in Veterinary Science, 9*. doi:10.3389/fvets.2022.979548

Le, V. H., von Mettenheim, H.-J., Goutte, S., & Liu, F. (2022). News-based sentiment: Can it explain market performance before and after the Russia–Ukraine conflict? *The Journal of Risk Finance,* doi:10.1108/JRF-06-2022-0168

Leakey, R. R. B., Mabhaudhi, T., & Gurib-Fakim, A. (2021). African lives matter: Wild food plants matter for livelihoods, justice, and the environment—A policy brief for agricultural reform and new crops. *Sustainability (Switzerland), 13*(13). Scopus. doi:10.3390/su13137252

Lei, L., & Shimokawa, S. (2020). Promoting dietary guidelines and environmental sustainability in China. *China Economic Review, 59*. doi:10.1016/j.chieco.2017.08.001

Leung, K. K. Y., & Hon, K. L. (2023). Lessons from Animal Culling during Human Pandemics: Is Vaccination a Viable Option for Animals? *Current Pediatric Reviews, 19*(1), 2–4. doi:10.2174/1573396318666220316124155

Lewandowska-Tomaszczyk, B. (2022). Online comments as inwards and outwards-directed acts in Polish and English The Russia-Ukraine conflict. *International Review of Pragmatics, 14*(2), 252–275. doi:10.1163/18773109-01402005

Lewtak, K., Kanecki, K., Tyszko, P., Gorynski, P., Kosinska, I., Poznanska, A., Rzad, M., & Nitsch-Osuch, A. (2022). Hospitalizations of Ukrainian Migrants and Refugees in Poland in the Time of the Russia-Ukraine Conflict. *International Journal of Environmental Research and Public Health, 19*(20), 13350. doi:10.3390/ijerph192013350 PMID:36293932

Li, N., Chen, M., & Huang, D. (2023). How Do Logistics Disruptions Affect Rural Households? Evidence from COVID-19 in China. *Sustainability (Switzerland), 15*(1). Scopus. doi:10.3390/su15010465

Li, Y., Zhuang, Q., Jiang, L., Jiang, W., Peng, C., Jiang, N., Zhang, F., Yu, X., Wang, S., Yuan, L., Hou, G., Liu, S., Wang, J., Yu, J., Li, J., Zhao, C., Huang, B., Liu, H., & Wang, K. (2021). Traceable surveillance and genetic diversity analysis of coronaviruses in poultry from China in 2019. *Virus Research, 306*. doi:10.1016/j.virusres.2021.198566

Lialina, A. V., & Morachevskaya, K. A. (2022). Economic Access to Food and COVID-19: New Challenges for the Russian Exclave. *Regional Research of Russia, 12*(3), 335–349. Scopus. doi:10.1134/S2079970522700289

Lian, T., Haq, S. U., Shahbaz, P., Zhao, L., Nadeem, M., & Aziz, B. (2023). Changing Food Patterns during the Pandemic: Exploring the Role of Household Dynamics and Income Stabilization Strategies. *Sustainability (Switzerland), 15*(1). Scopus. doi:10.3390/su15010123

Liang, Y., & Zhong, T. (2023). Impacts of community-level grassroots organizations on household food security during the COVID-19 epidemic period in China. *International Journal of Disaster Risk Reduction, 85*. Scopus. doi:10.1016/j.ijdrr.2022.103490

Li, J., & Song, Z. (2022). Dynamic Impacts of External Uncertainties on the Stability of the Food Supply Chain: Evidence from China. *Foods, 11*(17), 2552. doi:10.3390/foods11172552 PMID:36076737

Limon, G., Gamawa, A. A., Ahmed, A. I., Lyons, N. A., & Beard, P. M. (2020). Epidemiological Characteristics and Economic Impact of Lumpy Skin Disease, Sheeppox and Goatpox Among Subsistence Farmers in Northeast Nigeria. *Frontiers in Veterinary Science, 7*. doi:10.3389/fvets.2020.00008

Ling, J., Xue, Y., Yang, C., & Zhang, Y. (2023). Effect of Farmers' Awareness of Climate Change on Their Willingness to Adopt Low-Carbon Production: Based on the TAM-SOR Model. *International Journal of Environmental Research and Public Health, 20*(1). Scopus. doi:10.3390/ijerph20010619

Lioubimtseva, E., & Henebry, G. M. (2012). Grain production trends in Russia, Ukraine and Kazakhstan: New opportunities in an increasingly unstable world? *Frontiers of Earth Science, 6*(2), 157–166. Scopus. doi:10.1007/s11707-012-0318-y

Liu, M., Yao, H., & Matsui, H. (2022). Marked rebound of agricultural fire emissions in Asia after the outbreak of COVID-19. *Environmental Research Letters, 17*(11). Scopus. doi:10.1088/1748-9326/ac9e69

Liu, Y., & Rabinowitz, A. N. (2021). The impact of the COVID-19 pandemic on retail dairy prices. *Agribusiness, 37*(1), 108–121. doi:10.1002/agr.21687

Liu, Y., Liu, S., Ye, D., Tang, H., & Wang, F. (2022). Dynamic impact of negative public sentiment on agricultural product prices during COVID-19. *Journal of Retailing and Consumer Services, 64*. doi:10.1016/j.jretconser.2021.102790

Loebnitz, N., Frank, P., & Otterbring, T. (2022). Stairway to organic heaven: The impact of social and temporal distance in print ads. *Journal of Business Research, 139*, 1044–1057. doi:10.1016/j.jbusres.2021.10.020

Lopes, M. S., de Freitas, P. P., de Carvalho, M. C. R., Silva, U. M., & Lopes, A. C. S. (2022). The COVID-19 pandemic in a Brazilian metropolis: Repercussion on food prices. *Cadernos de Saude Publica, 38*(4). Scopus. doi:10.1590/0102-311XEN166721

Lopez-Ridaura, S., Sanders, A., Barba-Escoto, L., Wiegel, J., Mayorga-Cortes, M., Gonzalez-Esquivel, C., Lopez-Ramirez, M. A., Escoto-Masis, R. M., Morales-Galindo, E., & García-Barcena, T. S. (2021). Immediate impact of COVID-19 pandemic on farming systems in Central America and Mexico. *Agricultural Systems, 192*. doi:10.1016/j.agsy.2021.103178

Losso, J. N., Losso, M. J. N., Toc, M., Inungu, J. N., & Finley, J. W. (2021). The Young Age and Plant-Based Diet Hypothesis for Low SARS-CoV-2 Infection and COVID-19 Pandemic in Sub-Saharan Africa. *Plant Foods for Human Nutrition, 76*(3), 270–280. doi:10.1007/s11130-021-00907-6

Lusk, J. L., & Chandra, R. (2021). Farmer and farm worker illnesses and deaths from COVID-19 and impacts on agricultural output. *PLoS ONE, 16*(4 April 2021). Scopus. doi:10.1371/journal.pone.0250621

Lutaladio, N., & Prakash, A. (2010). The potato: Its history and economic development. *Cahiers de Nutrition et de Dietetique, 45*(6), S5–S16. doi:10.1016/S0007-9960(10)70003-1

Macias, P., Stelmasiak, D., & Szafranek, K. (2022). Nowcasting food inflation with a massive amount of online prices. *International Journal of Forecasting*. Scopus. doi:10.1016/j.ijforecast.2022.02.007

Madzorera, I., Ismail, A., Hemler, E. C., Korte, M. L., Olufemi, A. A., Wang, D., Assefa, N., Workneh, F., Lankoande, B., Chukwu, A., Ourohire, M., Mattei, J., Soura, A., Berhane, Y., Sie, A., Oduola, A., & Fawzi, W. W. (2021). Impact of COVID-19 on nutrition, food security, and dietary diversity and quality in Burkina Faso, Ethiopia and Nigeria. *American Journal of Tropical Medicine and Hygiene, 105*(2), 295–309. Scopus. doi:10.4269/ajtmh.20-1617

Mafruchati, M. (2020). Broiler Chicken vs. Turkey Meat; which One Has the Least Bad Fat to Avoid Positive Case of COVID-19? *Systematic Reviews in Pharmacy, 11*(10), 799–802. doi:10.31838/srp.2020.10.119

Mahmud, K., Panday, D., Mergoum, A., & Missaoui, A. (2021). Nitrogen losses and potential mitigation strategies for a sustainable agroecosystem. *Sustainability (Switzerland), 13*(4), 1–23. Scopus. doi:10.3390/su13042400

Mahran, H. A. (2022). The impact of the Russia–Ukraine conflict (2022) on volatility connectedness between the Egyptian stock market sectors: Evidence from the DCC-GARCH-CONNECTEDNESS approach. *The Journal of Risk Finance*. doi:10.1108/JRF-06-2022-0163

Mai, R., Hoffmann, S., & Balderjahn, I. (2021). When drivers become inhibitors of organic consumption: The need for a multistage view. *Journal of the Academy of Marketing Science, 49*(6), 1151–1174. doi:10.1007/s11747-021-00787-x

Mallet, Y., Pivette, M., Revest, M., Angot, E., Valence, M., Dupin, C., Picard, N., Brelivet, G., Seyler, T., Ballet, S., Le Tertre, A., & Guillois, Y. (2021). Identification of Workers at Increased Risk of Infection During a COVID-19 Outbreak in a Meat Processing Plant, France, May 2020. *Food and Environmental Virology, 13*(4), 535–543. doi:10.1007/s12560-021-09500-1

Manian, M. A., Khoshbakht, K., Mahmoudi, H., & Liaghati, H. (2022). Dynamic Conservation in Risk Society: A Case Study of COVID-19 Pandemic Risk in Kashan Qanat Irrigated Agriculture. *Frontiers in Public Health, 10*. Scopus. doi:10.3389/fpubh.2022.882943

Manolova, V., & Kolev, K. (2012). Economic results from growing of cherry in different levels of intensification. *Acta Horticulturae*, *981*(1), 719–724. https://doi.org/10.17660/actahortic.2013.981.115

Manyong, V., Bokanga, M., Akonkwa Nyamuhirwa, D.-M., Bamba, Z., Adeoti, R., Mwepu, G., Cole, S. M., & Dontsop Nguezet, P. M. (2022). *COVID-19 outbreak and rural household food security in the Western Democratic Republic of the Congo. World Development Perspectives, 28.* Scopus., doi:10.1016/j.wdp.2022.100469

Maples, J. G., Thompson, J. M., Anderson, J. D., & Anderson, D. P. (2021). Estimating COVID-19 Impacts on the Broiler Industry. *Applied Economic Perspectives and Policy, 43*(1), 315–328. doi:10.1002/aepp.13089

Márquez-Berber, S. R., Torcuato-Calderón, C., Almaguer-Vargas, G., Colinas-León, M. T., & Gardezi, A. K. (2012). Cactus pear (Opuntia albicarpa and O. megacantha) production system in axapusco, Estado de Mexico. Problems and alternatives. *Revista Chapingo Serie Horticultura*, *18*(1), 81–93.

Martey, E., Etwire, P. M., Adzawla, W., Atakora, W., & Bindraban, P. S. (2022). Perceptions of COVID-19 shocks and adoption of sustainable agricultural practices in Ghana. *Journal of Environmental Management, 320.* Scopus. doi:10.1016/j.jenvman.2022.115810

Martin, W., & Minot, N. (2022). The impacts of price insulation on world wheat markets during the 2022 food price crisis. *Australian Journal of Agricultural and Resource Economics, 66*(4), 753–774. Scopus. doi:10.1111/1467-8489.12498

Martinho, V. J. P. D., Cunha, C. A. D. S., Pato, M. L., Costa, P. J. L., Sánchez-Carreira, M. C., Georgantzís, N., Rodrigues, R. N., & Coronado, F. (2022). Machine Learning and Food Security: Insights for Agricultural Spatial Planning in the Context of Agriculture 4.0. *Applied Sciences (Switzerland), 12*(22). Scopus. doi:10.3390/app122211828

Martinho, V. J. P. D. (2022). Impacts of the COVID-19 Pandemic and the Russia–Ukraine Conflict on Land Use across the World. *Land (Basel), 11*(10), 10. Advance online publication. doi:10.3390/land11101614

McBurney, M., Tuaza, L. A., Ayol, C., & Johnson, C. A. (2021). Land and livelihood in the age of COVID-19: Implications for indigenous food producers in Ecuador. *Journal of Agrarian Change, 21*(3), 620–628. doi:10.1111/joac.12417

McClure, E. S., Vasudevan, P., Bailey, Z., Patel, S., & Robinson, W. R. (2020). Racial capitalism within public health-how occupational settings drive covid-19 disparities. *American Journal of Epidemiology, 189*(11), 1244–1253. doi:10.1093/aje/kwaa126

Medina, R. B., Cantuarias-Avilés, T. E., Angolini, S. F., & Da Silva, S. R. (2018). Performance of 'emerald' and 'jewel' blueberry cultivars under no-chill incidence. *Pesquisa Agropecuaria Tropical, 48*(2), 147–152. Scopus. doi:10.1590/1983-40632018v4852093

Meekins, D. A., Gaudreault, N. N., & Richt, J. A. (2021). Natural and Experimental SARS-CoV-2 Infection in Domestic and Wild Animals. *Viruses, 13*(10), 10. Advance online publication. doi:10.3390/v13101993 PMID:34696423

Mendi, A. F. (2022). Blockchain for Food Tracking. *Electronics (Switzerland), 11*(16). Scopus. doi:10.3390/electronics11162491

Merchant, E. V., Fatima, T., Fatima, A., Maiyo, N., Mutuku, V., Keino, S., Simon, J. E., Hoffman, D. J., & Downs, S. M. (2022). The Influence of Food Environments on Food Security Resilience during the COVID-19 Pandemic: An Examination of Urban and Rural Difference in Kenya. *Nutrients, 14*(14). doi:10.3390/nu14142939

Meunier, C., Casagrande, M., Rosiès, B., Bedoussac, L., Topp, C. F. E., Walker, R. L., Watson, C. A., & Martin, G. (2022). Interplay: A game for the participatory design of locally adapted cereal–legume intercrops. *Agricultural Systems, 201*. Scopus. doi:10.1016/j.agsy.2022.103438

Meyfroidt, P., Schierhorn, F., Prishchepov, A. V., Müller, D., & Kuemmerle, T. (2016). Drivers, constraints and trade-offs associated with recultivating abandoned cropland in Russia, Ukraine and Kazakhstan. *Global Environmental Change, 37*, 1–15. Scopus. doi:10.1016/j.gloenvcha.2016.01.003

Miao, Y., & Fei, X. (2022). The international monetary system: Evolution and revolution. *China Economic Journal, 15*(3), 235–252. doi:10.1080/17538963.2022.2117193

Mignogna, C., Costanzo, S., Ghulam, A., Cerletti, C., Donati, M. B., De Gaetano, G., Iacoviello, L., & Bonaccio, M. (2022). Impact of Nationwide Lockdowns Resulting from the First Wave of the COVID-19 Pandemic on Food Intake, Eating Behaviors, and Diet Quality: A Systematic Review. *Advances in Nutrition, 13*(2), 388–423. doi:10.1093/advances/nmab130

Mihai, I. F., Anita, D., Dorneanu, O. S., Luca, C. M., Manciuc, C. D., Budacu, C. C., Roşu, F. M., Savuta, G., Anita, A., & Vâţă, A. (2022). Seroprevalence of Anti-Hepatitis E Virus Antibodies among Patients from a Tertiary Hospital from Northeast Romania. *Medicina (Lithuania), 58*(8). doi:10.3390/medicina58081020

Mila, F. A., Nahar, A., Amin, M. R., Culas, R. J., & Ahmed, A. (2022). Empirical assessment of onion supply chain constraints in Bangladesh: A pre-covid to covid situation. *Journal of Agriculture and Food Research, 10*. doi:10.1016/j.jafr.2022.100418

Miller, L. C., Neupane, S., Joshi, N., Lohani, M., Sah, K., & Shrestha, B. (2022). Dairy Animal Ownership and Household Milk Production Associated with Better Child and Family Diet in Rural Nepal during the COVID-19 Pandemic. *Nutrients, 14*(10). Scopus. doi:10.3390/nu14102074

Mishra, A., & Shukla, A. (2022). *Gyan Fresh: Digital Transformation of Dairy Business with Resilience and Technology Innovation. FIIB Business Review*. Scopus., doi:10.1177/23197145221113374

Mnisi, C. M., Marareni, M., Manyeula, F., & Madibana, M. J. (2021). A way forward for the South African quail sector as a potential contributor to food and nutrition security following the aftermath of COVID-19: A review. *Agriculture and Food Security, 10*(1). doi:10.1186/s40066-021-00331-8

Mohamed Sala, S., Otieno, D. J., Nzuma, J., & Mureithi, S. M. (2020). Determinants of pastoralists' participation in commercial fodder markets for livelihood resilience in drylands of northern Kenya: Case of Isiolo. *Pastoralism, 10*(1). doi:10.1186/s13570-020-00166-1

Mohite, J., Sawant, S., Pandit, A., & Pappula, S. (2022). Impact of lockdown and crop stubble burning on air quality of India: A case study from wheat-growing region. *Environmental Monitoring and Assessment, 194*(2). Scopus. doi:10.1007/s10661-021-09723-6

Molinero-Gerbeau, Y. (2021). The Problem is not Covid-19, it's the Model! Industrial Agriculture and Migrant Farm Labour in the EU. *EuroChoices, 20*(3), 69–74. Scopus. doi:10.1111/1746-692X.12308

Morávková, M., Veselá, L., & Kubíčková, L. (2022). Changes in households' stances on wasting food during the COVID-19 pandemic. *Waste Forum, 3*, 161–178.

Morell, M., & Söderhäll, M. (2019). Smallholders' and large estates' reaction to changed market conditions 1860–1910. *Scandinavian Economic History Review, 67*(3), 312–331. doi:10.1080/03585522.2019.1620850

Mo, S., & Wang, T. (2022). Synergistic Effects of International Oil Price Fluctuations and Carbon Tax Policies on the Energy-Economy-Environment System in China. *International Journal of Environmental Research and Public Health, 19*(21), 14177. doi:10.3390/ijerph192114177 PMID:36361057

Mubeena, M., Muthukumar, M., & Barbuddhe, S. B. (2021). Trends and seasonal variations in broilers prices. *Veterinary Practitioner*, *22*(2), 165–171.

Mueller, V., Grépin, K. A., Rabbani, A., Navia, B., Ngunjiri, A. S. W., & Wu, N. (2022). Food insecurity and COVID-19 risk in low- and middle-income countries. *Applied Economic Perspectives and Policy, 44*(1), 92–109. Scopus. doi:10.1002/aepp.13200

Muhammad, A., & Kilmer, R. L. (2008). The impact of EU export subsidy reductions on U.S. dairy exports. *Agribusiness, 24*(4), 557–574. doi:10.1002/agr.20179

Müller, A., Schmidhuber, J., Hoogeveen, J., & Steduto, P. (2008). Some insights in the effect of growing bio-energy demand on global food security and natural resources. *Water Policy, 10*(Suppl. 1), 83–94. doi:10.2166/wp.2008.053

Mumena, W. A. (2020). Impact of COVID-19 curfew on eating habits, food intake, and weight according to food security status in Saudi Arabia: A retrospective study. *Progress in Nutrition, 22*(4). Scopus. doi:10.23751/pn.v22i4.9883

Munonye, J., Osuji, E., Olaolu, M., Okoisu, A., Obi, J., Eze, G., Ibrahim-Olesin, S., Njoku, L., Amadi, M., Izuogu, C., & Azuamairo, G. (2022). Perceived Effects of COVID-19 Pandemic on Food Security in Southeast Nigeria. *Frontiers in Sustainable Food Systems, 6*. Scopus. doi:10.3389/fsufs.2022.936157

Murithi, T. (2022). The Failure of the United Nations Security Council in Creating the Framework Conditions for Mediation in the Russia-Ukraine Crisis. *Strategic Review for Southern Africa*, *44*(1), 65–92. doi:10.35293rsa.v44i1.4082

Murphy, B., Martini, M., Fedi, A., Loera, B. L., Elliott, C. T., & Dean, M. (2022). Consumer trust in organic food and organic certifications in four European countries. *Food Control, 133*. doi:10.1016/j.foodcont.2021.108484

Murungi, M. K., Muloi, D. M., Muinde, P., Githigia, S. M., Akoko, J., Fèvre, E. M., Rushton, J., & Alarcon, P. (2021). The Nairobi Pork Value Chain: Mapping and Assessment of Governance, Challenges, and Food Safety Issues. *Frontiers in Veterinary Science, 8*. doi:10.3389/fvets.2021.581376

Mu, S., Huang, G., Li, P., & Hou, Y. (2022). A Study on Volatility Spillovers among International Stock Markets during the Russia-Ukraine Conflict. *Discrete Dynamics in Nature and Society*, *4948444*, 1–8. doi:10.1155/2022/4948444

Muthamilarasan, M., & Prasad, M. (2021). Small Millets for Enduring Food Security Amidst Pandemics. *Trends in Plant Science, 26*(1), 33–40. Scopus. doi:10.1016/j.tplants.2020.08.008

Myeki, L. W., Nengovhela, N. B., Mudau, L., Nakana, E., & Ngqangweni, S. (2022). Estimation of Technical, Allocative, and Economic Efficiencies for Smallholder Broiler Producers in South Africa. *Agriculture-Basel, 12*(10), 1601. doi:10.3390/agriculture12101601

Nakandala, D., Tsang, Y. P., Lau, H., & Lee, C. K. M. (2022). An Industrial Blockchain-Based Multi-Criteria Decision Framework for Global Freight Management in Agricultural Supply Chains. *Mathematics, 10*(19), 3550. doi:10.3390/math10193550

Nasir, M. A., Nugroho, A. D., & Lakner, Z. (2022). Impact of the Russian–Ukrainian Conflict on Global Food Crops. *Foods, 11*(19). Scopus. doi:10.3390/foods11192979

Nautiyal, S., & Lal, C. (2022). Product knowledge as a facilitator of organic purchase intention in emerging markets: Empirical evidence from India. *Journal of Cleaner Production, 372*. doi:10.1016/j.jclepro.2022.133782

Naveen, J., Saikia, M., Borah, N., Pathak, K., & Das, R. (2021). Influence of nutrient management and moisture conservation practices on Growth and development of organic baby corn (Zea mays L.) in Assam. *Indian Journal of Agricultural Research, 55*(4), 497–500. doi:10.18805/IJARe.A-5531

Neganova, V. P., & Chistyakov, Yu. F. (2020). Development of agricultural foreign trade in the post-soviet countries. *Economy of Region, 16*(2), 597–611. Scopus. doi:10.17059/2020-2-20

Nekmahmud, M. (2022). Food consumption behavior, food supply chain disruption, and food security crisis during the COVID-19: The mediating effect of food price and food stress. *Journal of Foodservice Business Research*. Scopus. doi:10.1080/15378020.2022.2090802

Nelles, W. (2023). Academic network-building for agroecology and sustainable agri-food systems in Southeast Asia: Critical reflections on a regional initiative. *Agroecology and Sustainable Food Systems, 47*(1), 126–155. Scopus. doi:10.1080/21683565.2022.2134955

Nematollahi, M., Tajbakhsh, A., & Mosadegh Sedghy, B. (2021). The reflection of competition and coordination on organic agribusiness supply chains. *Transportation Research Part E: Logistics and Transportation Review, 154*. doi:10.1016/j.tre.2021.102462

Nerlinger, M., & Utz, S. (2022). The impact of the Russia-Ukraine conflict on energy firms: A capital market perspective. *Finance Research Letters, 50*, 103243. doi:10.1016/j.frl.2022.103243

Niaz, A., Asad, M. M., Abdulmuhsin, A. A., Shavanov, M., & Churi, P. (2022). Risk factors for the rice crop farming community in China: A documentary analysis of the challenges during post COVID-19. *Asian Education and Development Studies, 11*(2), 298–310. Scopus. doi:10.1108/AEDS-11-2020-0257

Nichols, M., Gollarza, L., Sockett, D., Aulik, N., Patton, E., Francois Watkins, L. K., Gambino-Shirley, K. J., Folster, J. P., Chen, J. C., Tagg, K. A., Stapleton, G. S., Trees, E., Ellison, Z., Lombard, J., Morningstar-Shaw, B., Schlater, L., Elbadawi, L., & Klos, R. (2022). Outbreak of Multidrug-Resistant Salmonella Heidelberg Infections Linked to Dairy Calf Exposure, United States, 2015-2018. *Foodborne Pathogens and Disease, 19*(3), 199–208. doi:10.1089/fpd.2021.0077

Nikonenko, U., Shtets, T., Kalinin, A., Dorosh, I., & Sokolik, L. (2022). Assessing the Policy of Attracting Investments in the Main Sectors of the Economy in the Context of Introducing Aspects of Industry 4.0. *International Journal of Sustainable Development and Planning, 17*(2), 497–505. Scopus. doi:10.18280/ijsdp.170214

Nikooyeh, B., Rabiei, S., Amini, M., Ghodsi, D., Rasekhi, H., Doustmohammadian, A., Abdollahi, Z., Minaie, M., Sadeghi, F., & Neyestani, T. R. (2022). COVID-19 epidemic lockdown-induced remarkable decrease in dairy products consumption of Iran population: Does it really matter? National Food and Nutrition Surveillance. *BMC Nutrition, 8*(1). Scopus. doi:10.1186/s40795-022-00612-w

Nili, H., Bouzari, M., Attaran, H. R., Ghalegolab, N., Rabani, M., & Mahmoudian, A. (2022). Hyper-Immune Bovine Milk as an Immunological and Nutritional Supplement for COVID-19. *Frontiers in Nutrition, 9*. Scopus. doi:10.3389/fnut.2022.868964

Nohra, R. G., Naim, E., Chaaban, T., & Rothan-Tondeur, M. (2022). Nurses' eating habits in Lebanon during the economic and health crises: A cross-sectional study. *Archives of Public Health, 80*(1). Scopus. doi:10.1186/s13690-021-00775-1

Nugroho, A. D. (2021). Does covid-19 have effects on the indonesian horticultural subsector? *Bulgarian Journal of Agricultural Science, 27*(5), 865–874.

Nurlaelah, S., Harnianti, Asnawi, A., & Sirajuddin, S. N. (2022). Income analysis on broiler chicken farming in partnership system during the covid-19 pandemic in Tanralili district, Indonesia. *Online Journal of Animal and Feed Research, 12*(4), 210–216. doi:10.51227/ojafr.2022.28

O'Meara, L., Turner, C., Coitinho, D. C., & Oenema, S. (2022). Consumer experiences of food environments during the Covid-19 pandemic: Global insights from a rapid online survey of individuals from 119 countries. *Global Food Security, 32*. Scopus. doi:10.1016/j.gfs.2021.100594

OECD. (2023). *Agriculture—OECD Data*. TheOECD. https://data.oecd.org/agriculture.htm

Okello, W. O., MacLeod, E. T., Muhanguzi, D., Waiswa, C., Shaw, A. P., & Welburn, S. C. (2021). Critical Linkages Between Livestock Production, Livestock Trade and Potential Spread of Human African Trypanosomiasis in Uganda: Bio-economic Herd Modeling and Livestock Trade Analysis. *Frontiers in Veterinary Science, 8*. doi:10.3389/fvets.2021.611141

Okolie, C. C., & Ogundeji, A. A. (2022). *Effect of COVID-19 on agricultural production and food security: A scientometric analysis. Humanities and Social Sciences Communications, 9(1)*. Scopus., doi:10.105741599-022-01080-0

Okoli, T. O. (2021). Coping strategies adopted by female-headed households in combating food insecurity during corona virus-19 (Covid-19) lockdown in rural areas in anambra state. [Scopus.]. *Journal of Home Economics Research, 28*(2), 132–141.

Olaizola, A. M., Ameen, F., & Manrique, E. (2015). Potential strategies of adaptation of mixed sheep-crop systems to changes in the economic environment in a Mediterranean mountain area. *Livestock Science, 176*, 166–180. doi:10.1016/j.livsci.2015.03.012

Olejar, K. J., Vasconcelos, M. C., King, P. D., Smart, R. E., Ball, K., & Field, S. K. (2021). Herbicide reduction through the use of weedmat undervine treatment and the lack of impact on the aromatic profile and volatile composition of Malbec wines. *Food Chemistry, 343*. doi:10.1016/j.foodchem.2020.128474

Olipra, J. (2020). Leading properties of GDT auctions for dairy prices. *British Food Journal, 122*(7), 2303–2328. doi:10.1108/BFJ-06-2018-0404

Omotayo, A. O., Omotoso, A. B., Daud, S. A., Omotayo, O. P., & Adeniyi, B. A. (2022). Rising Food Prices and Farming Households Food Insecurity during the COVID-19 Pandemic: Policy Implications from SouthWest Nigeria. *Agriculture (Switzerland), 12*(3). Scopus. doi:10.3390/agriculture12030363

Onduso, R., Onono, J. O., & Ombui, J. N. (2020). Assessment of structure and performance of cattle markets in western Kenya. *Tropical Animal Health and Production, 52*(2), 725–732. doi:10.1007/s11250-019-02062-2

Ordofa, G., Zemedu, L., & Tegegne, B. (2021). Structure conduct and performance of dairy market in Ada'a Berga district, Ethiopia. *Cogent Food and Agriculture, 7*(1). doi:10.1080/23311932.2021.1918878

Ott, H. (2014). Volatility in cereal prices: Intra- versus inter-annual volatility. *Journal of Agricultural Economics, 65*(3), 557–578. doi:10.1111/1477-9552.12073

Pampori, Z. A. (2021). Saga of Hunger in India- Challenges, Chances and Resolutions: A Review. *Asian Journal of Dairy and Food Research, 40*(4), 398–407. Scopus. doi:10.18805/ajdfr.R-2183

Pan, D., Yang, J., Zhou, G., & Kong, F. (2020). The influence of COVID-19 on agricultural economy and emergency mitigation measures in China: A text mining analysis. *PLoS ONE, 15*(10). Scopus. doi:10.1371/journal.pone.0241167

Panta, R., Paswan, V. K., Kanetkar, P., Bunkar, D. S., Rose, H., & Bakshi, S. (2023). Exploring trade prospects of Chhurpi and the present status of Chhurpi producers and exporters of Nepal. *Journal of Ethnic Foods, 10*(1). Scopus. doi:10.1186/s42779-022-00165-0

Paria, B., Mishra, P., & Behera, B. (2022). Climate change and transition in cropping patterns: District level evidence from West Bengal, India. *Environmental Challenges, 7*. Scopus. doi:10.1016/j.envc.2022.100499

Parry, M. L., Rosenzweig, C., Iglesias, A., Livermore, M., & Fischer, G. (2004). Effects of climate change on global food production under SRES emissions and socio-economic scenarios. *Global Environmental Change, 14*(1), 53–67. doi:10.1016/j.gloenvcha.2003.10.008

Patel, S. K., Surve, J., Parmar, J., Ahmed, K., Bui, F. M., & Al-Zahrani, F. A. (2022). Recent Advances in Biosensors for Detection of COVID-19 and Other Viruses. *IEEE Reviews in Biomedical Engineering*, 1–16. Scopus. doi:10.1109/RBME.2022.3212038

Pathirana, R., & Carimi, F. (2022). Management and Utilization of Plant Genetic Resources for a Sustainable Agriculture. *Plants, 11*(15). Scopus. doi:10.3390/plants11152038

Patil, S., Verma, A., Patil, I., Singh, R., & Gaur, R. (2021). Bibliometric Survey on Effect of Socio-Economic Factors on Spread of Corona virus (COVID-19). [Scopus.]. *Library Philosophy and Practice*, *2021*, 1–16.

Patyka, N., Khodakivska, O., Mohylnyi, O., & Pugachov, M. (2021). Ukraine's Agrarian Sector in the Conditions of COVID-19 Distribution and Restrictive Quarantine Measures: Methodological Principles of Empirical Evaluation. *Scientific Horizons, 24*(12), 55–69. Scopus. doi:10.48077/scihor.24(12).2021.55-69

Peng, L., Lu, G., Pang, K., & Yao, Q. (2021). Optimal farmer's income from farm products sales on live streaming with random rewards: Case from China's rural revitalisation strategy. *Computers and Electronics in Agriculture, 189*. doi:10.1016/j.compag.2021.106403

Pereira, P. S., Sarmento, R. A., Galdino, T. V. S., Lima, C. H. O., dos Santos, F. A., Silva, J., dos Santos, G. R., & Picanço, M. C. (2017). Economic injury levels and sequential sampling plans for Frankliniella schultzei in watermelon crops. *Pest Management Science, 73*(7), 1438–1445. doi:10.1002/ps.4475

Pérez-Cutillas, P., Pérez-Navarro, A., Conesa-García, C., Zema, D. A., & Amado-Álvarez, J. P. (2023). *What is going on within google earth engine? A systematic review and meta-analysis. Remote Sensing Applications: Society and Environment, 29*. Scopus., doi:10.1016/j.rsase.2022.100907

Perrin, A., & Martin, G. (2021). Resilience of French organic dairy cattle farms and supply chains to the Covid-19 pandemic. *Agricultural Systems, 190*. Scopus. doi:10.1016/j.agsy.2021.103082

Philippidis, G., Ferrer Pérez, H., Gracia de Rentería, P., M'barek, R., & Sanjuán López, A. I. (2021). Eating your greens: A global sustainability assessment. *Resources, Conservation and Recycling, 168*. doi:10.1016/j.resconrec.2021.105460

Phillips, H. N., & Heins, B. J. (2022). Alternative Practices in Organic Dairy Production and Effects on Animal Behavior, Health, and Welfare. *Animals, 12*(14). doi:10.3390/ani12141785

Phukamchanoada, P. (2022). Community-based tourism product development based on the community identity of Klong-Lad-Ma-Yom floating market Bangkok, Thailand. *Turyzm/Tourism, 32*(1), 59–76. Scopus. doi:10.18778/0867-5856.32.1.03

Pieter, L. A. G., Utomo, M. M. B., Suhartono, S., Sudomo, A., Sanudin, S., Fauziyah, E., Widyaningsih, T. S., Palmolina, M., Hani, A., & Siagian, C. M. (2022). The Nexus of COVID-19 Pandemic and Rural Agroforestry Farmers' Livelihoods in Tasikmalaya Regency, East Priangan, Indonesia. *Forest and Society, 6*(1), 335–354. doi:10.24259/fs.v6i1.18773

Pilvere, I., Upite, I., Muska, A., Zdanovskis, K., Nipers, A., & Janmere, L. (2021). Resilience of milk supply chains during and after the CoviD-19 crisis in Latvia. *Rural Sustainability Research, 45*(340), 53–64. doi:10.2478/plua-2021-0007

Polukhin, A. A., & Panarina, V. I. (2022). *Financial Risk Management for Sustainable Agricultural Development Based on Corporate Social Responsibility in the Interests of Food Security. Risks, 10(1)*. Scopus., doi:10.3390/risks10010017

Popescu, G. C., & Popescu, M. (2022). COVID-19 pandemic and agriculture in Romania: Effects on agricultural systems, compliance with restrictions and relations with authorities. *Food Security, 14*(2), 557–567. Scopus. doi:10.1007/s12571-021-01239-8

Popović, G., Erić, O., & Bjelić, J. (2020). Factor analysis of prices and agricultural production in the European union. *Economics, 8*(1), 73–81. doi:10.2478/eoik-2020-0001

Pourghazi, F., Eslami, M., Ehsani, A., Ejtahed, H.-S., & Qorbani, M. (2022). Eating habits of children and adolescents during the COVID-19 era: A systematic review. *Frontiers in Nutrition, 9*. Scopus. doi:10.3389/fnut.2022.1004953

Pramod, R. K., Nair, A. V., Tambare, P. K., Chauhan, K., Kumar, T. V., Rajan, R. A., Mani, B. M., Asaf, M., & Pandey, A. K. (2021). Reverse zoonosis of coronavirus disease-19: Present status and the control by one health approach. *Veterinary World, 14*(10), 2817–2826. doi:10.14202/vetworld.2021.2817-2826

Prasad, A., & Kothari, N. (2022). Cow products: Boon to human health and food security. *Tropical Animal Health and Production, 54*(1). Scopus. doi:10.1007/s11250-021-03014-5

Priyadarshini, P., & Abhilash, P. C. (2021). Agri-food systems in India: Concerns and policy recommendations for building resilience in post COVID-19 pandemic times. *Global Food Security, 29*, 100537. doi:10.1016/j.gfs.2021.100537 PMID:35155096

Qin, T., Wang, L., Zhou, Y., Guo, L., Jiang, G., & Zhang, L. (2022). Digital Technology-and-Services-Driven Sustainable Transformation of Agriculture: Cases of China and the EU. *Agriculture (Switzerland), 12*(2). Scopus. doi:10.3390/agriculture12020297

Qingbin, W., Liu, C.-Q., Zhao, Y.-F., Kitsos, A., Cannella, M., Wang, S.-K., & Han, L. (2020). Impacts of the COVID-19 pandemic on the dairy industry: Lessons from China and the United States and policy implications. *Journal of Integrative Agriculture, 19*(12), 2903–2915. Scopus. doi:10.1016/S2095-3119(20)63443-8

Quandt, R. E. (1960). Tests of the Hypothesis that a Linear Regression System Obeys Two Separate Regimes. *Journal of the American Statistical Association, 55*(290), 324–330. doi:10.1080/01621459.1960.10482067

Quigley, D. D., Qureshi, N., Gahlon, G., & Gidengil, C. (2022). Worker and employer experiences with COVID-19 and the California Workers' Compensation System: A review of the literature. *American Journal of Industrial Medicine, 65*(3), 203–213. Scopus. doi:10.1002/ajim.23326

Quilliot, D., Gérard, M., Bonsack, O., Malgras, A., Vaillant, M.-F., Di Patrizio, P., Jaussaud, R., Ziegler, O., & Nguyen-Thi, P.-L. (2021). Impact of severe SARS-CoV-2 infection on nutritional status and subjective functional loss in a prospective cohort of COVID-19 survivors. *BMJ Open, 11*(7). Scopus. doi:10.1136/bmjopen-2021-048948

Qureshi, A., Rizwan, M. S., Ahmad, G., & Ashraf, D. (2022). Russia-Ukraine war and systemic risk: Who is taking the heat? *Finance Research Letters, 48*, 103036. doi:10.1016/j.frl.2022.103036

Rad, A. K., Shamshiri, R. R., Azarm, H., Balasundram, S. K., & Sultan, M. (2021). Effects of the covid-19 pandemic on food security and agriculture in iran: A survey. *Sustainability (Switzerland), 13*(18). Scopus. doi:10.3390/su131810103

Radojević, V., Tomaš Simin, M., Glavaš Trbić, D., & Milić, D. (2021). A profile of organic food consumers—Serbia case-study. *Sustainability (Switzerland), 13*(1), 1–22. doi:10.3390/su13010131

Rahmat, Z. S., Islam, Z., Mohanan, P., Kokash, D. M., Essar, M. Y., Hasan, M. M., Hashim, H. T., & Basalilah, A. F. M. (2022). Food Insecurity during COVID-19 in Yemen. *American Journal of Tropical Medicine and Hygiene, 106*(6), 1589–1592. Scopus. doi:10.4269/ajtmh.22-0059

Rajpoot, K., Singla, S., Singh, A., & Shekhar, S. (2022). Impact of COVID-19 lockdown on prices of potato and onion in metropolitan cities of India. *Journal of Agribusiness in Developing and Emerging Economies, 12*(3), 386–399. Scopus. doi:10.1108/JADEE-04-2021-0099

Rasul, G. (2021). A Framework for Addressing the Twin Challenges of COVID-19 and Climate Change for Sustainable Agriculture and Food Security in South Asia. *Frontiers in Sustainable Food Systems, 5*. Scopus. doi:10.3389/fsufs.2021.679037

Rawtani, D., Gupta, G., Khatri, N., Rao, P. K., & Hussain, C. M. (2022). Environmental damages due to war in Ukraine: A perspective. *The Science of the Total Environment, 850*, 157932. doi:10.1016/j.scitotenv.2022.157932 PMID:35952889

Recchia, D., Rollet, P., Perignon, M., Bricas, N., Vonthron, S., Perrin, C., Méjean, C., & Surfood-Foodscapes Working Group. (2022). Changes in Food Purchasing Practices of French Households During the First COVID-19 Lockdown and Associated Individual and Environmental Factors. *Frontiers in Nutrition, 9*. Scopus. doi:10.3389/fnut.2022.828550

Rejeb, A., Rejeb, K., Abdollahi, A., Zailani, S., Iranmanesh, M., & Ghobakhloo, M. (2022). Digitalization in food supply chains: A bibliometric review and key-route main path analysis. *Sustainability (Switzerland), 14*(1). Scopus. doi:10.3390/su14010083

Rejeb, A., Rejeb, K., Appolloni, A., Iranmanesh, M., Treiblmaier, H., & Jagtap, S. (2022). Exploring Food Supply Chain Trends in the COVID-19 Era: A Bibliometric Review. *Sustainability (Switzerland), 14*(19). Scopus. doi:10.3390/su141912437

Rezitis, A. N., & Tremma, O. A. (2022). The linkage between international dairy commodity prices and volatility: A panel-GARCH analysis. *Journal of Agribusiness in Developing and Emerging Economies*. doi:10.1108/JADEE-02-2022-0027

Rivera-Ferre, M. G., López-i-Gelats, F., Ravera, F., Oteros-Rozas, E., di Masso, M., Binimelis, R., & El Bilali, H. (2021). The two-way relationship between food systems and the COVID19 pandemic: Causes and consequences. *Agricultural Systems, 191*, 103134. doi:10.1016/j.agsy.2021.103134

Rocchi, B., Romano, D., & Hamza, R. (2013). Agriculture reform and food crisis in Syria: Impacts on poverty and inequality. *Food Policy, 43*, 190–203. doi:10.1016/j.foodpol.2013.09.009

Rojík, S., Zámková, M., Chalupová, M., Pilař, L., Prokop, M., Stolín, R., Malec, K., Appiah-Kubi, S. N. K., Maitah, M., Dziekański, P., & Prus, P. (2022). Pre-COVID-19 Organic Market in the European Union—Focus on the Czech, German, and Slovak Markets. *Agriculture (Switzerland), 12*(1). doi:10.3390/agriculture12010082

Ronaghi, M. (2022). Effects of COVID-19 on Iran's Livestock and Meat Market. *Journal of Agricultural Science and Technology, 24*(5), 1017–1028.

Rosegrant, M. W., Cline, S. A., Li, W., Sulser, T. B., & Valmonte-Santos, R. A. (2005). Looking ahead long-term prospects for Africa's agricultural development and food security. Food, Agriculture, and the Environment Discussion Paper, 41, 1–71.

Rosen, K., Llonch, P., & Langford, F. M. (2020). The welfare of culled dairy cattle sent to market and its impact on economic return. *Animal Welfare, 29*(4), 359–370. doi:10.7120/09627286.29.4.359

Rude, J., & An, H. (2013). Trans-pacific partnership: Implications for the canadian industrial dairy sector. *Canadian Public Policy, 39*(3), 393–410. doi:10.3138/CPP.39.3.393

Rukasha, T., Nyagadza, B., Pashapa, R., & Muposhi, A. (2021). Covid-19 impact on Zimbabwean agricultural supply chains and markets: A sustainable livelihoods perspective. *Cogent Social Sciences, 7*(1). Scopus. doi:10.1080/23311886.2021.1928980

Salama, H. H., Trif, M., Rusu, A. V., & Bhattacharya, S. (2022). Application of Functional and Edible Coatings and Films as Promising Strategies for Developing Dairy Functional Products—A Review on Yoghurt Case. *Coatings, 12*(6). Scopus. doi:10.3390/coatings12060838

Saleh, M. M., de Carvalho, A. M., de Andrade, N. M., Fonseca, A. B. M., Keller, L. A. M., Duarte, M. C. K. H., & Franco, R. M. (2023). Bacteriological quality and antimicrobial resistance of Staphylococcus spp. And Escherichia coli isolated from organic and conventional fresh cheese. *Food Science and Technology (Brazil), 43*. Advance online publication. doi:10.1590/fst.54922

Salo, T., Eskelinen, J., Jauhiainen, L., & Kartio, M. (2007). Reduced fertiliser use and changes in cereal grain weight, test weight and protein content in Finland in 1990-2005. *Agricultural and Food Science, 16*(4), 407–420. doi:10.2137/145960607784125375

Sama-Berrocal, C., & Martínez-Azúa, B. C. (2022). How Has the COVID-19 Pandemic Affected the Different Branches of the Agri-Food Industry in Extremadura (Spain)? *Land, 11*(6). Scopus. doi:10.3390/land11060938

Sánchez, M. V., Cicowiez, M., & Ortega, A. (2022). Prioritizing public investment in agriculture for post-COVID-19 recovery: A sectoral ranking for Mexico. *Food Policy, 109*. Scopus. doi:10.1016/j.foodpol.2022.102251

Sánchez-Hidalgo, M., Bravo, V., & Gallo, C. (2020). Behavior and Health Indicators to Assess Cull Cow's Welfare in Livestock Markets. *Frontiers in Veterinary Science, 7*. doi:10.3389/fvets.2020.00471

Sanders, K. F. (2005). Orange harvesting systems review. *Biosystems Engineering, 90*(2), 115–125. doi:10.1016/j.biosystemseng.2004.10.006

Sangha, K. K., Ahammad, R., Mazahar, M. S., Hall, M., Owens, G., Kruss, L., Verrall, G., Moro, J., & Dickinson, G. (2022). An Integrated Assessment of the Horticulture Sector in Northern Australia to Inform Future Development. *Sustainability (Switzerland), 14*(18). Scopus. doi:10.3390/su141811647

Sarker, M. A. (2021). The art and myth of organic agriculture for nature conservation and sustainable food production: Perspective of Bangladesh. *Agricultural Technology (Thailand), 17*(6), 2293–2302.

Satama, M., Singaña Tapia, D. A., & Paul, C. (2022). Adoption of sustainable agricultural practices during the pandemic: The case of Ecuadorian family farming systems. *Journal of Agribusiness in Developing and Emerging Economies*. Scopus. doi:10.1108/JADEE-04-2022-0067

Schaalan, M., Abou Warda, A. E., Osman, S. M., Fathy, S., Sarhan, R. M., Boshra, M. S., Sarhan, N., Gaber, S., & Ali, A. M. A. (2022). The Impact of Sociodemographic, Nutritional, and Health Factors on the Incidence and Complications of COVID-19 in Egypt: A Cross-Sectional Study. *Viruses, 14*(3). doi:10.3390/v14030448

Schat, K. A. (2022). Northeastern Conference on Avian Diseases from 1928-2021: 93 Years of Contributions to Organized Avian Medicine. *Avian Diseases, 66*(1), 7–19. 10.1637/aviandiseases-21-00074

Schilde, K. (2017). European Military Capabilities: Enablers and Constraints on EU Power? *Journal of Common Market Studies*, *55*(1), 37–53. doi:10.1111/jcms.12444

Scopus. (2023). *Scopus Database*. https://www.scopus.com/search/form.uri?display=basic#basic

Scopus. (2023). *Scopus Database*. Scopus. https://www.scopus.com/search/form.uri?display=basic#basic

Scown, M. W., Brady, M. V., & Nicholas, K. A. (2020). Billions in Misspent EU Agricultural Subsidies Could Support the Sustainable Development Goals. *One Earth, 3*(2), 237–250. Scopus. doi:10.1016/j.oneear.2020.07.011

Selim, T., & Eltarabily, M. G. (2022). Impact of COVID-19 lockdown on small-scale farming in Northeastern Nile Delta of Egypt and learned lessons for water conservation potentials. *Ain Shams Engineering Journal, 13*(4). Scopus. doi:10.1016/j.asej.2021.11.018

Shahabi Ahangarkolaee, S., & Gorton, M. (2021). The effects of perceived regulatory efficacy, ethnocentrism and food safety concern on the demand for organic food. *International Journal of Consumer Studies, 45*(2), 273–286. doi:10.1111/ijcs.12619

Shahini, E., Skuraj, E., Sallaku, F., & Shahini, S. (2022). The Supply Shock in Organic Fertilizers for Agriculture Caused by the Effect of Russia-Ukraine War. *Scientific Horizons, 25*(2), 97–103. Scopus. doi:10.48077/scihor.25(2).2022.97-103

Shenggen Fan, D. H. (2010). Reflections on the global food crisis: How did it happen? How has it hurt? And how can we prevent the next one? *Research Report of the International Food Policy Research Institute, 165*, 1–140.

Shen, L., & Hong, Y. (2023). Can geopolitical risks excite Germany economic policy uncertainty: Rethinking in the context of the Russia-Ukraine conflict. *Finance Research Letters, 51*, 103420. doi:10.1016/j.frl.2022.103420

Shepard, J. U., & Pratson, L. F. (2022). The myth of US energy independence. *Nature Energy, 7*(6), 462–464. doi:10.103841560-022-01053-2

Shi, J., Wen, Z., Zhong, G., Yang, H., Wang, C., Huang, B., Liu, R., He, X., Shuai, L., Sun, Z., Zhao, Y., Liu, P., Liang, L., Cui, P., Wang, J., Zhang, X., Guan, Y., Tan, W., Wu, G., … Bu, Z. (2020). Susceptibility of ferrets, cats, dogs, and other domesticated animals to SARS-coronavirus 2. *Science, 368*(6494), 1016–1020. doi:10.1126/science.abb7015

Shyaka, A., Quinnell, R. J., Rujeni, N., & Fèvre, E. M. (2022). Using a Value Chain Approach to Map the Pig Production System in Rwanda, Its Governance, and Sanitary Risks. *Frontiers in Veterinary Science, 8*. doi:10.3389/fvets.2021.720553

SIMA. (2023). *Several Statistics and Information*. https://www.gpp.pt/index.php/sima/sima-2018

Singh, A., Patel, R., & Singh, H. (2022). Recalibration of priorities: Investor preference and Russia-Ukraine conflict. *Finance Research Letters, 50*, 103294. doi:10.1016/j.frl.2022.103294

Sjursen, H., & Rosén, G. (2017). Arguing Sanctions. On the EU's Response to the Crisis in Ukraine. *Journal of Common Market Studies, 55*(1), 20–36. doi:10.1111/jcms.12443

Skotnicka, M., Karwowska, K., Kłobukowski, F., Wasilewska, E., & Małgorzewicz, S. (2021). Dietary habits before and during the covid-19 epidemic in selected European countries. *Nutrients, 13*(5). doi:10.3390/nu13051690

Sodji, K. (2022). Estimating the link between trade uncertainty, pandemic uncertainty and food price stability in Togo: New evidence for an asymmetric analysis. *Review of Development Economics*. Scopus. doi:10.1111/rode.12963

Sodri, N. I., Mohamed-Yassin, M.-S., Nor, N. S. M., & Ismail, I. A. (2021). Rickets due to severe vitamin d and calcium deficiency during the covid-19 pandemic in Malaysia. *American Journal of Case Reports, 22*(1). Scopus. doi:10.12659/AJCR.934216

Sohag, K., Islam, M. M., Tomas Žiković, I., & Mansour, H. (2022). Food inflation and geopolitical risks: Analyzing European regions amid the Russia-Ukraine war. *British Food Journal*. doi:10.1108/BFJ-09-2022-0793

Soliman, I. (2021). A model for prediction of the buffalo and cattle male calves' live weight. *Journal of Buffalo Science, 10*, 67–74. doi:10.6000/1927-520X.2021.10.10

Spearman, C. (1904). The Proof and Measurement of Association between Two Things. *The American Journal of Psychology, 15*(1), 72–101. doi:10.2307/1412159 PMID:3322052

Spies, M., Schick, A., Karomatov, S., Bakokhoja, B., Zikriyokhon, K., Jobirov, S., Bloch, R., & Ibisch, P. L. (2021). Adapting a Participatory and Ecosystem-Based Assessment Impacted by the Pandemic: Lessons Learned With Farmers in Tajikistan. *Frontiers in Sustainable Food Systems, 5*. Scopus. doi:10.3389/fsufs.2021.750252

Spósito, M. B., & Filho, F. D. A. A. M. (2003). 'Tahiti' lime fruit set related to gibberellic acid application on out-of-season flowering and the accumulation of degree days. *Fruits, 58*(3), 151–156. doi:10.1051/fruits:2003003

Srairi, M. T. (2021). Rethinking the agricultural development model in Morocco in the post Covid-19 era. *Cahiers Agricultures, 30*(6). Scopus. doi:10.1051/cagri/2021003

Sridhar, A., Balakrishnan, A., Jacob, M. M., Sillanpää, M., & Dayanandan, N. (2022). *Global impact of COVID-19 on agriculture: Role of sustainable agriculture and digital farming. Environmental Science and Pollution Research*. Scopus., doi:10.100711356-022-19358-w

Stamatopoulou, M., & Tzimitra-Kalogianni, I. (2022). Consumers' profile analysis for chicken meat, during the first wave of COVID-19 pandemic: Case of northern Greece. *Agraarteadus, 33*(1), 162–176. doi:10.15159/jas.22.01

Stata. (2023). *Statistical software for data science | Stata*. https://www.stata.com/

Stata. (2023). *Statistical software for data science*. https://www.stata.com/

Stata. (2023). *Statistical software for data science*. Stata. https://www.stata.com/

StataCorp. (2017a). *Stata 15 Base Reference Manual*. Stata Press.

StataCorp. (2017b). *Stata Statistical Software: Release 15*. StataCorp LLC.

Stewart, H., & Blayney, D. P. (2011). Retail dairy prices fluctuate with the farm value of milk. *Agricultural and Resource Economics Review, 40*(2), 201–217. doi:10.1017/S1068280500008017

Stojkov, J., von Keyserlingk, M. A. G., Duffield, T., & Fraser, D. (2020a). Fitness for transport of cull dairy cows at livestock markets. *Journal of Dairy Science, 103*(3), 2650–2661. doi:10.3168/jds.2019-17454

Stojkov, J., von Keyserlingk, M. A. G., Duffield, T., & Fraser, D. (2020b). Management of cull dairy cows: Culling decisions, duration of transport, and effect on cow condition. *Journal of Dairy Science, 103*(3), 2636–2649. doi:10.3168/jds.2019-17435

Štreimikienė, D., Baležentis, T., Volkov, A., Ribašauskienė, E., Morkūnas, M., & Žičkienė, A. (2022). Negative effects of covid-19 pandemic on agriculture: Systematic literature review in the frameworks of vulnerability, resilience and risks involved. *Economic Research-Ekonomska Istraživanja, 35*(1), 529–545. doi:10.1080/1331677X.2021.1919542

Sturm, R., & Datar, A. (2011). Regional price differences and food consumption frequency among elementary school children. *Public Health, 125*(3), 136–141. doi:10.1016/j.puhe.2010.11.016

Suarez, D. L., Pantin-Jackwood, M. J., Swayne, D. E., Lee, S. A., DeBlois, S. M., & Spackman, E. (2020). Lack of Susceptibility to SARS-CoV-2 and MERS-CoV in Poultry. *Emerging Infectious Diseases, 26*(12), 3074–3076. doi:10.3201/eid2612.202989 PMID:33219803

Subramaniam, S., & Rajoo, M. (2022). Review on Vietnamese Refugees, Resettlement and Mental Health: From Pulau Bidong, a Malaysian Experience. *Clinical Social Work and Health Intervention, 13*(2), 38–44. doi:10.22359/cswhi_13_2_10

Sun, F., Qian, P., Cao, S., Chen, Y., & Feng, Z. (2022). The impact of crop specialization on nutritional intake: Evidence from farm households in China. *PLoS ONE, 17*(8 August). Scopus. doi:10.1371/journal.pone.0272347

Sun, Z., Scherer, L., Zhang, Q., & Behrens, P. (2022). Adoption of plant-based diets across Europe can improve food resilience against the Russia-Ukraine conflict. *Nature Food, 3*(11), 905–910. doi:10.103843016-022-00634-4

Swinkels, J. M., Rooijendijk, J. G. A., Zadoks, R. N., & Hogeveen, H. (2005). Use of partial budgeting to determine the economic benefits of antibiotic treatment of chronic subclinical mastitis caused by Streptococcus uberis or Streptococcus dysgalactiae. *Journal of Dairy Research, 72*(1), 75–85. doi:10.1017/S0022029904000603

Sy, M. A. T., Massaoud, W., Oumarou, D. H., Issoufou, B., Simeoné, C., & Abdouramane, B. (2022). Impact of cereal production and cereal prices on food consumption and changes in household livelihoods in Burkina Faso. *African Journal of Food, Agriculture, Nutrition and Development, 22*(1), 19252–19270. doi:10.18697/ajfand.106.20715

Talukder, B., vanLoon, G. W., & Hipel, K. W. (2022). Planetary health & COVID-19: A multi-perspective investigation. *One Health, 15*, 100416. doi:10.1016/j.onehlt.2022.100416 PMID:35892119

Tanaka, T., & Guo, J. (2020). *International price volatility transmission and structural change: A market connectivity analysis in the beef sector. Humanities and Social Sciences Communications, 7(1).* doi:10.105741599-020-00657-x

Tansuchat, R., Suriyankietkaew, S., Petison, P., Punjaisri, K., & Nimsai, S. (2022). Impacts of COVID-19 on Sustainable Agriculture Value Chain Development in Thailand and ASEAN. *Sustainability (Switzerland), 14*(20). Scopus. doi:10.3390/su142012985

Temesgen, F., Wakjira, M., & Abirham, A. (2022). Assessing the Economic Impact of the Coronavirus Pandemic on the Production and Distribution of Livestock Across Value Chain Approach: The Case of Kellem Wollega Zone, Oromia National Regional State, Ethiopia. *International Journal of Rural Management, 18*(1), 39–55. Scopus. doi:10.1177/0973005221993369

Teng, Y., Pang, B., Zhang, M., & Guo, X. (2022). Driving mechanism of farmers' green production behavior under normalization of COVID-19 prevention and control: A case study in China. *Frontiers in Public Health, 10.* Scopus. doi:10.3389/fpubh.2022.826846

Tizard, I. R. (2020). Vaccination against coronaviruses in domestic animals. *Vaccine, 38*(33), 5123–5130. doi:10.1016/j.vaccine.2020.06.026

Torres-Reyna, O. (n.d.). *Getting Started in Factor Analysis (using Stata) (ver. 1.0 beta/draft).* https://www.princeton.edu/~otorres/Stata/Factor

Torres-Reyna, O. (n.d.). *Getting Started in Factor Analysis (using Stata) (ver. 1.0 beta/draft).* Princeton. https://www.princeton.edu/~otorres/Stata/Factor

Tosun, O. K., & Eshraghi, A. (2022). Corporate decisions in times of war: Evidence from the Russia-Ukraine conflict. *Finance Research Letters, 48*, 102920. doi:10.1016/j.frl.2022.102920

Tripathi, H. G., Smith, H. E., Sait, S. M., Sallu, S. M., Whitfield, S., Jankielsohn, A., Kunin, W. E., Mazibuko, N., & Nyhodo, B. (2021). Impacts of covid-19 on diverse farm systems in tanzania and south africa. *Sustainability (Switzerland), 13*(17). Scopus. doi:10.3390/su13179863

Tsvakirai, C. Z., Liebenberg, F., & Kirsten, J. F. (2018). Does research and development (R&D) investment lead to economic growth? Evidence from the South African peach and nectarine industry. *African Journal of Science, Technology, Innovation and Development, 10*(4), 463–472. doi:10.1080/20421338.2018.1473826

Uçak, H., Yelgen, E., & Arı, Y. (2022). The Role of Energy on the Price Volatility of Fruits and Vegetables: Evidence from Turkey. *Bio-Based and Applied Economics, 11*(1), 37–54. doi:10.36253/bae-10896

Uddin, M. M., Akter, A., Khaleduzzaman, A. B. M., Sultana, M. N., & Hemme, T. (2021). Application of the Farm Simulation Model approach on economic loss estimation due to Coronavirus (COVID-19) in Bangladesh dairy farms—Strategies, options, and way forward. *Tropical Animal Health and Production, 53*(1). doi:10.1007/s11250-020-02471-8

Umar, Z., Polat, O., Choi, S.-Y., & Teplova, T. (2022). The impact of the Russia-Ukraine conflict on the connectedness of financial markets. *Finance Research Letters*, *48*, 102976. doi:10.1016/j.frl.2022.102976

Usman, M., Farooq, M., Jellali, S., Charabi, Y., Al-Sadi, A. M., & Al-Badi, A. (2021). COVID-19: The Challenges and Opportunities for Water, Air, Agriculture and Energy Sectors. *International Journal of Agriculture and Biology, 25*(5), 1085–1095. Scopus. doi:10.17957/IJAB/15.1767

Uwishema, O., Adriano, L. F., Chalhoub, E., Onyeaka, H., Mhanna, M., David, S. C., Nasrallah, Y., Ribeiro, L. L. P. A., & Berjaoui, C. (2021). Bird flu outbreak amidst COVID-19 pandemic in South Africa: Efforts and challenges at hand. *Journal of Medical Virology, 93*(10), 5676–5679. doi:10.1002/jmv.27124

Vall Castelló, J., & Lopez Casasnovas, G. (2021). The effect of lockdowns and infection rates on supermarket sales. *Economics and Human Biology, 40*. doi:10.1016/j.ehb.2020.100947

van Eck, N. J., & Waltman, L. (2022). *Manual for VOSviewer version 1. 6.18*. CWTS, Universiteit Leiden. https://www.vosviewer.com/documentation/Manual_VOSviewer_1.6.18.pdf

van Eck, N. J., & Waltman, L. (2010). Software survey: VOSviewer, a computer program for bibliometric mapping. *Scientometrics, 84*(2), 523–538. doi:10.100711192-009-0146-3 PMID:20585380

van Vliet, N., Sirén, A., Paemelaere, E. A. D., Rios, E. Y. C., Correa, M. S. A., Williams, T. B., & Quintero, S. (2022). Immediate effects of the COVID-19 pandemic on the use of wildlife as food among indigenous people and local communities in South America. *Ecology and Society, 27*(3). Scopus. doi:10.5751/ES-13570-270342

Vargas, R., Fonseca, C., Hareau, G., Ordinola, M., Pradel, W., Robiglio, V., & Suarez, V. (2021). Health crisis and quarantine measures in Peru: Effects on livelihoods of coffee and potato farmers. *Agricultural Systems, 187*. Scopus. doi:10.1016/j.agsy.2020.103033

Varshney, D., Kumar, A., Mishra, A. K., Rashid, S., & Joshi, P. K. (2021). COVID-19, Government Transfer Payments, and Investment Decisions in Farming Business: Evidence from Northern India. *Applied Economic Perspectives and Policy, 43*(1), 248–269. Scopus. doi:10.1002/aepp.13144

Vatta, K., Bhogal, S., Green, A. S., Sharma, H., Petrie, C. A., & Dixit, S. (2022). COVID-19 Pandemic-Induced Disruptions and Implications for National Food Security and Farm Incomes: Farm-Level Evidence from Indian Punjab. *Sustainability (Switzerland), 14*(8). doi:10.3390/su14084452

Verdial, M. F., Neto, J. T., Minami, K., Scarpare Filho, J. A., Christoffoleti, P. J., Scarpare, F. V., Barela, J. F., Del Aguila, J. S., & Kluge, R. A. (2007). Vernalization on five cultivars of strawberry. *Ciência Rural, 37*(4), 976–981. doi:10.15900103-84782007000400009

Vicente-Vicente, J. L., Doernberg, A., Zasada, I., Ludlow, D., Staszek, D., Bushell, J., Hainoun, A., Loibl, W., & Piorr, A. (2021). Exploring alternative pathways toward more sustainable regional food systems by foodshed assessment – City region examples from Vienna and Bristol. *Environmental Science and Policy, 124*, 401–412. Scopus. doi:10.1016/j.envsci.2021.07.013

Vik, J. (2020). The agricultural policy trilemma: On the wicked nature of agricultural policy making. *Land Use Policy, 99*. Scopus. doi:10.1016/j.landusepol.2020.105059

Vincent, J. E. (1971). *Factor Analysis in International Relations: Interpretation, Problem Areas, and an Application*. University of Florida Press.

Violet, A. O., Elizabeth, A. O., & Cynthia, E. N. (2022). Enterprise Training Needs of Poultry Farmers During Covid-19 Pandemic in Enugu State, Nigeria. *Journal of Agricultural Extension, 26*(2), 129–137. doi:10.4314/jae.v26i1.16S

Vo, D. H. (2020). Sustainable agriculture & energy in the U.S.: A link between ethanol production and the acreage for corn. *Economics and Sociology, 13*(3), 259–268. Scopus. doi:10.14254/2071-789X.2020/13-3/16

VOSviewer. (2023). *VOSviewer—Visualizing scientific landscapes—Version 1.6.18*. VOSviewer. https://www.vosviewer.com//

Wang, C.-M. (2021). Performing and counter-performing organic food markets in East Asia: The role of ahimsa, scientific knowledge and faith groups. *Geographical Journal, 187*(4), 361–372. doi:10.1111/geoj.12338

Wang, Y., Peng, D., Yu, L., Zhang, Y., Yin, J., Zhou, L., Zheng, S., Wang, F., & Li, C. (2020). Monitoring Crop Growth during the Period of the Rapid Spread of COVID-19 in China by Remote Sensing. *IEEE Journal of Selected Topics in Applied Earth Observations and Remote Sensing, 13*, 6195–6205. Scopus. doi:10.1109/JSTARS.2020.3029434

Wang, Y., Wang, J., & Wang, X. (2020). COVID-19, supply chain disruption and China's hog market: A dynamic analysis. *China Agricultural Economic Review, 12*(3), 427–443. doi:10.1108/CAER-04-2020-0053

Web of Science. (2023). *Web of Science Core Collection*. https://www.webofscience.com/wos/woscc/basic-search

Web of Science. (2023). *Web of Science Core Collection*. WOS. https://www.webofscience.com/wos/woscc/basic-search

Wegerif, M. (2022). The impact of Covid-19 on black farmers in South Africa. *Agrekon, 61*(1), 52–66. Scopus. doi:10.1080/03031853.2021.1971097

Wegner, G. I., Murray, K. A., Springmann, M., Muller, A., Sokolow, S. H., Saylors, K., & Morens, D. M. (2022). Averting wildlife-borne infectious disease epidemics requires a focus on socio-ecological drivers and a redesign of the global food system. *EClinicalMedicine, 47*, 101386. doi:10.1016/j.eclinm.2022.101386 PMID:35465645

White, R. J., & Razgour, O. (2020). Emerging zoonotic diseases originating in mammals: A systematic review of effects of anthropogenic land-use change. *Mammal Review, 50*(4), 336–352. Scopus. doi:10.1111/mam.12201

Windsor, P. A. (2021). Progress With Livestock Welfare in Extensive Production Systems: Lessons From Australia. *Frontiers in Veterinary Science, 8*. doi:10.3389/fvets.2021.674482

Wojciechowska-Solis, J., Kowalska, A., Bieniek, M., Ratajczyk, M., & Manning, L. (2022). Comparison of the Purchasing Behaviour of Polish and United Kingdom Consumers in the Organic Food Market during the COVID-19 Pandemic. *International Journal of Environmental Research and Public Health, 19*(3). doi:10.3390/ijerph19031137

Wu, Y., & Turvey, C. G. (2020). The impact of the China–USA trade war on USA Chapter 12 farm bankruptcies. Agricultural Finance Review, 81(3), 386–414. doi:10.1108/AFR-05-2020-0076

Wudil, A. H., Usman, M., Rosak-Szyrocka, J., Pilař, L., & Boye, M. (2022). Reversing Years for Global Food Security: A Review of the Food Security Situation in Sub-Saharan Africa (SSA). *International Journal of Environmental Research and Public Health, 19*(22). Scopus. doi:10.3390/ijerph192214836

Xu, X., & Mola-Yudego, B. (2021). Where and when are plantations established? Land-use replacement patterns of fast-growing plantations on agricultural land. *Biomass and Bioenergy, 144*. doi:10.1016/j.biombioe.2020.105921

Yan, W., Cai, Y., Lin, F., & Ambaw, D. T. (2021). The Impacts of Trade Restrictions on World Agricultural Price Volatility during the COVID-19 Pandemic. *China and World Economy, 29*(6), 139–158. doi:10.1111/cwe.12398

Yang, B., Asche, F., & Li, T. (2022). Consumer behavior and food prices during the COVID-19 pandemic: Evidence from Chinese cities. *Economic Inquiry, 60*(3), 1437–1460. Scopus. doi:10.1111/ecin.13067

Yang, R., Sun, H., Gao, F., Luo, K., Huang, Z., Tong, Q., Song, H., Han, Q., Liu, J., Lan, Y., Qi, J., Li, H., Chen, S., Xu, M., Qiu, J., Zeng, G., Zhang, X., Huang, C., Pei, R., … Liu, J. (2022). Human infection of avian influenza A H3N8 virus and the viral origins: A descriptive study. *The Lancet Microbe, 3*(11), e824–e834. doi:10.1016/S2666-5247(22)00192-6

Yang, X., & Liu, W. (2022). Agricultural Production Networks and Upgrading from a Global-Local Perspective: A Review. *Land (Basel), 11*(10), 1864. doi:10.3390/land11101864

Yazbeck, N., Mansour, R., Salame, H., Chahine, N. B., & Hoteit, M. (2022). The Ukraine-Russia War Is Deepening Food Insecurity, Unhealthy Dietary Patterns and the Lack of Dietary Diversity in Lebanon: Prevalence, Correlates and Findings from a National Cross-Sectional Study. *Nutrients, 14*(17), 3504. doi:10.3390/nu14173504 PMID:36079761

Yazlcl, D., Fersahoğlu, M. M., Fersahoğlu, T., Bulut, N. E., Çiğiltepe, H., Çeler, Ö., Sancak, S., Sulu, C., Durcan, E., Şahin, S., Polat Korkmaz, Ö., Bozdoğan Polat, S. H., Taşkln, E., İşeri, C., Gürsoy, E., Küçük Yetgin, M., Kaya, T., Özdemir, F., Mestanoğlu İstanbullu, N., … Yumuk, V. D. (2022). Status of Weight Change, Lifestyle Behaviors, Depression, Anxiety, and Diabetes Mellitus in a Cohort with Obesity during the COVID-19 Lockdown: Turk-Com Study Group. *Obesity Facts, 15*(4), 528–539. doi:10.1159/000522658

Yilmaz, B., Sharma, H., Melekoglu, E., & Ozogul, F. (2022). Recent developments in dairy kefir-derived lactic acid bacteria and their health benefits. *Food Bioscience, 46*. Scopus. doi:10.1016/j.fbio.2022.101592

Yovchevska, P. (2021). Bulgarian agriculture in the focus of representative statistical surveys. [Scopus.]. *Bulgarian Journal of Agricultural Science, 27*(5), 859–864.

Yuan, Y.-S., Cao, Z.-Y., Chen, Y.-T., Gong, P.-L., Huang, G.-H., & He, L. (2022). Efficiency Analysis of the Crop Production in China in 2019 and 2020: Role of Uncertainty Perceptions in COVID-19. *Discrete Dynamics in Nature and Society, 2022*. Scopus. doi:10.1155/2022/7044474

Zámková, M., Rojík, S., Prokop, M., Činčalová, S., & Stolín, R. (2022). Czech Consumers' Preference for Organic Products in Online Grocery Stores during the COVID-19 Pandemic. *International Journal of Environmental Research and Public Health, 19*(20). Scopus. doi:10.3390/ijerph192013316

Zargaraan, A., Nasrabadi, F. M., Mehr, A. R., Eskandari, S., Salmani, Y., & Esfarjani, F. (2021). A Mixed-Method Study of Chicken Meat Safety in Iran during the COVID-19 Pandemic: SWOT Analysis. *Poultry Science Journal, 9*(2), 233–243. doi:10.22069/psj.2021.19049.1690

Zawadka, J., Jęczmyk, A., Wojcieszak-Zbierska, M. M., Niedbała, G., Uglis, J., & Pietrzak-Zawadka, J. (2022). Socio-Economic Factors Influencing Agritourism Farm Stays and Their Safety during the COVID-19 Pandemic: Evidence from Poland. *Sustainability (Switzerland), 14*(6). doi:10.3390/su14063526

Zhao, W., Chou, J., Li, J., Xu, Y., Li, Y., & Hao, Y. (2022). Impacts of Extreme Climate Events on Future Rice Yields in Global Major Rice-Producing Regions. *International Journal of Environmental Research and Public Health, 19*(8). Scopus. doi:10.3390/ijerph19084437

Zheng, Y., Fan, Q., & Jia, W. (2022). Article How Much Did Internet Use Promote Grain Production? Evidence from a Survey of 1242 Farmers in 13 Provinces in China. *Foods, 11*(10). doi:10.3390/foods11101389

Zielińska-Chmielewska, A., Mruk-Tomczak, D., & Wielicka-Regulska, A. (2021). Qualitative research on solving difficulties in maintaining continuity of food supply chain on the meat market during the COVID-19 pandemic. *Energies, 14*(18). Scopus. doi:10.3390/en14185634

Zmami, M., & Ben-Salha, O. (2019). *Does oil price drive world food prices? Evidence from linear and nonlinear ARDL modeling. Economies, 7(1)*. doi:10.3390/economies7010012

Zmrhal, V., Svoradova, A., Batik, A., & Slama, P. (2022). Three-Dimensional Avian Hematopoietic Stem Cell Cultures as a Model for Studying Disease Pathogenesis. *Frontiers in Cell and Developmental Biology, 9*. doi:10.3389/fcell.2021.730804

Zollo, L., Carranza, R., Faraoni, M., Díaz, E., & Martín-Consuegra, D. (2021). What influences consumers' intention to purchase organic personal care products? The role of social reassurance. *Journal of Retailing and Consumer Services, 60*. doi:10.1016/j.jretconser.2020.102432

Zysk, E., Mroczkowski, T., & Dawidowicz, A. (2022). Small Farms as "Data Producers" for the Needs of Agricultural Management Information System. *Geomatics and Environmental Engineering, 16*(3), 79–109. Scopus. doi:10.7494/geom.2022.16.3.79

About the Author

Vítor João Pereira Domingues Martinho is Coordinator Professor with Habilitation at the Polytechnic Institute of Viseu, Portugal, and holds a Ph.D. in Economics from the University of Coimbra, Portugal. He was President of the Scientific Council, President of the Directive Council and President of the Agricultural Polytechnic School of Viseu, Portugal, from 2006 to 2012. He was an Erasmus student in the Faculty of Economics from the University of Verona, Italy, participated in various technical and scientific events nationally and internationally, has published several technical and scientific papers, is referee of various scientific and technical journals and participates in the evaluation of national and international projects.(d739eb72-5649-49d3-a648-ac43386cb903)

Index

M

N

O

P

R

S

T

www.ingramcontent.com/pod-product-compliance
Lightning Source LLC
LaVergne TN
LVHW081249011026
849162LV00006BA/1235